Vehicular ad hoc Networks

Claudia Campolo • Antonella Molinaro
Riccardo Scopigno
Editors

Vehicular ad hoc Networks

Standards, Solutions, and Research

Editors
Claudia Campolo
University Mediterranea
 of Reggio Calabria
Reggio Calabria, Italy

Antonella Molinaro
University Mediterranea
 of Reggio Calabria
Reggio Calabria, Italy

Riccardo Scopigno
Istituto Superiore Mario
Boella (ISMB), Torino, Italy

ISBN 978-3-319-15496-1   ISBN 978-3-319-15497-8 (eBook)
DOI 10.1007/978-3-319-15497-8

Library of Congress Control Number: 2015936909

Springer Cham Heidelberg New York Dordrecht London
© Springer International Publishing Switzerland 2015
This work is subject to copyright. All rights are reserved by the Publisher, whether the whole or part of the material is concerned, specifically the rights of translation, reprinting, reuse of illustrations, recitation, broadcasting, reproduction on microfilms or in any other physical way, and transmission or information storage and retrieval, electronic adaptation, computer software, or by similar or dissimilar methodology now known or hereafter developed.
The use of general descriptive names, registered names, trademarks, service marks, etc. in this publication does not imply, even in the absence of a specific statement, that such names are exempt from the relevant protective laws and regulations and therefore free for general use.
The publisher, the authors and the editors are safe to assume that the advice and information in this book are believed to be true and accurate at the date of publication. Neither the publisher nor the authors or the editors give a warranty, express or implied, with respect to the material contained herein or for any errors or omissions that may have been made.

Printed on acid-free paper

Springer International Publishing AG Switzerland is part of Springer Science+Business Media (www.springer.com)

Preface

Editing a book is a rewarding experience, but also an endeavor which steals a lot of time—especially your personal time; even more, you can hardly know everything from the beginning and the final outcome is somehow a bet.

Consequently, you have to throw yourself wholeheartedly into something which is not completely guaranteed. This is not infrequent in the working life, but it already challenges your motivation.

So, when given the opportunity to take up this experience, these, indeed, were our sensations: pride for being given the possibility to coordinate and contribute to such a hot topic as vehicular communications are, but, at the same time, manifold doubts about the contents to include, the authors to invite, their possible acceptance, the possible final result, the targets, and many other initial dilemmas.

Then, firstly, we three sat around a table (virtually, since we are at the opposite sides of Italy) and we wondered whether it could be worth writing one more book on vehicular communications—there are already some of them in the market. We debated rationally, coming to a full consensus (it may not be obvious even in two), supported by several arguments.

Indeed, these arguments, together with a presentation of the contents, constitute the preface that we chose to write: if we managed to fulfill our initial aim—we hope so—our motivations are the best way to introduce the cut and the purpose which we agreed for this volume.

The first motivation is the one which probably triggered also Springer to contact us: vehicular ad hoc networks (VANETs) constitute a hot and alive topic, a still-becoming one. Hence, it is a topic of interest both from a commercial and from a scientific point of view.

Concerning the topicality, VANETs are about to be deployed: at the time of this writing, both European Countries and the United States are making their final decisions on the rulemaking for vehicular networking technologies in newly sold cars, while more and more field trials are running worldwide. Even more, there are great expectations set on VANETs for improving road safety and transport efficiency and as an enabler of value-added services for passengers and drivers.

As a consequence, not only wireless scientists and engineers are involved in the future of VANETs, but also people working in the automotive sector, service providers, stakeholders of the smart cities and of the highways, and the list could be made longer and longer.

All of them set inside our intended audience.

So, related to the previous item, decisions about the contents to include were influenced by our brainstorming on the addressees of the book.

We envisaged a book requiring very few prerequisites and providing as many details as possible on the foundations of vehicular communications (spanning from a historical perspective to the wireless phenomena), their peculiarities and the relevant standardization activities (i.e., presenting solutions highlighting their differences to the previous art), the tools to study them, and the emerging solutions.

All in all, we made our best to write a comprehensive guide—but not an encyclopedia—digging as much as needed to keep a scientific approach (the reader can skip some details) but, at the same time, preserving the overall readability. Cross-references among chapters ensure a self-contained work, and rich per-chapter bibliographies also provide the reader with hints for further readings.

Then, there is the still evolving nature of VANETs and this is related to a second strong motivation. We are all scientists in this field and participated in the evolution of the topic since its dawn, maturing a nice picture of it as a whole; and, importantly for us, we got sufficient contacts to fill our gaps—we are the first to distrust who "knows" everything.

Over the years, we have contributed to the research, by theoretical studies and experiments, to the dissemination (by IEEE tutorials) and, in little part, supporting the standardization. All in all, we are so involved in VANETs and so keen on the topic that the idea of editing a book let our enthusiasm run wild—we have to acknowledge it.

We have mentioned our contacts, so it is time to spend some words on who contributed to the book: all the coauthors are top-ranking ones, a real list of who's who in the respective topics of each chapter (please take time to go through the list). What is more, the list of authors includes a rich and diversified number of contributors, many from the standardization and the Industry (this is not obvious for a book: we have rarely seen together so many of them), and, obviously, some from the Academia; they are from different world regions, so as to prevent a partial, polarized view.

In the happy attitude which is characterizing our final wrap-up, we can say that, in the end, we have closer relationships with such eminent fellow travelers in our book editing adventure: we hope we have managed to coordinate and support their work and, whenever they had doubts (or even something more than doubts), we may have reassured them enough. Also in this case, we did our best to be credible and patient head coaches.

Now, some words about the characterizing contents of the book.

In addition to the general-purpose scope and the extensiveness—which have already been mentioned—our aims, agreed with all the authors, included the avoidance of technological integralisms and a significant differentiation from other books on the same topics: the volume does not intend to replace existing literature but to complement it by covering still unaddressed topics and by updating covered but already outdated contents.

In this book, you will find details on the day-1 development, on evolution in the nearby future, on the work and research which is under way, on the tools to study vehicular communications and, what is even more uncommon for a book on VANETs, also on alternative or complementary technology solutions (so far we are from integralisms).

Every time we all tried to specify the actual feasibility, the timeline, and an objective analysis, trying not to give an opinion for granted a priori, with no proof.

The book consists of 18 chapters, organized into five sections briefly outlined in the following:

- Part I provides introductory material with an historical overview of vehicular communications (Chap. 1) and of the standardization and harmonization activities pursuing a global solution (Chap. 2). Standardization is a continuous running process; hence the reader is advised to consult source documents for successive updates.
- Part II is about the fundamentals of VANETs: the IEEE 802.11-based physical (PHY) and medium access control (MAC) layer (respectively, Chaps. 3 and 4), including phenomena, protocols, and packet formats; then the message sets for vehicular communications (Chap. 5); and finally two topics still related to the PHY-MAC layers: Decentralized Congestion Control (DCC) techniques conceived to counteract safety channel congestion (Chap. 6) and the multichannel operation (Chap. 7). With the exception of DCC (which is in Part II for coherence), this Part includes all subjects which refer to the day-1 of VANETs (i.e., the nearest future).
- With Part III, the focus moves to additional features and upper layers, pertaining to the vehicular networking and the security. Chapter 8 is about forwarding (geonetworking), Chap. 9 is about the adoption of IPv6 for Cooperative Intelligent Transport System (C-ITS), and Chap. 10 sheds light on the security and privacy aspects.
- Part IV is of primary interest for both scientists and newcomers: it overviews the main tools used for the performance study of VANETs: mobility models (Chap. 11), channel models (Chap. 12), simulation tools (Chap. 13), and field operational tests (Chap. 14).
- The last part (Part V) is about the evolution of VANETs, both in terms of technology improvements or complementation by other communication technologies and in terms of new paradigms and applications. More in detail, here we find: Chap. 15, about the possible evolutions of PHY and MAC layers and about complementary solutions, excluded Long-Term Evolution (LTE), to which is devoted an entire chapter (Chap. 16); Chap. 17 is about the emerging paradigm of information-centric networking to disseminate contents in VANETs, and, finally, Chap. 18 looks into some possible future vehicular applications.

In wishing a comfortable reading, for those who are not familiar with the topic, we would like to leave some impressions and to motivate the reading, sharing our passion in VANETs by intentionally instilling some conundrums whose answers can be found in the book.

- *Can you imagine how challenging it may be to set up wireless distributed communications in high mobile outdoor environments?*
- *What are the main issues related to the outdoor environment and to the radio propagation in urban environments at vehicular speed?*
- *What happens if hundreds of nodes should start transmitting simultaneously? How to prevent that too many collisions occur while guaranteeing a timely update of the kinematics information for cooperative vehicular applications?*
- *How would it be possible to forward information in such a dynamic context with rapidly varying topology?*
- *What is the state of the art in the implementation, standardization, and deployment worldwide?*
- *Which are the best candidate technologies to support vehicular communications? And how to use complementary wireless technologies?*
- *Which are the main reference standards for VANETs?*

These and many more questions will be answered throughout the book. From a scientific point of view, the field was very fascinating, challenging, and open to innovation: it subtended so many issues to be solved that it was like a mother lode in a researcher's perspective.

And we can hardly imagine what future will be possible when VANETs' penetration will be extended to all the cars (even to autonomously driven ones) but also to bikers, cyclists, and pedestrians. It is not science fiction; rather it is a process which is already started.

We really hope that this book may help a little bit and acquaint new followers with the topic.

***

The preface has come to its end and we can eventually state our acknowledgments:

- The first thanks are for our families. We have stolen many evenings and weekends to achieve our goal. We do not dare promise that it was the last time ;-)
- Our editorial contacts in Springer, Rebecca R. Hytowitz, Mary James, and Brett Kurzman, who have supported us with prompt and professional advices and carefully followed the formation of this book.
- Then we want to mention the coauthors. They are as diversified as the humankind is, and their characters have emerged just through their way of working: each of them has delivered, in our view, a precious artifact which returns part of their knowledge, filtered by their experience—this indeed what we appreciated the most.

Preface

- We wish also to thank IEEE, ISO, and ETSI for their work and their documents which were necessarily a continuous reference for us.
- Last but not least, the reader. Thanks for your trust in our work. Should you have any idea on how to enrich and improve our future editions, please just contact us.

We wish you a pleasant reading.

Reggio Calabria, Italy	Claudia Campolo
Reggio Calabria, Italy	Antonella Molinaro
Torino, Italy	Riccardo Scopigno

Contents

Part I Introduction

1. **The History of Vehicular Networks** 3
   Marco Annoni and Bob Williams

2. **Standardization and Harmonization Activities Towards a Global C-ITS** 23
   Hans-Joachim Fischer

Part II Fundamentals of Vehicular Communications

3. **The Physical Layer of VANETs** 39
   Riccardo M. Scopigno, Alessia Autolitano, and Weidong Xiang

4. **The MAC Layer of VANETs** 83
   Claudia Campolo, Antonella Molinaro, Riccardo Scopigno, Serkan Ozturk, Jelena Mišić, and Vojislav B. Mišić

5. **Message Sets for Vehicular Communications** 123
   Lan Lin and James A. Misener

6. **Decentralized Congestion Control Techniques for VANETs** 165
   Dieter Smely, Stefan Rührup, Robert K. Schmidt, John Kenney, and Katrin Sjöberg

7. **Multi-Channel Operations, Coexistence and Spectrum Sharing for Vehicular Communications** 193
   Jérôme Härri and John Kenney

Part III Vehicular Networking and Security

8 Forwarding in VANETs: GeoNetworking 221
Andrea Tomatis, Hamid Menouar, and Karsten Roscher

9 The Use of IPv6 in Cooperative ITS: Standardization Viewpoint 253
Fernando Pereñiguez, José Santa, Pedro J. Fernández,
Fernando Bernal, Antonio F. Skarmeta, and Thierry Ernst

10 Security and Privacy for ITS and C-ITS 283
Scott W. Cadzow

Part IV Evaluation of Vehicular Networks

11 Mobility Models for Vehicular Communications 309
Pietro Manzoni, Marco Fiore, Sandesh Uppoor,
Francisco J. Martínez Domínguez, Carlos Tavares Calafate,
and Juan Carlos Cano Escriba

12 Channel Models for Vehicular Communications 335
Mate Boban and Wantanee Viriyasitavat

**13 Simulation Tools and Techniques for Vehicular
Communications and Applications** 365
Christoph Sommer, Jérôme Härri, Fatma Hrizi,
Björn Schünemann, and Falko Dressler

14 Field Operational Tests and Deployment Plans 393
Yvonne Barnard, François Fischer, and Maxime Flament

Part V The Evolution of Vehicular Networks

15 Insights into Possible VANET 2.0 Directions 411
Xinzhou Wu, Junyi Li, Riccardo M. Scopigno,
and Hector Agustin Cozzetti

16 LTE for Vehicular Communications 457
Christian Lottermann, Mladen Botsov, Peter Fertl, Robert Müllner,
Giuseppe Araniti, Claudia Campolo, Massimo Condoluci,
Antonio Iera, and Antonella Molinaro

17 Information-Centric Networking for VANETs 503
Peyman TalebiFard, Victor C.M. Leung, Marica Amadeo,
Claudia Campolo, and Antonella Molinaro

18 Future Applications of VANETs 525
Cristofer Englund, Lei Chen, Alexey Vinel, and Shih Yang Lin

Part I
Introduction

Chapter 1
The History of Vehicular Networks

Marco Annoni and Bob Williams

Abstract The chapter begins by providing a short historical description of the evolution of the technologies and the standards enabling a vehicle to communicate with other vehicles and the surrounding environment and become part of an extended intelligent transportation system (ITS) communication system able to support a wide range of services by using different communication media. In the following, the evolution from the V2I (vehicle to infrastructure) toward the V2X (vehicle to any) scenarios is discussed as an extension of the original vehicular ad hoc network (VANET) concept. The reference architecture of the ITS-station is then introduced by highlighting the roles and the contributions of the main standard development organizations involved in the development and consolidation of the concept. Finally, some consideration on the role of the regulatory environment and the related open issues are reported.

Keywords VANET • ISO • ETSI • CEN • IEEE • WAVE • CALM • ITS • C-ITS • V2V • V2I • V2X • ITS Station • Cooperative ITS • C-ITS

1.1 Motivation and History of VANETs

The general definition of the term VANET (vehicular ad-hoc network) refers to the possibility of having a communication node on-board a vehicle able to establish a wireless communication with other surrounding communication nodes visible in the radio range. Another implicit concept is that the vehicles are, by definition, mobile objects and, as a consequence, the network topology is randomly variable in time even if, in this particular scenario, some predictions can be made on the motion of communication nodes since any vehicle is supposed to be moving along predefined trajectories (i.e., roads).

M. Annoni (✉)
Telecom Italia SpA, Via Reiss Romoli 274, Turin 10148, Italy
e-mail: marco.annoni@telecomitalia.it

B. Williams
CSI UK Ltd., Nottinghamshire, UK
e-mail: bw_csi@fastmail.fm

© Springer International Publishing Switzerland 2015
C. Campolo et al. (eds.), *Vehicular ad hoc Networks*,
DOI 10.1007/978-3-319-15497-8_1

The concept of using radio communications to communicate from a vehicle in order to improve the safety has been around well before the advent of the digital radio communications we are familiar with today. One example is the patent "Radio Warning Systems for use on Vehicles" submitted on 1922 and issued in 1925 [8], based on the concept of peer-to-peer radio communication between equal devices installed on two different vehicles (see Fig. 1.1).

In its simplicity, the proposed solution anticipated some of the requirements for vehicular safety that have driven the demand for development of communication among vehicles in more recent years. It was not yet a wireless networking technology as we know today, but was aiming at one of the very similar needs, which, it turns out, we are still considering today and which motivated the development and the consolidation of the VANET technology.

Outside of the military, nothing much happened however, for nearly half a century, until the 1980s and 1990s, by which time most vehicles were already sold with a "radio-set" as standard. Of course, by this time, the radio provided broadcasts through local radio stations, advising motorists of weather conditions and major incidents, but simply via the voice of the presenter. So it was not networking of any sort, just a broadcast.

Radio data system (RDS), a communication protocol standard for embedding small amounts of digital information in conventional Frequency Modulation (FM) radio broadcasts, in 1984, became the first digital infrastructure to vehicle (I2V) communication, and was introduced in the USA as radio broadcast data system (RBDS) a few years later. In 1990, RDS became a European Standard.

Both RDS and RBDS carry data at 1,187.5 bits per second on a 57 kHz subcarrier, so there are exactly 48 cycles of subcarrier during every data bit. The RBDS/RDS subcarrier was set to the third harmonic of the 19 kHz FM stereo pilot tone to minimize interference and inter-modulation between the data signal, the stereo pilot and the 38 kHz Double-sideband suppressed-carrier (DSB-SC) stereo difference signal. The stereo difference signal extends up to $38\,\text{kHz} + 15\,\text{kHz} = 53\,\text{kHz}$, leaving 4 kHz for the lower sideband of the RDS signal.

The data is sent with error correction. RDS defines many features including how private (in-house) or other undefined features can be "packaged" in unused program groups. However, it is unidirectional, and not a network.

Around 2005, following long trials, RDS was enhanced to provide RDS-Traffic Message Channel (TMC). Each traffic incident is binary-encoded and sent as a TMC message. Each message consists of an event code, location code, expected incident duration, affected extent and other details.

The message is coded according to the Alert C standard and contains a list of up to 2,048 event phrases that can be translated by the receiver into the user's language. Some phrases describe individual situations such as a crash, while others cover combinations of events such as construction causing long delays.

RDS-TMC is also a low-bandwidth system, with each RDS-TMC message comprising 37 data bits sent at most 1–3 times per second, using a basic data channel primarily designed for FM radio tuning and station name identification. Compressing traffic incident descriptions in multiple languages into 16 bits for a location,

1 The History of Vehicular Networks 5

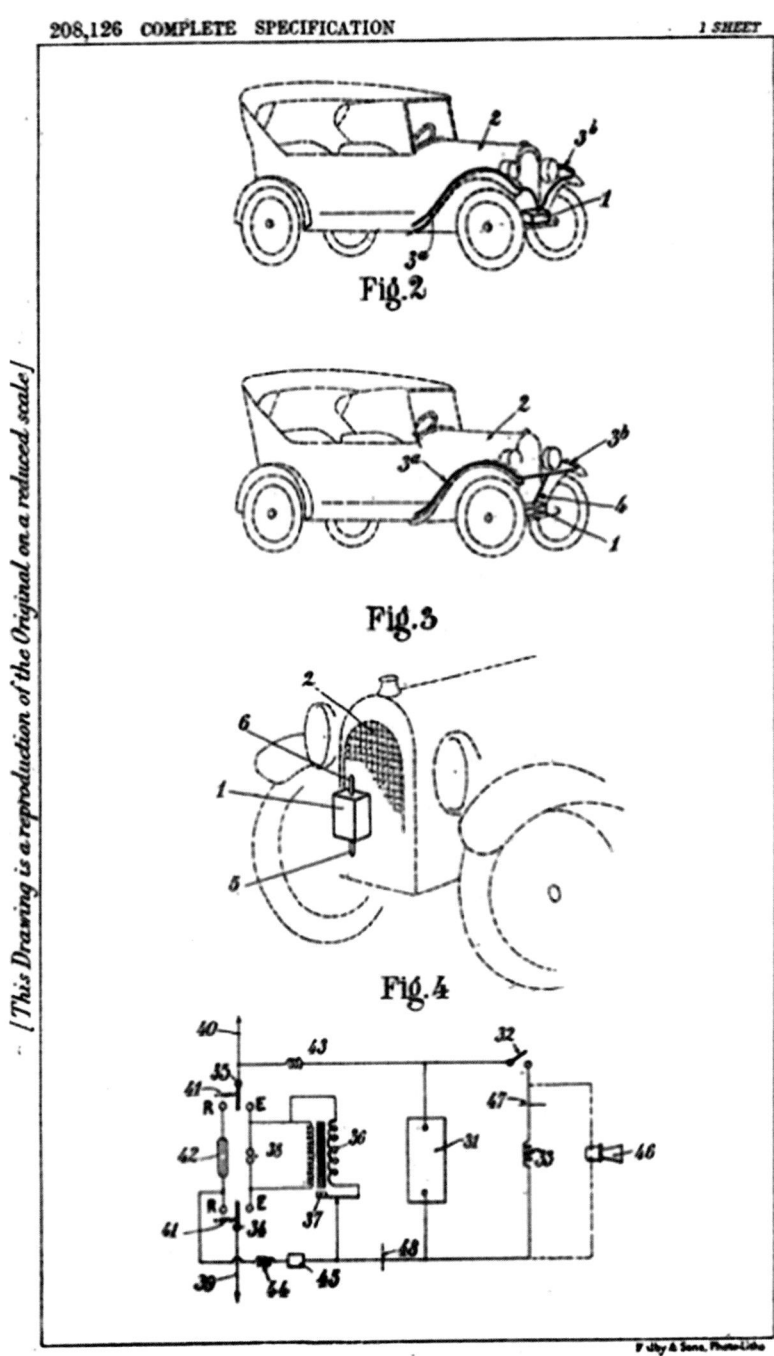

Fig. 1.1 Drawing from "Radio Warning Systems for use on Vehicles" [8]

11 bits for an event code, plus 5 bits for an extent and a few more bits for the duration and system management was necessary due to constraints in the RDS standard.

Sources of traffic information typically include police, traffic control centres, camera systems, traffic speed detectors, floating car data, winter driving reports and roadwork reports.

But these communications remained unidirectional, and not networks.

The first effective bidirectional systems came in the 1980s with tolling systems introducing Radio Frequency IDentification (RFID) tags into vehicles (initially at 2.45 GHz, then migrating to 5.8 GHz in Europe and subsequently, much of the world, and at 915 MHz in the USA). These were the first bidirectional, if primitive, communications, in which, in its simplest form, the infrastructure interrogated RFID tags passing under beacons and the tag responded with its identification. More complex systems, such as that devised by Philips, could store and transmit entry and exit data as well.

It was first envisioned that the 5.8 GHz dedicated short range communications (DSRC) system, invented by Philips in the early 1990s as an adaptation of its 2.45 GHz system, would become the basis for what was then (and probably still more appropriately) called "Intelligent vehicle-highway systems" (IVHS) communications. The DSRC In-vehicle system was able to make different types of transactions, so in the limited telephony and internet world of the 1990s it was envisioned that it would become the obvious means of IVHS service delivery. The concept of an infrastructure controlled "Network" evolved.

In this architecture, a network of beacons, operating at 500 kbit/s / 250 kbit/s, would hold short communications sessions with vehicles as they passed within the short (2.5–10 m) range of the beacons, and the controlling infrastructure would ask what services were required, and pass on the benefit of the updated information it was constantly receiving throughout its centrally connected and managed network.

But the cars did not communicate with each other so there was not even a vehicle network, let alone an ad-hoc network, and the expensive and fixed nature of the downlink beacons meant that there was nothing "ad-hoc" about that infrastructure either. The technology was infrastructure driven, in a master–slave relationship.

And the bandwidth was too limited, and the range too short, and by then the three principal developers of the Comité Européen de Standardization (CEN) Standards for DSRC had got embroiled in trying to lock the technology into their proprietary protocols (at least two of the three, with the third fighting for non-proprietary protocols).

But the fundamental weakness lay in the business case. There was no business case, other than road tolling, that could bear the required infrastructure cost.

Following one of the many legal disputes, and the eventual compromise and flawed standard, a couple of the competitors, together with one of the lead consultants at the time, sat down to work out how to avoid this mess in the future world of ITS communications.

They came up with an idea which was, at the time evolutionary, but, unknown to them, similar design issues were facing internet developers, and the eventual

1 The History of Vehicular Networks

solution they conceived, was similar to those that have become the norm for the internet as well—i.e., to separate the applications from the communications means, and introduce an on-board communication and network management function. The advantage of this system is that it would work with any standardized wireless media, and all that was needed was a tailored "Service Access Point" for each wireless medium, which controlled the opening, management and closing of each communication session. It was then realized that the protocols had to be standardized, and so they adopted the IPv4 (later IPv6) protocols, and introduced the concept to International Standards Organization (ISO) TC204 (Wide Area Communications). From 2000–2013 ISO, working with IEEE, and latterly also with ETSI [6], developed a set of standards to manage these communication sessions.

Early developments had considered the issues of Infrastructure: vehicle communications (V2I) being an extension of the 5.8 GHz DSRC master/slave relationships with vehicle/vehicle (V2V) being peer/peer relationships. But in the ITS world, a police car can be a vehicle at one moment, and after an incident become a node of the infrastructure. The communication structure turned out, in fact, to be more simple, and as with the rest of the new architecture, the application, and the "roles" of the application had to be separated from the communications architecture. All ITS communications, were, as with any mobile communications between actors, peer/peer communications.

By 2013 the ISO Standards had evolved to the concept of ITS-station communications as shown in Fig. 1.2.

This peer-to-peer relationship could involve two or any number of ITS-stations in peer-to-peer, broadcast, or unicast communications (see Fig. 1.3). Communications could also "hop", extending the range of an end-to-end communication.

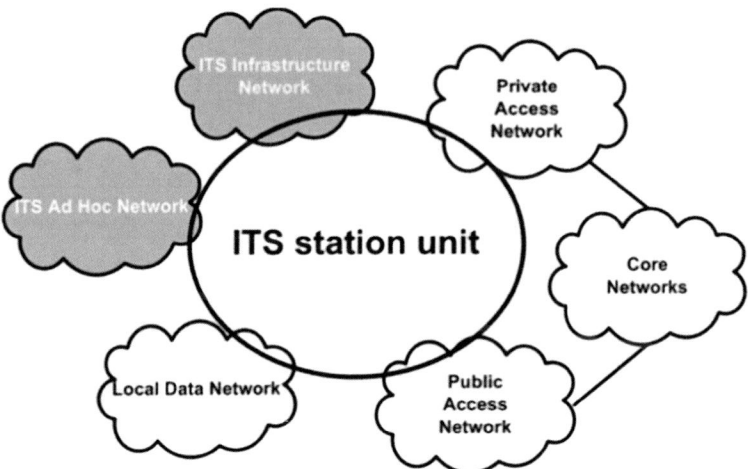

Fig. 1.2 Networking view of ITS communications [10]

Fig. 1.3 Peer-to-peer, broadcast, unicast ITS-station communications (*source* CSI (UK) Ltd)

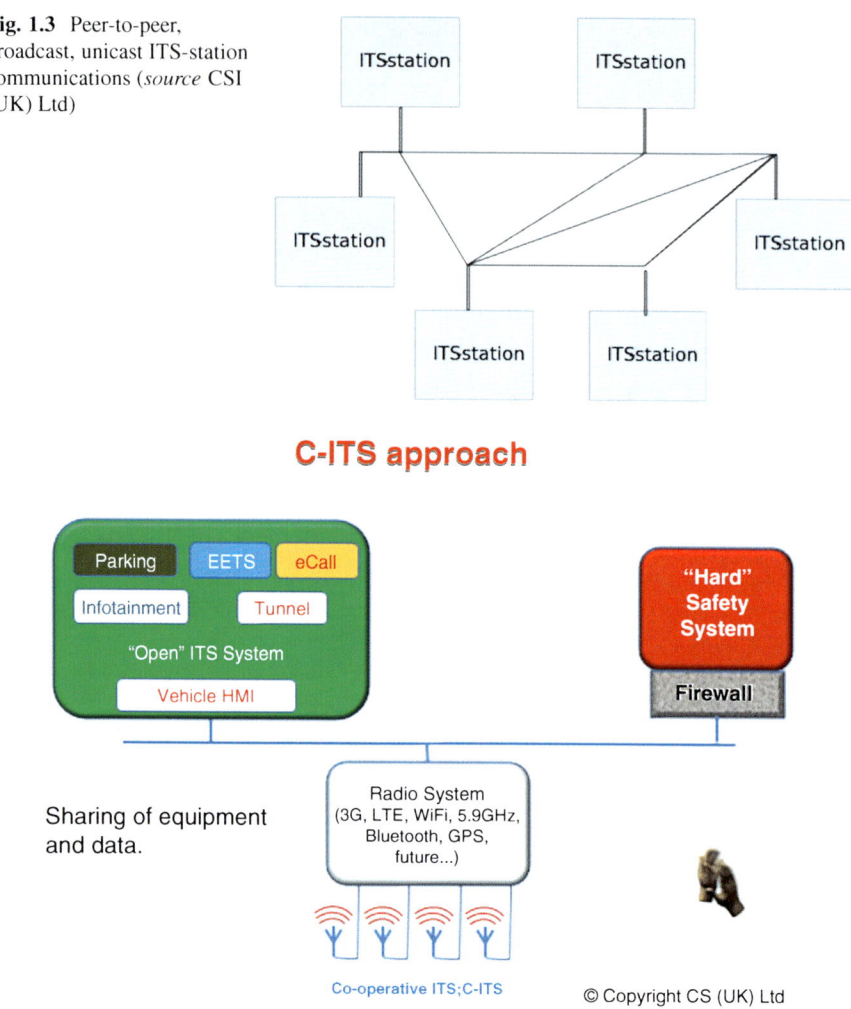

Fig. 1.4 ITS-station with multiple media (*source* ESF GmbH)

The role of any actor in an ITS-stations network (infrastructure, vehicle, street furniture, etc.) was unimportant at the communications level. Furthermore, the medium, though convenient if the same wireless medium, need not be, and the use of multiple media expanded capacity significantly. This approach is represented in Fig. 1.4.

The high level architecture view of each ITS-station was therefore defined as shown in Fig. 1.5.

Security is of course a key feature, especially for safety systems (but also with respect to privacy) and so ISO 21217 [10] espouses the concept of the "Bounded Secure Managed Domain" (BSMD).

Fig. 1.5 Simplified ITS-S reference architecture [10]

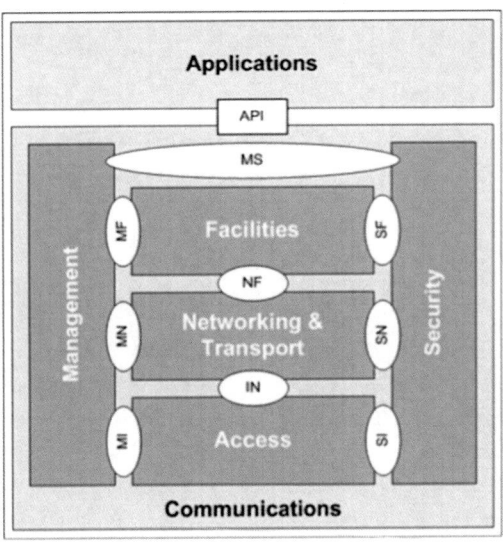

However, while some transactions, particularly important safety transactions, need to operate within a BSMD, that is not true for all transactions. Experience from interactive sat-nav systems with traffic information, RDS-TMC, and Variable Message Sign (VMS), has shown that drivers quickly and easily discount redundant and out of date information, even if they find it annoying. An ice alert, for example, carries little downside if it continues to show after the ice has melted, or even if indeed that data was inaccurate. The car and/or driver is simply more alert and cautious to this risk for a while. On the other hand, for collision avoidance, ramp access control, etc., security and faith in the data received, is of time critical, crucial and paramount importance.

Figure 1.6, taken from ISO 17427-1 [9], shows the different levels of security required within a cooperative ITS architecture, managed within the ISO 21217 concepts.

The ISO multiple media supporting ITS-station concept can also be portrayed, as displayed in ISO 21217 as in Fig. 1.7, showing examples of adoption in the vehicle, road side infrastructure, the service centres and the personal devices.

The ISO standards, by now, constitute a comprehensive suite where several management issues are identified:

- Local station management including interference and channel congestion management (ISO 24102-1)
- Access technology management (ISO 21218)
- Remote station management (ISO 24102-2)
- Station-internal management communications (ISO 24102-4)
- Application management

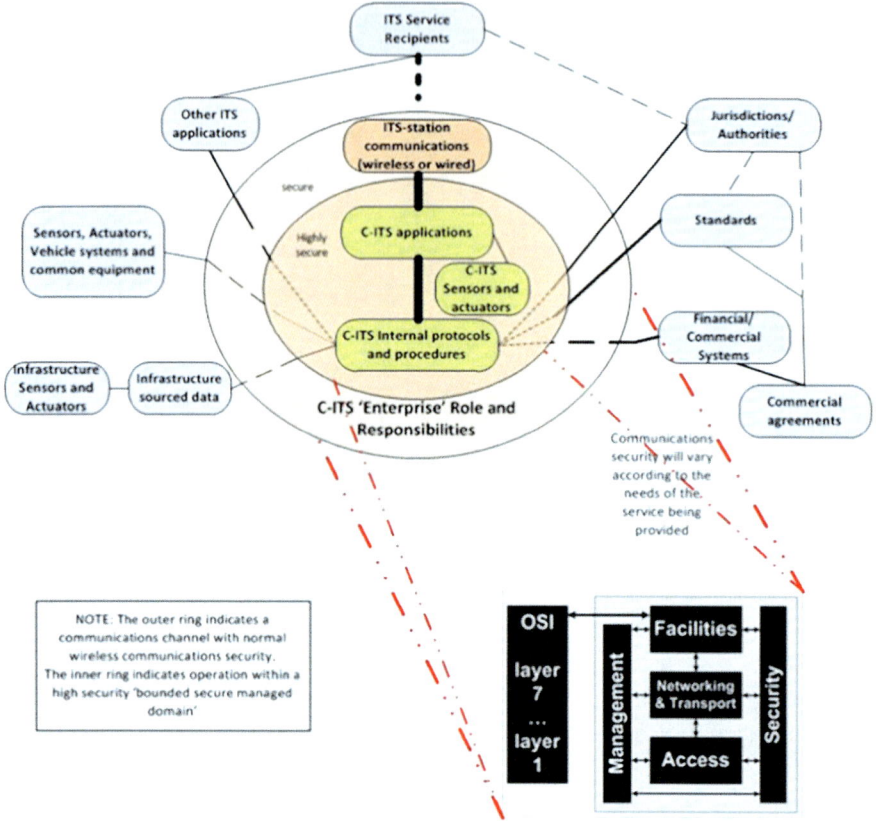

Fig. 1.6 C-ITS Roles and Responsibilities mixed security requirements architecture [11]

- Selection of communication profiles (CEN/ISO TS 17423)
- Path and flow management (ISO 24102-6)
- Service advertisement (ISO 24102-5)
- Management service access points (IS0 24102-3)

Early implementations do not need to implement all of this, and management standards are continuously extended.

* * *

Here, from its first conception around 2000, we have the first true VANET architecture. Each node, be it car or infrastructure, was simply an ITS-station, and it could and would communicate, within its firewalls, with any other compatible node to form an ad-hoc network.

Of course those nodes acting as the infrastructure would perform additional, largely broadcasting to the network, tasks, and provide a link to other non-ITS systems for all vehicles within their range, but this becomes an application level activity, and the basic peer-to-peer network provided the opportunity for vehicles to make and utilize ad-hoc networks.

1 The History of Vehicular Networks

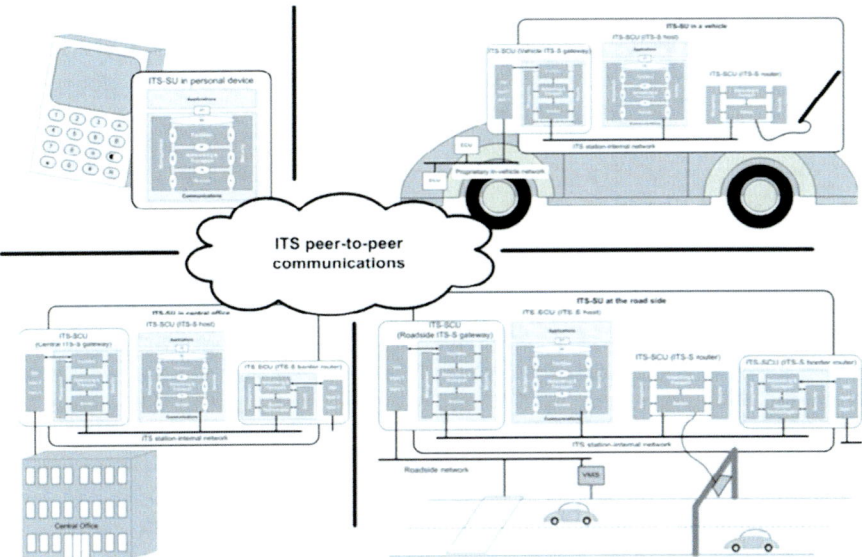

Fig. 1.7 Examples of implementations of ITS-station units [10] and [7]

* * *

But life is not a simple singular and straightforward path, so in parallel, but somewhat separate from these developments, automotive researchers had meanwhile determined that with more bandwidth at 5.9 GHz, they could use the longer range (500 m as opposed to 5 m) system and provide important safety systems, such as collision avoidance, ramp access control, and in the shorter term ice, fog, and obstacle alerts.

Facing the issue of the 5.8 GHz infrastructure cost, considered unfeasible, they thought that due to the extended range provided by the 5.9 Ghz technology a cost reduction factor of 10x would have been achievable. The beacons would not be anymore the source of identification of vehicle location (Global Navigation Satellite System, GNSS, has moved this on) and simple unidirectional antennas would have been used in most cases. In their view, the infrastructure cost would become viable.

But US Department of Transport (DoT), with its crumbling and underfunded road structure, and road authorities in Europe, with the heavy infrastructure costs of the largely non-tolled roads in Europe, combined with heavy adverse budget pressures, soon disabused them that there would be any widescale roll-out of such an infrastructure network.

However, the researchers argued, if the networks were operating in a peer-to-peer fashion, why did they need the infrastructure at all? Particularly as they could, at least for some nodes, connect to previously infrastructure dependent service provision via Wi-Fi from within the vehicle, or via cellular communications? In the places where ITS was needed most, there was usually heavy density of traffic, indeed, always another vehicle within 500 m. Messages and data could not just be simply be "hopped" from one car to the next, but through the network could be "multi-hopped" to another vehicle maybe many kilometres away. Following this logic, there was no need for infrastructure investment! *The world would finally be that of V2V VANETS!*

Throughout the 2000s such systems were developed, demonstrated and evolved to the next level. In 2005 at ITS world congress, BMW demonstrated a VANET based skid alert system between vehicles. Mercedes and GM demonstrated VANET based collision warning, passing car warning and similar systems, proving that VANET technology was feasible in practice. By 2010 the large automotive companies of USA had lobbied and persuaded the US DoT to recommend to introduce these systems mandatorily, at 5.9 GHz. Further US DoT conceded that there would be a business case for infrastructure investment at "hot-spots".

All of the ISO Standards were being developed in an open, Intellectual Property Rights (IPRs)-free environment. However, the automotive companies R&D developments were understandably locked into their IPR, and so inconsistencies between the various Standard Development Organizations (SDO) emerged. The researchers reached near impasse, particularly as the scalability of some of the patent based systems became exposed.

While Europe, i.e., the European Telecommunications Standards Institute (ETSI), has developed a comprehensive *cooperative awareness message* (CAM) and *decentralized environmental notification message* (DENM), IEEE in the USA developed the simpler *basic service message* (BSM), which has formed the basis for the most extensive test and trials that have been enacted.

Further, by this time, an EU-US task force was in place to internationalize ITS and cooperative ITS developments and government strategies. EU/US Harmonization Task Groups (HTG) 1 & 3, in a joint study, advised that some of the choices made were "unfortunate", and that in any event, as US 5.9 GHz trials had been expanded, it had become clear that if all ITS services were to be loaded through this 5.9 GHz channel, even with the "lighter" IEEE BSM, the network would soon become overcrowded, even at relatively low traffic densities, let alone 6 and 12 dual direction lane highways at rush hour!

But technology once again moved faster than ITS research and development. For while the ITS community has spent one and a half decades developing these future systems, and had still not made any significant commercial implementations, the world of mobile communications had sprinted forward. The humble analogue cellphone had been ditched in the mid 1990s and the now all-digital, cellphones had moved to Global System for Mobile Communication (GSM) and most to Universal Mobile Telecommunication Systems (UMTS). Both modes supported digital data as well as voice, moving from General Packet Radio Service (GPRS), with theoretical transfer speed of max. 50 kbit/s (40 kbit/s in practice) to Enhanced Data Rates for GSM Evolution (EDGE), with a theoretical transfer speed of max. 250 kbit/s (150 kbit/s in practice), and over the cell ranges achieved for all wireless mobile telecommunications.

Meanwhile, commercial vehicle fleet management systems and some interactive sat-nav systems had long ago ditched the idea of expensive 5.9 GHz unproven

1 The History of Vehicular Networks

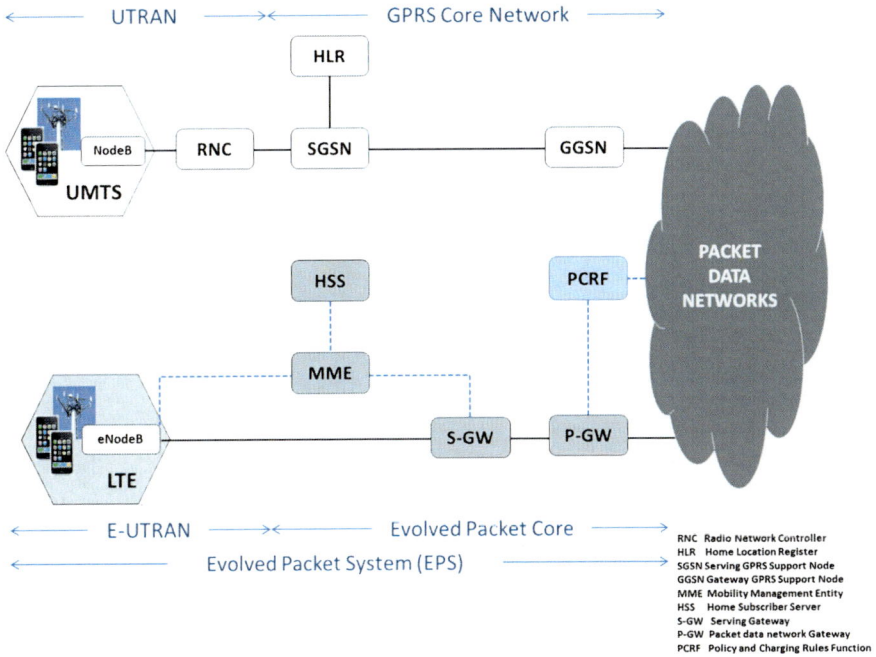

Fig. 1.8 Evolution of the public mobile networks for packet services

technologies, to satisfactorily using GRPS/EDGE. Then in 2012, so-called 3.9G or 4G networks started to be introduced (a variant having already been used in Japan for several years).

These Evolved-Universal Terrestrial Radio Access Network (E-UTRAN), also known in Europe and by the Third Generation Partnership Project (3GPP) as "Long Term Evolution" (LTE) are fundamentally different in that whereas previous generations are "circuit switched networks", these systems are "packet switched networks", totally digital and capable of rapid and fast evolving data rates, offering peak download rates up to 299.6 Mbit/s and upload rates up to 75.4 Mbit/s depending on the user equipment category, and continuing to improve. Figure 1.8 shows the differences among the 2.5G and the LTE networks.

Now a schism developed between the marketing departments of automotive manufacturers and their Research departments.

Nearly everyone now carries a smartphone. In any car there is now likely to be at least as many smartphones as occupants. Nearly all of these smartphones support Bluetooth, a Ultra-High Frequency (UHF) wireless radio system in the Industrial, Scientific and Medical (ISM) band from 2.4 to 2.485 GHz, from fixed and mobile devices, which builds personal area networks (PANs). Invented by telecom vendor Ericsson in 1994, it was originally conceived as a wireless alternative to RS-232 data cables. Bluetooth can connect several devices, overcoming problems of synchronization and making phone calls is just one of their functions. They have become the entertainment and communication centres of their owners.

Because of regulations forbidding manual use of cellphones while driving, almost all vehicles now provide Bluetooth or plugged capability to support hands-free use of the cell phone. Car owners now pressured automotive manufacturers to use this link to also provide infotainment linked to their smartphone. The marketing departments have pressurized the automotive product designers to provide these facilities.

It is now only one step further to use this technology, whether linked by Bluetooth to the car owners phone, or embedded in the vehicle, for ITS service provision.

Additionally, in Europe, regulation for eCall, a post incident "silo" system to link affected vehicles to the emergency services, will legally require a cellphone Universal Subscriber Identity Module (USIM) in every light vehicle, so the basic infrastructure will already be in place.

On another front, Wi-Fi is now supported on all smartphones, and Wi-Fi operators want the 5.9 GHz band for shared use. They argue that while the ITS sector has implemented just a few hundred research units at 5.9 GHz, in the past 15 years, they have implemented several billion active users, including for safety applications. They are currently arguing for shared use of the bandwidth, if granted this will fundamentally affect the current approaches to provision of critical safety services dependent on using 5.9 GHz.

But it is clear that, for advanced ITS safety systems, a preselected 5.9 GHz or similar "fast" ITS dedicated communication system will still be required.

1.2 V2X Communication Scenarios and Requirements

V2X communication is a very general term that includes all possible forms of communications involving a vehicle and the external environment. It is the natural extension of the VANET concept, where the vehicle it is not anymore the only communication node involved, but the vehicle becomes part of a larger system where many elements are involved together. It belongs to the family of the Cooperative Intelligent Transport Systems (C-ITS) communications.

Automotive researchers have developed a number of application and service scenarios, which have been considered and specified based on this concept, and can be considered as tightly related to the VANET technology.

In order to be actually deployed in the operational environment, most of the application scenarios require the adoption of some standardized solution (e.g. the vehicles need to speak some "common language") and a high level of adoption by the vehicles on the roads. Even if it is agreed that an increasing number of vehicles will be equipped with V2V communication capability in the coming years, the level of market penetration of these technologies will have to increase gradually with rate that will be proportional to the vehicle replace rate. As a matter of fact, several decades will be needed before an adequate level of V2V technology penetration in the market will be reached able to guarantee a reasonable operational service availability for those services, such as collision avoidance and ramp management, that require a high population penetration to enable them to operate safely or effectively.

For this reason, it was soon realized that the V2V communication technology alone was not enough to enable a fast and effective deployment of most of the application scenarios proposed by the ITS community and pushed by the policy makers. Therefore, the extension of the same communications principles and technology was considered by realizing that the vehicle should not restrict its capability to communicate to other vehicles only, but should become able to connect with many different communication nodes that can be deployed or used in the roadside infrastructure, remote service centres, pedestrians, bikers, smart devices, etc. (the ITS-station concept). So the concept of V2I and V2V evolves to a more general extension to the vehicle-to-anything (V2X).

Nowadays, in the ITS community, it is taken for granted that the vehicle communication capabilities will enable the future vehicles to become one of the many co-operative communication nodes of a distributed ITS ecosystem supporting an increasing variety of services and applications by means of the use of the most appropriate communication media.

This vision is not restricted to vehicular transportation only, but it will also include other transportation modes that will equally benefit from a gradual adoption of ICT in order to make the mobility of people and freight safer, more efficient, environmentally and economically sustainable. This vision is represented by Fig. 1.9 showing the vehicle as one of the elements of a fully integrated multi-modal ITS communication ecosystem.

The only way to make possible the establishment of such extended integrated communication environment is to develop and adopt a standardization approach. At the beginning many proprietary technological solutions have been developed able to quickly answer to specific market or operational requirements. These solutions, even if sometimes elegant from a technical standpoint, often miss to fulfill objectives for scalability, massive deployment and interoperability. The result has been the creation of a very fragmented market for ITS solutions which are often very difficult to integrate or anti-economical to deploy and operate.

ITS is a very global market involving a large number of private and public stakeholders. Different specific geographic constraints apply from both regulatory and policy standpoints (e.g., regional frequency bands regulations, communication licensing rules, regulated services, strategic industrial priorities, etc.). A global standardization process is the fundamental tool needed to convert these general constraints into technically usable requirements.

A number of SDO have been involved in the process and one result has been the formalization and publication of standards based optimizing aspects of specific communication stacks. When trying to integrate the different standards into a global ITS scenario, some inconsistencies and gaps have evolved and the following phase, currently in progress, is devoted to the creation and increasing involvement of the different SDOs and ad-hoc task forces into a joint harmonization process which today is achieving a worldwide footprint.

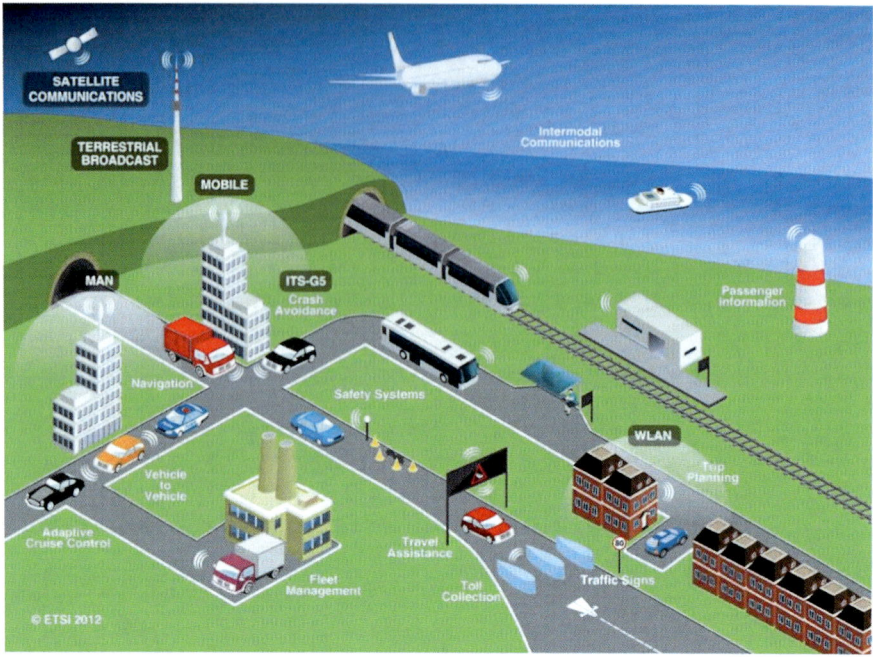

Fig. 1.9 The extended ITS scenario (*source* ETSI ITS)

As it becomes clear that the future world of ITS will be multi-media, in order for VANETs to function efficiently and usefully, we need first to ensure that whenever applicable/possible the service should be provided with a "media-independent" approach, this means that the application standardization groups, largely ISO and CEN, need to better ensure the provision of applications should be media independent. This also means that the providers of communications standards, largely IEEE and ETSI, should ensure that their standards provide usable application-unaware communications and not application biased nor application-centric communications standards. Here the evolution of smartphone "apps" provides a model that can be followed. Temptation to add "features" into heartbeats and awareness messages, because the developers can see its usefulness in specific applications, loads and burdens the communications standards with overhead that is not used by most, and clutters up the network, largely uselessly.

Also, while the focus of interest of course lies with the most safety critical systems (collision avoidance, ramp access management for example), those systems require a high population penetration to work, and has already been observed, with a car park replenishment rate of about 5 % per annum, may not be fully workable for 20 years. Designing such systems to operate using only the current 5.9 GHz band, with current radio technology, is therefore fraught with danger. Put yourself back 25 years in your mind to the world of analogue mobile communications, with handsets the size of bricks and struggling to make even a reliable phone call most of the time. Could you have imagined then the state of the art today for mobile communications? Can you therefore realistically imagine today the radio environment in which VANETS will operate 20 years time from today? Perhaps researchers should focus more on what can be achieved in the very near future with low population penetration services, than struggle with the more difficult long-term challenges, albeit that such applications are intellectually and technically far more interesting. And of course, it is these high profile services which attract the attention and sympathy of regulators and Departments of Transport. A conundrum indeed.

The concepts of the BSMD is critical for many of the safety critical applications, but it too can be an unnecessary overhead where security is a lesser threat. Developers of application standards need to be more selective in their propositions for when it is appropriate. Perhaps the answer may be quite simple. It is clear that for most safety critical applications, time is of the essence. Taking the examples of collision avoidance and ramp access management (managing ingress and egress onto/from busy highways), these are time critical, and position critical, applications. They carry far more risk than, for example, an ice alert or a pothole warning. A fog alert has a loose location sensitivity, whilst a pothole alert is very location specific. Neither, in the context of life saving communications, are very time or security critical. Neither have a downside in the event of malicious hacking. Neither need a BSDM. But the collision avoidance and ramp access systems both need rapid communication and security. The malicious hacking of a ramp entry system or collision avoidance system is potentially devastating, and, once in place, highly likely to be the target of terrorists.

All of the more recent developments in ITS wireless communications recognize the desirability of using an IPv4/IPv6 approach for interoperability, but the need, with current radio technology, for more rapid communications for these safety critical systems, is also well understood.

Perhaps the simple solution that we alluded to above would be to limit use of the 5.9 GHz band to safety critical "fast" applications and require all other systems, for example safety systems such as fog and pothole alerts, even cooperative traffic efficiency and active road safety, to use a different wireless medium. Technology developers and regulators need to sit together, perhaps in the framework of the EU/US HTGs, to develop a practicable, workable and politically acceptable solution.

1.3 The Architecture of ITS-Station (ISO, CEN, ETSI)

The reference architecture of what is now referred by most of the standards as the *ITS-Station* (ITS-s) is the result of a long evolution and joint harmonization effort among the many organizations involved in the process.

The initial approach was necessarily to study and develop independent specialized communication stacks able to allow a peer-to-peer communication between two peers.

In most of the cases, the original concept of the proposed communication architecture was based on the adoption of specialized units (e.g. vehicle unit, roadside unit, etc.) conceived for specific implementation.

Multiple SDOs have been and are active in standardizing the specific communication stacks. But most of the physical R&D activity was actually prototyping, testing and further developing proprietary applications and then trying to get their proprietary solutions adopted by the standardization process. This has not been productive nor helpful.

Due to the international nature of the ITS standardization effort, an increasing co-operation has developed among the international standardization organizations involved in the process, such as ISO, ETSI, CEN, IEEE, SAE, Association of Radio Industries and Businesses (ARIB) Japan, Telecommunications Technology Association (TTA) Korea, IETF and International Telecommunications Union (ITU) with the aim of achieving internationally deployed and harmonized standards and worldwide interoperability.

The starting point for harmonization has been the reference architecture of the generic ITS-station.

ISO/TC204 [9]—ITS was established in 1993 and is responsible for the overall system aspects, infrastructure aspects and application aspects of ITS. In particular, its working group "WG18—Cooperative Systems" is focused on C-ITS and, in the frame of its activity started developing the concept of a general architecture for a generic node of the C-ITS network able to accommodate different communication stacks. ISO TC204's communication stacks are developed by its working group 16 "Wide Area Communications", but, with the exception of millimetre wave (60 GHz) and infrared communications, are largely based on adapting other available wireless network technologies to support ITS. In respect of 5.9 GHz, ISO started out to develop its own protocols, but transferred its efforts to collaboration with IEEE to ensure that its IEEE 802.11p and IEEE 1609 standards met its requirements, and have developed its 5.9 GHz communications around these. The basic idea is that this architecture is able to include many different options (i.e., stacks) that can be selected and adopted, whenever applicable, in specific implementations.

CEN/TC278 [1] was established in 1991, predating ISO 204 by a little less than 2 years, and operates, at European level, in tight synergy with ISO/TC204 which manages the corresponding standardization at global level. These days its works jointly with ISO TC204, and works under its lead in global aspects of ITS standardization, and concentrates TC278's remaining efforts on European-specific requirements, largely associated with the EU and the single market.

ETSI/TC-ITS [7] was established with the approval of its Term of Reference by the ETSI Board#64 in 2007 and the related Technical Committee started its activity in January 2008 with the objective to carry out the development and maintenance of Standards, Specifications and other deliverables to support the development and implementation of ITS Communications provision across the network, for transport networks, vehicles and transport users, including interface aspects and multiple modes of transport and interoperability between systems. In general, ETSI produces globally-applicable standards for ICT.

The European Commission in order to speed up the process of deployment of the ITS solutions in the European Market and to steer the standard harmonization process in the domain took a number of actions. The European Commission published an Action plan for the deployment of ITS in Europe in 2008 [3, 4]. This was followed in 2009 by a request to the European standardization organizations to develop harmonized standards for ITS implementation, in particular regarding cooperative systems. This request was formalized in the standardization Mandate M/453 [2]. ETSI and CEN jointly accepted the mandate that was then carried out by ETSI/TC- ITS and CEN/TC278 and finalized in 2013 with the finalization of the "Release1" of the ITS standards.

1.4 Regional Regulation

In respect of regulatory aspects on the use of VANETs, and indeed ITS services in general, approaches have differed throughout the globe.

In Europe, the ITS Directive, the ICT Rolling Plan and the current and subsequent year "Annual Program for Standardisation" encompass the EU aims for regulation and standardization of ITS. VANETs are not specifically mentioned in any of these documents, but of course many of the applications promoted will or may use VANETS.

So far regulation and regulation proposals in Europe that affect ITS are restricted to four areas: Electronic Fee Collection (EFC), eCall, HGV Tachographs and HGV Weigh in motion. Within the car, there has been far more action, for example requiring electronic traction control systems, but ITS is, by its nature, between a vehicle and other parties outside of the vehicle, so in-vehicle systems are not included here.

In respect of EFC there are regulations determining that in the situation you employ EFC toll collection in Europe, the manner that it shall be done. However, take-up/compliance in this area is poor.

HGV tachograph remote read and weigh in motion are still future regulations, but the EC is minded to use the 1990s 5.8 GHz DSRC technology to do this. Now that there are more capable alternatives, the logic behind this decision may be questioned, but if it is the will of the EC, no doubt will and resource will be found, to provide the work required.

eCall remains a special case. It can in no way be described as a VANET—in its current inception it is better described as a "silo" system. However there were good reasons for this as it was originally conceived as an extension of the public E112 pan-European system already deployed by the European mobile operators in their network and is based on the use of a circuit-switched voice channel to deliver both the data (provided within the "Minimum Set of Data"—MSD) coming from the vehicle involved in a road accident and the actual voice call. Regrettably, the time elapsed since the completion of the standardization work and the decision about the actual operational deployment has been postponed several times and, in the meanwhile the mobile network technology has evolved introducing different technologies which will gradually replace 2G. Therefore, the EC is looking for the means to migrate to packet switched data at an appropriate time and has already mandated to the involved standardization organizations (namely CEN and 3GPP) the related analysis for the future migration.

Anyway, eCall does bring a USIM into a vehicle which could be used to also support C-ITS functionality—if the network operators and application designers can use this opportunity. However the current eCall modem is a very limited and constrained beast, and it may well be that eCall migrates to support over an ITS-station over a period of time.

Another important aspect to enable the deployment of ITS at global level is related to the spectrum allocation and to the relevant regulations. In Europe, the EU Decision 2008/671/EC [3] established the use of the 5,855–5,925 MHz band for ITS safety related applications [5]. The deployment in this initial ITS band is in progress with the channel 176 used as control channel. The European ITS channels are compatible with the US-DRSC channelization (see Fig. 1.10) and are close to the radio band used by Wi-Fi devices. This radio band is referred to as the Unlicensed National Information Infrastructure (U-NII) and, in particular, the 5,850–5,925 MHz spectrum is called U-NII-4 and is being studied by the FCC and the NTIA for possible extensions of the spectrum available for Wi-Fi connectivity. In general, all devices operating in any U-NII band must ensure to be able to prevent harmful interference.

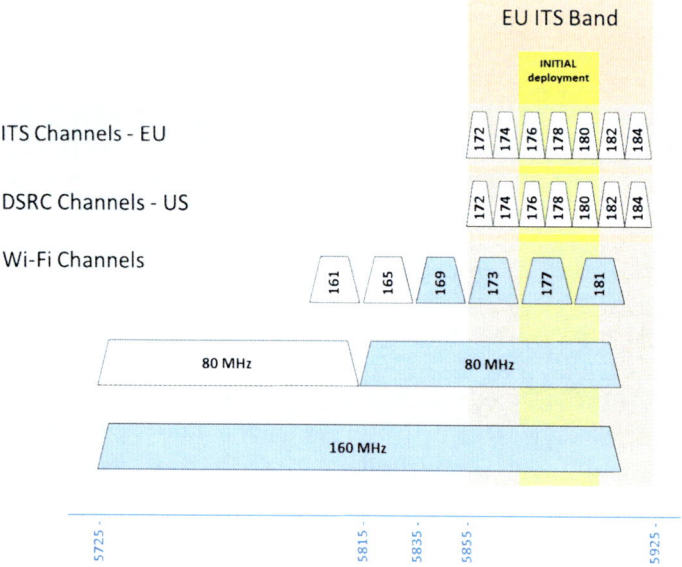

Fig. 1.10 5.9 GHz spectrum allocation

The current proposal coming from some radio local area network (RLAN) stakeholders for sharing spectrum and ensuring coexistence with ITS services consists of migrating the ITS control channel beyond the boundary of U-NII-4, from channel 176 to channel 180. This would not be possible in Europe where the 176–180 channels can be used without restrictions. Therefore, rearranging spectrum as proposed by some parties in the US is not feasible in Europe due to the different spectrum allocations. This is just an example of the issues to be addressed when considering spectrum regulation at global level. In general, some global harmonized spectrum sharing solution would be needed as it will become difficult to control movements of equipment across regions.

In terms of policy, in order to speed up and support the EU goals for the achievement of a competitive and resource-efficient transport system, the European Commission issues specific standardization mandates requests to the European standardization organizations (ESOs) to finalize coherent set of guidelines, specifications and standards to support the different aspects ITS deployment. For example, the Mandate M/453 [2], successfully created the conditions for a joint standardization activity among ETSI ITS and CEN TC278. New mandates are expected soon on strategic areas such as the ITS deployment in urban areas.

References

1. CEN (2014) CEN/TC278 - Intelligent transport systems. Retrieved from http://www.itsstandards.eu/
2. EC - DG ENTR D4 (2009) Standardization mandate addresse to CEN, CENELEC and ETSI in the field of information and communication technologies to support the interoperability of co-operative systems for intelligent transport in the European Community. European Commission
3. EC (2008) Action plan for the deployment of intelligent transport systems in Europe. European Commission. Commission of European Communities, Brussels
4. EC (2009) Action plan on urban mobility. Commission of the European Communities, Brussels
5. ECC (2008) ECC Decision of 14 March 2008 on the harmonised use of the 5875–5925 MHz frequency band for intelligent transport systems (ITS). Electronic Communications Committee. ECC/DEC/(08)01
6. ETSI (2014) Intelligent transport systems. Retrieved from $http://www.etsi.org/technologies-clusters/technologies/intelligent-transport$
7. ETSI TC ITS (2010) ETSI EN 302 665 V1.1.1 (2010-09) - intelligent transport systems (ITS); communications architecture. ETSI
8. Flurscheim H (1925). Patent No. (US 1612427 A) 28522/23. UK, 1925
9. ISO (2014). ISO/TC 204 Intelligent transport systems. Retrieved from $http://www.iso.org/iso/iso_technical_committee?commid=54706$
10. ISO (n.d.) ISO 21217 - Intelligent transport systems - communications access for land mobiles (CALM) - architecture. ISO
11. ISO/NP 17427-1 - Intelligent transport systems – cooperative ITS – Part 1: roles and responsibilities in the context of co-operative ITS architecture(s)

Chapter 2
Standardization and Harmonization Activities Towards a Global C-ITS

Hans-Joachim Fischer

Abstract Standardization plays a crucial role in facilitating adoption of a technology while enabling the interoperability between products of different manufacturers. Standardization of intelligent transport systems (ITS) involves stakeholders in different regions. For enabling a globally interoperable ITS system, EU, Japan, and the USA closely work together towards harmonized standards. A compact overview on a consistent set of important standards to build ITS station units for vehicular ad hoc networks (VANETs) and cooperative-ITS (C-ITS) and, in general for ITS, is provided in this chapter. To a large extent, references are made to standards from CEN, ETSI, IEEE, ISO, and SAE.

Keywords VANET • ISO • ETSI • IEEE • WAVE • CALM • CEN • C-ITS • ITS-Station • IEEE 1609 • Standardization • Harmonization

2.1 Introduction

Vehicular ad-hoc network[1] (VANET) is a term mainly used in scientific papers, presentations, and white-papers rather than in standards. However the term is associated with technologies (architecture, data, and protocols) developed and standardized under the work title of *intelligent transport systems* (ITS) and cooperative ITS (C-ITS).

VANETs differ from other networks, mainly due to the dynamic changes of topology of communication nodes. Their applications are related to road safety, traffic efficiency, and comfort, but include also commercial services. ETSI TC ITS summarized use case specifications for "day 1 deployment" in TS 102 637-1 [13]. Examples are:

---

[1]The term "ad-hoc networking" refers to a network without infrastructure nodes.

H.-J. Fischer (✉)
ESF GmbH, Fichtenweg 9, 89143 Blaubeuren, Germany
e-mail: HJFischer@fischer-tech.eu

- Active road safety—driving assistance (to be implemented in the European C-ITS corridor The Netherlands–Germany–Austria)
  - cooperative awareness
  - road hazard warning
- Cooperative traffic efficiency
  - speed management
  - navigation
- Cooperative local services
  - Point of interest notification
  - automatic access control and parking management
  - ITS local electronic commerce
  - media downloading
- Global Internet services
  - insurance and financial services
  - fleet management
  - loading zone management
  - others

A special characteristic of VANETs is that the nodes in the network typically are changing their position quite rapidly (vehicles), and by this the network topology and potential links between nodes also change. In such a context multi-hop communications are difficult to be achieved and even the setup of communications with neighbor stations becomes tricky.

However, essential in VANETs is that information can be transported either to next neighbor stations or to defined *geo-areas* and disseminated there. This is the case of road safety and traffic efficiency use cases that are designed to operate on top of an IEEE 802.11 radio with non-IP communications (typically single-hop), whilst the other use cases may benefit from an IPv6 over a cellular link. It is to be noted that geo-dissemination of information in VANETs can also be implemented using IPv6. Usage of both communication technologies by a single application may be beneficial or even necessary.

Standardization of ITS and C-ITS is done in various governmental and non-governmental Standard Development Organizations (SDOs), as illustrated in Fig. 2.1. The purpose of standardization is to enable implementation of protocols in an interoperable way such that, in the case of C-ITS, communication nodes with different architectures and from different vendors "speak the same language", i.e., can exchange and interpret messages based on common syntax and semantics.

Whilst two decades ago, at the start of ITS standards development, harmonization between SDOs was a "private activity" of experts involved in different SDOs, the situation has changed with the more formal request for cooperation and harmonization expressed by the European Commission and the United States Department of Transportation (DoT).

Fig. 2.1 Standard Development Organizations developing ITS

Harmonization of standards will reduce costs of hardware and software development and maintenance, and allow for a global market.

The chapter serves the purpose to guide the interested reader in the complex standardization landscape.

The remainder of the chapter is organized as follows. The main standardization bodies along with their scope will be presented in Sect. 2.2. Sections 2.3 and 2.4, respectively, introduce the reference architecture with a focus on the more general ITS station and communication architecture developed in ISO and standardized in ISO 21217/EN 302 665, but also informing about the highly optimized WAVE (Wireless Access in Vehicular Environment) device architecture. Section 2.5 shortly summarizes the ongoing harmonization efforts. Finally, Sect. 2.6 concludes the chapter, by providing an overview of main standards for C-ITS.

2.2 Main Standardization Bodies

Among governmental SDOs, the main advocate on ITS standards is the International Standards Organization (ISO) Technical Committee (TC) 204 "Intelligent Transport Systems."

ISO/TC 204, created in 1992,[2] with its 18 working groups (WGs), is responsible for the overall system and infrastructure aspects of ITS, taking into account the work of existing international standardization bodies.

ISO TC2 04 cooperates with the Comité Européen de Standardization (CEN) TC 278 *ITS*,[3] developing jointly CEN/ISO standards.

---

[2]http://www.iso.org/iso/iso_technical_committee?commid=54706.

[3]http://www.itsstandards.eu/index.php?option=com_content&view=article&id=47&Itemid=27.

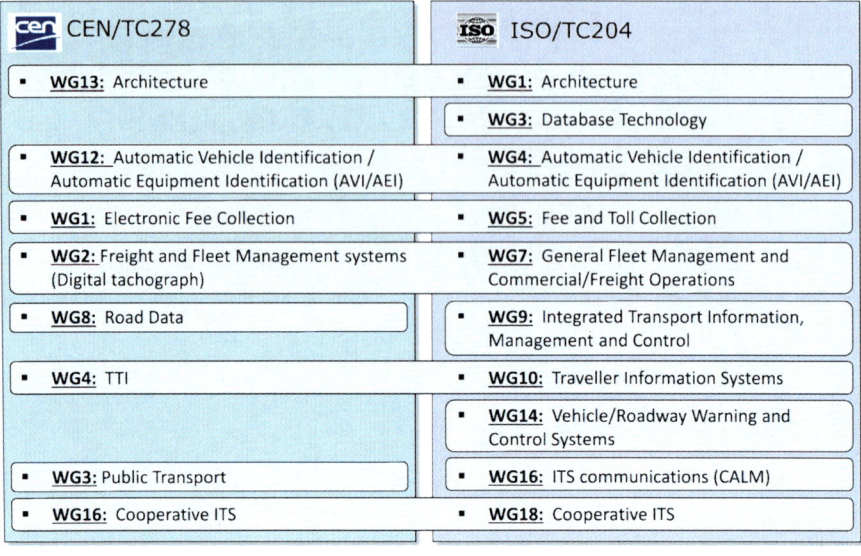

Fig. 2.2 Joint development of standards by CEN TC278 and ISO TC204

CEN/ISO are working on all layers and entities in the ITS protocol stack with a global scope serving the needs of all stakeholders (i.e., road operators, city authorities, car makers, cellular network operators). The alignment of the WGs is shown in Fig. 2.2.

The International Telecommunication Union (ITU) recently just tried to start creation of a unit for ITS. No standards are available so far.

Among non-governmental SDO, as an initiative of members of ISO TC204 WG16 "Wide Area Communications"—also known under the work title of "Communications Access for Land Mobiles" (CALM), the European Telecommunications Standards Institute (ETSI) TC ITS was founded in 2007[4] in order to benefit from faster processes in ETSI and the ETSI competence center on testing. In practice, ETSI followed a car-centric approach focused on a single protocol stack for car-to-car and car-to-roadside communications with C-ITS applications for road safety and traffic efficiency.

Institute of Electrical and Electronics Engineers (IEEE) standards such as IEEE 802.11 are used for ITS, and IEEE 1609 WG[5] developed a set of standards under the work title "Wireless Access in Vehicular Environment" (WAVE), which is an optimized sub-system of ITS with a specific own architecture and a major focus on IEEE 802.11 access technology.

---

[4]http://www.etsi.org/technologies-clusters/technologies/intelligent-transport.

[5]http://standards.ieee.org/develop/wg/1609_WG.html.

A further organization developing standard-like specifications is the Society of Automotive Engineers (SAE)[6] in the USA with a focus on specifications of data dictionaries, also referred to as message sets. SAE cooperates with CEN, ETSI, and ISO.

In Japan, Association of Radio Industries and Businesses (ARIB), Japan's standardization organization, performs investigative studies, research and development on C-ITS. ARIB standards are already deployed in large quantities.

2.3 The ITS Station

2.3.1 Reference Architecture

CEN, ETSI, and ISO share a common *ITS station and communication reference architecture* specified in ISO 21217 [21] and illustrated in Fig. 2.3. This architecture is also presented in the ETSI standard [11] that was written on the basis of the already published ISO 21217.

The ITS station reference architecture uses a simplified and extended Open Systems Interconnection (OSI) model. The communication layers are the *ITS-S access layer* (OSI layers 1 and 2), the *ITS-S Networking & Transport layer* (OSI layers 3 and 4), and the *ITS-S Facilities layer* (OSI layers 4, 5, and 6). On top of the communications layers is the *ITS-S Applications* entity.

The *ITS-S Management entity* and the *ITS-S Security* entity provide management and security features. These layers and entities are interconnected with Service access points (MI, MN, MF, MA, MS, SI, SN, SF, IN, NF and FA), with MA, FA, and SA being implemented in an application programming interface (API).

This architecture supports:

- any kind of access technology for station-internal and station-external communications (e.g., infrared, microwave, millimeter wave, 2G/3G) to provide seamless connection to the end users and applications;
- any kind of networking and transport protocol. In addition to the Internet Protocol version 6 (IPv6) gluing together all different access technologies, to support some ITS applications for which IPv6 may fail due to, for example, the rapidly changing environment, geonetworking [10], and a low overhead network and transport layer protocol, called CALM Non-IP networking [34], have been developed;
- a large variety of facilities (e.g., message handling, Local Dynamic Map, publish-subscribe mechanisms for standardized data and messages);

---

[6]http://www.sae.org/standardsdev/dsrc/.

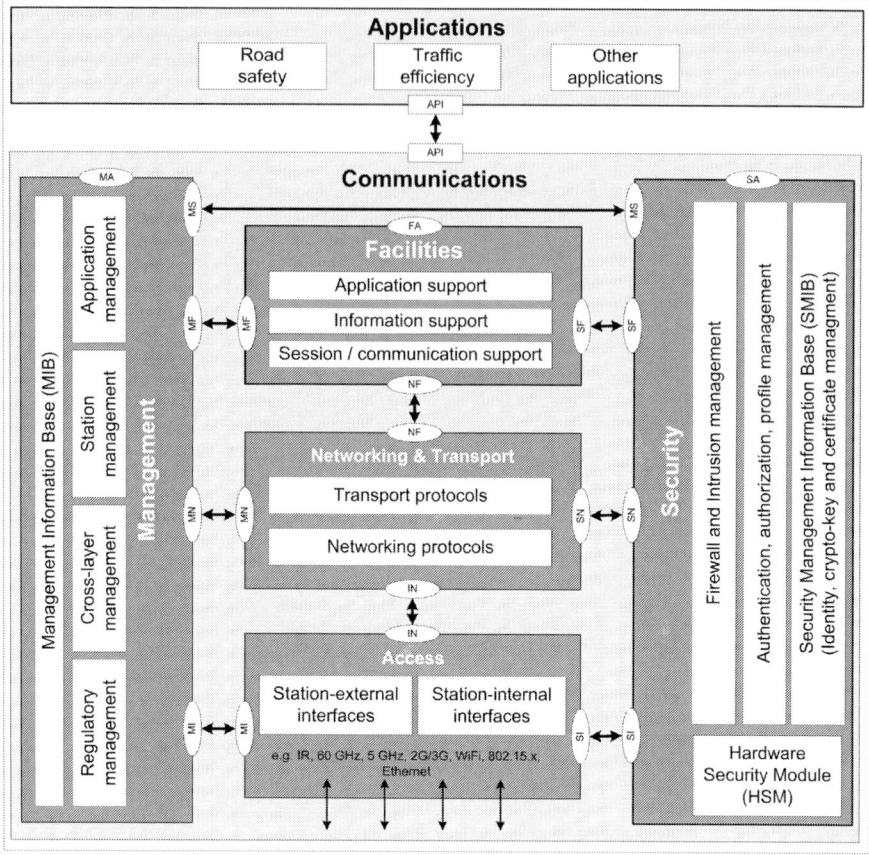

Fig. 2.3 ITS station reference architecture

enabling various communication protocol stacks for non-IP and IPv6 communications according to the needs of C-ITS applications (e.g., road safety, traffic efficiency, comfort, and commercial purposes).

The ITS station is defined as a bounded secured managed domain (BSMD) in order to provide trust to the users. Trust can be achieved with certificates and a public key infrastructure (PKI) by authentication of ITS station units and ITS applications ("Is the station unit authorized that this software is being installed?", "Is this an authorized software that I can accept to be installed in my station unit?"), and by authentication of message sources (is the station unit authorized to send me this type of message?).

Essential elements and procedures to ensure trust are illustrated and standardized in CEN/ISO 17419 [1].[7]

---

[7] A number of standards related to security details are already published or in preparation. Major work is done at IEEE and ETSI.

2.3.2 Abstraction of Applications from Communications

A further essential characteristic of an ITS station is the abstraction of applications from communications illustrated in Fig. 2.4, i.e., ITS-S application processes [21] do not need to know anything about the communication tools in an ITS station unit, but inform the ITS station management about their functional requirements for communications as specified in CEN/ISO 17423 [2].

Subsequently the ITS station management, in a continuous process, maps the communication flows of an application to be the best suited available communication protocol stack for the given location of the ITS station unit as specified in ISO 24102-6 [27] using the communication protocols' status, the ITS-S applications' requirements and objectives, and a set of rules (e.g., regulations and policies) as illustrated in Fig. 2.4.

This allows applying efficiently hand-over mechanisms in the context of the rapidly changing VANET network topology. Dedicated hand-over protocols for, e.g., change of channel, change of access technology, change of networking protocol, are under preparation at ISO TC204.

Multiple access technologies (OSI layers one and two) are supported by means of adaptation layers specified in ISO 21218 [28].

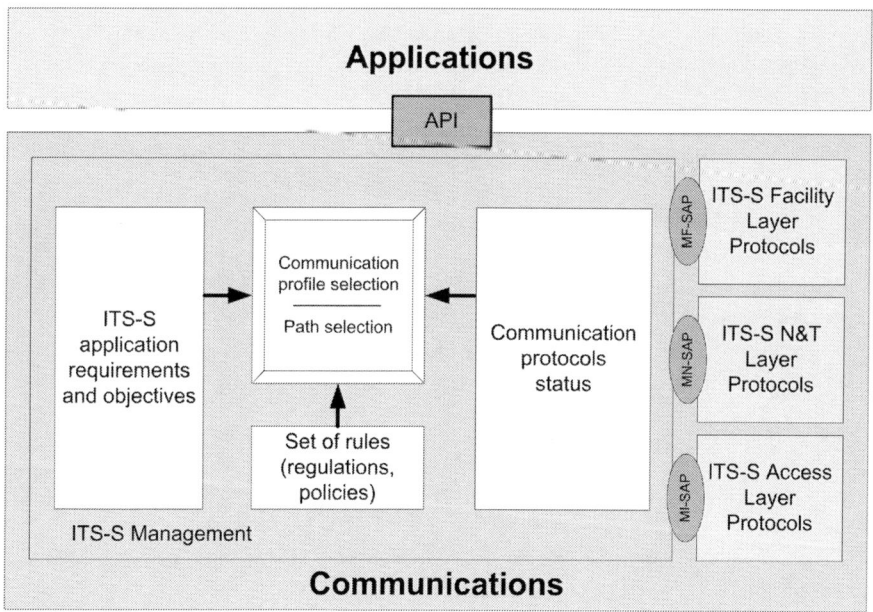

Fig. 2.4 Abstraction of applications from communications

2.3.3 Messages and Data

Management of data to be disseminated at specific geo-locations, and probably to different ITS-S application processes in the same ITS station unit which is an essential characteristic of C-ITS, is enabled by basically two complementary mechanisms:

- a direct publish-subscribe mechanism specified in CEN/ISO 17429 [3] for the various messages, e.g., basic safety messages (BSM) [36], cooperative awareness messages (CAM) [8], decentralized environmental messages (DENM) [9], signal phase and timing (SPAT) [27], In-vehicle Information (IVI) [5];
- the Local Dynamic Map specified in CEN/ISO 18750 [4] within the ITS station facilities layer acting as a data store within the ITS Station for standardized geo-referenced and time-stamped data elements.

Besides the dissemination of data, there is also the need to provide services in sessions whilst a vehicle is passing a roadside installation.

For this purpose, a roadside ITS station unit acting as a service announcer periodically transmits *service advertisement messages* (SAMs) with the fast single-hop *push* service advertisement protocol FSAP (fast service advertisement protocol, specified in [26]) over FNTP in the *Service Advertisement Channel*, inviting vehicle ITS station units acting as service users to perform the session on the indicated *Service Channel* (SCH). Note that the standards also allow a mobile station to be a service provider, and any station being simultaneously a service provider and a service user.

2.4 The WAVE Architecture

In addition to the presented ITS station architecture specified in ISO 21217 [21], the WAVE device architecture is standardized in IEEE 1609.0 [17]. This architecture by intention only provides a sub-set of the functionality of an ITS station.

WAVE is built on top of IEEE Std 802.11 [20], and optimized for fast reliable broadcast of safety messages.

The 1609 family of standards includes:

- *IEEE Std. 1609.0, Architecture*. It describes how the IEEE 1609 standards work together.
- *IEEE Std 1609.2, Security Services for Applications and Management Messages*. It defines secure message formats and processing.
- *IEEE Std 1609.3, Networking Services*. It defines network and transport layer services, including the "WAVE Short Message Protocol" (WSMP) for efficient single-hop null-networking communications, and ordinary IP addressing and routing (without the mobility features standardized in ISO).

- *IEEE Std 1609.4, Multi-Channel Operation.* It provides enhancements to the IEEE 802.11 Media Access Control (MAC) to support multichannel WAVE operations.
- *Draft IEEE P1609.6, Remote Management Services.* It provides inter-operable services to manage WAVE devices that follow IEEE Std 1609.3.
- *IEEE Std 1609.11, Over-the-Air Electronic Payment Data Exchange Protocol for ITS.* It defines the services and secure message formats in support of secure electronic payments.
- *IEEE Std 1609.12, Identifier Allocations.* It indicates identifier values that have been allocated for use by WAVE systems, including the Provider Service Identifier (PSID) allocations harmonized with the ITS Application Identifier (ITS-AID) used in ISO, CEN, and ETSI, ISO, CEN, and ETSI.

2.5 Harmonization in C-ITS

In Europe, a more synchronized development of standards started in October 2009 with mandate M/453 of the European Commission [37]. Standards dedicated to C-ITS and the M/453 were developed at CEN TC278 WG16 and ETSI TC ITS. CEN TC278 WG16 works jointly together with ISO TC204 WG18.

The term C-ITS was created in the context of this mandate and points to a feature of ITS, i.e. the sharing of data between different ITS applications.

In November 2009, EC/DGINFSO and USDOT/RITA signed a Joint Declaration of Intent on Research Cooperation in Cooperative Systems [14], which finally lead to the EU-US Cooperative Systems Standards Harmonization Action Plan (HAP) [15] in 2011, where also Japan is actively contributing, and other regions are preparing to join this approach.

The joint EU/US standardization approach linked SAE, CEN, ETSI, and ISO together. The attempt was to achieve harmonized solutions for message sets directly from the beginning.

For already existing standards, the Harmonization Task Groups HTG 1 and HTG 3 investigated in gaps and overlaps of communications and security standards, and provided recommendations to SDOs on missing standards and standards to be harmonized.[8]

HTG3 published a recommendation [16] to harmonize the over-the-air message formats of two protocol sets from ISO and IEEE, i.e.,:

- the message formats of the *messaging protocols* FNTP (Fast Networking & Transport Layer Protocol) standardized in ISO 29281-1 [34] with WSMP standardized in IEEE 1609.3 and

---

[8]The deliverables of these two HTGs are online at, e.g., http://ec.europa.eu/digital-agenda/en/news/progress-and-findings-harmonisation-eu-us-security-and-communications-standards-field.

- the message formats of the "service advertisement protocols" FSAP standardized in ISO 24102-5 with WSA (WAVE Service Advertisement) protocol standardized in IEEE 1609.3 [18]. The technical approach on how to harmonize WSMP with FNTP is agreed upon.

There is also a proposal to enable the ETSI GeoNetworking functionality [10], developed to enable geo-dissemination in the European Commission's GeoNet project (http://www.geonetproject.eu/), in this harmonized message format by using the GeoNetworking headers and related procedures.

It has to be noted that:

- the series of GeoNetworking standards from ETSI is covered by Intellectual Property Rights (IPRs) from several organizations,
- there are significant problems related to GeoNetworking as identified by the EU/US harmonization task groups 1 and 3 [35],
- deployment projects such as the CONVERGE project (http://www.converge-online.de/) in Germany expressed the urgent need to have a common geo-dissemination protocol suitable for different communication profiles (e.g., single-hop 5.9 GHz communications and LTE) instead of a GeoNetworking protocol. Such a geo-dissemination protocol preferably is located in the ITS station (ITS-S) facilities layer specified in ISO 21217 [21].

In the meanwhile HTGs up to number eight are identified. HTG2 had the task to work on BSM/CAM harmonization. HTG 4 and 5 are in progress to address the harmonization of messaging standards development (i.e., SPaT, MAP, IVI). HTG 6 is working on security standards.

2.6 Conclusion

The standardization of vehicular communications is a quite complex and slow process involving several stakeholders.

A plethora of standards have been published so far covering different aspects, e.g., the overall architecture, the station and communication architecture, messages, management, access technologies, networking, security, and facilities aspects.

Most important standards of the CEN/ISO/ETSI Release 1 under the European Commission's mandate M/453 specifying protocols for VANETs with ITS stations and of the IEEE/WAVE family are summarized in Table 2.1. For the reader's convenience, the book chapters covering aspects addressed by standard specifications are also indicated. This Release 1 already is under revision and extension towards a Release 2. Release 2 can be considered as the first real deployment release for C-ITS. The standardization approach at CEN/ISO, to a large extent, enables adding new features without the need to update standards. Such features are made publicly known by means of registries.

Table 2.1 Summary of relevant standard documents by the main SDOs

Category	Specification number	Content	Reference	Main related chapter(s)
Architecture	ISO 21217	ITS Communication Architecture	[21]	1, 3, 4, 5, 6, 7,
	ETSI EN 302 665	Content aligned with ISO 21217. To be replaced by a reference to ISO 21217	[11]	8, 9
	1609.0	WAVE architecture	[17]	
Access technologies	ISO 21218	General technical details related to the usage of access technologies in an ITS station unit	[28]	3, 4, 16
	ISO 21215	General requirements and guidelines on how to use the IEEE 802.11 access technology in an ITS station unit	[29]	
	IEEE 802.11	Wireless LAN Medium Access Control (MAC) and Physical Layer (PHY) Specifications	[20]	
	ETSI ES 202 663	European Profile Standard for the IEEE 802.11 access technology operating in the 5 GHz frequency band	[12]	
	ISO 17515 series	LTE for ITS	[30]	
Networking	ISO 21210	IPv6 networking	[31]	4, 7, 8, 9
	ETSI TS 102 636	GeoNetworking protocol	[10]	
	ISO 16789	ITS IPv6 Optimization	[33]	
	ISO 29281-1:2013	Fast Networking & Transport Layer Protocol (FNTP) for non IP single-hop and N-hop broadcast communications	[34]	
	IEEE 1609.3	WAVE Networking services	[18]	
Applications	CEN ISO 17423	Application requirements for selection of communication profile	[2]	5, 10, 18
	ETSI TS 102 637-1	Basic set of applications—use case descriptions	[13]	
	ISO 24102-5	Fast Service Advertisement Protocol (FSAP)	[26]	

(continued)

Table 2.1 (continued)

Category	Specification number	Content	Reference	Main related chapter(s)
Management and security	ISO/TS 17419:2014	Globally unique addresses and identifiers used for ITS station management	[1]	10
	ISO 24102-1/2/3/4/5/6	ITS station and communication management	[22–27]	
	IEEE 1609.2 - 2013	WAVE Security Services for Applications and Management Messages	[19]	
	ISO 16788	ITS IPv6 Security	[32]	
Messages and facilities	ETSI EN 302 637-3	Decentralized Environmental Message	[9]	5, 7
	ETSI EN 302 637-2	Cooperative Awareness Message	[8]	
	SAE J2735	DSRC Message Set Dictionary (BSM)	[36]	
	CEN/ISO TS 19321	In-vehicle signage data specification	[5]	
	CEN/ISO TS 19091	(SPaT) Message	[6, 7]	
	CEN ISO 19321	Dictionary of in-vehicle information (IVI) data structures	[5]	
	CEN/ISO 18750	Local Dynamic Map	[4]	
	EN ISO 17429	Message handling	[3]	

Worldwide efforts are underway to facilitate harmonization of similar standards enabling a global market and a single ITS allowing at the same time for regional specialities.

References

1. CEN ISO 17419:2014, Intelligent transport systems – cooperative systems – classification and management of ITS applications in a global context
2. CEN ISO 17423:2014, Intelligent transport systems – cooperative systems – ITS application requirements and objectives for selection of communication profiles
3. CEN ISO 17429:2014, Intelligent transport systems – cooperative systems – profiles for processing and transfer of information between ITS stations for applications related to transport infrastructure management, control and guidance (Title likely will be changed. Publication expected for Q4/2014 or Q1/2015.)
4. CEN ISO 18750:2014, Intelligent transport systems – cooperative systems – definition of a global concept for local dynamic maps

5. CEN ISO 19321, Intelligent transport systems – cooperative systems – dictionary of in vehicle information (IVI) data structures
6. CEN ISO 19091, Intelligent transport systems – cooperative systems – using V2I and I2V communications for applications related to signalized intersections
7. CEN ISO 19091:2014, Intelligent transport systems – cooperative systems – signal phase and timing data structures
8. ETSI EN 302 637-2, Intelligent transport systems (ITS); vehicular communications; basic set of applications; Part 2: specification of cooperative awareness basic service
9. ETSI EN 302 637-3, Intelligent transport systems (ITS); vehicular communications; basic set of applications; Part 3: specifications of decentralized environmental notification basic service
10. ETSI EN 302 636-4-1, Intelligent transport systems (ITS); vehicular communications; GeoNetworking; Part 4: geographical addressing and forwarding for point-to-point and point-to-multipoint communications; Sub-part 1: media-independent functionality
11. ETSI EN 302 665, Intelligent transport systems (ITS); communications architecture
12. ETSI ES 202 663, Intelligent transport systems (ITS); European profile standard for the physical and medium access control layer of intelligent transport systems operating in the 5 GHz frequency band
13. ETSI TS 102 637-1, Intelligent transport systems (ITS); vehicular communications; basic set of applications; Part 1: functional requirements
14. EU-U.S (2009) Joint declaration of intent on research cooperation in cooperative systems, EC/DGINFSO and USDOT/RITA, 13 Nov 2009
15. EU-US (2011) Cooperative systems standards harmonization action plan (HAP), EC DG INFSO and USDOT RITA JPO, 30 June 2011
16. Feedback to ITS Standards Development Organizations Communications, EU-US ITS Task Force - Standards Harmonization Working Group - Harmonization Task Group 3, Document HTG3-3, http://ec.europa.eu/digital-agenda/en/news/progress-and-findings-harmonisation-eu-us-security-and-communications-standards-field
17. IEEE 1609.0 - 2013, IEEE standard for wireless access in vehicular environments (WAVE) – architecture
18. IEEE 1609.3 - 2010, IEEE standard for wireless access in vehicular environments (WAVE) – networking services
19. IEEE 1609.2 - 2013, IEEE standard for wireless access in vehicular environments (WAVE) – security services for applications and management messages
20. IEEE Standard 802.11-2012 - IEEE standard for information technology–telecommunications and information exchange between systems local and metropolitan area networks – specific requirements Part 11: wireless LAN medium access control (MAC) and physical layer (PHY) specifications
21. ISO 21217:2014, Intelligent transport systems – communications access for land mobiles (CALM) – architecture
22. ISO 24102-1:2013, Intelligent transport systems – communications access for land mobiles (CALM) –ITS station management – Part 1: local management
23. ISO 24102-2:2013, Intelligent transport systems – communications access for land mobiles (CALM) – ITS station management – Part 2: remote management
24. ISO 24102-3:2013, Intelligent transport systems – communications access for land mobiles (CALM) – ITS station management – Part 3: Service access points
25. ISO 24102-4:2013, Intelligent transport systems – communications access for land mobiles (CALM) – ITS station management – Part 4: ITS station-internal management communications protocol (IICP)
26. ISO 24102-5:2013, Intelligent transport systems – communications access for land mobiles (CALM) – ITS station management – Part 5: Fast Service Advertisement Protocol (FSAP)
27. ISO 24102-6, Intelligent transport systems – communications access for land mobiles (CALM) – ITS station management – Part 6: path and flow management
28. ISO 21218:2013, Intelligent transport systems – communications access for land mobiles (CALM) – access technology support

29. ISO 21215:2010, Intelligent transport systems – communications access for land mobiles (CALM) – M5
30. ISO 17515, Intelligent transport systems – communications access for land mobiles (CALM) – evolved universal terrestrial radio access network (E-UTRAN)
31. ISO 21210:2012, Intelligent transport systems – communications access for land mobiles (CALM) – IPv6 networking
32. ISO 16788, Intelligent transport systems – communications access for land mobiles (CALM) – IPv6 networking security
33. ISO 16789, Intelligent transport systems – communications access for land mobiles (CALM) – IPv6 optimization
34. ISO 29281-1:2013, Intelligent transport systems – communications access for land mobiles (CALM) – non-IP networking – Part 1: fast networking & transport layer protocol (FNTP)
35. Observations on GeoNetworking, EU-US ITS Task Force - Standards Harmonization Working Group - Harmonization Task Groups 1 & 3, Document HTG1&3-3, http://ec.europa.eu/digital-agenda/en/news/progress-and-findings-harmonisation-eu-us-security-and-communications-standards-field
36. SAE J2735, Dedicated short range communications (DSRC) message set dictionary
37. Standardisation mandate addressed to CEN, Cenelec and ETSI in the field of information and communication technologies to support the interoperability of co-operative systems for intelligent transport in the European Community, European Commission, Enterprise and Industrial Directorate-General, Brussels, 6 Oct 2009

Part II
Fundamentals of Vehicular Communications

Chapter 3
The Physical Layer of VANETs

Riccardo M. Scopigno, Alessia Autolitano, and Weidong Xiang

Abstract This chapter is about the physical layer (PHY) of VANETs and will present all its main features with a bottom-up approach, starting from the relevant physical propagation and reception phenomena; afterwards, the *Orthogonal Frequency Division Multiplexing* (OFDM)-Wi-Fi standard, as adapted to the vehicular environment, will be introduced; then, the architecture of transceivers will be presented, so to match theoretical solutions with practical implementations and a hands-on perspective; eventually, some open research areas will be introduced, highlighting how much the physical layer could be improved in a standard-compliant way. More innovative solutions will be presented in the following chapters.

Keywords VANET • PHY • IEEE 802.11p • ITS-G5 • OFDM • Antennas • Noise figure • Transceiver • Fading • Coherence time • Coherence bandwidth • Inter-symbol interference • Inter-carrier interference • ISI • ICI

3.1 Physical Layer Phenomena Relevant to VANETs

Before entering into details about how the physical layer of VANETs works, it is important to understand the rationale for each mechanism foreseen by the transceiver design and, even before, the challenges which such mechanisms are intended to cope with. For this reason, in this section the most relevant phenomena affecting radio propagation and reception will be shortly reviewed, highlighting their impact on the vehicular environment. The analysis is split into two subsections: Sect. 3.1.1 about the phenomena pertaining the events from the propagation to the

---

R.M. Scopigno (✉) • A. Autolitano
Istituto Superiore Mario Boella, Via P.C. Boggio 61, Torino, Italy
e-mail: scopigno@ismb.it; autolitano@ismb.it

W. Xiang
University of Michigan-Dearborn, Dearborn, MI, USA
e-mail: xwd@umd.umich.edu

antenna coupling, also including quantities such as interference and noise levels; Sect. 3.1.2 will focus on additional channel phenomena and characteristics, such as the coherence time and the coherence bandwidth, which, under some conditions, may significantly affect the received signal.

3.1.1 Signal-to-Noise Ratio at the Antenna Terminal

VANETs are radio systems operating in the GHz range. It is well known from the *fundamental theory* [19] that the received power by a radio system can be computed by the Friis formula (3.1):

$$P_R = P_T * \underbrace{PL_\alpha}_{\substack{\text{Atten.}+ \\ \text{Fading}}} * \underbrace{G_T(\theta_T,\phi_T) * G_R(\theta_R,\phi_R) * \left(1-|\Gamma_T|^2\right) * \left(1-|\Gamma_R|^2\right) * |\mathbf{a}_T \cdot \mathbf{a}_R|^2}_{\text{Effect of the antennas}}.$$

(3.1)

From left to right, the formula states that the received power P_R can be computed as the transmitted power (P_T), reduced by the attenuation factor PL_α and by a series of factors depending on the antennas (G_T, G_R, Γ_T, Γ_R, $\mathbf{a}_T$, and $\mathbf{a}_R$). Attenuation-related and antenna-related factors are, respectively, discussed in the following paragraphs.

3.1.1.1 Signal Attenuation and Fading

In Eq. (3.1) the term PL_α accounts for the propagation phenomena. It is referred to as *attenuation* or *path loss*, but it actually includes also the *fading* term: even if attenuation and fading are usually kept distinct, they are two faces of the same coin (in fact, attenuation is also known as *large scale fading*). The two will be here shortly presented, from an academic perspective: their more realistic modeling will be further discussed in Chap. 12.

In all our analysis we will suppose to be sufficiently apart from the transmitter to adopt the hypothesis of **Fraunhofer Region**—*which takes place when the receiver is at a distance d significantly higher than D^2/λ (where D represents the antenna dimensions and λ the wavelength, which is 5 cm in the case of VANETs).*[1]

---

[1] The Fraunhofer distance $d = D^2/\lambda$ is usually adopted as cut-off between the near- and far-field; in far-field, you can also suppose that wave-fronts are spherical and that the field is TEM (transverse electromagnetic (TEM) modes: neither electric nor magnetic field in the direction of propagation).

3 The Physical Layer of VANETs

It is well known [6] that, in case of *free space* conditions, and under ideal reception conditions (that is, the factors G_T, G_R, and the multiplicative corrective term $\left(1 - |\Gamma_T|^2\right) * \left(1 - |\Gamma_R|^2\right) * |\mathbf{a}_T \cdot \mathbf{a}_R|^2$ equal to 1), Eq. (3.1) becomes:

$$P_R = \underbrace{\frac{P_T}{4\pi d^2}}_{\text{Isotropically radiated power density}} * \underbrace{\frac{\lambda^2}{4\pi}}_{A_\alpha - \text{Antenna aperture}} = P_T * \left(\frac{\lambda}{4\pi d}\right)^2, \quad (3.2)$$

where d is the distance between transmitter and receiver and the antenna aperture,[2] A_α represents the power which can be captured by the receiving antenna[3]; hence, *with free space propagation, the power decays with the power 2 of distance*; additionally, as an effect of the antenna aperture, the received power will also decrease with the square of the frequency being used.

In case you consider also reflections from the ground, you get the so-called *Two-Ray Ground-reflection* model (2RG). Despite simplistic, the 2RG model lays the foundations for more complex attenuation models: it considers the interfering effect of two main rays, a direct one and one reflected by the ground, as depicted in Fig. 3.1. Still under the hypothesis of an ideal and isotropic antennas, and with a perfect reflection by the ground (for an angle of incidence $\theta_i \to 0$, the reflection is almost perfect [7]), the interference at the receiver is purely due to the phase difference Δ between the direct ray and the reflected one, that is:

$$\Delta = \frac{2\pi}{\lambda}\left[\sqrt{d^2 + (h_T + h_R)^2} - \sqrt{d^2 + (h_T - h_R)^2}\right] \approx \frac{4\pi}{\lambda}\frac{h_T h_r}{d}, \quad (3.3)$$

(the approximation holds if d is sufficiently greater than the height of the receiving antenna h). Considering Δ, the received power can then be written [33]:

$$P_R = \left(\frac{\lambda}{4\pi d}\right)^2 * \left(2\sin\left(\frac{2\pi}{\lambda} \cdot \frac{h_T h_R}{d}\right)\right)^2 * P_T \approx \frac{h_T^2 h_R^2}{d^4} * P_T. \quad (3.4)$$

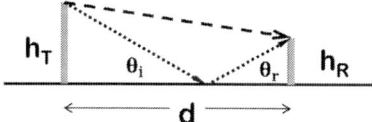

Fig. 3.1 The scenario and quantities which the two-ray ground-reflection model refers to

---

[2]The antenna aperture is measured as a surface (square meters) but is not directly related to the physical size of the antenna.
[3]The aperture of the receiving antenna is $A_{\alpha,R} = G_R * \lambda^2/(4\pi)$, as demonstrated in the antenna theory, for instance in [6]; here, we suppose $G_R(\theta_T, \phi_T) = 1$.

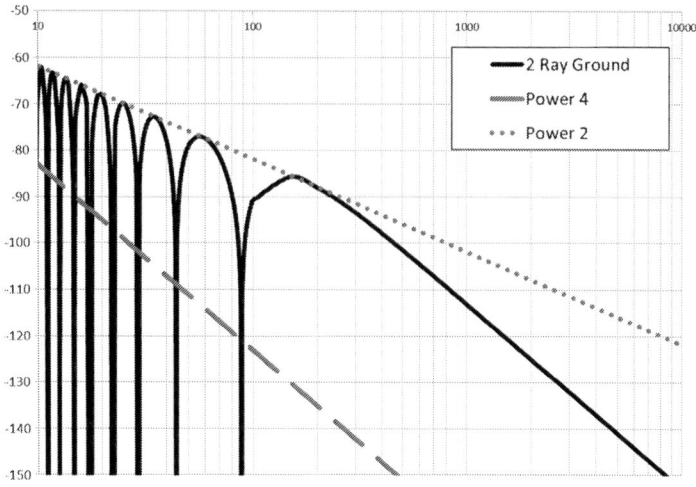

Fig. 3.2 The pattern of received power according to the two-ray model

The first part of Eq. (3.4)—before the approximation—represents a function whose graph (logarithmic scale) is displayed in Fig. 3.2 (for $h_T = h_R = 1.5$ m). The graph shows a cut-off distance at about 120 m which highlights two main phenomena: (1) before the cut-off there are several fluctuations which are interpolated by a decay with slope 2 (20 dB/decade); (2) after the cut-off the fluctuations stop and the plot becomes a line with slope 4 (40 dB/decade)[4]—and this is confirmed by the approximations in the latter part of Eq. (3.4). Fluctuations will be interpreted by fading, while the slope embodies the attenuation: the 2RG model witnesses that it makes sense to model it with a dual slope exponential decay (remember that the graph in Fig. 3.2 is logarithmic).

In *real environments* the attenuation is much more complex: the reflections may be manifold and are not ideal. As a result, the community has at length relied on empirical models and several measurements have sustained the use of a generalized dual-slope attenuation (for instance, [25]), defined heuristically: a cut-off distance at about 100 m, initial slope around 2 and depending on the environment (frequently 1.9), final slope smaller than 4 (typically 3.8). These are the so-called *multi-log models* and have also been adopted by official documents such as [15]: they typically set the maximum reception for VANETs (in line-of-sight) around 200–300 m (Fig. 3.4). Recently, novel models have emerged, including a detailed (and non-statistical) analysis of the environment, as further discussed in Chap. 12, leading also to the analysis of the effect of obstructions (Chap. 15).

Already in the simplistic model of 2RG (which considers only two rays between transmitter and receiver) the power of the received signal oscillates in the initial

---

[4]The attenuation with power 4 seems to violate the principle of energy conservation but, actually, it does not: due to reflections, the power is only differently distributed; if one considers the power received at a height h_T growing with distance, the power would still decay with power 2.

region, as a result of the phase in the composition of the rays. In the target frequency of VANETs you have a λ of about 5 cm: this implies that all the shapes of magnitude comparable (or higher than) λ will cause a significant reflection. Consequently, with VANETs you will meet signals which are the total of several reflected rays, some even after multiple reflections; as a result, the received power may be expected to spread in a quite irregular way, with fluctuations of a λ-scale around the values estimated by the attenuation. *Fading* considers such phenomena and describes them by statistical models (the most frequent are *Rician*, for a strong line-of-sight contribution, *Rayleigh* for non-line-of-sight and *Nakagami*, the most used in literature, addressing more complex cases): they will be further studied in Chap. 15 but need to be considered since now, so to explain some mechanisms which VANETs put in place at physical layer. Fading may account for fluctuations as high as 10–20 dB (or more) on a λ-scale, meaning also that if an object (the receiver or a scatterer) moves over a λ-distance, the signal might suddenly drop. Figures 3.3 and 3.4 show two different qualitative visualizations of fading: the former displays the fluctuations; the latter the effect of fading superimposed to attenuation.

From a practical point of view, fading has two main drawbacks: (1) the performance of reception can vary also within the same frame; (2) one cannot count on the received signal strength (RSS) to estimate mutual distance between cars.

3.1.1.2 Antenna Phenomena

The antenna phenomena would require themselves a volume and their explanation is out of the scope of this book: a rapid overview is here provided just to recall or briefly introduce them. For this purpose we will shortly step back to the formula (3.1); the terms which refer to the antennas (transmitting and receiving) are:

- **Antenna gain**. $G(\theta, \phi)$ characterizes how well an antenna can transmit (respectively receive) a wave coming from a given direction: the reciprocity theorem [38] states that for any antenna, the receiving and transmitting gain are the same. For instance, the reception gain can be defined as the ratio between the electric power produced by the antenna from a far-field source on the antenna's beam axis and the one produced by the same wave on a hypothetical lossless isotropic antenna (equally sensitive to signals from all directions). As such, gain also takes into account the efficiency of the antenna, that is, the fraction of the input power dissipated in losses such as resistance. An example of antenna gain is available in Fig. 3.5. In case of VANETs, for example, it will not make sense to have high gains in the vertical directions (waves will arrive from around).
- **Impedance match**. It refers to the proper adaptation of the impedance of the antenna and of the electronic circuitry, so to maximize the power transfer (or minimize signal reflection) between them. In Eq. (3.1) the additional factors that take the effect of the impedance mismatch into account include Γ_T and Γ_R, i.e. the reflection coefficients of the transmit and receive antennas.
- **Polarization**. In the Friis formula $\mathbf{a}_T$ and $\mathbf{a}_R$ are the polarization vectors of the transmit and receive antennas. The electromagnetic radiations are composed

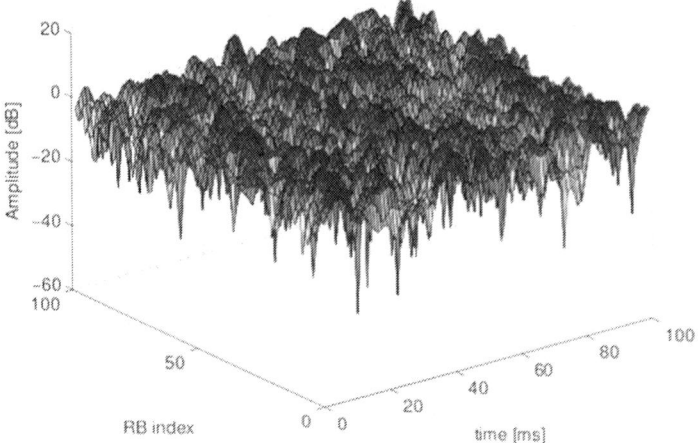

Fig. 3.3 A 3D representation of the signal fading at vehicular speed according to the 3GPP scenarios (i.e., pedestrian, vehicular, and urban) as defined in [1, Annex B.2]) (actually, referred to LTE) and implemented over the open-source Network-Simulator NS-3 (an RB index corresponds to a frequency sub-range). *Source*: http://www.nsnam.org/docs/models/html/lte-user.html

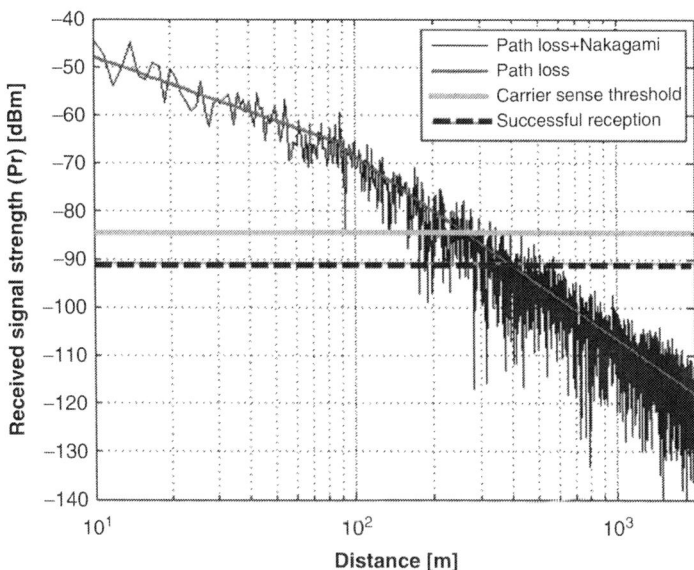

Fig. 3.4 A graph showing some concepts related to propagation and reception: dual-slope attenuation, superimposed fading, reception threshold, carrier sensing threshold. The reference scenario is that of a 20 dBm signal and a 3 Mb/s transfer rate. The threshold for Carrier Sensing (see algorithm CSMA/CA in Chap. 4.) can be set higher than the reception threshold to permit more simultaneous transmissions at a reasonable distance. Taken from [15]—Courtesy of *European Telecommunications Standards Institute* (ETSI)

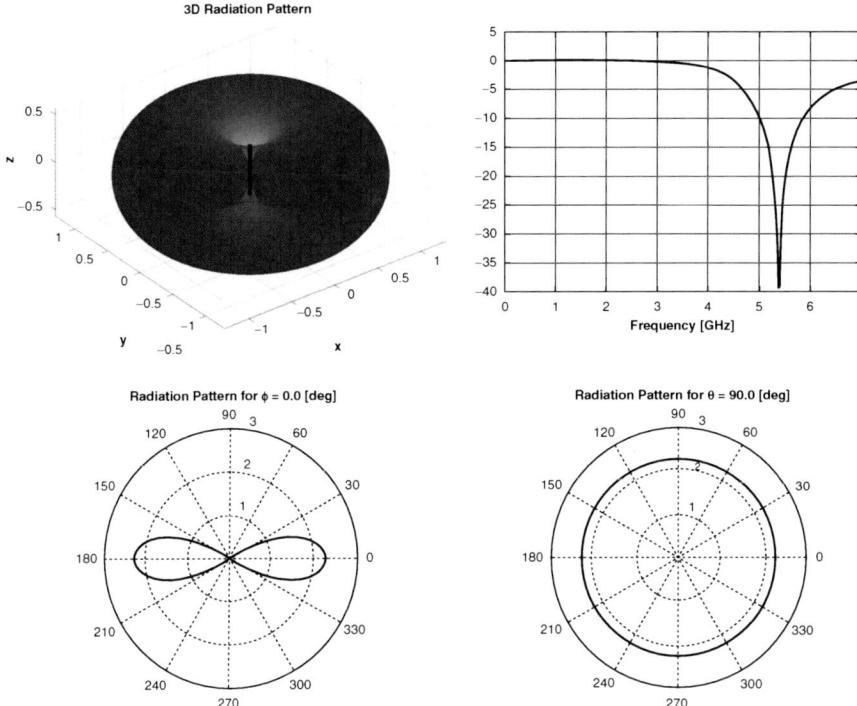

Fig. 3.5 An example of antenna diagram for the gain $G(\theta, \phi)$. (*Top, from left to right*) 3D diagram and frequency response of the reflection factor (notch in the target frequency); (*Bottom, from left to right*) 2D Azimuth and elevation diagrams: the antenna is omni-directional in the Azimuth plane

of electric and magnetic fields: the electric one determines the direction of polarization of the wave (which may be linear—in any directions—or elliptic). When an antenna extracts energy from a passing radio wave, maximum pickup will result when the antenna has the same polarization as the electric field (that is, has the same polarization of the transmitting antenna and $|\mathbf{a}_T \cdot \mathbf{a}_R| = 1$). As already mentioned, in the far-field, the polarization will be basically TEM; additionally, in case of vertical dipoles (commonly used for Wi-Fi and in VANETs) it will be linear and vertical. Importantly, an antenna defines the polarization of the transmitted signal, but reflections can affect it, hence further influence the resulting received signal.

Altogether, once the waves are received at the antenna, after attenuation and reflections, the reception can be facilitated or impeded, depending on the antenna design and on the vectorial properties of the signal: this may have also positive perspectives, as further discussed in Sects. 3.4.1 and 3.4.2.

3.1.1.3 Interference, Noise, and Sensitivity

Considering the propagation phenomena and the reception at the antenna, a generic signal reaches the receiver and is ready for decoding. Intuitively, the stronger the signal and the lower the interference and noise are, the better the reception is. This is actually confirmed by two parameters characterizing the reception. The first one is the *Signal to Interference and Noise Ratio* (SINR) and is defined in Eq. (3.5):

$$\text{SINR} = \frac{P_R}{\sum_i P_I^i + P_N}; \quad (3.5)$$

P_R represents the power of the received signal being decoded (affected by attenuation, fading, and antenna phenomena); $\sum_i P_I^i$ is the sum of the interfering signals (if any) in the target frequency (they may be simultaneous transmissions by other nodes or disturbs); finally, P_N is the power of noise (or *noise floor*) coming from the environment which, often, can be mostly ascribed to the environmental thermal noise [49] and assimilated to white noise (with flat spectral density), with average power $N_t = k \times T_0 \times B$, where k is Boltzmann's constant (1.38×10^{-23} J/K), T_0 is the temperature, and B the bandwidth. At 290 K, $kT_0 B$ is -114 dBm/MHz; with a 10 MHz channel, a noise of -104 dBm is found; depending on the environmental conditions it is typically set in the range $(-104, -99)$ dBm.

From a different perspective, it has also been proposed [12] to define the so called *background noise* as the sum of noise and of the average of all the interfering signals which are not strong enough to be perceived by CSMA/CA as competing frames (below the sensing range—for instance, the interference coming from far apart nodes): as such, the background noise would account for the specific environment and also for the congestion state of a network.

After studying the specific case of VANET transceivers, it will be shown (Sect. 3.3.3) that SINR is indeed a significant parameter directly linked to the probability of correctly receiving a frame.

The other parameter acting on reception lies instead in the receiver side and can be alternatively described by the *Internal Noise* (of the Receiver), by its *Noise Figure* or *Sensitivity* [18].

If one considered only the effect of SINR, he would find that, until the received power sufficiently exceeds the environmental noise, regardless its absolute power, it can be received. Suppose that an SINR of 8 dBm is sufficient (e.g., for 6 Mb/s rate): in this case, a signal of -96 dBm could be correctly received; conversely, devices cannot receive signals so weak. This is explained by the sensitivity of the receiver which is defined as the minimum magnitude of input signal required to produce a specified output signal. Sensitivity also accounts for the internal noise figure (F) of the receiver [18]: each component/block in the receiver performs its intended function but also degrades SINR. The Noise Figure F is the ratio between the available SINR at the input and at the output of the receiver and is in the range $[1, +\infty[$, where $F = 1$ represents the ideal case and [5, 13] the most frequent cases. If G is the gain of the receiver (which amplifies both signal and noise), and

3 The Physical Layer of VANETs

indexes i and o refer respectively, to input and output of the receiver, then the following equations hold:

$$F = \frac{S_i/N_i}{S_o/N_o} = \frac{N_o}{G * N_i} = \frac{N_o}{G * kT_0B} = \frac{GkT_0B + N_R}{GkT_0B}. \quad (3.6)$$

Importantly, in case there is a cascade of multiple stages in a transceiver, numbered from 1 to N (from the first input to the last output) the overall noise figure will be: $F = F_1 + \frac{F_2-1}{G_1} + \frac{F_3-1}{G_1G_2} + \cdots + \frac{F_N-1}{G_1G_2...G_N}$: hence, the noise figure of the first stage is the most critical for the overall performance.

Equation (3.6) implies that the received power must be sufficiently higher than the environmental noise *and* the internal noise caused by active and passive components. One has two possible ways to keep track of the internal noise: (1) either to evaluate frame reception based on $S_o/N_o = S_i/(N_i * F)$ (instead of S_i/N_i) [41], or (2) to introduce the concept of sensitivity [11, 41].

Concerning the latter, we will see in Sect. 3.3.3 that, at a given transfer rate, there is a threshold (depending on frame length ℓ)—$SINR_{min(rate,\ell)}$—which a signal needs to exceed in order to be decoded.

The sensitivity threshold is defined as $S_m(\text{rate}, \ell) = SINR_{min(rate,\ell)} * FkT_0B$. The dependence on frame length ℓ is often neglected, as shown in Table 3.1: for this reason, using internal noise or figure noise is in general more precise. The sensitivity is also indicated (as a line) in Fig. 3.4 for a 3 Mbps transfer rate; possible values for the transfer rates of VANETs are included in Table 3.1.

3.1.2 Physical Layer Challenges

In Sect. 3.1.1 the wireless reception has been presented as a matter which is controlled by the SINR ratio. However, reception may be degraded, in spite of a good signal-to-noise ratio: the reason is that several additional phenomena degrading the signal exist and a receiver must deal with them. They are addressed in this section.

Table 3.1 Possible sensitivity thresholds for the transfer rates of VANETs (see Sect. 3.2)

Data rate (Mbit/s)	Minimum sensitivity (dBm)	Data rate (Mbit/s)	Minimum sensitivity (dBm)
3	−85	12	−77
4.5	−84	18	−73
6	−82	24	−69
9	−80	27	−68

The levels here indicated are those adopted by the simulator NS-2 according to the model proposed in. Regardless of the absolute values here listed, it is significant that different rates have different sensitivity threshold, coherently with what discussed in [11]. Absolute sensitivity values will actually depend on the specific implementation of the receivers

The main critical challenges of VANET propagation are due to the outdoor environment and to the high-speed mobility: both leave significant traces in the signal being propagated. The outdoor environment itself makes fading harsh (see the measurements from the field trials in [4, 47]) and is even worsened by mobility: coherence time and coherence bandwidth effectively summarize their main effects.

Before discussing them, a brief foreground on Doppler displacement is provided.

Mobility, Doppler Displacement, and Spread Supposing an absolute speed of 150 km/h, the maximum relative speed of two cars would be 300 km/h, which is far from being negligible. *Doppler displacement* is the most known phenomenon related to mobility and can be quantified, in the simplest case, by the following formula:

$$\Delta f = f - f_0 = \left(1 + \frac{v_{sr}}{c}\right) \cdot f_0 - f_0 = \frac{v_{sr}}{c} f_0, \quad (3.7)$$

with v_{sr} the relative speed between the sender and the receiver, c the speed of light, and f_0 the nominal frequency of 5.9 GHz. Supposing v_{sr} as high as 2×200 km/h, the displacement is still lower than the guard-band of IEEE 802.11p [28]. So, the Doppler displacement is not itself an issue; however, mobility leads also to *Doppler Spread*, that is the broadening, in the frequency domain, of the scattered components.

The occurrence of Doppler spread is qualitatively depicted in Fig. 3.6: the mutual speed between cars causes a perception of the frequencies characterizing the wavefronts slightly different from the nominal f_0 and the receiver perceives this as a frequency broadening. A deep theoretical and practical analysis is available, for instance, in [34].

Considering Eq. (3.7) it is clear that the contribution of scattering by a vehicle D approaching or by a vehicle A getting far from vehicle B and driving in the opposite direction than the receiver C will result in a frequency displacement, respectively, positive or negative (see Fig. 3.6); there will be also components due to still scatterers. As a result, the overall displacement will be a distribution in which each component causes its own frequency displacement, depending on the mutual speeds of the transmitter, of the scatterer(s), and of the receiver.

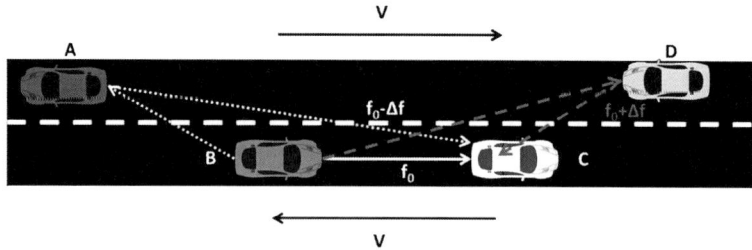

Fig. 3.6 How Doppler broadening occurs: vehicle getting closer or farther causes the perception of more/less frequent wave fronts

3.1.2.1 Coherence Time

Doppler broadening impacts on fading, as studied in [42]. A qualitative explanation of this effect comes from some insights into Rayleigh fading: in fact, Rayleigh model assumes a node *moving* at speed $\vec{v}$, and receiving a large number (N) of reflected waves (*multipath*). We will outline a synthetic demonstration.

Firstly, under the hypothesis of a still transmitted and a moving receiver at speed $\vec{v}$, the Doppler displacement (Δf) will depend on the angle of arrival (AoA) of the ray (ϑ), so that $(\Delta^{\vartheta} f) = f \cdot v/c \cdot \cos\vartheta = \Delta f \cos\vartheta$ (see Fig.3.7a).

If the signal transmitted is $s(t)$ and is composed by the baseband component $u(t)$ and by the carrier of frequency f_0, it can be written as:

$$s(t) = \Re\{(u(t) \cdot e^{j2\pi f_0 t})\}. \tag{3.8}$$

Under our analysis it is reasonable to suppose $u(t)$ so slow to assume an almost constant transmitted power ($u(t) \simeq u_T$). The received power $x(t)$ under mobility, will be the composition of the signals due to several reflections (*taps*), each independent of the others (the Doppler shift is different on each angle and the phase/delay are independent and related to the position of the scatterer), so that:

$$x(t) = \Re\left\{e^{j2\pi \overset{\circ}{f} t} \sum_{n=1}^{N} \alpha_n e^{j\varphi_n} e^{-j2\pi(\Delta^{\vartheta} f_0)t}\right\} \tag{3.9}$$

$$= \cos(2\pi f_0 t)\underbrace{\left(\sum_{n=1}^{N} \alpha_n \cos(\varphi_n - 2\pi(\Delta^{\vartheta} f_0)t)\right)}_{\text{In-Phase Component } \Gamma_{\angle}(t)} - \sin(2\pi f_0 t)\underbrace{\left(\sum_{n=1}^{N} \alpha_n \sin(\varphi_n - 2\pi(\Delta^{\vartheta} f_0)t)\right)}_{\text{Quadrature Component } \Gamma_{\perp}(t)}.$$

Under our assumptions, for $s(t)$ the only dependence on time added by the reflections is due to the AoA-dependent Doppler displacement.

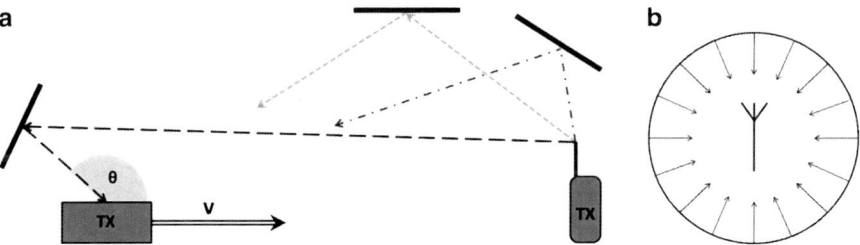

Fig. 3.7 (a) Scenario representing Rayleigh fading, with still transmitter and moving receiver, with the indication of the angle ϑ for a specific ray: actually, differently from what depicted, Rayleigh better fits non-line-of-sight scenarios, due to the uniform scattering hypothesis; (b) representation of the hypothesis of uniform angular distribution of rays

For the two components $\Gamma_\angle(t)$ and $\Gamma_\perp(t)$ of the fading process (which act on the signal envelope), the autocorrelation $\Phi_{xx}(\tau)$ and cross-correlation $\Phi_{xy}(\tau)$ can be computed in terms of *expectations* $E(\cdot)$

$$\Phi_{xx}(\tau) = E\{\Gamma_\angle(t) \cdot \Gamma_\angle(t+\tau)\} = E\{\Gamma_\perp(t) \cdot \Gamma_\perp(t+\tau)\} \qquad (3.10)$$

$$= \underbrace{\left(\frac{1}{2} \sum_{n=1}^{N} E\{\alpha_n^2\}\right)}_{\text{Defined} := \Omega} \cdot E_\vartheta \{\cos(2\pi \cdot (\Delta f_0) \cos \vartheta \cdot \tau)\} \qquad (3.11)$$

$$= \Omega \frac{1}{2\pi} \int_{-\pi}^{+\pi} \cos(2\pi \cdot \Delta f_0 \cos \vartheta \cdot \tau) \, d\vartheta \qquad (3.12)$$

$$= \Omega \, J_0(2\pi \cdot \Delta f_0 \cdot \tau) . \qquad (3.13)$$

The previous computation subtends some assumptions: in (3.11) each directional subcomponent (i.e., each reflected component) is autocorrelated only to itself (not to the other ones) since it is characterized by a different frequency, destroying correlation ($\Delta^\vartheta f$ shows a frequency which depends on the angle ϑ), and a casual amplitude and phase φ_n. As a result the autocorrelation is computed only as the sum of the autocorrelation of each reflected wave. Additionally, still in (3.11), the autocorrelation of each reflected wave is computed using the Werner's trigonometric formulas[5] that of a $\cos(\cdot)$ term which, for $\tau = 0$ is the energy of the function ($\Omega_n = \alpha_n^2/2$). In the step (3.12), ϑ can be supposed to be uniformly distributed in $[0, 2\pi]$, so that a non-weighted integration permits to get the ϑ-independent expectation. Finally, in (3.13) a change of variable is used so that $\int_{-\pi}^{+\pi} \cos(2\pi f \tau \cos \vartheta) = 2 \int_0^{+\pi} \cos(2\pi f \tau \sin \vartheta)$, and this is, indeed, the definition of Bessel's function of order 0.

So, the autocorrelation is J_0, the zeroth-order Bessel function, with Δf the maximum Doppler spread: it does not depend on the time t, hence the two components are *wide-sense stationary* (WSS).

With similar steps, one shows that the cross-correlation is null for $\tau = 0$ (hence the phase and quadrature components are not mutually correlated):

$$\Phi_{xy}(\tau)\big|_{\tau=0} = E\{\Gamma_\angle(t) \cdot \Gamma_\perp(t+\tau)\}\big|_{\tau=0} \qquad (3.14)$$

$$= \left(\frac{1}{2} \sum_{n=1}^{N} E\{\alpha_n^2\}\right) \cdot E_\vartheta \{\sin(2\pi \cdot \Delta f_0 \cos \vartheta \cdot \tau)\}\big|_{\tau=0} = 0 \qquad (3.15)$$

---

[5] In fact $\alpha^2/2T \int_{-T}^{+T} \cos(t)\cos(t+\tau)dt = \alpha^2/4T \int_{-T}^{+T} \cos(2t+\tau)dt + \alpha^2/4T \cos(\tau)\int_{-T}^{+T} dt$; when $t \longrightarrow \infty$ the first terms decreases to 0 (due to the divisor $4T$) and the second keeps constant.

For those interested, the null cross-correlation is helpful to demonstrate the *Probability Density Function* (PDF) of Rayleigh fading.[6]

Eventually, it is possible to compute the Fourier transform of the autocorrelation, which represents the power spectrum (theorem of Wiener–Kinchin[7]). For a zeroth order Bessel function, it is in the form:

$$S(f) = \begin{cases} \frac{1}{4\pi \Delta f \cdot \sqrt{1-\left(\frac{f-f_0}{\Delta f_0}\right)^2}} & \text{if } |f - f_0| < \Delta f_0 \\ 0 & \text{otherwise.} \end{cases} \quad (3.16)$$

From a practical point of view, the formula (3.16) states that the power spectrum, due to the different angular displacements, has a U-shape with peaks at the maxima displacements $\pm \Delta f$. It is known as Clarke or Jakes' model [27].

Applying back the Heisenberg's theorem or Uncertainty Principle,[8] one can say that the broadening (B_D, related to Δf) is inversely proportional to the coherence time of the signal. In the Rayleigh case, for instance, you get that the time over which a signal keeps an autocorrelation > 0.5 is $T_c = 9/(16\pi \Delta f) \propto f_0/v$.

The physical interpretation of the coherence time is that the higher the speed, the more significant the differences in the angular Doppler displacements: as a result, less time is required to have a significantly different phase (hence a different signal envelope), compared to what initially estimated.

When considering a generic signal, the Rayleigh fading would be multiplicative and, in the frequency domain, it would result in a convolution causing a comparable widening, due to the properties of convolution.[9] Finally, for the sake of precision, the theory holds only to still transmitters but it is also applied when both the transmitter and the receiver are moving, considering their mutual speed: in principle, more rigorously, one should consider also the angular distribution of the Doppler displacements at the transmitter.

In general, one will talk of *fast fading*, when the symbol period is longer than the coherence time; otherwise, it will be called *slow fading. As further discussed in Chap.15, slow fading can still be dangerous for OFDM, for the problem of equalization throughout a frame (coherence time may be longer than a frame duration)*.

---

[6]Rayleigh fading defines a PDF which is in the form $PDF(\rho) = \rho/\sigma^2 e^{-\rho/2\sigma^2}$, leveraging the central limit theorem (all the components are independent)—supposing that the two independent components are both gaussian.

[7]The Wiener–Kinchin theorem states that the Fourier transform of the autocorrelation function of a wide-sense-stationary random process, represents its power spectrum.

[8]The uncertainty principle applied to the Fourier's transforms states that, if $h(t)$ is a normalized function and $H(f)$ its Fourier transform, then $\sigma_h^2 \cdot \sigma_H^2 \geq 1/4\pi$. The equality holds only if $h(t)$ is a gaussian function.

[9]The Titchmarsh convolution theorem states that if φ and ψ are two functions in the domain $\mathbb{R}$ whose supports are compact (are, respectively, non-null in that compact sets supp φ and supp φ), then supp $\phi * \psi \subset$ supp ϕ + supp ϕ.

All in all: *The coherence time and Doppler broadening are two inversely related entities providing information about the time-varying nature of the channel.*

3.1.2.2 Coherence Bandwidth

A receiver will have to decode a signal resulting from multiple scatterers: it will not be a sharp and sharply synchronized signal with clear edges, but rather the sum of delayed copies of the same signal (even with some displacement in the frequency,as previously discussed). At a given time and position the signal received will be something like [4]:

$$h(t) = \sum_k A_k \cdot s(t - \tau_k), \quad (3.17)$$

being A_k the complex amplitude contribute coming for kth scatterer. The function $h(t)$ represents the impulse response and, in general it will be time-variable; a parameter characterizing $h(t)$, independently of its specific profile, is the *root mean square (RMS) delay spread,* σ_τ, that is, the RMS-duration of the impulse response as a result of the multiple, delayed reflections (something different from the *maximum* excess delay). Such delayed copies lead to the definition of the coherence bandwidth.

In this case the modeling would be more complex than for the coherence time, so we will skip it; however, with the purpose of a qualitative explanation of the underlying concepts, some considerations follow.

Moving in the *dual* frequency domain, we can reason on the effect of σ_τ delay at two different frequencies f_1 and f_2. The number of cycles which the frequency f_1 encompasses during σ_τ is $N_1 = f_1 \cdot \sigma_\tau$; in the same way, $N_2 = f_2 \cdot \sigma_\tau$ is the number subtended by a wave whose frequency is f_2. Then, the difference between the two becomes significant (and relevant for the pattern of composition of the scatterings), when $N_2 - N_1$ approaches some non-negligible portion of a cycle. Before that conceptual *threshold* is exceeded, the signal composition will be almost independent of the frequency (i.e., the fading of the channel will be flat).

So, we can identify the cut-off condition for the flat fading with: $(f_2 - f_1)\sigma_\tau \sim$ Threshold; frequently (for instance, for Rayleigh fading) Threshold $= 1/(2/pi)$.

In a more formal way, the coherence bandwidth is the bandwidth for which the auto covariance of the signal amplitudes at two extreme frequencies reduces from 1 to 0.5. Consequently, this would require to move into *dual* frequency domain and to study the covariance of the Fourier transform of the signal:

$$H(f) = \mathcal{F}_t\{h(t)\} = S(f) \sum_k A_k \cdot e^{-j2\pi f \tau_k}. \quad (3.18)$$

All in all, the RMS value of $H(f)$ is B_c, that is the *coherence bandwidth* and is inversely proportional to the RMS delay spread (there is a constant which depends on the shape of the impulse response). From a conceptual point of view, B_c represents the frequency range over which the channel response is similar (in [4] the authors propose that $|H(f)| \approx |H(f + \Delta f)|$ for $|\Delta f| < B_c/2$).

Delay spread and coherence bandwidth are inversely related parameters which describe the time dispersive nature of the channel in an area and at a given time.

As in the case of time one defines fast and slow fading, in the frequency domain one will talk of *flat fading* and *frequency-selective fading* not in an absolute way for a given channel, but rather compared to a specific signal.

We will see that in the OFDM of VANETs, each sub-carrier experiences flat fading, but the distance between pilots exceeds coherence bandwidth—hence equalization will undergo the phenomena of frequency-selective fading.

3.1.2.3 Some Reflections on the Physical Layer Challenges

Altogether, there are potential issues with coherence bandwidth and coherence time: to concisely assert it, wideband radio channels in high mobility are typically referred to as *doubly selective*, meaning that they are exposed to a time-varying frequency-selective fading. Problems with doubly selective channels arise, respectively, when (1) the coherence bandwidth of the channel is narrower than the signal bandwidth and if (2) the coherence time gets shorter than a symbol or, in some cases than a frame duration (see Table 3.2).

So it is important to quantify these parameters through the empirical data from measurements. They are actually available from multiple sources, for example in [32, 36, 45, 47]. In general, throughout the literature, you can find RMS delay spread as high as 200 ns (worse spreads for non-line-of-sight conditions) and Doppler Spreads between 200 and 600 Hz (they are worse in highways, due to speed). They respectively correspond to a coherence bandwidth lower than 1 MHz and a coherence time of about 300 μs. The same authors of [32] mention other measurements by trials and conclude that it is better to focus on the distribution of the RMS delay and Doppler Spread and to consider the most frequent cases, rather than the worst. These are just some initial numbers to fix ideas, but this analysis will be recalled and continued in Chap. 15, where the weaknesses of VANETs and the possible future solutions will be discussed.

For now, it is sufficient to wrapup this theoretical discussion, by highlighting some points which may be relevant to OFDM—just mentioning that it makes use of coherent digital modulations on close-by, partially overlapped sub-carriers.

- The coherence time and coherence bandwidth are relevant to the definition of the OFDM system: in order to correctly identify the sub-carriers, a frame should not last more than the coherence time (so that the channel estimations by the initial *pilot* signals can be held-on) and the spacing between continuous pilot sub-carriers does not exceed the coherence bandwidth (details in Sect. 3.2).
- Related to the previous item, in an OFDM system, the concepts of *Inter-Carrier Interference* (ICI) and *Inter-Symbol Interference* (ISI) assume a specific contextual meaning:

Table 3.2 Physical Layer parameters

Parameter	Definition	Relationship	Requirement	OFDM requirements
Time variance				
Coherence time T_c	Time interval over which channel responses are highly correlated	$T_c \propto 1/B_D$	$T_c > T_s$ T_s is the symbol time	$T_c > T_f$ T_f is the frame duration[a]
Doppler spread B_D	Measure of the spectral broadening	$B_D \propto 1/T_c$	$B_D \ll B_s$ B_s is the signal band	$B_D \ll \Delta f_c$ Δf_c is the sub-carriers spacing
Time dispersion				
Coherence bandwidth B_c	Range of frequencies over which the channel response is flat	$B_c \propto 1/\sigma_\tau$	$B_c \gg B_s$	$B_c > \Delta f_p$ Δf_p is the pilots spacing[b]
Delay spread σ_τ	RMS-duration of the overall response by all the scatterers	$\sigma_\tau \propto 1/B_c$	$\sigma_\tau \ll T_s$	$\sigma_\tau \ll T_s$

[a] After T_c the channel estimation performed at the beginning of the frame (and the equalization performed according to it) may lead to a wrong decoding

[b] Channels whose coherence bandwidth is much smaller than the signal bandwidth are B_s *frequency selective*. In the time domain, this manifests as inter-symbol interference (since the RMS delay spread is larger than the symbol period). With OFDM (Sect. 3.2.3) this statement achieves a slightly different embodiment: the symbols over each OFDM carriers are robust to ISI by the *same* carrier, thanks to a sufficient guard-time, but not by ISI by multiple carriers. In fact, the estimation by pilots cannot be sufficient, due to their spacing compared to the coherence bandwidth. In this case B_s will refer to the pilot spacing

- in OFDM the sub-carriers partially overlap: ICI occurs when the *orthogonality* between sub-carriers does not hold at the receiver, either due to Doppler displacement and/or for a wrong channel (and frequency) estimation. To prevent this, an OFDM frame should be smaller than the coherence time (so that the frequency of the sub-carriers can be precisely identified) and the spacing between sub-carriers less than the coherence bandwidth[10]; additionally, there are techniques meant to cancel ICI at the receiver side, leveraging once more the knowledge of the channel;
- ISI happens when consecutive OFDM symbols overlap. To avoid it there should be a guard interval between symbols and it should exceed the channel maximum delay spread; also in this case, digital filters can support the cancellation;

---

[10]In VANETs' OFDM (Sect. 3.2.3) the spacing between pilots is about 3.2 MHz which may be more than the coherence bandwidth met outside: this may be counteracted by the initial denser pilots held-on over a time shorter than the coherence time, but, indeed, it represents a challenge (see Chap.15).

- altogether: ICI can be influenced by the coherence time (due to a wrong estimation of sub-carriers) and can be partially prevented if the spacing between pilots is narrower than the coherence bandwidth; ISI, instead, is related to the maximum delay spread;
- the 10 MHz spacing of OFDM channel is the most frequently mentioned setting for VANETs (Sect. 3.2): it is proved to offer a good trade-off and to prevent most of the mentioned issues at physical layer. However, the IEEE 802.11p foresees also a 20 MHz option and, there are authors [45] who propose to use 20 MHz: shortened symbols would decrease channel congestion and counteract coherence time issues, while the guard time (halved) should still be sufficient.

- current simulators cannot integrate a detailed physical layer *and* a flexible MAC/network engine; as a result, you will find either studies based on a detailed physical modeling but neglecting network-level simulations (as, for example, those by Matlab-Simulink ©) or a flexible modeling of upper-layer protocols, even with enhanced physical models, but disregarding some phenomena such as Doppler, coherence time, and coherence bandwidth.

3.2 The Physical Layer of VANETs

In the present section the physical layer of VANET is presented step-by-step, starting with the available channels (Sect.3.2.1), then motivating the adoption of OFDM (Sect.3.2.2) (and including some recalls on *Discrete Fourier Transform* (DFT) for OFDM in Sect.3.2.2.1), afterward introducing the modulation and encoding formats (Sect.3.2.3)—also comparing them to the ones of indoor Wi-Fi—and eventually discussing the typical reception performance of VANET receivers (Sect. 3.3.3).

The information reflects what specified in Europe by ETSI [14, 16] and in USA by IEEE [26]: it may be worth recalling that the PHY of VANETs has been for a long time called IEEE 802.11p, from the name of the amendment to the IEEE 802.11, but currently is simply mentioned as IEEE 802.11.[11]

---

[11] According to the rules of the IEEE Standards Association, there is only one current standard which, for Wi-Fi, is denoted by IEEE 802.11 followed by the date of its publication. At the time of our writing, IEEE 802.11-2012 is the only version in publication and has integrated the previous IEEE 802.11p amendment; next version is expected to be the IEEE 802.11-2015. The standards are updated by means of amendments, which are created by task groups (TG). Both the task group and their finished document are denoted by 802.11 followed by a letter (or a couple of letters), such as IEEE 802.11a and IEEE 802.11ac. For the creation of a new stable version, task group m (TGm) combines the previous version of the standard and all the published amendments not subsumed yet. New versions of the IEEE 802.11 were published in 1999, 2007, and 2012.

3.2.1 The Channelization and Power Constraints of VANETs

The wide range of possible applications of the vehicular technology, both for safety and non-safety purposes, raised the concern of many organizations of standardization or governmental authorities, that allocated specific bands for the vehicular communications. ETSI defined the European profile standard [14] for ITS communication in the 5 GHz band. The functionality described in the standard is named "ITS-G5." In Europe for the ITS applications a band of 50 MHz around the frequency of 5.9 GHz is reserved. This is divided in to two portions:

- ITS-G5A is a 30 MHz-wide band, from 5.875 to 5.905 GHz, intended to support safety applications;
- the remaining band of 20 MHz, ITS-G5B, includes the frequencies from 5.855 to 5.875 GHz and is dedicated to non-safety applications;

Given this availability of band and with a *channel spacing of 10 MHz*, it is possible to allocate five channels in *ITS-G5A and ITS-G5B bands*: one physical channel is classified as G5CC (*ITS-G5 Control Channel*), four fixed are identified as G5SC (*ITS-G5 Service Channel*) [44]. The *ITS-G5C*, from 5.470 to 5.725 GHz, is a 255 MHz-wide band used in the Radio Local Area Network (RLAN), that can be also employed for ITS applications. The channel spacing can be *10 or 20 MHz*.

So, the most significant case, for EU VANETs, is that of 10 MHz-wide channels.

The European channel allocation for ITS applications is shown in Table 3.3, specifying the default data rate and the power limits for each channel. For each ITS-G5 channel, the corresponding channel identification number, according to the IEEE 802.11p allocation, is pointed out. As regards the G5SC band, the default rate is not fixed, but rather depends on the channel spacing. Besides, such frequency range is shared by the RLAN devices, so the ITS communication on this band suffers interference with RLAN transmissions. In order to reduce the interference, methods of *Dynamic Frequency Selection* (DFS) are applied and the data exchange occurs in a master (ITS roadside station)/slave (vehicular ITS station) fashion. The maximum

Table 3.3 European channel allocation

Band	Channel type ETSI/IEEE	Center frequency (MHz)	Default data-rate (Mb/s)	TX power max (dBm EIRP)	TX density power max (dBm/MHz)
G5A	G5CC/180	5,900	6	33	23
G5A	G5SC2/178	5,890	12	23	13
G5A	G5SC1/176	5,880	6	33	23
G5B	G5SC3/174	5,870	6	23	13
G5B	G5SC4/172	5,860	6	0	−10

transmission power is equal to 30 dBm *Effective Isotropic Radiated Power* (EIRP)[12] or 17 dBm/MHz for a DFS master, while for a DFS slave a power limit of 23 dBm EIRP or 10 dBm/MHz is defined.

The G5CC is intended to the exchange of messages related to safety and traffic efficiency applications and for the announcement of the services that will be offered on the G5SC1-G5SC5. Given its safety purposes, the Control Channel is placed at lower bound of the frequency range, so to benefit from less interference sources from nearby channels (the lowest part of the range is used as guard-band). For the same reason, GCC is followed by GSC2, rather than GSC1 (the two are inverted), since nodes are allowed to transmit only at 13 dBm, rather than 23 dBm, in GSC2.

The G5SC1 and GSSC2 are dedicated to ITS road safety and traffic efficiency applications, the other types of ITS user applications shall be provided in the channels G5SC3–G5SC5.

*****

As regards the US standardization, in 1999 the *Federal Communication Commission* (FCC) established the regulations for the *Dedicated Short Range Communications* (DSRC) operations in the USA, reserving a band of 75 MHz from 5.850 to 5.925 GHz for the vehicular technology [30]. In this spectrum seven 10 MHz-wide communication channels are available (numbered with even numbers from 172 to 184): one *Control Channel* (CCH) and six *Service Channels* (SCHs). A guard-band of 5 MHz is introduced in the lower portion of the DSRC band [17]. Indeed, the standard [26] defines operations on channels of 5, 10, and 20 MHz for the OFDM systems. However, some experimental tests [5], performed in the USA, proved that the bandwidth of 10 MHz is the one most suited to support an extended set of applications and to counteract the delay and Doppler spreads, physical phenomena that affect the vehicular networks. So, *the channel spacing of 10 MHz is commonly used for DSRC communications.*

For each channel a transmission power limit is fixed, typically the maximum EIRP permitted is 33 dBm. However, for many applications, it is recommended to limit the power in the range from 10 to 20 dBm in order to mitigate interference at long distances. In particular, four Classes for the transmitted power (or better, the EIRP) have been defined and are listed in Table 3.4.

It is possible to join two adjacent 10 MHz Service Channels, in order to operate on 20 MHz-wide channels (the rationale has already been discussed in the last paragraph of Sect. 3.1.2, as from [45]). Channel 178 is used as Control Channel, for the transmission of safety information and advertisements about the services provided on the SCHs. The Service Channels (identified by the numbers 174, 176, 180, and 182) are exploited for infotainment, commercial and traffic management applications. The channels 172 and 184 were intended for other generic service applications. The channel allocation for DSRC communication is shown in Fig. 3.8.

---

[12]The EIRP corresponds to the power that an isotropic antenna should radiate in order to produce the peak power density observed in the direction of maximum gain for the antenna.

Table 3.4 Power classes defined for WAVE (USA)

Power class	Max. (reference) transmit power (mW)	Maximum permitted EIRP (dBm)
Class A	1	23
Class B	10	23
Class C	100	33
Class D	760	33 for non government, 44.8 for government

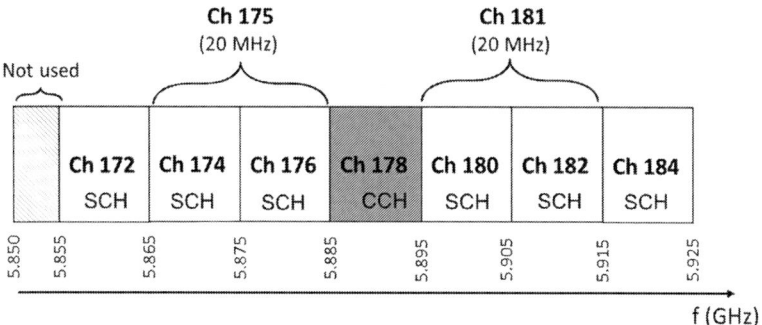

Fig. 3.8 DSRC channel allocation

In 2006, after receiving many requests from the automotive industries, the FCC specified more clear rules for the operations on these SCHs. In particular, the Channel 172 is dedicated to the vehicle-to-vehicle safety communication, where messages are exchanged between vehicles (and involving in some cases also RSUs) in order to avoid or reduce the risk of road accidents. The Channel 184 shall be used for the high-power long-distance communications that support public safety applications. One of them is the signal preemption, whose aim is to facilitate the trip of the emergency vehicles, intended to the safety of life and property. For such operations, a higher transmit power (up to 40 dBm) is required in order to reach longer distances.

The standard 802.11p at the physical layer defines four power classes (A, B, C, D) up to 44.8 dBm (30 W) to support long ranges, as shown in Table 3.4. Each class specifies the maximum EIRP allowed.

*****

Still about channelization, there are some additional facts to mention or to summarize, for the sake of clarity.

- Europe and USA opted for two different solutions for the multichannel, as further discussed in Chap. 7.
    - Europe decided from the beginning to enforce the safety purposes and to require that VANET receivers are always connected to the GCC. If one intended to exchange also data on a GSC, he needs to have a double receiver.

3 The Physical Layer of VANETs

- At the beginning in the USA the approach proposed for the channel coordination was based on the channel switching. All DSRC devices were forced to tune to the CCH during a fixed time interval, in order to receive safety alerts and also the announcements about the services provided on the SCHs. Since the main target of the DSRC technology is the exchange of traffic data safety, the automotive industries and the research groups investigated on alternative solutions, based on the full-time usage of a channel for the safety messages.
  - Currently, in the USA, the Service Channel 172 is continuously used only for vehicle-to-vehicle safety communications (accident avoidance or mitigation and public emergency applications). A vehicle interested in safety and non-safety information shall have two dual radios: one always tuned to the channel 172 for the safety data, the other one switching between channels.
  - Altogether, both in EU and in the USA, the VANETs will be exposed to adjacent interference. Even if the transceivers are compatible to the emission masks (see next item and Fig. 3.9), the interference might become significant, as studied in some papers [9, 10].
- The standards specified strict spectrum emission masks compared to IEEE 802.11a, perhaps to cope with higher transmitted power and the expected background noise due to several simultaneous transmissions. Such masks depend on the power class of the device (the higher the transmitted power, the more severe the drop in the spectral density of power). IEEE 802.11p also adjusts the value of the *Adjacent Channel Rejection* (ACR) parameter in order to satisfy the requirements of the outdoor vehicular environment: in particular, ACR is 12 dB higher for adjacent channels (and 10 dB for non-adjacent ones), in the respective modulations. These spectral requirements are shown in Fig. 3.9.
- Even if the standard foresees also other channel widths, with the exception of some channels where there are optional settings, the current and fixed channel width for VANET is 10 MHz.

3.2.2 OFDM for VANETs: Motivations and Background

Summing up: the VANET channels are 10 MHz-wide and in the vehicular environment—due to the multipath fading and the high mobility—the channel bandwidth is *doubly selective* (in the frequency and in the time domain).

- The channel response, during a symbol period, is represented by a set of delayed delta functions and results in a fluctuating power but also in a propagation delay, and the delay spread, indeed, leads to ISI.
- Also the frequency response of the channel is not flat, but shows many fluctuations, leading to ICI.
- In order to achieve a flat channel response in both domains, the bandwidth of the signal must be narrower than the coherence bandwidth of the channel, the

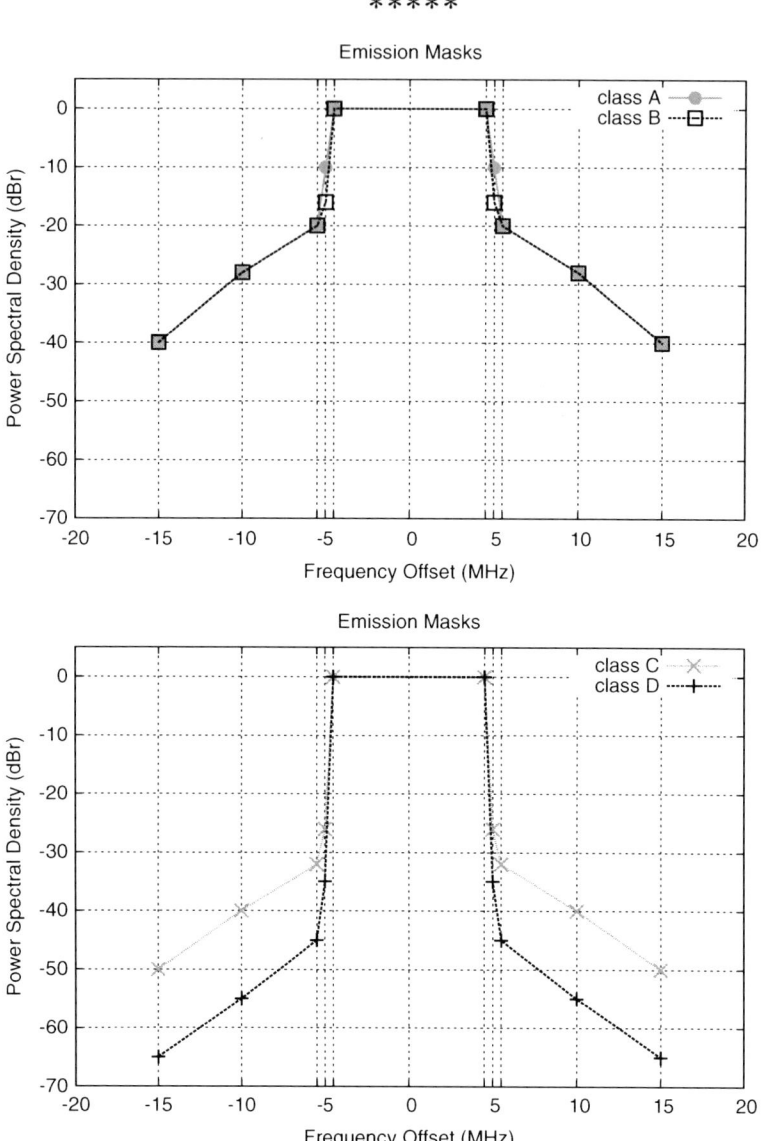

Fig. 3.9 Transmit spectral mask for devices of Class A and B *(top)*, C and D *(bottom)*

coherence time must be longer than the frame time and a guard interval is needed to prevent the ISI due to the delay spread.
- Applying a single bearer modulation, these conditions cannot be easily satisfied, because it is required to carry out channel equalization and estimation.

3 The Physical Layer of VANETs

Given these premises, the OFDM modulation technique was considered a good candidate to counteract the effects of the multipath fading and to meet the expected bit-rate with a 10 MHz channel. For an extensive analysis of OFDM techniques, several reference texts are available (for instance, [37]): here below, only the main ideas are shortly recalled. OFDM subtends two main ideas: (1) to distribute the information across many sub-carriers at narrow band (equally spaced in frequency) and (2) to use *orthogonal* carriers.

Thanks to the multiple sub-channels, smaller portions of the overall data can be transferred by each bearer using traditional modulation scheme (e.g., from BPSK, *Binary Phase Shift Keying*, to 64QAM, *64-point Quadrature Amplitude Modulation*). Furthermore, each sub-carrier will behave as a non-frequency-selective and time-invariant channel and the duration of each symbol is relatively long, thus reducing the ISI.

Concerning the orthogonality, even if the sidebands of the sub-carriers overlap, the cross-talk will be eliminated, with no need for inter-carrier guard-bands: sub-carriers are placed closely, side by side, dramatically improving the spectrum efficiency (from 33.3 % to 50 %). Specifically, the term orthogonal indicates a mathematical property in the frequency domain: the carriers are all synchronous and arranged so that, in the time domain, all of them contain a whole number of cycles in the symbol period T. This is sufficient to prevent cross-talk in the integration by the DFT in an OFDM digital receiver.

The setting is depicted in Fig. 3.10: in the frequency domain, the orthogonality occurs when the mutual spacing between carriers is a multiple of $1/T$, so that each DFT frequency will capture exclusively the power of a single carrier (even more, its peak).[13]

Moving back in the time domain, at each period of time T, the overall (serialized) OFDM symbol will result from the sum of all the amplitude and phase modulated sinusoid functions corresponding to all the OFDM sub-carriers. In Fig. 3.11a the baseband OFDM signal is shown: the constructive sum of sinusoid functions leads to *sudden* and rapidly varying peaks; in Fig. 3.11b, the effect of up-modulation in the target channel is shown (the baseband signal is highlighted as an envelope).

3.2.2.1 OFDM and Fast-Fourier-Transform Techniques

Far from the intention of providing an extensive explanation of the theoretical foundations of OFDM modems, this section is meant to recall the main concepts of DFT/FFT (*Fast Fourier Transform*), for those who may need to understand the

---

[13]The orthogonality exploits the well-known properties of $\text{sinc}(f) = \sin(f)/f$, the Fourier transform of the rectangular function in the time domain ($\text{rect}(t)$). The same properties hold also when you take a segment of the $\sin()$ function (it can be meant as the product $\sin(f_0 t) \cdot \text{rect}(t)$); in this case, the transform would be $\text{sinc}(f - f_0)$.

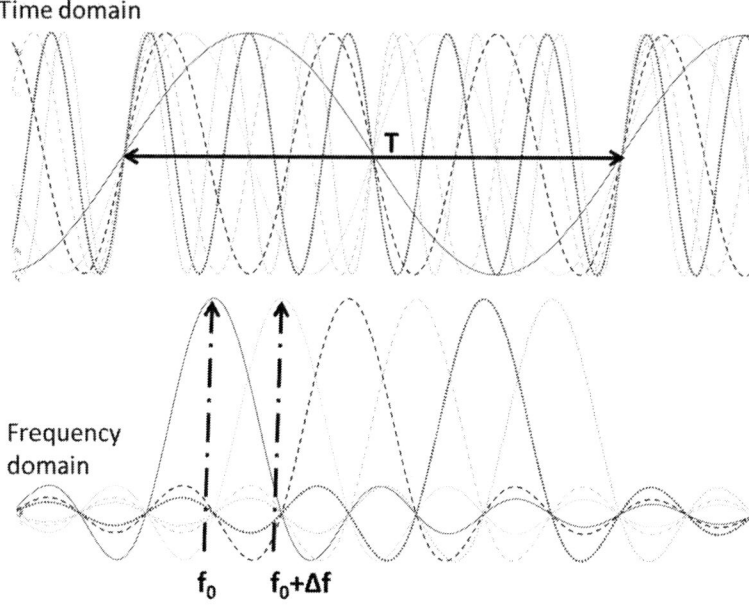

Fig. 3.10 How the time and frequency domains are related for regular segments of sin( ) functions with an integer number of periods in the sampling period

underlying principles. A more detailed presentation is available in several dedicated volumes, such as [46].

An OFDM signal consists of orthogonal sub-carriers modulated by parallel data streams. One baseband OFDM symbol (without a cyclic prefix) multiplexes N sub-carriers in the following way:

$$s(t) = \frac{1}{N} \sum_{k=0}^{N-1} X_k e^{j2\pi f_k t}, \qquad 0 < t < NT \tag{3.19}$$

being X_k the complex coefficient of the modulation (amplitude and phase of the modulation scheme, such as 64QAM) and NT the duration of an OFDM symbol; consequently, the equally spaced OFDM carriers are $f_k = k/NT$.

The resulting multi-carrier signal $s(t)$ of Eq. (3.19), before the advent of DFT, would have been received by a bank of matched filters. Fortunately, we have the Fourier transform, which is a powerful tool to analyze the signals and construct them to and from their frequency components. If the signal is discrete in time (it is sampled), one uses the DFT, which converts samples into a set of coefficients for the discrete frequency spectrum. To reduce the mathematical operations used in the calculation of DFT and IDFT (*Inverse DFT*) one uses the FFT algorithm. In other words, supposing that *the sub-carriers are still orthogonal* and that the *receiver*

3 The Physical Layer of VANETs

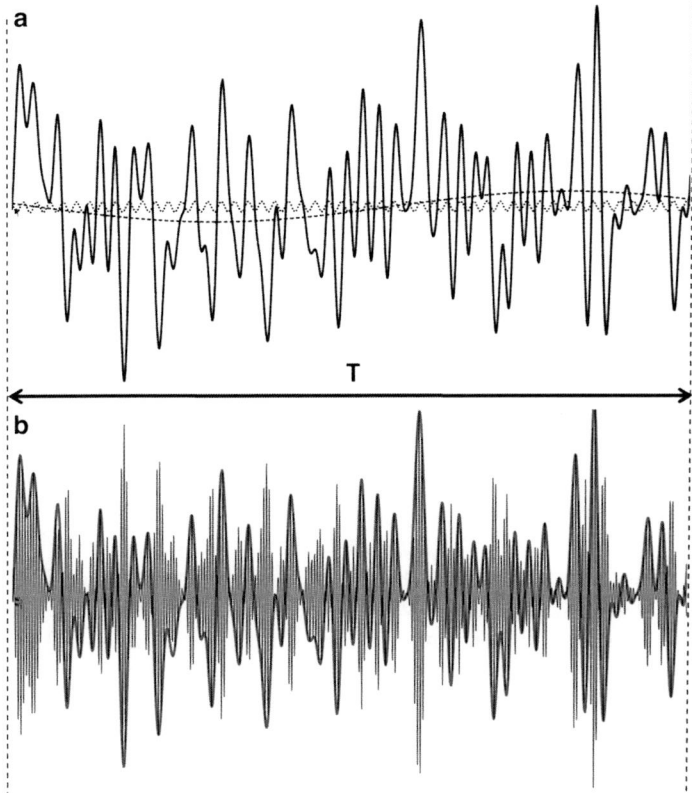

Fig. 3.11 An OFDM symbol (in the time domain) built with 48 64 QAM modulated sub-carriers + pilots, as from Wi-Fi standard: (**a**) the baseband signals with *dotted* the lowest-frequency and the highest-frequency sub-carriers; (**b**) the up-modulated signals in the target frequency

perfectly knows the frequency, phase and amplitude of the sub-carriers, then the signal can be demodulated back in the following way:

$$X(m) = X(e^{j2\pi f_k t})\big|_{k=m} = \sum_{0}^{N-1}, \qquad (3.20)$$

meaning that, if one samples the multi-carrier signal $s(t)$ exactly at the frequency f_m and he sums all the samples, he will find only a complex term which is equal—but a multiplicative factor (estimated thanks to the equalization)—to X_m; this is indeed the symbol modulated by the sub-carrier addressed. This result is one of the key properties of OFDM, proposed for the first time in [48].

So far, we have discussed about demodulating the signal. However the signal needs first to be modulated and, in multi-carrier modulation, separate oscillators tuned to different frequencies would be required. An IFFT (*Inverse FFT*) block removes also this potential issue and was the key for OFDM success: the inverse transform can be in fact used to convert the discrete frequency form into the discrete time form—the first step to modulate the signal. This can be easily performed by the following computation on the digital signal at the time $t = NT$.

$$s^*(n) = s^*(t = NT) = \sum_{k=0}^{N-1} X(k) \cdot e^{j\frac{2\pi nk}{N}}. \qquad (3.21)$$

Now the signal is ready for *Digital to Analogue Conversion* (DAC), as further discussed in Sect. 3.3.1.

3.2.3 The OFDM Settings for VANETs

OFDM is such a powerful technique that has spread across several telecommunications media: it is used in *Digital Video Broadcasting* (DVB), *Wi-Fi*, *WiMax*, *Long Term Evolution* (LTE), xDSL (ADSL, VDSL), powerline communications, etc. In particular, OFDM was already implemented in the PHY layer of the radio devices based on IEEE 802.11a, that was designed for high data rate communications among slow moving devices in indoor environments. Compared to IEEE 802.11a, IEEE 802.11p introduces minimal variations, aimed at addressing the peculiarities of VANET communications. These differences are shortly recalled in Table 3.5 and will constitute a track for our present analysis.

First of all, the 802.11a devices operate in a frequency range around 5 GHz, while for 802.11p a portion of spectrum around 5.9 GHz is allocated.

Further, in 802.11a, the spectrum is divided in to 20 MHz-wide channels. Each channel includes 64 sub-carriers: 48 are used as data carriers, four are pilots, and the remaining 12 are null carriers (in 802.11 they act as guard-bands, since they are located at the borders of the channel).

Due to the high relative speed of the nodes, the vehicular communication is more vulnerable to the Doppler Effect. Besides, with a channel bandwidth of 20 MHz, as in 802.11a, the community was worried that the guard interval between

Table 3.5 Main differences between 802.11p and 802.11a

Parameter	IEEE 802.11p	IEEE 802.11a
Frequency band	5.85–5.95 GHz	5.15–5.35 GHz; 5.725–5.835 GHz
Data rate	Max 27 Mb/s	Max 54 Mb/s
Channel bandwidth	10 MHz	20 MHz
Number of channels	7	12
OFDM signal duration	8.0 μs	4.0 μs
Guard time	1.6 μs	0.8 μs
FFT period	6.4 μs	3.2 μs
Preamble duration	32 μs	16 μs

OFDM symbols might not be long enough to avoid ISI. So, the physical layer of the VANETs promotes the use 10 MHz-wide channels (even if the 20 MHz still constitutes a possible setting, the 10 MHz option is the one most frequently mentioned).

Halving the bandwidth, all the time parameters get doubled, compared to the corresponding ones in 802.11a, improving the robustness against the effects of mobility; in particular, the doubled guard interval reduces the inter-symbol interference caused by multipath propagation—the signal becomes robust to maximum delay spreads as high as twice those compatible with IEEE 802.11a.

Each IEEE 802.11 frame starts with a preamble (which in IEEE 802.11p lasts 16 μs) which is aimed at facilitating the work of receiver in detecting the signal in the air, performing the automatic gain control (needed for the equalization and the consequent QAM demodulation), achieving synchronization (so to keep the orthogonality in the reception), the correct phasing (needed for the demodulation as well) and for manifold other channel evaluation tasks. Typically, these low-level mechanisms are not addressed (and taken for granted) by network simulators such as NS-2,[14] NS-3,[15] Omnet++,[16] Qualnet[17], etc.[18]

Giving the critical impact of these initial tasks, they are enforced by the so-called *pilots*. Pilot sub-carriers are employed to identify the exact frequency and phase of the signal at the receiver and to facilitate the estimation of the channel coefficients for the equalization of subchannels *Pilot-symbol Aided Channel Estimation* (PACE). The use of the pilot sub-carriers to compensate the effects of the fading is known as *Pilot Symbol Assisted Modulation* (PSAM).

Pilots are massively transmitted in the preamble[19] and, afterwards, continuously transmitted on four dedicated sub-channels, in order to increase the resistance to the fading. The frequency spacing among the pilots must be carefully fixed [40] and it should be sufficiently small to identify and track all the channel variations in terms of frequency. In 802.11p, the spacing between the pilots is about 2 MHz (about eight subchannels). Details on the preamble, the pilots tones, the guard-bands, and the guard-intervals are provided in Fig. 3.12.

---

[14]NS-2, the network simulator (Open Source) available at http://www.isi.edu/nsnam/ns/.

[15]NS-2, the network simulator (Open Source) available at http://www.nsnam.org/.

[16]Omnet++, available at http://www.omnetpp.org/ or (commercial edit.) at http://www.omnest.com/.

[17]Qualnet, available at http://web.scalable-networks.com/content/qualnet.

[18]The main purpose of network simulators is the testing of protocol and, consequently, they do not focus on physical layer modeling. As a result they typically model physical layer through statistical model, not considering the physical phenomena which may lead to events such as wrong equalization or OFDM misalignment.

[19]As shown in Fig. 3.12, during the first 16 μs there is a short training sequence where only 12 sub-carriers are used (for pilots) in BPSK; then, after the GI, the long training sequence uses all the 52 sub-carriers as pilots (+DCC); eventually, data follow. The short sequence is used for signal detection and coarse tuning; the long sequence for fine-tuning.

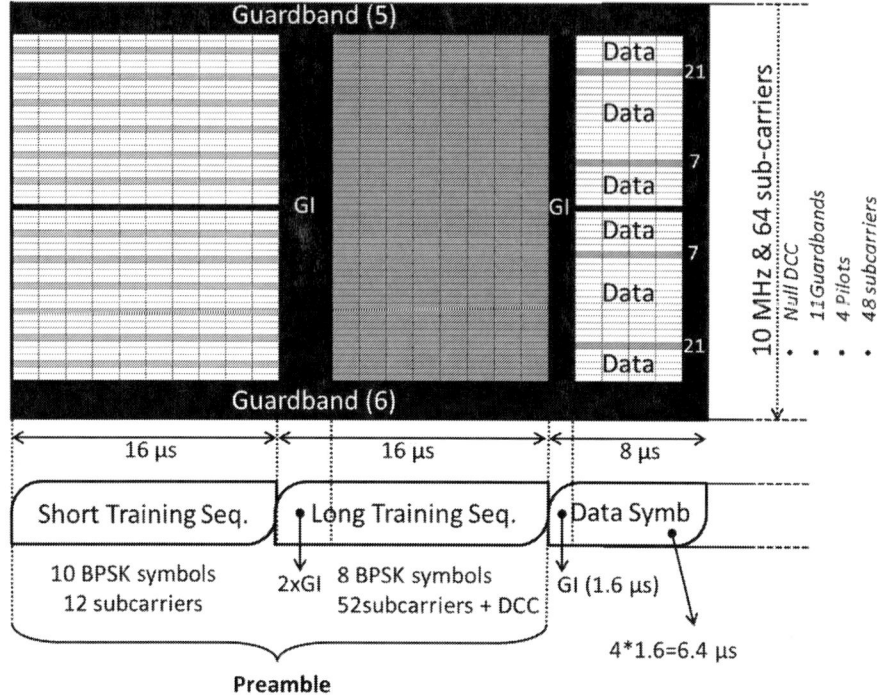

Fig. 3.12 Pilots and preamble in the IEEE 802.11p: *short training sequence* for coarse tuning by 12 sub-carriers; *long training sequence* (after a double guard interval—2xGI) for fine-tuning, involving all the sub-carriers (including the Direct Current Carrier—DCC components); data symbols follow one another, symbol by symbol, each preceded by a GI, as explained in Chap. 4

Then, data can start and this may be transmitted at different rates (with the exception of some additional PLCP fields sent at BPSK, as explained in next chapter).

Similarly to IEEE 802.11a, IEEE 802.11p transmits on each single sub-carrier using one of the following digital modulation schemes: BPSK, *Quadrature Phase Shift Keying* (QPSK), *16-point Quadrature Amplitude Modulation* (16-QAM), and 64-QAM.

Furthermore, both IEEE 802.11a and IEEE 802.11p adopt a *Forward Error Correction* (FEC) to the messages, so to increase the probability to recover some errors and decode the data successfully, at the price of a reduction of the actual bit-rate.

The resulting data rates are determined by the couple coding rate (of the FEC) and modulation type, as shown in Table 3.6. Since in 802.11p the values of the timing parameters are doubled (compared to the ones used in 802.11a), the possible data rates are halved and vary from 3 to 27 Mbps.

Summing up, the physical layer of 802.11p exploits the OFDM technique as 802.11a, but some modifications directed to the management of some issues related

3 The Physical Layer of VANETs

Table 3.6 Data rates and channel width in IEEE 802.11 OFDM

Modulation scheme	Coding rate	Data rate (Mbit/s) 10 MHz	Data rate (Mbit/s) 20 MHz	Bits per symbol and sub-carrier
BPSK	1/2	3	6	0.5
BPSK	3/4	4.5	9	0.75
QPSK	1/2	6	12	1
QPSK	3/4	9	18	1.5
16-QAM	1/2	12	24	2
16-QAM	3/4	18	36	3
64-QAM	2/3	24	48	4
64-QAM	3/4	27	54	4.5

to the high dynamism of the vehicular scenario. However, still the solution could be further improved, since some physical phenomena could be better handled and require additional studies, as suggested, for example, in [52]. These aspects will be addressed in Chap. 15.

3.3 The VANET Transceivers

Thanks to the long-lasting experience on Wi-Fi systems, VANET transceivers constitute a consolidated topic: OFDM radio transceivers have been developed for years and the Standard [26] specifies both the signal characteristics and the signal processing blocks in great detail.

As a result, today not only industrial components are available, but also lab-made implementations, leveraging *Software-Defined Radio* (SDR) tools (as, for instance, [8, 50]) and paving the way to the experimental testing of new solutions (both backward-compatible and disruptive ones), as the ones mentioned in Chap. 16.

For those who intend to approach the implementation or just to better understand how the physical layer of VANETs works, the next subsections, respectively, introduce the main building blocks of transmitters and receivers. Some further tricks of IEEE 802.11 will be so explained.

3.3.1 The Blocks of a Wi-Fi Transmitter

So far, several mechanisms which the OFDM techniques subtend have been introduced. It is time to analyze orderly all the main blocks which a VANET transmitter is made of. In the following they will be mentioned and shortly explained, following the conceptual chain which is depicted in Fig. 3.13.

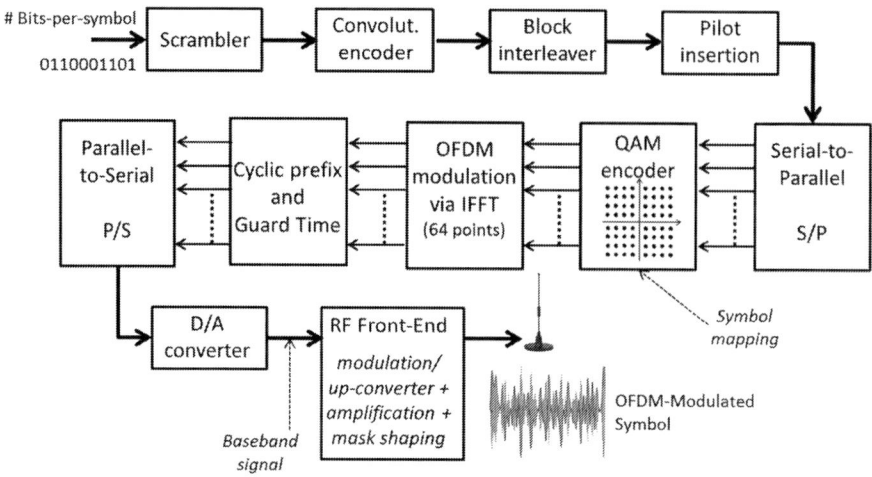

Fig. 3.13 The building blocks of a IEEE 802.11 transmitter

Input Firstly, the input to be encoded. With the exception of beginning of the frame (the preamble which has been presented before and is depicted in Fig. 3.12), the transmission concerns the binary data received by the upper logical layers (the PLCP sub-layer, presented in Chap. 4). In order to be encoded, they will be processed in chunks made of as many bits as the bits-per-symbol of the target encoding (they are recalled in Table 3.6).

Scrambler It processes all the information bits, except a field (the Service Field, as discussed in next, Chap. 4) and is aimed at preventing long streams of ones or zeros. It consists of a pseudo-random binary sequence (PRBS) by a polynomial ($S(x) = x^7 + x^4 + 1$), which generates a cyclic sequence of length 127 and is randomly initialized at the beginning of the transmission. The receiver will estimate the initial state of the scrambler by observing the first seven bits of the Service field.

Convolutional Encoder The principle of convolutional encoding is that the input stream undergoes a convolution with the encoder's impulse response, aiming at adding some controlled redundancy, so to achieve a more reliable data transfer (it adds parity bits with error correction capability). The L input bits are mapped into M bits by the convolutional encoder; $R = L/M$ represents the coding rate which is displayed in Table 3.6: the more redundancy is introduced, the more robust but slower the resulting transfer rate is. The IEEE 802.11 convolutional encoding supports the 1/2, 2/3, and 3/4 coding rates by *puncturing* the data prior to transmission (puncturing is the process of removing some of the parity bits after encoding with an error-correction code).

Block Interleaver The purpose of interleaving is to make the forward error correction (by the convolutional encoding) robust against burst errors: every encoded group of bits is interleaved by an algorithm which maps adjacent bits over non-adjacent sub-carriers and adjacent encoded bits alternatively onto less and more significant bits of the constellation. The size of the interleaver is determined by the number of the coded bits per OFDM.

Pilots, S/P and QAM These three steps are grouped because there is not much to add to what already discussed: pilots are inserted according to the scheme which is shown in Fig. 3.12; then, the serial-to-parallel conversion takes place and equal chunks of bits feed the QAM encoding of the 48 sub-carriers used for data transmission. Once more, the available digital encoding techniques (from BPSK to 64QAM) are listed in Table 3.6.

OFDM Modulation The step of the actual OFDM modulation, as a multi-carrier modulation, would require separate oscillators precisely tuned to the respective sub-carrier frequencies. An IFFT block simplifies indeed this step, thanks to the properties of OFDM orthogonality. This has been shortly recalled in Sect. 3.2.2.1.

Cyclic Prefix and Guard-Time Addition At this stage the signal processing still works in parallel: in principle, two consecutive symbols, due to their likely different phase and amplitude (by QAM modulation) might not join up and this discontinuity would cause a *spectral regrowth* (i.e., bandwidth broadening). Besides, in order to cope with the maximum propagation delay, a guard interval (GI) needs to be added (see Fig. 3.12). The two aspects have a common solution: the GI is inserted and is used also to prevent the mismatch between the phase and amplitude of consecutive symbols. In fact, during the GI the consecutive symbols are smoothly linked: a prefix is inserted and a suffix window appended (cyclic extensions) on which the symbols are extrapolated and matched each other, following precise mathematical relationships (*windowing*). This also improves the spectrum emission mask (SEM) and helps identifying the borders of each symbol (through self-correlation).

P/S, DAC, and Radio Front-End At this point of time, the data is OFDM modulated and can be serialized (P/S) and ready to be transmitted. A Digital-to-Analogue Converter (DAC) is used to transform the time domain digital data to time-domain analogue data (an IQ RF modulator). This leads to the baseband signal which is shown in Fig 3.11a: in some cases the highest peaks which may occur can get clipped (that is, the envelope soft-limited top a saturation) so to help the linear amplification at the receiver. RF modulation is then performed and the signal gets up-converted to the target transmission frequency, leading to the final signal to be sent (Fig. 3.11b) transmitted, after an adequate amplification and shaping of the spectral mask, so as to meet the requirements shown in Fig. 3.9. The signal, eventually, reaches the antenna.

3.3.2 The Blocks of a Wi-Fi Receiver

At the receiver, in principle, the same processing steps of the transmissions should be performed in reverse order. However, the receiving itself takes place after the signal has been subjected to a significant worsening, due to the propagation phenomena. Besides, being a reverse function, it cannot count on the knowledge of some variables which are well known at the transmitter, such as: the beginning-end of a frame, the exact synchronization and duration of each symbol, the exact phase and spacing of each pilot and the minimum/maximum amplitude of the QAM of the different sub-carriers (which may experience slightly different propagation phenomena). As a result, the receiver needs also to implement some extra functions (e.g., frame and symbol synchronization, detection of sub-carriers' phase and frequency, equalization): actually, these tasks will further motivate some strategies which have been mentioned at the transmitter side.

A conceptual block diagram of a WAVE receiver is depicted in Fig. 3.14. Also in this case it will support our analysis: we will skip all those functions which are nothing but the same of the transmitting side, in reverse order; instead, we will pinpoint the peculiarities of the reception process.

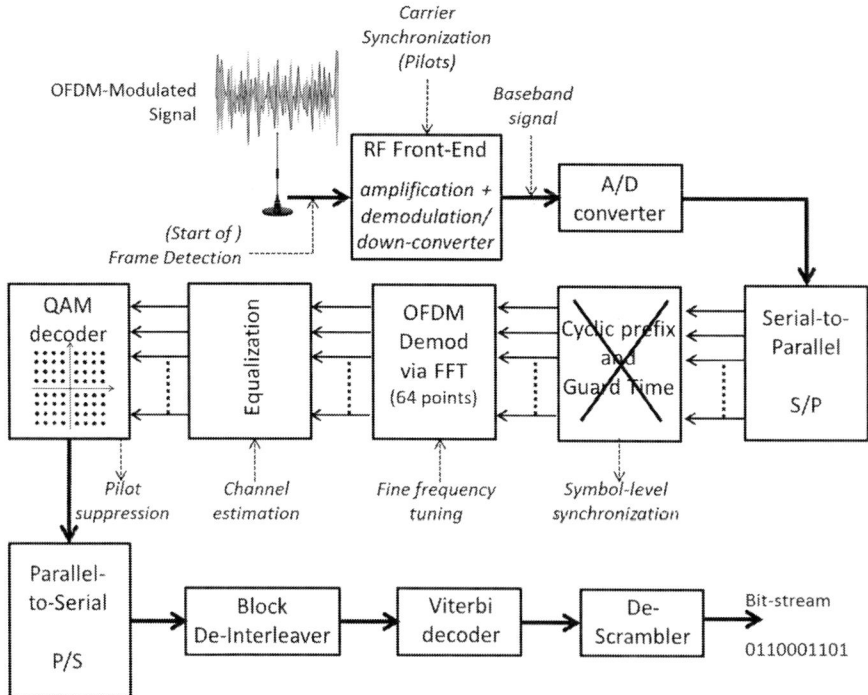

Fig. 3.14 The building blocks of an IEEE 802.11 receiver

Firstly, the phenomena in analogue electronic stages (*Analogue Front-End*, AFE) tend to get neglected but these, indeed, play a significant role. For example, in the radio front-end, after the antenna coupling phenomena which have already been discussed in Sect. 3.1.1, the amplification assumes a different and critical relevance, compared with the *amplifier* at the transmitter. In fact, if in the first amplification stage a low noise amplifier were not adopted, the overall noise figure of the receiver would get dramatically worsened. Actually, inside the radio front-end, there are several components (a cascade of amplifiers and the demodulation chain conducting to the baseband signal).

The following non-obvious task in the receiver chain is the *detection of the incoming frame*: for this purpose, the *short preamble* is continuously fetched. A frequent strategy is that, if a new preamble is detected while a frame is being received, a new reception starts even if the previous has not been completed, because the new one is supposed to have a stronger power (the preamble has been received in spite of the interference by the previous one).

At the same time, thanks to short preamble, the receiver roughly estimates the carrier frequency offset; the frequency offset error is then corrected more precisely using the information in the long preambles (*frequency recovery*). These steps are the keys to separate the OFDM subchannels (S/P). At the same time, thanks to the cyclic extension during the GT, one can perform self-correlation and straightforwardly identify the beginning of an OFDM symbol (*symbol-level synchronization*).

Furthermore, again thanks to the long preamble symbols, the *Channel Impulse Response* (CIR) coefficients can be estimated. The CIR can be used for two different purposes: on the one hand, to feed a digital filter which would cancel mutual and self-interference; on the other hand, to support the *equalization*. In order to correctly QAM demodulate the OFDM components, one needs to recover the precise amplitude and phase of each sub-carrier but, due to propagation phenomena, each of them may undergo different some attenuation and delay: equalization copes with this problem, but it requires to estimate CIR with a sub-carrier granularity, so to get its parameters.

For this purpose Wi-Fi has two types of pilots: *short* pilots covering all the sub-carriers but sent only at the beginning of the frame (long training sequence); pilots continuously sent in each symbol. The former permits a finer tuning but fails with fast fading (the initial estimation is not correct anymore at the end of the frame); the latter works, especially to hold on the former estimation, but the spacing between the pilots needs to be less than the coherence bandwidth. The two types of pilots, together, reach, in most of the cases, a sufficient performance.

The other steps, more or less, represent the dual case of the transmitter chain. Just one note to mention that the digital decoding (after the de-interleaving) will tightly integrate QAM decoding and *Viterbi decoding* (for the error correction) and the final decisions on the decoded symbols could be either a hard or soft Viterbi decision. Descrambling will complete the process, delivering the actual bitstream.

3.3.3 Reception Probability

The OFDM are quite complex systems and their optimal configuration requires a trade-off involving several variables. For instance, at the beginning of Wi-Fi and Wi-Fi-VANETs development, the community had to make joint decisions on:

- the number of sub-carriers, their width and spacing (hence the channel width);
- the digital modulation techniques;
- the redundancy codes.

All these things together lead, from a service-level point of view, to the available transfer-rate (and indirectly to the scalability of the network), to the number of available channels (given the overall bandwidth available), and to the communication robustness. The other way around, the target bit-rate, the expected number of channels and the expected reliability (*Packet Error Ratio*, PER) influenced back all the variables.

In this sense it may be useful to reason for a while on the reception probability. The researchers working on source coding (how to encode the information in order to cope at best with the transmission conditions) typically reason in terms of *Bit Error Ratio* (BER) vs SNR: they plot logarithmic diagrams where they study how a given digital encoding reacts to growing SNR, in terms of bit-error rate. As an intuitive rule of thumb, the lower the rate, the higher the robustness, thanks to the higher energy transmitted and/or the redundancy introduced.

However, when studying vehicular communications in a network-oriented perspective, one is interested more in packet-reception rather than in bit-error rate. For this reason, network simulators typically adopt two parameters: reception and sensitivity thresholds (both have already been introduced in Sect. 3.1.1), which obviously depend on the transfer rate.

More precisely, the reception of frames in presence of noise is not a deterministic process: at a given SINR, one has a certain probability (depending on his receiver implementation) to receive. For this reason, recently, this logic has been included in network simulators [2, 35]: in particular, with NS-3, you have to evaluate frame reception as a stochastic process, considering also the case of variable SNR throughout a frame reception. As already for the sensitivity, regardless of the absolute values, in Fig. 3.15 some possible diagrams of the PER vs SNR for VANET PHY are depicted (adopting the simulation/experimental model proposed in [2]). All the curves are almost equal and just displaced: at 3 Mb/s the PER drops to 0 when SINR is about 2 dBm; at 27 Mb/s, the ratio needs to be 20 dB higher.

Actually, the graphs in Fig. 3.15 are for a specified packet length (400 bytes); this is because frame reception will also depend on frame length: the longer the frame, the more probable that some bits get misunderstood by the receiver and that the errors are so many that the redundancy codes cannot recover them.

When analyzing the graphs, one sees that at 3 Mb/s the effect is almost negligible and the reception scarcely depends on the frame length (graph not shown here);

3 The Physical Layer of VANETs   73

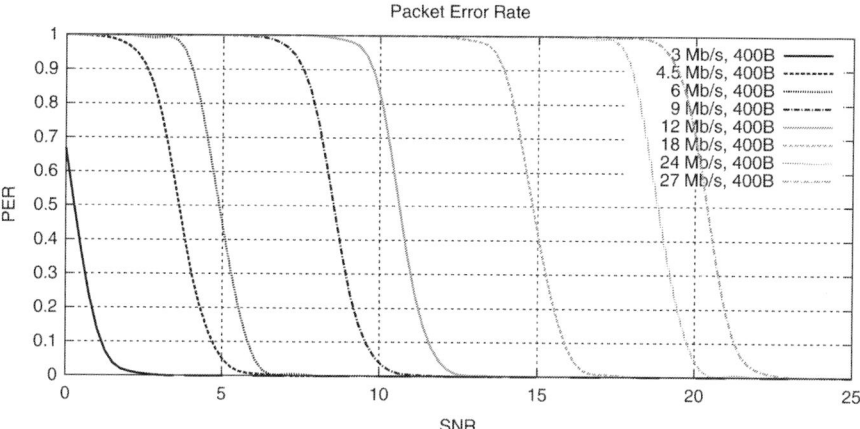

Fig. 3.15 The reception rate as a function of SNR, for different transfer rates (frame length is fixed at 400 bytes). Following the model proposed in [2]

even more, the reception rate abruptly drops from 1 to 0, independently of the frame length. This is because, vice versa, when the SNR lowers, the encoding is so robust which can recover almost all the errors until they are not too many.

Instead, as displayed in Fig. 3.16, at 27 Mb/s the transition is more gradual and is significantly parametrized by frame length.

Importantly, in all these frame-level reception models, the challenges which have been pointed out in Sect. 3.1.2 are not taken into account: one does not consider, in fact, how mobility can create problems on ICI or to the correct equalization, due to the short coherence time; typically, only a stochastic variable modeling fading gets considered. This is also our last remark: one should always pay attention to the approximation which are being adopted in his study.

3.4 Emerging Issues and Techniques Relevant to Physical Layer

An entire chapter (Chap. 15) is devoted to the analysis of possible future evolution of VANETs—including the evolution of physical layer; however, it will deal more with possible standard evolutions. There are also possible standard-compliant techniques which still deserve deeper investigations. Among them, the topic of antenna positioning and antenna diversity are two research branches which have not been exhaustively discussed. The two following subsections, respectively, introduce them.

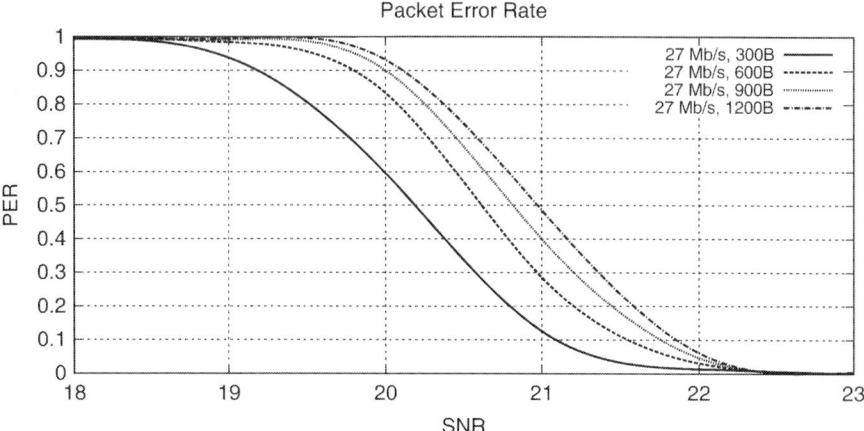

Fig. 3.16 The reception rate as a function of SNR, for different frame lengths (transfer rate is fixed at 27 Mb/s). Following the model proposed in [2]

3.4.1 Antenna-Related Topics

We have already discussed the role of antennas in VANETs in Sect. 3.1.1. Those preliminary theoretical considerations can be enriched by some practical issues which we personally met in our experience. Far from providing exhaustive answers to the problem of the best antenna design, this section is meant to highlight some design constraints. Our synthetic examination will concern the following aspects:

- the impact of the antenna diagram;
- the impact of the car body on the antenna diagram;
- the relevance of the antenna position;
- the mutual interference between nearby antennas.

Despite their relevance these topics have not been extensively studied in VANET literature: only few papers have been found on VANETs antennas [24, 29, 31, 39].

First of all, the antenna diagram may strongly influence the behavior of VANETs: depending on the preferred antenna gain direction, one car might preferably receive frames from cars in front of it or from following ones. Intuitively, this may also lead to a deeper awareness on a specific direction (the same for dissemination, thanks to the reciprocity theorem) and, the other way around for the directions where the antenna gain is lower. However the decision is not obvious: one might think that it is better to know what is happening in front of himself so to be able to react; however, this is not always true on roads and constitutes only a relative gain: it would also imply that cars in the back would receive less frames and would not be able to react at best (if one gains in one direction, loses in the others).

3 The Physical Layer of VANETs 75

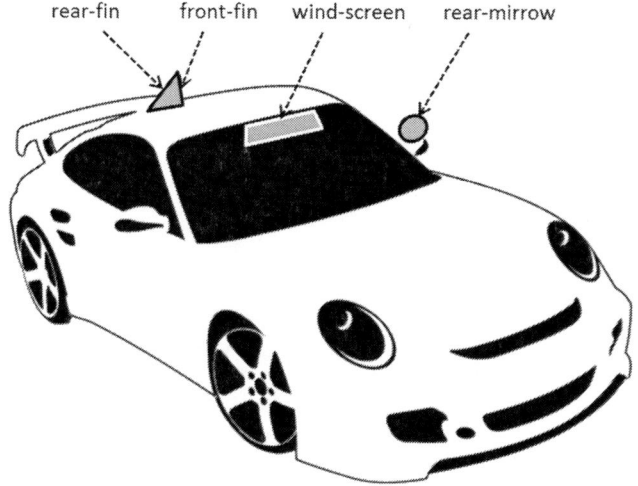

Fig. 3.17 Some possible positions for the VANETs antennas, as suggested in [24]

Altogether, also due to the lack of extensive simulation tools modeling all the involved communication aspects, it is hard today to conclude what the preferred shape of antenna gain should be. As general rule it seems sensible that:

- the antenna gain should preferably have a minimum in the *zenith* direction, since no *rays* are expected from that direction (except, for example, for some road-side unit). A possible diagram is depicted in Fig. 3.5: it would be beneficial to frame reception, with some useful lobes in the horizontal plane;
- the shape of the car body should be taken into account: if one mounts on the back of the car roof an antenna with a strong gain in the forth direction, the antenna gain might be canceled by the metallic surface of the roof (which is expected not to be perfectly flat).

As discussed in [31], the antenna gain patterns of omni-directional antennas become asymmetric in many mounting positions, showing distortions as high as 15 db: in that specific case, to preserve omni-directional characteristics, the antenna had to be mounted in the center of the vehicle.

A preliminary study on the effect of antenna position on frame reception is available in [24]: the authors have carried out a comparative analysis on the effect of 4 antenna positions, see Fig. 3.17, (windscreen, rear-view mirror, front of the roof-fin, back of the roof-fin), in different scenarios (LOS, forest, urban intersection) and evaluating *Received Signal Strength Indication* (RSSI) and throughput. In their preliminary conclusions they point out that, in their tests, no antenna shows a significantly better or worse performance, but, rather, each has its points of strength and weaknesses[20]; in the hypothesis of aiming at a diversity solution, the most effective compensation is likely to occur between the rear-fin antenna and the front-fin or the windscreen one. Obviously, these conclusions strongly depend on the type of antenna and on the type of car.

This analysis lets us introduce also some other criteria for the antenna placement, that is (1) the position will depend also on cabling constraints which may not be obvious on board; (2) two or more VANETs antennas could be mounted (at least at a λ-distance) so to benefit from diversity techniques (see Sect. 3.4.2) and, in order for diversity to effectively work; (3) the cabling of the two antennas should have similar attenuation—otherwise one of the two contributions would become useless.

As a sub-case of the cabling strategies, someone proposes to co-locate (with the antenna) the receiver or an input amplifier.

Last but not least, there is the issue of coexistence among antennas. Frequently, to optimize cabling, antennas are co-located. This already happens, for instance in the fin-shaped packaging which is mounted on the top of cars roofs: there you can already find a GPS and an LTE[21] antenna, and VANET antennas are likely to be mounted there as well. While it is true that each antenna has a frequency response so that it is supposed to capture waves at a given carrier frequency (see Fig. 3.5), however it will be able to capture also spurious frequencies (for example, in the figure the reflection is not complete outside the target frequency): even if the gain may be low, due to the close proximity, the antenna could catch a lot of disturbs. This may be indeed troublesome.

Altogether, when talking about physical layer, issues related to antennas are always neglected but, conversely, the antenna design and integration should not be taken as a straightforward and simple task.

3.4.2 Signal Diversity, Beamforming, Channel Estimation and Beyond

The idea of using multiple antennas is quite intuitive: fading sharply affects signals in a *lambda*-scale and, by combining the signal received by two different antennas, one gains the possibility to have, on average, a stronger signal ready for decoding, by different combining possibilities (including minimum mean squared error or maximum ratio combining).

---

[20]For instance, the performance of the antenna on the rear-view mirror strongly depended on the position of the transmitter (left or right); the windscreen and the front-fin antennas achieved similar performance under LOS; the rear-fin antenna performed the best for negative coordinates.

[21]This applies if the mobile terminals are assumed to be located inside the vehicle and coupled to the vehicle's roof-top antenna in order to avoid wireless signal degradation due to the penetration loss of the vehicle body.

3 The Physical Layer of VANETs

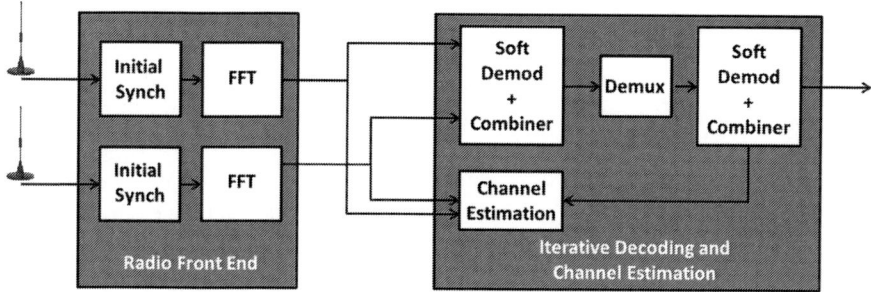

Fig. 3.18 The architecture proposed in [4] for joint turbo decoding and signal diversity

For instance, in [29, 39] the authors propose a frame-based diversity technique: they collect the frames received from all antennas and discard the duplicates: this method, despite very simple, already provides a 10–25 % gain in frame reception rate and 2–5 dB gain in the received power, also in strong line-of-sight (where the gain would be expected to be less significant).

The authors of [3, 32] came to a solution for diversity with a more theoretical approach. Already in 2007 they demonstrated that the outdoor urban environment is particularly hostile (as discussed in Sect. 3.1.2) and is characterized by heavy multi-path, with strong delay spreads which cause conventional Wi-Fi receivers to fail under ISI and ICI (being designed for indoor). The authors ascribed the sub-optimal performance to the time-varying effects altering parameters (for example, the equalization), even throughout a frame: even the channel estimates obtained by the continuous pilot sub carriers (due to their separation) are not sufficient to extend the estimates of the initial pilot symbols.

For this reason they proposed to adopt two parallel processing streams: an *early decoder* estimating channel and preventing interferences (hence addressing ISI and ICI); a *main decoder* (delayed of two symbols), which performs maximum ratio combining of two adjacent symbols, followed by ISI and ICI cancellations, thanks to the information provided by the *early decoder*. In their simulations this leads to a dramatic improvement of the reception rate at low SINR (much more than the diversity used in [29, 39]). The proposed technique falls in the area of iterative channel estimation and decoding, hence represents an application of the so-called turbo-principle for channel estimation (see for instance [51, 53]).

Coming to *diversity*, the same authors of [3, 32] present in [4] a new architecture, leveraging the signals conveyed by two antennas. In this case they can perform a doubled continuous estimation of the channel parameters and softly demodulate the two antenna signals combining them based on the channel estimation (Fig. 3.18); the other way around, the tentative decoding feeds back the channel estimation. The authors show how, during several field trials, the proposed solution outperformed COTS devices, leading to a better reception rate, with ranges 200 m wider and good reception at SINR 6 to over 20 db lower.

After diversity, the next step in the development of Wi-Fi indoor solutions was the introduction of MIMO[22] techniques (for the IEEE 802.11n amendment, now merged in IEEE 802.11 [26]). So one may expect the same for VANETs.

In [13] the authors mention the manifold advantages which VANETs might benefit from, in the perspective of adopting MIMOs: extended range, increased data-rates, dynamic cross-layer configuration of MIMO depending on the traffic and environmental scenario. However, in spite of all these expected benefits, very few papers have investigated this possibility so far: the reason is that, at the current stage of the technology, MIMO are too challenging for VANETs. In fact, MIMO requires a pre-coding which is based on the precise knowledge of channel state (the Channel State Information, a.k.a. as CSI, which is typically based on the Channel Impulse Response or CIR); this is sensible for relatively static conditions and, unfortunately, not for VANETs and for their highly dynamic channel conditions under mobility.

However, if not for pure MIMO solutions, multiple antennas can be used also for *beamforming* and this seems closer to a concrete exploitation. For instance, in [43] it is proposed to optimize the forwarding of broadcasted traffic by leveraging the heading of the car and the angle of the beamforming; even with a simpler approach, one could have a directional beamforming (e.g., forth/back) depending on the relevance of the message being propagated (its target area). This would reduce the interfering energy sent to nodes other than the ones in the intended direction. And, indeed, interference reduction is one of the aims which is proposed also by [13] where, in case of unicast traffic, they propose that the receiver's *Channel State Information* (CSI) can be used by the transmitter to orientate its beam.

There are other studies involving the knowledge of the channel response (for instance of CIR) which are still challenging and interesting but pertain applications other than wireless communications. For instance, there is an interesting research branch about the positioning based on the Time Difference of Arrival (TDoA) [20–23]: they estimate the TDoA of different rays, by a MIMO Wi-Fi equipment, through the evaluation of the preamble. All these studies (by the same authors) would not require a significant *a priori* built-in knowledge of the environment (such as fingerprinting) and should work both in LOS and NLOS conditions (detected direct path—DDP—or undetected direct path—UDP)[23]; besides, still thanks UMP applied to the OFDM preamble, they achieve a sub-nanosecond identification of TDoA for a multipath propagation [21] (in indoor pedestrian environment, with 20 MHz OFDM). Finally [22] they claim that the overall estimation of TDoA may lead, especially with multiple antennas, to an accurate indoor positioning

---

[22]MIMO means multiple-input and multiple-output and refers to the use of multiple antennas to improve communication performance: they may be used just for a diversity gain (counteracting fading) or even to increase the channel capacity by spatial multiplexing (using the different antennas as parallel channels). MIMO techniques have been used by Wi-Fi (802.11n and 802.11ac), LTE and Wimax, for instance.

[23]By considering the mean excess time delay (τ_{MED}) which a node can estimate over the multipath channels and thanks to the algorithm UMP (Unitary Matrix Pencil) [20], they can distinguish on the fly between UDP and DDP [23].

(in the range of some centimeters). Even if this is something which needs a deeper understanding and a detailed analysis for the vehicular environment, also the analysis of TDoA, for positioning or for channel estimation, is certainly an area which deserves our attention: it might further enforce vehicular safety. Definitely it represents another challenging but interesting research area also for VANETs.

3.5 Conclusions

After analyzing all the main phenomena affecting VANETs propagation and the characteristics of the current standards at physical layer, it can be stated that, from this point of view, VANETs are ready for exploitation.

As further discussed in Chap. 15, there are some areas which could be improved—especially to advance them upon their current performance—and other which may hold beneficial surprises—as in case of an optimal antenna positioning or of solutions adopting antenna diversity. Altogether, something will certainly evolve, however stable foundations have been set.

Acknowledgements The FP7 project GLOVE (joint GaliLeo Optimization and VANET Enhancement Grant Agreement 287175) has partially supported this work.

The authors thank also Giorgio Giordanengo (ISMB-LACE) for the antenna diagrams.

References

1. 3GPP (2007) Evolved universal terrestrial radio access (e-utra); base station (bs) radio transmission and reception. 3GPP TS 36.104, 3rd Generation Partnership Project, Sophia Antipolis
2. Abrate F, Vesco A, Scopigno R (2011) An analytical packet error rate model for wave receivers. In: Vehicular technology conference (VTC Fall), 2011 IEEE, pp 1–5. doi:10.1109/VETECF.2011.6093093
3. Alexander P, Haley D, Grant A (2007) Outdoor mobile broadband access with 802.11. IEEE Commun Mag 45(11):108–114. doi:10.1109/MCOM.2007.4378329
4. Alexander P, Haley D, Grant A (2011) Cooperative intelligent transport systems: 5.9-ghz field trials. Proc IEEE 99(7):1213–1235. doi:10.1109/JPROC.2011.2105230
5. Bai F, Stancil D, Krishnan H (2010) Towards understanding characteristics of dedicated short range communications (dsrc) from a perspective of vehicular network engineers. In: Proceedings of the sixteenth annual international conference on mobile computing and networking (MobiCom), pp. 329–340. ACM, New York. doi:10.1145/1859995.1860033. http://doi.acm.org/10.1145/1859995.1860033
6. Balanis CA (2005) Antenna theory: analysis and design, 3rd edn. Wiley, Hoboken (2005)
7. Barton D (1998) Radar technology encyclopedia. Artech, Boston
8. Bloessl B, Segata M, Sommer C, Dressler F (2013) Towards an open source IEEE 802.11p stack: a full sdr-based transceiver in gnu radio. In: Vehicular networking conference (VNC), 2013 IEEE, pp 143–149. doi:10.1109/VNC.2013.6737601
9. Campolo C, Molinaro A (2013) Multichannel communications in vehicular ad hoc networks: a survey. IEEE Commun Mag 51(5):158–169

10. Campolo C, Cozzetti HA, Molinaro A, Scopigno RM (2012) Overhauling ns-2 phy/mac simulations for IEEE 802.11 p/wave vehicular networks. In: 2012 IEEE international conference on communications (ICC). IEEE, Ottawa, pp 7167–7171
11. Chen Q, Schmidt-Eisenlohr F, Jiang D, Torrent-Moreno M, Delgrossi L, Hartenstein H (2007) Overhaul of IEEE 802.11 modeling and simulation in ns-2. In: Proceedings of the 10th ACM symposium on modeling, analysis, and simulation of wireless and mobile systems, MSWiM '07. ACM, New York, pp 159–168. doi:10.1145/1298126.1298155. http://doi.acm.org/10.1145/1298126.1298155
12. Cozzetti H, Vesco A, Abrate F, Scopigno R (2010) Improving wireless simulation chain: impact of two corrective models for vanets. In: Vehicular networking conference (VNC), 2010 IEEE, pp 323–329. doi:10.1109/VNC.2010.5698253
13. El-Keyi A, ElBatt T, Bai F, Saraydar C (2012) Mimo vanets: research challenges and opportunities. In: 2012 International conference on computing, networking and communications (ICNC), pp 670–676. doi:10.1109/ICCNC.2012.6167507
14. ETSI (2009) Intelligent transport systems (its); European profile standard for the physical and medium access control layer of intelligent transport systems operating in the 5 GHz frequency band. ETSI Draft - ES 202 663 v1.1.0. European Telecommunication Standards Institute, Sophia Antipolis
15. ETSI (2011) Intelligent transport systems (its); on the recommended parameter settings for using stdma for cooperative its; access layer part. ETSI TR 102 861. European Telecommunication Standards Institute, Sophia Antipolis
16. ETSI (2012) Intelligent transport systems (its); harmonized channel specifications for intelligent transport systems operating in the 5 GHz frequency band. ETSI TR 102 724. European Telecommunication Standards Institute, Sophia Antipolis
17. FCC (2003) Dedicated short range communications (dsrc) report and order. FCC R.O. FCC 03-324, U.S. Federal Communications Commission
18. Friis HT (1945) Discussion on "noise figures of radio receivers". Proc IRE 33(2):125–127. doi:10.1109/JRPROC.1945.233208
19. Friis H (1946) A note on a simple transmission formula. Proc IRE 34(5):254–256. doi:10.1109/JRPROC.1946.234568
20. Gaber A, Omar A (2012) A study of tdoa estimation using matrix pencil algorithms and IEEE 802.11ac. In: Ubiquitous positioning, indoor navigation, and location based service (UPINLBS), pp 1–8. doi:10.1109/UPINLBS.2012.6409772
21. Gaber A, Omar A (2012) Sub-nanosecond accuracy of tdoa estimation using matrix pencil algorithms and IEEE 802.11. In: 2012 International symposium on wireless communication systems (ISWCS), pp 646–650. doi:10.1109/ISWCS.2012.6328447
22. Gaber A, Omar A (2014) Recent results of high-resolution wireless indoor positioning based on IEEE 802.11ac. In: Radio and wireless symposium (RWS), 2014 IEEE, pp 142–144. doi:10.1109/RWS.2014.6830157
23. Gaber A, Alsaih A, Omar A (2014) Udp identification for high-resolution wireless indoor positioning based on IEEE 802.11ac. In: 2014 11th workshop on positioning, navigation and communication (WPNC), pp 1–6. doi:10.1109/WPNC.2014.6843303
24. Gavilanes G, Reineri M, Brevi D, Scopigno R, Gallo M, Pannozzo M, Bruni S, Zamberlan D (2013) Comparative characterization of four antennas for vanets by on-field measurements. In: 2013 IEEE international conference on microwaves, communications, antennas and electronics systems (COMCAS), pp 1–5. doi:10.1109/COMCAS.2013.6685290
25. Grau G, Pusceddu D, Rea S, Brickley O, Koubek M, Pesch D (2010) Vehicle-2-vehicle communication channel evaluation using the cvis platform. In: 2010 7th International symposium on communication systems networks and digital signal processing (CSNDSP), pp 449–453
26. IEEE (2012) Standard for information technology—telecommunications and information exchange between systems local and metropolitan area networks—specific requirements part 11: wireless LAN medium access control (MAC) and physical layer (Phy) specifications. IEEE 802.11-2012. Institute of Electrical and Electronics Engineers - Standard Association
27. Jakes WC (1994) Microwave mobile communications. Wiley, New York

28. Jiang T, Chen HH, Wu HC, Yi Y (2010) Channel modeling and inter-carrier interference analysis for v2v communication systems in frequency-dispersive channels. Mobile Netw Appl 15(1):4–12. doi:10.1007/s11036-009-0177-2. http://dx.doi.org/10.1007/s11036-009-0177-2
29. Kaul S, Ramachandran K, Shankar P, Oh S, Gruteser M, Seskar I, Nadeem T (2007) Effect of antenna placement and diversity on vehicular network communications. In: 4th Annual IEEE Communications Society conference on sensor, mesh and ad hoc communications and networks, 2007. SECON '07, pp 112–121. doi:10.1109/SAHCN.2007.4292823
30. Kenney J (2011) Dedicated short range communications (DSRC) standards in the united states. In: Proceedings of the IEEE, pp 1162–1182. doi:10.1109/JPROC.2011.2132790
31. Klemp O (2010) Performance considerations for automotive antenna equipment in vehicle-to-vehicle communications. In: 2010 URSI international symposium on electromagnetic theory (EMTS), pp 934–937. doi:10.1109/URSI-EMTS.2010.5637361
32. Letzepis N, Grant A, Alexander P, Haley D (2011) Joint estimation of multipath parameters from ofdm signals in mobile channels. In: 2011 Australian communications theory workshop (AusCTW), pp 106–111. doi:10.1109/AUSCTW.2011.5728746
33. Liberti J, Rappaport T (1996) A geometrically based model for line-of-sight multipath radio channels. In: IEEE 46th vehicular technology conference, 1996. Mobile technology for the human race, vol 2, pp 844–848. doi:10.1109/VETEC.1996.501430
34. Mecklenbrauker C, Molisch A, Karedal J, Tufvesson F, Paier A, Bernado L, Zemen T, Klemp O, Czink N (2011) Vehicular channel characterization and its implications for wireless system design and performance. Proc IEEE 99(7):1189–1212. doi:10.1109/JPROC.2010.2101990
35. Mittag J, Papanastasiou S, Hartenstein H, Strom E (2011) Enabling accurate cross-layer phy/mac/net simulation studies of vehicular communication networks. Proc IEEE 99(7):1311–1326. doi:10.1109/JPROC.2010.2103291
36. Molisch A, Tufvesson F, Karedal J, Mecklenbrauker C (2009) A survey on vehicle-to-vehicle propagation channels. IEEE Wirel Commun 16(6):12–22. doi:10.1109/MWC.2009.5361174
37. Nee RV, Prasad R (2000) OFDM wireless multimedia communication, 1st edn. Artech House, Norwood
38. Neiman M (1943) The principle of reciprocity in antenna theory. Proc IRE 31(12):666–671. doi:10.1109/JRPROC.1943.233683
39. Oh S, Kaul S, Gruteser M (2009) Exploiting vertical diversity in vehicular channel environments. In: 2009 IEEE 20th international symposium on personal, indoor and mobile radio communications, pp 958–962. doi:10.1109/PIMRC.2009.5449728
40. Ozdemir MK (2007) Channel estimation for wireless ofdm systems. IEEE Commun Surv Tutorials 9(2):18–48. doi:10.1109/COMST.2007.382406
41. Scopigno R (2012) Physical phenomena affecting vanets: open issues in network simulations. In: 2012 14th International conference on transparent optical networks (ICTON), pp 1–4. doi:10.1109/ICTON.2012.6253776
42. Sklar B (1997) Rayleigh fading channels in mobile digital communication systems. I. Characterization. IEEE Commun Mag 35(7):90–100. doi:10.1109/35.601747
43. Soua A, Ben-Ameur W, Afifi H (2012) Broadcast-based directional routing in vehicular ad-hoc networks. In: 2012 5th Joint IFIP wireless and mobile networking conference (WMNC), pp 48–53. doi:10.1109/WMNC.2012.6416146
44. Strom E (2011) On medium access and physical layer standards for cooperative intelligent transport systems in Europe. Proc IEEE, 1183–1188. doi:10.1109/JPROC.2011.2136210
45. Strom E (2013) On 20 MHz channel spacing for v2x communication based on 802.11 ofdm. In: IECON 2013 - 39th annual conference of the IEEE Industrial Electronics Society, pp 6891–6896. doi:10.1109/IECON.2013.6700274
46. Sundararajan D (2001) The discrete fourier transform: theory, algorithms and applications. World Scientific, Singapore. http://books.google.it/books?id=54kTgFg5IVgC
47. Tan I, Tang W, Laberteaux K, Bahai A (2008) Measurement and analysis of wireless channel impairments in dsrc vehicular communications. In: IEEE international conference on communications, 2008. ICC 08. IEEE, Beijing, pp 4882–4888

48. Weinstein S, Ebert P (1971) Data transmission by frequency-division multiplexing using the discrete fourier transform. IEEE Trans Commun Technol 19(5):628–634. doi:10.1109/TCOM.1971.1090705
49. Wu ZD, Qin SY (2012) Effect study of spectrum analyzer noise floor on antenna noise temperature measurement. In: 2012 10th International symposium on antennas, propagation EM theory (ISAPE), pp 8–10. doi:10.1109/ISAPE.2012.6408688
50. Xiang W, Shan D, Yuan J, Addepalli S (2012) A full functional wireless access for vehicular environments (wave) prototype upon the IEEE 802.11p standard for vehicular communications and networks. In: 2012 IEEE consumer communications and networking conference (CCNC), pp 58–59. doi:10.1109/CCNC.2012.6181050
51. Ylioinas J, Juntti M (2007) An iterative receiver for joint detection, decoding, and channel estimation in turbo coded mimo ofdm. In: Conference record of the forty-first Asilomar conference on signals, systems and computers, 2007. ACSSC 2007, pp 1581–1585. doi:10.1109/ACSSC.2007.4487497
52. Zang Y, Stibor L, Orfanos G, Guo S, Reumerman H (2005) An error model for intervehicle communications in highway scenarios at 5.9GHz. In: Proceedings of the second ACM international workshop on performance evaluation of wireless ad hoc sensor and ubiquitous networks, PE-WASUN '05. ACM, New York, pp 49–56. doi:10.1145/1089803.1089966. http://doi.acm.org/10.1145/1089803.1089966
53. Zhao M, Shi Z, Reed M (2008) Iterative turbo channel estimation for ofdm system over rapid dispersive fading channel. IEEE Trans Wirel Commun 7(8):3174–3184. doi:10.1109/TWC.2008.070228

Chapter 4
The MAC Layer of VANETs

Claudia Campolo, Antonella Molinaro, Riccardo Scopigno, Serkan Ozturk, Jelena Mišić, and Vojislav B. Mišić

Abstract The design of the Medium Access Control (MAC) layer for vehicular networks should carefully cope with fast changing topologies caused by vehicle mobility, short connection lifetimes, harsh propagation environments, high node density, and heterogeneous traffic nature and quality demands. This chapter provides a detailed description of MAC functions (i.e., channel access rules, prioritization schemes, frame types, and formats), as specified by the IEEE 802.11p and ETSI ITS-G5 standards, and the multichannel operation. By discussing the main challenges which the MAC has to deal with, the chapter also presents the main evaluation tools (analytics, simulations, field-tests) that can be used to analyze the MAC performance.

Keywords VANET • MAC • IEEE 802.11p • ITS-G5 • OCB • Broadcast • Hidden terminals • Collisions • EDCA • 1609.4 • 1609.3 • WAVE • WSA • PLCP • Metrics

---

C. Campolo (✉) • A. Molinaro
University Mediterranea of Reggio Calabria, Reggio Calabria, Italy
e-mail: claudia.campolo@unirc.it; antonella.molinaro@unirc.it

R. Scopigno
Istituto Superiore Mario Boella, Torino, Italy
e-mail: scopigno@ismb.it

S. Ozturk
Erciyes University, Kayseri, Turkey
e-mail: serkan@erciyes.edu.tr

J. Mišić • V.B. Mišić
Ryerson University, Toronto, ON, Canada
e-mail: jmisic@ryerson.ca; vmisic@ryerson.ca

4.1 Introduction

Since the beginning of research on Vehicular ad-hoc networks (VANETs), in the early 1990s, different wireless technologies have been considered to support vehicle-to-vehicle (V2V) and vehicle-to-infrastructure (V2I) communications with the primary goal of enabling cooperative safety applications.

The MAC layer, whatever the access technology, plays a crucial role in regulating the access to the common wireless medium so that multiple stations share it efficiently. The most pressing requirement for a MAC in a vehicular environment was for a *decentralized* operation, without a coordinator and/or a central manager in charge of ruling the channel access or the resource assignment. This way, vehicular nodes are enabled to communicate without explicitly joining the network (i.e., associating with the access point or the base station playing the coordinator role).

Among available radio technologies, the simple and worldwide well-known IEEE 802.11 [27] technology was considered the best candidate to enable V2V and V2I wireless communications; notably, it also guaranteed the perspective of supporting services other than safety (such as automated tolling, enhanced navigation, and traffic management). Initially proposed as an amendment to the IEEE 802.11, 802.11p is now part of the last IEEE 802.11 standard [27] and represents the de facto solution for vehicular communications.

This decision was one of the few shared between the IEEE 1609 family of standards for Wireless Access in Vehicular Environments (WAVE) and the European profile standard of Intelligent Transport Systems (ITS) operating in the 5 GHz frequency band standardized by ETSI (ITS-G5) [28].

IEEE 802.11p inherits the main operating principles of the 802.11 technology, however, unlike 802.11, it has been specifically conceived to provide short-to-medium range connectivity (in a range up to 1,000 m) to fast moving vehicles in a very different environment compared to wireless office/home. Many other differences might (and will) be highlighted in Sect. 4.2.

Summing up:

- VANETs adopt the same PHY-MAC layers of IEEE 802.11;
- the PHY layer, relying on Orthogonal Frequency Division Multiplexing (OFDM) as discussed in Chap. 3, has been slightly modified for VANETs and, in most cases, it now suits the VANETs' requirements—with exception of those cases mentioned in Chap. 15;
- the adaptation of the MAC layer, based on Carrier Sense Multiple Access with Collision Avoidance (CSMA/CA), seemed challenging from the beginning, due to the deeply different context, underlying traffic and scenarios.

For the above reasons, the MAC layer requires some additional insight,

> which necessarily will exceed the boundaries of a chapter. For editorial decision, this chapter is about the MAC basics in VANETs and is focused on the access mechanisms (including elements of the multichannel operation), the frame formats and, eventually, the approaches available for the MAC analysis.
>
> Other topics related to the MAC are the Decentralized Congestion Control (Chap. 6), the detailed Multichannel operations (Chap. 7), and a recap on CSMA/CA weaknesses in VANETs with focus on the MAC evolution (Chap. 15).

Being the first part of this chapter on MAC operating principles and data structures (i.e., MAC frame formats and CSMA/CA basics), it was decided to briefly highlight in Sect. 4.2 the peculiarities of CSMA/CA when used for VANETs: this is meant to facilitate a rapid understanding of VANETs based on their differentiators. Then, for an exhaustive explanation of MAC features, the remainder of the chapter is organized as follows.

Frame types and formats are discussed in Sect. 4.3. The MAC protocol operations, spanning from the basic access rules to the prioritization mechanisms, as specified in IEEE 802.11 and inherited by 802.11p, are detailed in Sect. 4.4. Section 4.5 describes the MAC layer extension to support multichannel operation and interact with the IEEE 1609.4 specifications in the WAVE architecture [26]. Simulations, mathematical models and test-beds as tools for MAC analysis are introduced in Sect. 4.6. Finally, Sect. 4.7 concludes the chapter.

4.2 The Differentiators of VANETs and Their Effects on the MAC

In Chap. 3, the PHY layer of VANETs has been presented starting from the requirements of the vehicular environment. While the IEEE 802.11 PHY could be adapted quite effectively by down-clocking; its MAC layer was exposed to some deeper differences for the reasons which are shortly recalled in Table 4.1.

Let's go once more through the VANETs requirements and check how much they can be matched with IEEE 802.11p mechanisms:

- MAC protocols for VANETs should carefully face the quickly changing network topology due to vehicles' mobility, the short connection lifetime and the harsh effects of a hostile signal propagation environment: CSMA/CA intrinsically supports this feature, being decentralized and *connectionless*.
- IEEE 802.11p inherits the main rules of the baseline 802.11 MAC standard with its prioritized channel access, but it *simplifies* the operations for *authentication and association*, which were considered too time-consuming. This represents a successful adaptation to VANETs and is achieved, thanks to the definition of the new *Outside the Context of the Basic Service Set* (OCB) mode (Fig. 4.1).

Table 4.1 Differences between IEEE 802.11 in vehicular environment (802.11p) and legacy IEEE 802.11. The bale is split into: *constraints*, *solutions*, and *issues*, with some links to the other sections and chapters where the topics are discussed. In bold the topics which fall inside the present chapter

	802.11p	802.11
Constraints		
Environmental context	Outdoor with fast fading, mobility	Indoor, stationary or nomadic
Number of nodes	Large, unpredictable number	Often limited and a priori known
Traffic type	Heavy delay-sensitive traffic Mostly broadcast High percentage of periodic traffic + critical event-driven	Mostly browsing (unicast)
Traffic mode	Mostly OCB	Often within an *infrastructure* (BSS), or ad-hoc (IBSS)
Solutions		
PHY layer	(1) As IEEE 802.11a but narrower channels for slower and more robust encoding (Chap. 3) ⇒ Lower data rate (2) **PLCP sublayer (as 802.11)—Sect. 4.3.1**	OFDM as from IEEE 802.11a
MAC layer	**(1) MAC Frame formats: as 802.11 (but no beacons and other management frames) plus Timing Advertisement—Sect. 4.3.2 (2) MAC mechanisms: as 802.11 but no BSS setup (OCB enabled) and no centralized access—Sect. 4.4 Broadcast ⇒ No exponential back-off, no ACK, no RTS/CTS ⇒ see Issues (3) LLC frame formats: IEEE 802.2—Sect. 4.3.3**	(1) Data, Control and Management MAC frames (2) MAC mechanisms based on BSS-IBSS setup, distributed and centralized access (3) Standard LLC frame formats
Protocol architecture	**Differences in Fig. 4.2**	–
Issues		
PHY layer	Doubly selective radio channel (Chap. 3) ⇒ Future solutions (Chap. 15)	–
MAC layer	(1) Scalability ⇒ DCC (Chap. 6) ⇒ Future solutions (Chap. 15) (2) Hidden terminals ⇒ Future solutions (Chap. 15)	–

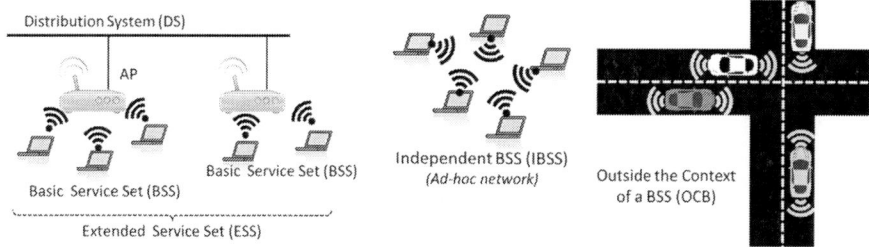

Fig. 4.1 IEEE 802.11 modes: *infrastructure mode* or basic service set (BSS), *ad-hoc mode* or independent BSS (IBSS), and the new outside the context of a BSS (OCB)—on the *right*. Nodes in a VANET are free to use any of them (BSS, IBSS, OCB); however, OCB is specifically defined to avoid all the setup times, particularly prejudicial for medium access with rapidly moving nodes

- Most vehicular traffic (primarily, safety) is *broadcast*—not sent to a specific station—this kills most of the CSMA/CA mechanisms that rely on the receiver's feedback; as a result, the prevention of collisions is not so effective as one could expect (*no exponential back-off, no prevention of hidden terminals*).
- Even worse, in VANETs the number of participating nodes is not always known and, more importantly, cannot be restricted: in principle, a vehicular MAC protocol must be properly scalable to account for *varying network density*.
  Sparse traffic conditions are common in off-peak hours, or expected in the initial VANET deployment phase due to a limited market penetration rate. High traffic conditions with several cars concentrated in a small area characterize instead urban crossroads in peak hours. In the latter scenario, the data traffic load can be severely heavy, especially when considering safety data, such as Cooperative Awareness Messages (CAMs) (see Chap. 5), transmitted by vehicles frequently (typically at every 100 ms). Such expected congestion further exacerbates the problem of collisions in VANETs: this threat led to the definition of Decentralized Congestion Control (DCC) mechanisms (see Chap. 6).
- From the point of view of frame *formats*, using the same formats as IEEE 802.11 may bring to some channel *inefficiencies*: in most cases, for instance, the use of multiple MAC addresses in the MAC frame is unnecessary (for broadcast transmissions the address of the sender would be sufficient); over 300-byte-long frames (the typical CAM size), this would result in a significant and useless overhead.
- The MAC operation should also be: (1) *real-time* (to guarantee low delivery-time of safety messages), (2) *reliable* and (3) *fair* (i.e., give all nodes at least one opportunity to access the channel within each time period). Actually, the first requirement (real-time transmission) is only apparently an issue: broadcast traffic—due to the lack of exponential back-off—is never delayed significantly [19], although channel access delay is unpredictable for CSMA/CA under heavy load conditions; conversely the number of collisions grows and this affects both reliability [12, 13, 15, 19] and fairness [3]. Once more, a partial solution is represented by DCC mechanisms.

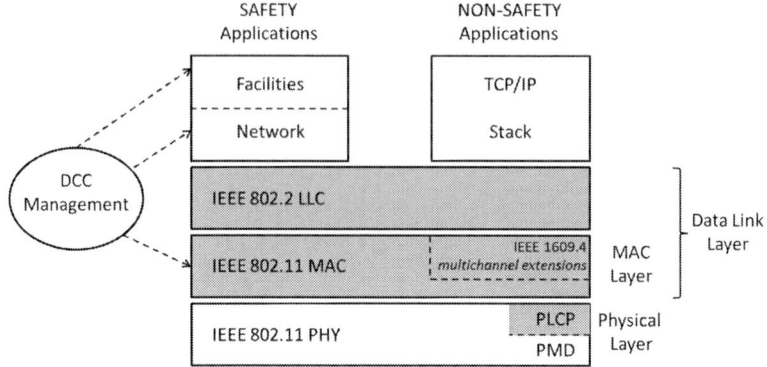

Fig. 4.2 A simplified stack highlighting the MAC related topics covered in this chapter (*grey*) and the main changes produced (either by ETSI or by IEEE) to improve and cope with VANET performance at MAC layer: cross-layer DCC (by ETSI)—acting also on upper layer—and multi-channel management (by IEEE)

All the aforementioned constraints and peculiarities of VANETs make the adaptation of MAC to vehicular environments a thorny task. Altogether, the IEEE 802.11-based MAC of VANETs has been extended along three main directions. Firstly, the definition of new optimized access parameter settings (e.g., EDCA settings), including a new operating mode preventing long association time. Such mode, called OCB, is further discussed in Sect. 4.4.

Second, specific DCC mechanisms have been defined to counteract the CSMA/CA poor performance in VANETs, due to the likely congestion and to the mostly broadcast traffic nature.

Finally, a specific protocol stack was defined for vehicular safety applications (see Fig. 4.2—a different view of what already discussed in Chap. 2). Such protocol architecture is optimized so to avoid the huge overheads of the Transport Control Protocol (TCP)/Internet Protocol (IP) stack, and to give the applications (the *Facility* Layer) the possibility to cope with MAC inefficiencies, for example by acting on the generation rate of safety messages.

4.3 Frame Types and Formats

In the previous chapter, the transmission and reception of IEEE 802.11 frames have been explained without talking about the framing, but just mentioning physical layer phenomena. However, information needs to be cut into frames in order to be sent on the radio interface, and the receiver has to accomplish several side-tasks in order to successfully decode a frame: it needs to know how long the frame is, what transfer-rate it is using, what it is meant for, what station it is addressed to, etc.

4 The MAC Layer of VANETs

These functions are split into layers, as shown in Fig. 4.2; bottom–up you meet: PMD, PLCP, MAC and LLC.

The Physical Medium Dependent (PMD) sublayer provides the actual *physical* transmission and reception, by interfacing directly the wireless medium, modulating and demodulating.

All the upper layers (first the PLCP, then the MAC and the LLC) will add their respective outer preambles and trailers—the other way around, LLC is carried within the MAC frame, which is nested into PLCP.

This process leads to the frame structure discussed in Sects. 4.3.1–4.3.3.

4.3.1 Formats: PLCP

IEEE 802.11 foresees multiple transfer rates and also permits to change packet fragmentation so to improve the overall throughput under changing conditions. The Physical Layer Convergence Procedure (PLCP) is aimed at signaling all this information and to send the auxiliary fields which are needed for a correct frame reception.

The PLCP information (Fig. 4.3) includes: a *preamble*, a *header* and a *trailer* (tail + padding). The MAC frame represents the upper-layer Service Data Unit (PLCP SDU, or PSDU): PLCP + PSDU constitute the so-called Physical Protocol Data Unit (PPDU).

Importantly, the PLCP portion of the frame is always sent at the lowest commonly supported data rate, to ensure the maximum reliability and compatibility with other stations. The PLCP carries the following information:

- the *preamble* is the one which has already been described in Chap. 3 about the correct pilot identification and the OFDM equalization;
- the *signal* brings several sub-fields:

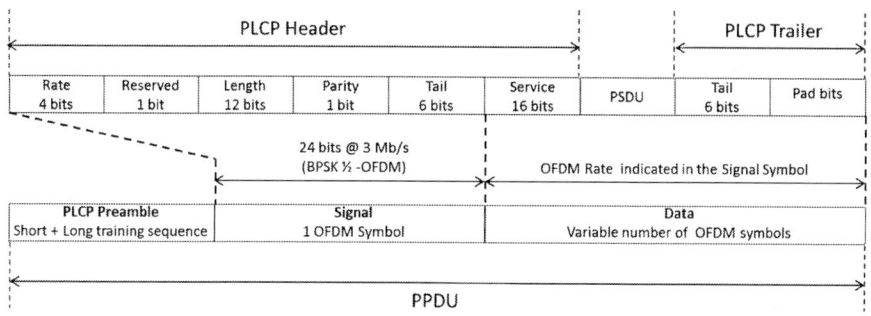

Fig. 4.3 PLCP frame format

- *rate*: these are 4 bits specifying what transfer rate is being used (in case of 10 MHz channels, among: 3, 4.5, 6, 9, 12, 18, 24 and 27 Mb/s);
- the next bit is a *reserved* bit set to 0;
- *length* indicates the number of octets in the PSDU (the first transmitted are the least significant);
- *parity* is an even parity bit for the preceding bits (0–16), used to detect errors;
- *tail* bits are 6 and are used to initialize the convolutional encoder (they are always set to 0);
- *service* field is already transmitted at the target rate—not at the basic one— and is set to all 0s. Being embedded in the MAC frame they also undergo the scrambling process (see Chap. 3): since they are *a priori* known, they are used to set the scrambler at the receiver; while the first 7 bits are used for the initialization of the scrambler, the last 9 are unused (*reserved*);

- *tail* includes 6 bits appended to the end of the MAC frame to smoothly stop the convolutional encoder (6 bits are required since the convolutional encoder has length 7);
- the final variable *pad* is necessary to have a integer number of symbols to be sent.

Thanks to the PLCP information, the PMD is managed and made transparent (transfer rate and frame length indication, initialization of scrambler and convolutional encoder, proper padding); additionally, also the required pilots are provided (preamble). The MAC frame, that is PSDU, can now be nested.

4.3.2 Formats: MAC Frames

The IEEE 802.11 MAC layer uses three types of frames: *data*, *control*, and *management frames*. Data frames carry user data from higher layers. Management frames (e.g., beacon) enable stations to establish and maintain communications; they are not forwarded to the upper layers.

Control frames, e.g., Request-to-Send (RTS), Clear-to-Send (CTS) and acknowledgement (ACK), assist in the delivery of data and management frames between stations.

MAC frames (whatever kind they are) consist of the following basic components: a MAC *header*; a variable length *Frame Body*; and a *Frame Check Sequence* (FCS).

- The MAC header format encompasses a set of fields that occur in a fixed order in all frames of a given type.
- The Frame Body field consists of the MAC service data unit (MSDU). The frame body is of variable size; its maximum length is determined by the maximum MSDU size plus any overhead from security encapsulation.
- FCS is a 4-byte field which contains a 32-bit cyclic redundancy code (CRC) used for detecting bit errors in the MAC frame. The CRC is computed over all the fields of the MAC header and the Frame Body.

4 The MAC Layer of VANETs

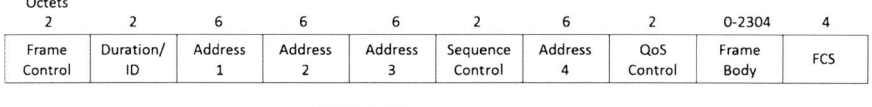

Fig. 4.4 MAC frame format

Figure 4.4 depicts the most common format for Data frames.

Among the header fields, *Frame Control*[1] includes rich information about the protocol version, the type and sub-type of frame (for instance, association, probe request, authentication or other management frame; otherwise RTS, CTS, ACK or other control frame; data frame), the number of retry, and if the MAC is destined for (or comes from) the *Distribution System* (DS)[2]; it includes also additional information, for an overall length of 2 bytes. For an extensive analysis, the standard represents the best reference [27].

Still in the header field, the *Sequence Control* (2 bytes) manages the fragmentation, while the *Duration/Id* field is typically used to update the Network Allocation Vector (NAV), since it includes the duration value (in microseconds) required to transmit the current frame, possibly including some overhead for its response, and appropriate Inter-Frame Spaces (IFSs). Finally, the *QoS Control* field involves details of Quality of Service (QoS) behavior for priority-based access to the channel.

About the *Address Fields*, they can be up to 4 in 802.11: this is because, in infrastructure mode, all the traffic needs to pass through the access point: hence, in principle, one should distinguish between *source* (SA) and *transmitter* (TA) address, and between *destination* (DA) and *receiver* (RA) address. Normally, SA and DA are the addresses of the endpoints of the communication, while TA and RA are the addresses of the intermediate nodes on the current wireless hop.

For this reason, one address is used to uniquely identify the basic service set (BSS), i.e., the network created by an access point, see Sect. 4.4, by the so-called BSS IDentification (BSSID). In a BSS the value of BSSID is the MAC address currently in use by the AP. In an independent BSS (IBSS) instead (ad-hoc operation), it is a locally administered IEEE MAC address formed from a 46-bit random number with the individual/group bit set to 0 and the universal/local bit set to 1. The value of all 1s is used to indicate the *wildcard BSSID*. A wildcard BSSID can be used in management frames of subtype probe request for example, and, according to the 802.11p amendment, also data frames with a wildcard BSSID are permitted when the *dot11OCBEnabled* flag is true and the OCB mode is enabled.

---

[1] For the sake of completeness, Frame Control includes the following bits: *Protocol version* (2 bit), *Type* (2 bit), *Subtype* (4 bit), *To DS* (1 bit), *From DS* (1 bit), *More fragments* (1 bit), *Retry* (1 bit), *Power management* (1 bit), *More data* (1 bit), *WEP* (1 bit), *Order* (1 bit).

[2] In IEEE 802.11 Distribution System is the LAN which the Access Point (AP) is connected to; one may have also a Wireless DS (WDS).

Table 4.2 Meaning of the address fields in 802.11 frames

To DS bit	From DS bit	Address1	Address2	Address3
0	0	DA	SA	BSSID
0	1	DA	BSSID	SA
1	0	BSSID	SA	DA

The number of address fields actually used depends on the frame type. The most frequent case foresees three of them: SA, DA, BSSID.

Things are only apparently complicated. In fact, with exception of the case when the frame comes from the DS and goes to another DS (from DS = 1, to DS = 1 in the Frame Control), the fourth address is never used and three of them are used according to the semantics of Table 4.2.

The first three fields (Frame Control, Duration/ ID, and Address 1) and the last field (FCS) in Fig. 4.4 constitute the minimal frame format and are present in all frame types. For instance, the commonly used control frames, i.e., RTS, CTS, ACK frames, include them, without carrying out a frame body.

Moving to the specific case of VANETs, three adaptations in the frame formats and definitions can be highlighted:

- Considering what said for OCB mode, broadcast frames in VANETs will have *FromDS* and *ToDS*=0; hence, *Address1* (destination) and *Address3* (BSSID) will be all to "1" and *Address2* will be the MAC of the transmitting station.
- IEEE 802.11p amendment inherits the same frames defined for baseline 802.11, with the main difference that the beacon frame foreseen in independent and infrastructure modes is replaced in the newly introduced OCB mode (Sect. 4.4) by the *Timing Advertisement* (TA) management frame. The TA frame can be used to distribute time synchronization information since it carries out a *Timestamp* field, which conveys the local time of the transmitting device, and the Time Advertisement information element, which contains data that can be used by recipients to estimate the Coordinated Universal Time (UTC).
- In addition, *Vendor Specific Action* (VSA) frames have been introduced to exchange management information such as WAVE Service Advertisements (WSAs). VSAs and TAs may be transmitted on any channel.

4.3.3 Formats: LLC Sublayer

Logical Link Control (LLC) represents one more proof that backward compatibility is intended to be kept in the design of VANETs. In fact, LLC is not necessary in principle for VANETs messages but it is kept to permit the multiplexing of multiple services—not only safety ones—on the VANETs' stack.

As shown in Fig. 4.2, LLC is the upper portion of the data link layer of the Open System Interconnection (OSI) Model and presents a uniform interface to the user of the data link service, usually the network layer. It has been standardized in [24].

4 The MAC Layer of VANETs

The LLC sublayer has the main purpose of carrying multiple protocols over the same MAC; this implies the following tasks:

- identifying the protocol of the nested protocol (i.e., TCP/IP in the WAVE stack: this accomplished by means of *logical addresses* of the network layer entity managing the message; this is referred as *Service Access Point* (SAP) in LLC terminology.
- providing a proper data-link service for such protocol conveyed inside.

Concerning the latter point, IEEE 802.2 provides three operational modes: two connectionless and one connection-oriented.

- *Type 1* is an unacknowledged connectionless mode and is compatible with *unicast*, *multicast* and *broadcast*;
- *Type 2* is connection-oriented and adopts sequence numbering and acknowledgments to ensure that all the frames are received and are correctly ordered; it is only for *unicast*;
- *Type 3* is an acknowledged connectionless service, and it also supports only unicast communications.

Obviously, VANETs need Type 1, which, however, does not only support multicast, but also typical TCP/IP flows.

In practice, LLC includes only an overhead made up of the following fields:

1. *Destination Service and Access Point (DSAP) address*: 1 byte;
2. *Source Service and Access Point (SSAP) address*: 1 byte;
3. *Control*: 1 or 2 bytes, conceptually related to High-Level Data Link Control (HDLC). It is required to manage sequence number, acks, flow control. The most used type is U-format, 1-byte long, for connectionless applications. It is identified by the least significant 2 bits set to 1.

VANET traffic cannot be encapsulated in IEEE 802.2 LLC frames without a Subnetwork Access Protocol (SNAP), due to the lack of a dedicated protocol type. Additionally, also IP traffic is not directly encapsulated because, even if there is a protocol type for IP, there is not for the Address Resolution Protocol (ARP).

In both cases, the SNAP is used inside IEEE 802.2 (it is aimed at distinguishing many more protocols than using of the 8-bit SAP fields); SNAP can be used with EtherType value and supports private protocol ID spaces. When both the DSAP and the SSAP are set to the hexadecimal 0xAA, the SNAP service is recalled.

The SNAP header consists of a 3-octet about the organization and is managed by IEEE (IEEE Organizationally Unique Identifier—OUI) followed by a 2-octet protocol ID.

As a result, the LLC values found in VANETs will be those indicated in Fig. 4.5.

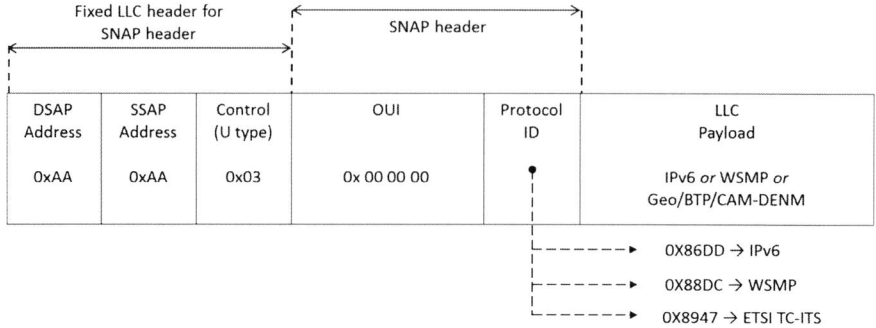

Fig. 4.5 The LLC/SNAP encapsulation as used in VANETs: the protocol types for IPv6, WSMP and ETSI-TC-ITS are indicated

4.4 IEEE 802.11 MAC Foundations

So far, we have not entered the analysis of the CSMA/CA as defined in IEEE 802.11, but just mentioned some features which are quite widely known. In this section the protocol mechanisms will be explained, paying attention to those differentiators which have been shortly mentioned in Sect. 4.2.

The analysis includes an overview of the BSS setup, of the MAC architecture and of the MAC mechanisms for distributed and prioritized channel access.

4.4.1 IEEE 802.11 BSS Setup

IEEE 802.11 networks are built upon the concept of BSS as a set of wireless stations (STAs) that agree to exchange data. Two types of BSS are specified by the standard: *infrastructure* and *independent BSSs* (Fig. 4.1-*left*).

In an infrastructure BSS an AP establishes and announces its own BSS by periodically sending *beacon* frames. An STA—listening beacons from an AP—can join the BSS through a number of interactive steps, including synchronization, authentication, and association.

Communications in an independent BSS (IBSS) (also referred to as *an ad hoc network*) directly occur among stations, without an AP. All STAs transmit beacon frames to announce the existence of the IBSS and its parameters and to allow synchronization of STAs.

Both infrastructure and independent BSSs are allowed by IEEE 802.11p. However, to better match the requirements of vehicular communications with very short-lived connections, a new operational mode has been introduced in 802.11p, referred to as OCB, that avoids the latency associated with establishing a BSS (Fig. 4.1-*right*).

Such a communication mode is activated by setting the *dot11OCBEnabled* flag to true. Differently from baseline 802.11, in OCB mode also STAs which are not member of a BSS are allowed to transmit data without preliminary authentication, and association signaling.

In OCB mode, beacon frames are used neither for BSS advertisement nor for synchronization purposes. Therefore, stations rely either on default parameter values or on information carried out in other frames, like the TA frame, introduced by 802.11p mainly to distribute time synchronization information.

Also, in OCB communications, any required authentication service would be provided by the upper layers to mutually identify communicating STAs.

4.4.2 IEEE 802.11 MAC Architecture

MAC mechanisms could be categorized as *contention-based* and *contention-free*. Contention-based approaches rely on carrier sensing, back-offs, and retry schemes, while contention-free approaches rely on time division multiple access and synchronization schemes.

IEEE 802.11 MAC defines the *Distributed Coordination Function* (DCF), the *Point Coordination Function* (PCF), the *Hybrid Coordination Function* (HCF), and the *Mesh Coordination Function* (MCF) [27].

DCF is the basic MAC protocol and it is based on the CSMA/CA mechanism; it is mandatory in all STAs. PCF is a centralized access that relies on a polling mechanism managed by a *coordinator*, usually co-located with AP.

The HCF augments DCF and PCF with QoS-specific mechanisms. HCF uses both a contention-based channel access method, called the *enhanced distributed channel access* (EDCA) mechanism for contention-based transfer and a controlled channel access, referred to as the *HCF controlled channel access* (HCCA) mechanism, for contention-free transfer.

DCF and EDCA are distributed contention-based channel access schemes, which can be used both in infrastructure and independent BSSs. PCF and HCCA, instead, can be only used in centrally coordinated infrastructure-based networks. The HCF, that has both a contention-based channel access and contention free channel access mechanism, is usable only in a Mesh BSS (MBSS).

A distributed channel access scheme is the best candidate for ad hoc communications in IEEE 802.11p, which, consequently, adopts the same core mechanism of EDCA in order to meet the prioritization requirements of different kinds of applications. Before EDCA, its precursor DCF is here shortly introduced.

4.4.2.1 Distributed Coordination Function

DCF relies on the CSMA/CA technique to reduce the collision probability among multiple stations accessing the medium. The CSMA basic idea is "to listen before talk", i.e., a station must probe the medium before transmission to determine whether it is busy or not.

Clear Channel Assessment (CCA) is the name for the channel monitoring and can be performed both through *physical* and *virtual* mechanisms to determine the state of the medium.

Physical carrier sensing is done by monitoring any channel activity caused by other sources, either by measuring the energy in the target frequency range (*Energy Detection*—ED) or by recognizing the frames on the air (by the preamble, realizing the *Carrier Sensing*—CS); in the former case, also interferences from other wireless standards may be caught.

Channel sensing is not sufficient to avoid collision in 802.11 networks, especially due to what is called the *hidden terminal* problem.

A potential hidden terminal is any STA which is located out of the range of the transmitting node S, but close enough to the destination D so that it can disturb an ongoing transmission towards D with its own simultaneous transmission towards any node in its coverage area.

Virtual sensing is, instead, implemented (1) by adding a preliminary (optional) *RTS/CTS* frame handshaking between the sender and the receiver to reserve the medium prior to actual data exchange, and (2) by including in all transmitted frames an indication of their expected duration so that the non-destination stations, by overhearing any of these frames, can be aware of the time interval during which the channel will remain busy.

A counter, called NAV, will be set accordingly by each STA to keep track of the channel status and it will be decremented by one slot-by-slot regardless of the sensed channel status. Once a station has set its NAV, its transmission is deferred until the channel becomes idle (i.e., the NAV is zero).

An immediate positive acknowledgment scheme for unicast frames allows the sending STA to realize that the transmission was successful. Upon a successful frame reception, the receiving STA sends back an ACK frame to the source. Like for a CTS frame, the intended receiver waits for a *Shorter Inter-Frame Space (SIFS)*, the shortest of the IFSs,[3] prior to transmit an ACK.

If the ACK is not received, the frame is considered as lost and retransmitted by the source node after a mandatory IFS and an additional random delay. Contrarily to unicast frames, broadcast and multicast frames are never acknowledged, because multiple simultaneous ACK transmissions would cause collisions at the sender.

Figure 4.6 shows the data exchange between a source and a destination node when the RTS/CTS handshaking is enabled and the NAV setting by other stations [27].

The source announces with the RTS that the channel will be busy until the reception of the ACK at the sender node. The CTS transmitted by the destination restates the same channel reservation, so that neighboring nodes which are not able to hear the source (e.g., hidden terminals) may set the NAV and refrain from accessing the channel.

---

[3]The 802.11 MAC layer mandates that a gap of a minimum specified duration, IFS, exists between contiguous frame sequences.

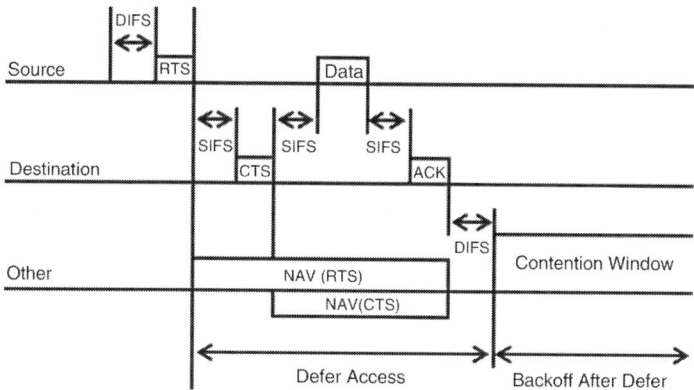

Fig. 4.6 Virtual carrier sensing through RTS/CTS frame handshaking before data transmission and NAV setting (adapted from [27])

To reduce the overhead incurred when the RTS/CTS access mode is enabled, the standard specifications allow an STA to use such an exchange only when the data frame size is longer than a given threshold indicated by the *dot11RTSThreshold* attribute. Longer packets are also more prone to channel errors and collisions; therefore, the RTS/CTS mechanism becomes more effective as the packet size increases.

Similarly to acknowledgment, the RTS/CTS mechanism cannot be used for broadcast/multicast frames, because multiple simultaneous CTS transmissions would cause collisions at the sender.

A station desiring to initiate a transmission using the DCF shall invoke the CS mechanism to determine the state of the medium. If the medium is idle for the duration of a *Distributed Interframe Space (DIFS)*[4] period, the station transmits immediately. If the medium is busy, the station shall defer until the medium is determined to be idle. When the medium is left idle for the duration of a DIFS period, the station shall then generate a random backoff period for an additional deferral time before transmitting. The basic access mechanism is illustrated in Fig. 4.7.

The backoff procedure minimizes collisions during contention between multiple stations that have been deferring to the same event. This procedure shall be performed for a station to transfer a frame when finding the medium busy as indicated in Fig. 4.8. To begin the backoff procedure, the station sets its *backoff timer* to:

$$Random() * aSlotTime \qquad (4.1)$$

---

[4] A DIFS is equal to a SIFS plus two slot times.

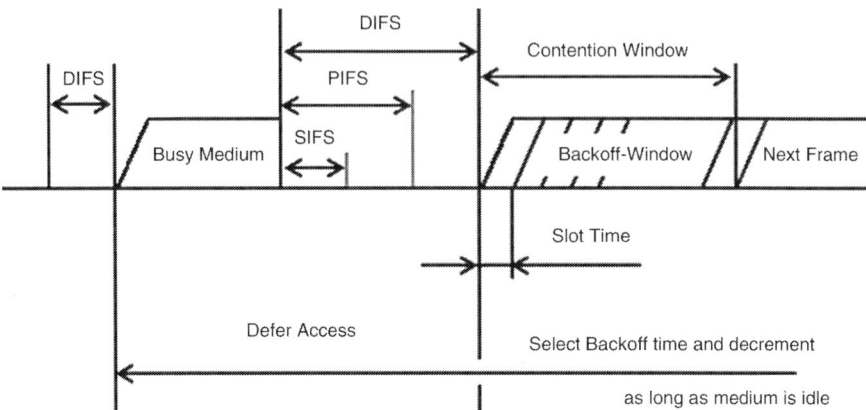

Fig. 4.7 Basic access mechanism (adapted from [27])

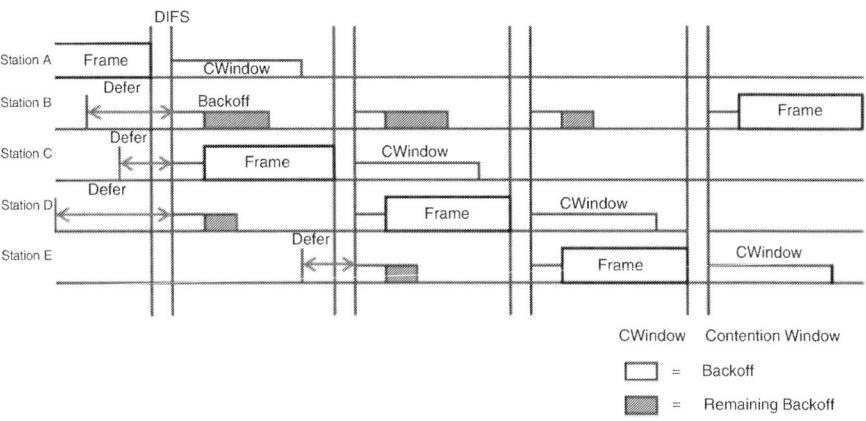

Fig. 4.8 Backoff procedure (adapted from [27])

where:

- *aSlotTime* is the slot time duration;
- Random() is a pseudo-random integer drawn from a uniform distribution over the interval *[0,CW-1]*, where *CW* is the *Contention Window*. CW takes *aCWmin* as an initial minimum value at the first transmission attempt and is doubled (exponential growth depicted in Fig. 4.9) at every failed transmission (with an upper limit equal to the maximum size *aCWmax*).

The values of the main MAC parameters in 802.11p are reported in Table 4.3.

If no medium activity is indicated for the duration of a particular backoff slot, then the backoff procedure shall decrement its backoff time by *aSlotTime*. If the

4 The MAC Layer of VANETs

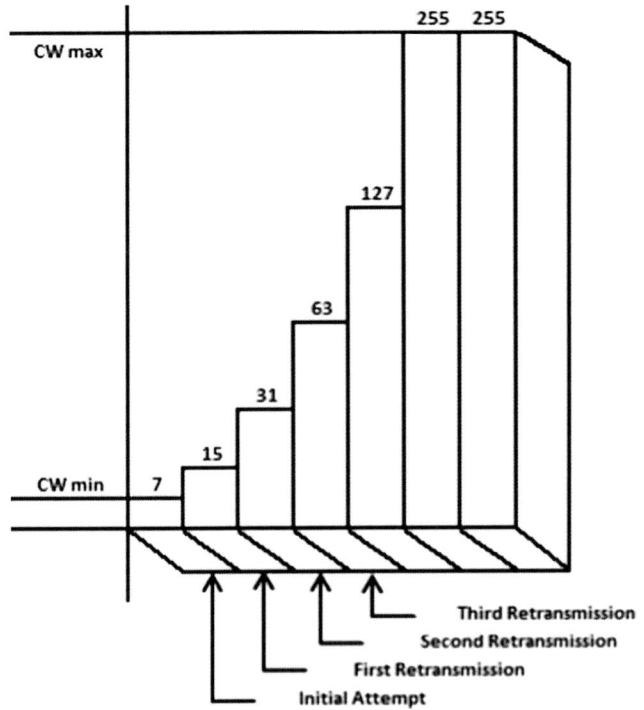

Fig. 4.9 Example of exponential increase of CW (adapted from [27])

Table 4.3 Main 802.11p MAC parameters

Parameter	Value
aCWmin	15
aCWmax	1,023
aSlotTime	13 μs
aSIFSTime	32 μs

medium is determined to be busy at any time during a backoff slot, then the backoff procedure is suspended until the medium is idle for the duration of a DIFS period. When the backoff timer reaches zero, transmission will start.

Since failed broadcast/multicast frames are never retransmitted, the backoff procedure will only be performed once per packet, consequently the size of the CW never changes, leading to poor performance due to collisions, especially if several nodes are contending for seizing the channel.

To compensate for the inherent unreliability of broadcast transmissions, broadcast frames are often transmitted at the lowest mandatory data rate (notoriously the most robust to channel errors).

4.4.2.2 EDCA

According to the DCF rules, all the STAs compete for accessing the channel with the same priority, i.e., they all use a DIFS as the mandatory idle channel period before selecting the random backoff, that is selected in the same range for each STA.

There is no differentiation mechanism that guarantees better service to traffic with more severe delivery requirements (such as safety traffic in the case of VANETs), and there is no mechanism that effectively differentiates multiple flows within an STA. To cope with these issues, the EDCA mechanism has been specified in 802.11 and adopted in 802.11p.

The EDCA mechanism supports service differentiation by assigning eight different user priorities (UPs), that are mapped into four access categories (ACs), namely voice, video, best effort, and background. The AC is derived from the UPs as shown in Table 4.4.

A model of the reference implementation is shown in Fig. 4.11 and illustrates a mapping from frame type or UP to AC, the four transmit queues and the four independent EDCA functions, one for each queue. The voice AC has the highest priority followed by video, best effort, and background ACs.

Each device has four ACs which are contending for the medium as per the CSMA/CA process of 802.11 DCF, with EDCA parameters replacing the parameters used in DCF. The priorities assigned to these categories depend on the following EDCA parameters: the *Arbitration IFS (AIFS)*, *CWmin* and *CWmax*, and *transmission opportunity (TXOP)*.

The AIFS idle duration time is not the constant value (DIFS) as defined for DCF, but is a distinct value for different ACs, proving to be an important factor in providing priority to ACs. It is the defined as the amount of time an STA senses the channel to be idle before randomly extracting the backoff or transmitting the data frame:

$$AIFS[AC] = aSIFSTime + AIFSN[AC] * aSlotTime \qquad (4.2)$$

where *AIFSN[AC]* is an integer AIFS number assigned to each AC. The higher priority ACs have shorter AIFS value and, hence, they have to wait for a shorter time to start the backoff procedure and would experience a shorter delay compared to lower priority ACs (Fig. 4.10).

The *CWmin* and *CWmax*, from which the random backoff is computed, are fixed in DCF. In EDCA, instead, different ACs have different values for them. The lower CW would cause the higher priority AC to choose a smaller random number for backoff which enables the highest priority AC to wait for the shortest period of time when the medium becomes idle as it has the shortest AIFSN value.

Collisions between contending EDCA functions (completing their respective backoff processes at the same time) within a station are internally resolved so that

4 The MAC Layer of VANETs

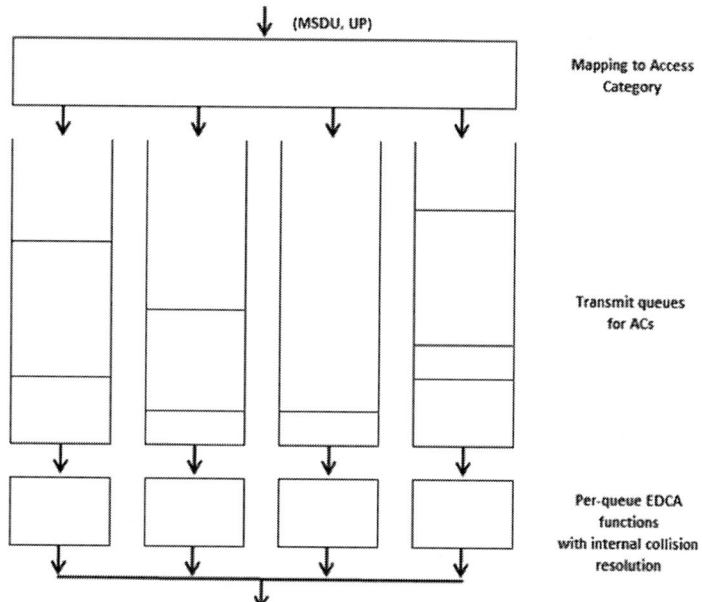

Fig. 4.10 EDCA differentiation

Table 4.4 UP to AC mappings in EDCA (adapted from [27])

Priority	UP	802.1D designation	AC	Designation
Lowest	1	BK	AC-BK	Background
	2	–	AC-BK	Background
	0	BE	AC-BE	Best effort
	3	EE	AC-BE	Best effort
	4	CL	AC-VI	Video
	5	VI	AC-VI	Video
	6	VO	AC-VO	Voice
Highest	7	NC	AC-VO	Voice

the frames from the higher priority AC receive the TXOP and the frames from the lower priority colliding AC(s) behave as if there was an external collision. This event is known as *virtual collision*.

TXOP is the time interval during which an STA may transmit a series of frames separated by a SIFS, after it has seized the channel.

The EDCA parameters set of each AC to be used in baseline 802.11 standard (when *dot11OCBEnabled* is false) is reported in Table 4.5, while values for 802.11p (when *dot11OCBEnabled* is true) are reported in Table 4.6. When *dot11OCBEnabled* is true, the TXOP limit is set to zero for each AC, indicating that only one frame can be transmitted after having seized the channel.

Table 4.5 Default EDCA parameters set if *dot11OCBEnabled* is false

ACI	AC	CWmin	CWmax	AIFSN	TXOP limit
0	Best effort	aCWmin	aCWmax	7	0
1	Background	aCWmin	aCWmax	3	0
2	Video	(aCWmin +1)/2 − 1	aCWmin	2	3.008 ms
3	Voice	(aCWmin +1)/4 − 1	(aCWmin +1)/2 − 1	2	1.504 ms

Table 4.6 Default EDCA parameters set if *dot11OCBEnabled* is true

ACI	AC	CWmin	CWmax	AIFSN
0	Best effort	aCWmin = 11	aCWmax = 511	9
1	Background	aCWmin = 15	aCWmax = 511	6
2	Video	(aCWmin +1)/2 − 1	aCWmin	3
3	Voice	(aCWmin +1)/4 − 1	(aCWmin +1)/2 − 1	2

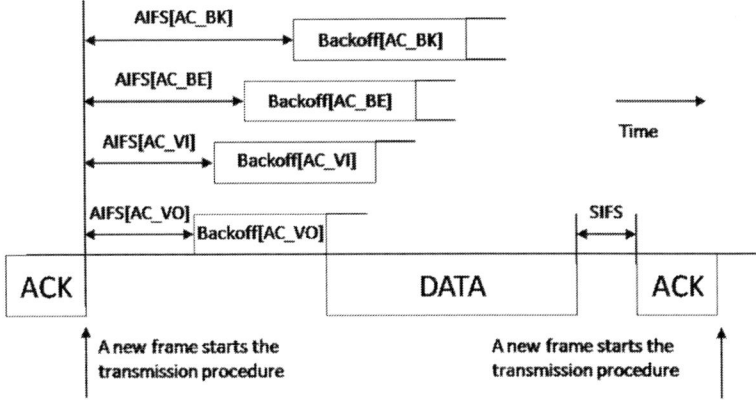

Fig. 4.11 Reference implementation model

4.5 IEEE 802.11p MAC: Multi-Channel Operations

The WAVE family of standards provides a communication protocol stack optimized for the vehicular environment, when considering multiple channels, specifically, one control channel (CCH) and many service channels (SCHs).

To operate over multiple wireless channels while in operation with the *OCBEnabled* mode (i.e., WAVE operation), there is a need to perform channel coordination. The latter one is specified in the IEEE 1609.4 standard [26] and consists of additional features for *OCBEnabled* operations in the MAC sublayer specified in the IEEE 802.11p standard [27].

1609.4 addresses routing and data transfer from the upper layers to the designated channel and queue at the MAC layer.

The basic reference model for the IEEE 1609.4 standard is depicted in Fig. 4.12. Both data plane and management plane features are specified. The MAC and PHY

4 The MAC Layer of VANETs

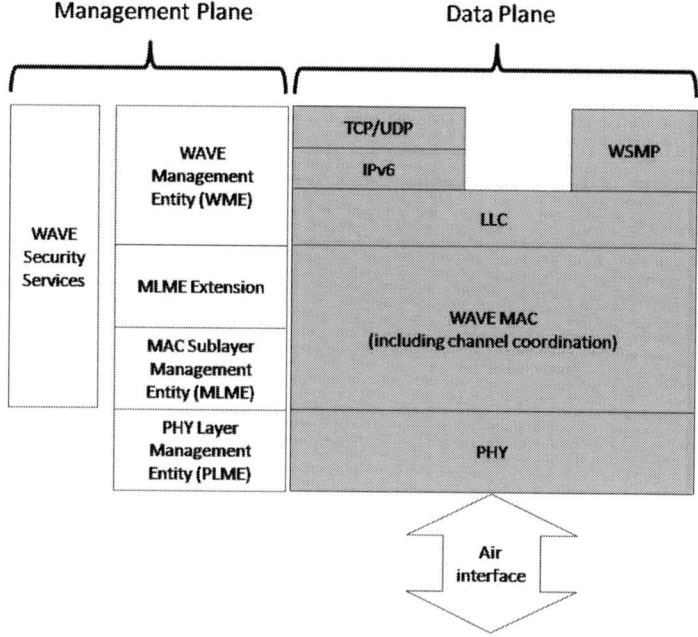

Fig. 4.12 Reference model for the IEEE 1609.4 standard, adapted from [26]

layers conceptually include management entities, called MAC sublayer management (MLME) and physical layer management entity (PLME), respectively.

The data and management planes, respectively, support:

- data plane:
  - channel coordination: the MAC sublayer coordinates channel intervals so that data packets are transmitted on the proper channel at the right time;
  - channel routing: the MAC sublayer handles outcoming higher layer data, by setting parameters (e.g., transmit power) for WAVE transmissions;
  - user priority: WAVE supports a variety of safety and non-safety applications with up to eight levels of priority mapped in EDCA parameters.

- management plane:
  - multi-channel synchronization with the objective of aligning channel intervals among communicating WAVE devices. The MLME provides the capability to generate TA frames to distribute system timing information and monitor received TA frames;
  - channel access to specific radio channels;
  - reception and generation of VSA frames;
  - maintenance of a management information base (MIB) containing configuration and status information.

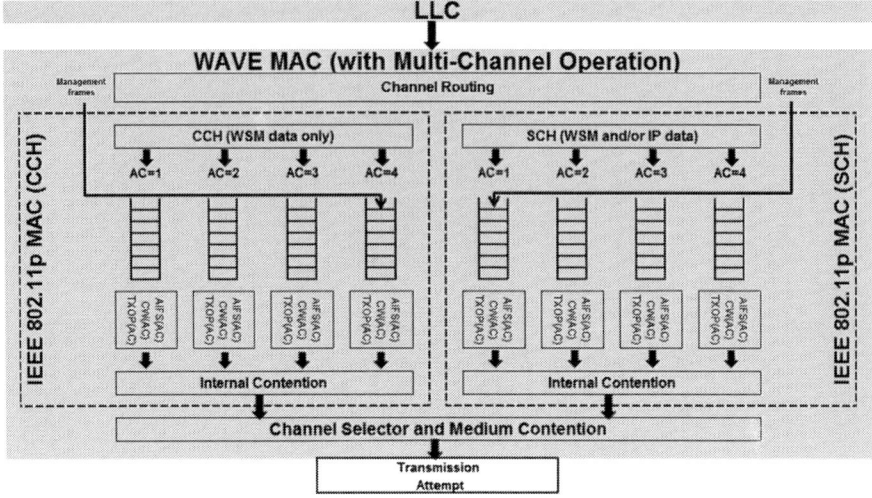

Fig. 4.13 MAC with channel coordination, adapted from [26]

Data, management, and control frames are supported as detailed in Sect. 4.3. For higher layer data exchanges, WAVE supports both IP (IPv6)- and non IP-based data transfers. Non IP-based data transfers are supported through the WAVE Short Message Protocol (WSMP) specified in IEEE 1609.3 standard [25].

Data frames containing WSMs may be exchanged among devices on either the CCH or an SCH; however, data frames containing IP datagrams are only allowed on service channels.

The WAVE multi-channel MAC internal architecture is shown in Fig. 4.13 [27].

This reference design architecture is used to specify the following transmit operations: channel routing, data queueing and prioritization, and channel coordination. There are two 802.11p MAC entities: one for the CCH and one for the SCH. Each channel has four independent queues, which have their own backoff process to provide aggressively differentiated priorities. Data is prioritized according to access category (directly related to user priority), as indicated by the queues shown in Fig. 4.13, which provide different contention and transmission parameters for different priority data frames.

For WSMs, the channel, transmit power, and data rate are set by higher layers on a per-message basis. For IP datagrams, the channel, transmit power, and data rate to be used are stored in a transmitter profile.

The default EDCA parameter set specified in IEEE Std 802.11p for *OCBEnabled* operation is optimized for short message transfer and is recommended to be used when operating on the CCH (see Sect. 4.4.2.2).

4 The MAC Layer of VANETs

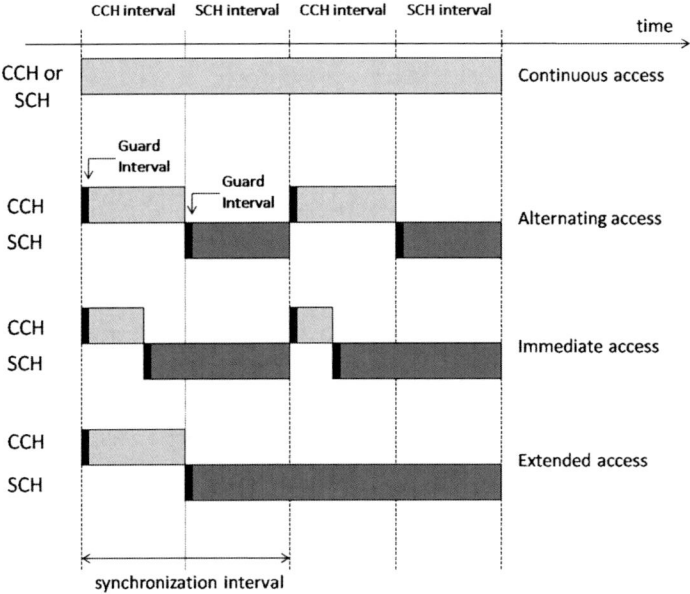

Fig. 4.14 Channel access options: continuous, alternating, immediate, and extended

4.5.1 Channel Coordination

Channel coordination is designed to support data exchanges involving one or more switching devices with concurrent alternating operation between the CCH and an SCH.

Channel access options include *continuous access*, *alternating* SCH and CCH access, *immediate* SCH access, and *extended* SCH access, as illustrated in Fig. 4.14 [26].

The continuous access allows a node to always stay on a given channel, either CCH and SCH.

The alternating access allows a node to switch between the CCH and the available SCHs at scheduled time intervals, *CCH interval* and *SCH interval*, respectively. The channel time is divided into synchronization intervals with a fixed length of 100 ms, consisting of a CCH interval and an SCH interval. While the synchronization interval duration is fixed in order to meet the requirements of safety applications which cannot tolerate latencies higher than 100 ms, CCH and SCH interval durations may be adaptable, although they are typically 50 ms-long intervals.

Immediate SCH access allows immediate communications access to the SCH without waiting for the next SCH interval, by avoiding the latency of the residual CCH interval. Extended SCH access allows communications access to the SCH without pauses for CCH access and is useful for services which require a huge amount of data to be transferred which takes several periods to be delivered.

Both immediate and extended access schemes have been designed to improve the delivery performance of bandwidth-demanding non-safety applications, requiring a huge amount of data to be transferred. However, they can be only beneficial to those vehicles which are not interested in safety applications such as a cooperative collision avoidance [7].

At the beginning of each channel interval, a *guard interval* accounts for the radio switching delay and timing inaccuracies in the devices. During the guard interval, a switching device is not available for communication due to the transition between channels. When the guard interval starts, all MAC activities are suspended and they start (or are resumed) at the end of the guard interval. To prevent multiple switching devices from attempting to transmit simultaneously at the end of a guard interval, the medium is declared as busy during the guard interval, so that all devices extract a random backoff before transmitting. This helps to limit, but it does not prevent, collisions between frames which were queued during the previous channel interval.

All 1609.4 channel access methods, except the continuous access, require time synchronization for switching channels on the channel interval boundaries.

The current WAVE standard assumes that coordination between channels may exploit a global time reference, such as the UTC, which is provided by a global navigation satellite system. This approach suffers from being centralized; an attack or failure in the global clock source could lead to widespread irrecoverable network failure. The TA frame has been introduced for synchronization purposes.

4.5.2 WSA Frames

VSA frames are used to exchange management information such as WSAs specified in IEEE Std 1609.3 [25] and containing information including the announcement of the availability of services.

WSAs contain all the information identifying the offered WAVE services and the network parameters necessary to join the BSS, such as the identifier of the BSS, the Provider Service Identifier (PSID) that identifies the provider application entity, the SCH where a given service is provided, timing information, the EDCA parameter sets.

The main fields of a WSA frame are depicted in Fig. 4.15. It includes header information and may include a series of variable-length *Service Info* segment for each service advertised in the WSA, a series of variable-length *Channel Info* segment for each SCH on which the advertised service is offered and WAVE routing advertisement (WRA), used for IPv6-based services to provide information about how to connect to the Internet, e.g., default gateway and domain name server address.

The *Change Count* may be used by the recipient to determine whether a WSA is a repeat of the previous one from the same source.

Since WSAs are broadcasted by providers without any feedback on their successful reception, sending more WSAs in the CCH interval provides greater

4 The MAC Layer of VANETs

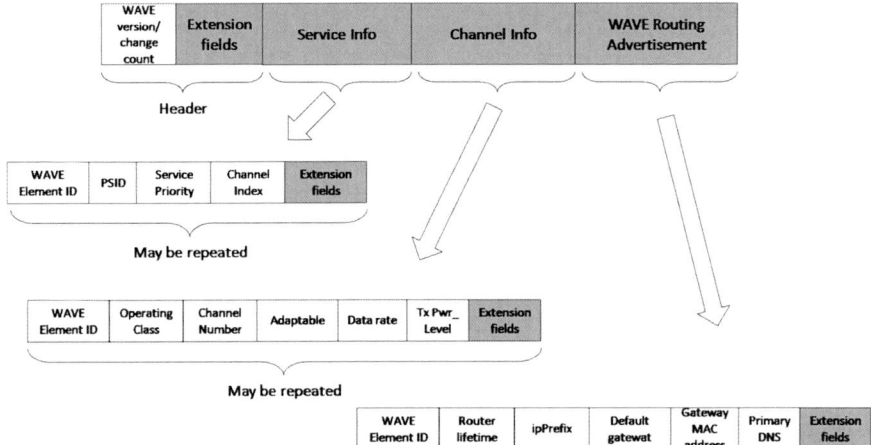

Fig. 4.15 WSA frame format, adapted from [25]

reliability. To this purpose, a *repeat rate* parameter is introduced that is the number of times the advertisement is transmitted every five seconds. Such a parameter is an optional field in the header extension, along with the transmission power used for the WSA and the location of the WSA transmit antenna.

4.6 Tools for MAC Analysis

From the point of view of variables related to wireless channel state (e.g., network load, transmission opportunities, reception rate, etc.) the behavior of VANETs is mostly ascribable to the PHY and MAC phenomena. Even more, both the PHY and the MAC phenomena are often non-easily foreseeable by simple models.

As a result, it is not straightforward answer to questions such as *How frequently is updated the information coming from a car, depending on the transmission rate and the distance?* or *Is it better to transmit more and overload the network or to rely on few successful transmissions?* and, to make one more example: *What happens to CSMA/CA when a large number of nodes are employed and some obstructions block propagation in certain directions?*

To answer such questions, one cannot rely just on intuition but may count on several tools, which include, from the most practical to the most theoretical: **measurements** and trials, **simulations**, and **analytical models**. Each of them has its own benefits, limitations, validity, and metrics. They are discussed in next sections. Measurements are a complex issue and as such require a precise methodology: they will be further discussed, devoting some space to the FESTA methodology, in Chap. 14. Also simulations will have a dedicated chapter (Chap. 13). Conversely, analytical models will be discussed only in the present section and, consequently, they are devoted some more pages.

4.6.1 Simulations

Just as an indicator of the appeal of simulators, we performed, at the time of this writing (November 2014), three searches over the items available through their metadata on the well-known portal IEEE Xplore[5] coming to the following results:

- SIMULATION AND (VANET OR VEHICULAR+WIRELESS):
  5,092 Conference publications + 1,129 Journals and Magazines
- ANALYTICAL + MODEL AND (VANET OR VEHICULAR+WIRELESS):
  746 Conference publications + 286 Journals and Magazines
- MEASUREMENT AND (TRIAL OR FIELD OR CAMPAIGN) (VANET OR VEHICULAR+WIRELESS):
  144 Conference publications + 33 Journals and Magazines

Additionally, for the sake of precision, we checked the relevance of some search results and the first category was the most coherent with the intended keys. Simulations are, from far, the most used approach for the study of VANETs.

We will discuss the weaknesses which simulations suffer from in Sect. 4.6.3: basically, if one is cautious and pays attention to simulation hypotheses, significant results can be achieved. But, the other way around, what are the reasons for the success of simulation tools?

We can answer with our point of view. Simulations have manifolds benefits: they are cheap (a lot of platforms are even open source) and more and more accurate; even a beginner can run simulation with a limited effort; heterogeneous scenarios can be tested, involving a large number of nodes, or testing very (or slightly) different settings by changing few script and in a repeatable way (repeatability in field tests may be an issue). Additionally, simulations benefit from the ever growing computational capacity and from the cloud computing [10] technology, so that several simulations can be launched simultaneously (by batch commands). These are all true, but there are two major conceptual benefits which are discussed here below: *heterogeneous modularity* and *flexible metric* definition.

4.6.1.1 Modularity

Simulators are typically modular; however, the modularity which we refer to is that *among* simulators. Figure 4.16 shows what we mean and will help us introduce some topics further discussed in other chapters (mainly in Part IV of the book).

In the perspective of evaluating MAC mechanisms, the preferred tool is a network simulator, which focuses its modeling on the protocol mechanism. Network simulators (see Chap. 13) typically adopt simplified propagation and reception models; however, some of them have integrated some refined models for the

---

[5]IEEE Xplore is the portal of all the IEEE publications (Conferences, Journals, Magazines) and is reachable at the URL: http://ieeexplore.ieee.org/.

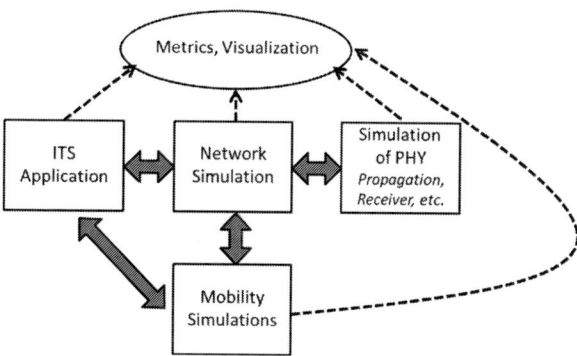

Fig. 4.16 Modular representation of the simulating blocks potentially involved in a VANET simulation

receivers (see Chap. 3) and improved propagation (see Chap. 12). As an alternative, some efforts have pursued the integration of two simulators (ray-tracing—for propagation—and network simulators): this would lead to very precise simulation results but, at the current stage, it has been demonstrated not to be computationally affordable [34].

Then there is the simulation of mobility, which is further discussed in Chap. 11. Concerning the mobility, it can be usually simulated *a priori* and feed the network simulator (providing the time-dependent position of each node). However, the most recent approaches (Chap. 13) tend to merge together (with a complete mutual feedback) the simulation of vehicular applications, the network (wireless) simulation and the mobility, so that: depending on the position, some messages will be received and sent; depending on the messages received the mobility pattern will change on the fly (e.g., to react to notified traffic accidents and road congestion events); depending on the mobility patterns the vehicular applications will be triggered.

In the end, in the most exhaustive simulation, you will have very complex results, depending on several variables, such as: topology, real propagation and obstructions, uniform/even distribution of nodes, position of the nodes, vehicles' maneuvers, etc. Here comes the issue of correctly interpreting the results in such a complex context.

One has two opportunities (Fig. 4.16): carefully analyze the output metrics—even defining new custom ones (Sect. 4.6.1.2)—or adopting a visual tool supporting data analytics, for example by animations showing how nodes move in the map, and how metrics may evolve (displaying them as labels or colormaps)—Fig. 4.17 shows the, so far unique, example as published in [5]. This is the last modular block that we will mention in the simulation chain.

It is difficult to explain the benefits of visualization without testing it. Just to give an example, by displaying the moving nodes in the map, with a color depending on their respective states, one could easily understand whether a complex multilayer mechanism, such as DCC (Chap. 6) is stable or not, if it depends on the node position, how frequently the mechanism is invoked, if it suffers from border effect at the edge of the simulated topology (in this case giving also the opportunity to refine metrics), etc. To say it simple, a visual approach would make the interpretation of results more immediate and less faulty.

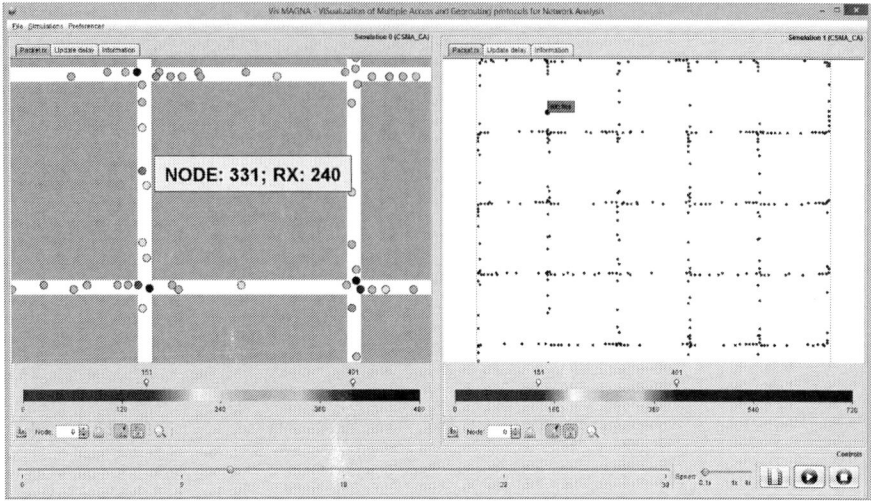

Fig. 4.17 The example of a visualization tool for VANET—screenshot by VisMagna [5]

All in all, the approach of visualization is very promising and, in spite of this, quite disregarded in the VANETs' community: certainly, it needs to be more extensively divulged.

4.6.1.2 Flexible Metrics

The intrinsic benefit of simulations stands in the nature of the occurring events: they are just variables which are triggered by other ones, through functions. This may seem obvious but is very powerful: sometimes it may take some time, but one always manages to go back up to the cause of such event.

Even more, when discovered the cause, one can recall the related variables and include them in the metrics.

From a practical point of view, one will have several opportunities, for example:

- split events into categories: for instance, simultaneous transmissions (collisions) can be split into their causes (simultaneous count-down, hidden terminals); this would be hardly feasible with tests;
- one could look at the distribution of internal states: for example, in DCC (Chap. 6) one could study the distribution of channel load throughout the nodes;
- it would be possible to perform some correlations between variables even over multiple nodes: for instance, how the reception rate of a given node is related to the transmissions by the nodes in the sensing range.

All in all, metrics can be defined in a very flexible way, considering both the events and the internal states of the nodes (including position and time).

VANETs represent a very novel and specific context and ever under development: they subtend new traffic paradigms (many-to-many, by broadcast), new requirements (e.g., regular periodic reception of messages by surrounding nodes, or geographical target for the forwarding of messages), and new environmental changes (deep fading and severe obstructions). So, the first message we desire to leave at the end of this analysis is that one should not hesitate to define a new performance metric.

Then, since this chapter is about MAC, we would like to shortly recall some MAC metrics which have been defined over time in the scientific literature, so to highlight how much VANETs may be specific and to present those metrics which the reader may meet more frequently in the literature. Given the purpose of this presentation, the metrics will not be defined as rigorously as they could be.

- *Legacy metrics* (non strictly related to VANETs)

    - **Transmission Time**-In most VANET simulation scenarios, the transmission time will be related to broadcast traffic (without re-transmission), hence corresponds to the time to access the channel (including backoff count-down and freezing) and send a frame.
    - **Packet** (*or* Frame) **Delivery Rate** (PDR)-This metric refers to the reception rate, usually depending on the distance from the transmitter under the hypothesis of uniformly distributed broadcast traffic sources. Notably, with broadcast transmissions will not puzzle things with retransmissions (and consequent congestion) depending on the mutual distance between transmitter and receivers. There may be additional flavors of the PDR, including:
        · *Reception Rate in a Region of Interest* (ROI): this becomes relevant, for example, when considering the effect of geo-forwarding;
        · *PDR Restricted to Certain Areas*: for example, PDR considering only the receivers in the center of the crossroads, so to emphasize the effect of hidden terminals.
    - **Protocol Overhead**-This metric is mentioned as one which someone uses but we do not agree it is really useful in a VANET scenario. In fact, the ultimate purpose of VANETs is to effectively forward some safety-related information, and this is related to the effective reception or, from a different perspective, to a smart spatial multiplexing of simultaneous transmissions (see Chap. 15). A protocol might involve significant overheads but lead to an ideal reception. So, the overhead itself can be an *indicator* of VANET performance, but should *not* be *a goal*.

- *VANET metrics* (more specifically, most of the metrics defined in this area refers to a scenario of periodic traffic, and more to CAMs rather than event-triggered Decentralized Environmental Messages, DENMs, considering the European standards, see Chap. 5)

- **Update Delay**-This is strictly related to periodic traffic and is defined as the time difference between two consecutive successfully received packets from the same transmitter. It can be further elaborated to show what is the probability that the time elapsed between the reception of two consecutive (periodically generated) messages between a transmitter A and a receiver B exceeds a certain threshold. This metric is very popular for the evaluation of VANET MAC effectiveness (and of DCC mechanisms).
- (Distribution of) the **Distance of Simultaneous** (i.e., overlapping) **Transmissions**-This is a very interesting metric which digs into the spatial multiplexing and is somehow related to the PDR. With periodic broadcast transmissions it works particularly well: for instance, on some node it could show the effect of re-iterated collisions.
- (Distribution of the) **Causes for Simultaneous Transmissions**-This is restricted, for instance, to CSMA/CA. When two nodes transmit simultaneously this may be due to several causes: hidden terminals, equal back-off, simultaneous count-down: distinguishing between the causes (depending on the distance) may help understanding how the MAC works.

- *Channel metrics*: for the sake of clarity, these are metrics which measure how the channel reacts to a given protocol, so they are again protocol metric cut in the perspective of channel.
    - **Channel Load** (CL)-This has emerged especially for DCC (where it becomes also a parameter of the algorithm) and represents the percentage of time that a channel is perceived non-free. This parameter has a protocol-dependent meaning: for instance, when CL exceeds 60 % in CSMA/CA, the number of collisions becomes significant, and the Update Delay worsens (see Chap. 6).
    - **Background Noise**-Not very popular, it can be defined as the average noise perceived during the correctly received frames and is also a measure of the interfering traffic which is sent sufficiently apart.

All the metrics may be expressed as functions of several parameters, such as the transmitter-receiver distance, the number of nodes or their local density, the transmission parameters (rate, power, etc.), the topology, the background traffic, etc., and many more metrics could be defined.

Altogether, simulations are very powerful and ever growing tools which can help us in our reasoning on real facts: they output a lot of possible and complex metrics, but they should be considered as a tool and not a goal.

4.6.2 Analytical Models

Analytical models represent a valuable means to evaluate networking protocols performance and, in particular, the behavior of the MAC protocols. They are typically less time- and resource-consuming than simulations. They allow to understand the interactions among the main processes and better focus on the behavior of the modeled phenomena. Performance can be predicted, so to make better decisions on adoption, adaptation (e.g., parameter settings), and improvement of standard and protocol solutions.

As a main drawback, models typically need to resort to simplifying and unrealistic assumptions to make the problem mathematically tractable. This is especially true in VANETs, where the complexity of the environment (i.e., mobility, time- and space-varying propagation conditions, variable node density) hinders the possibility to conceive accurate modeling. The accuracy of an analytical model should be validated either against simulations, mimicking the behavior of the modeled phenomena, or real experimentations, to understand how far the predictions are from the reality. In the following, basics for 802.11 MAC protocol modeling are provided, along with the new demands coming from VANETs.

4.6.2.1 Modeling the 802.11 MAC Protocol: Main Approaches

Modeling the 802.11p MAC protocol means dissecting the behavior of its backoff procedure. In the seminal paper in [4] a simple model is introduced that accounts for all the exponential backoff protocol details. It allows to compute the throughput performance of DCF for both standardized access mechanisms, with and without RTS/CTS handshake (and also for any combination of the two methods), under *saturation* conditions (i.e., when each station has immediately a packet available for transmission, after the completion of each successful transmission).

The saturation throughput is a fundamental performance figure defined as the *limit reached by the system throughput* as the offered load increases, and it represents the maximum load that the system can carry in stable conditions.

The model in [4] has been derived in the presence of a finite number of terminals and under the assumption of ideal channel conditions: (1) no channel errors, i.e., losses are only due packet collisions, and (2) a single collision domain, stations within the communications range of each other, i.e., no hidden terminals are considered.

The system time is broken down into *virtual time slots* where each slot is the time interval between two consecutive countdown of backoff timers by non-transmitting stations. It counts for the size of an empty slot time, the average time used for successful transmission, and the average time wasted by a packet collision. Hence, the model considers a bi-dimensional process $\{b(t), s(t)\}$ which is a discrete-time Markov chain.

$b(t)$ is the stochastic process representing the size of the backoff window for a given station at slot time t, and $s(t)$ is the stochastic process representing the backoff stage of the station at time t, as shown in Fig. 4.18.

The key approximation of the model is that the probability that a transmitted packet collides is independent on the state of the station (this is more accurate as the contention window size and the number of nodes are larger). In this condition, the bidimensional process is a discrete-time Markov chain, where the chain is frozen when the channel is detected as busy and the transition probabilities account for: the

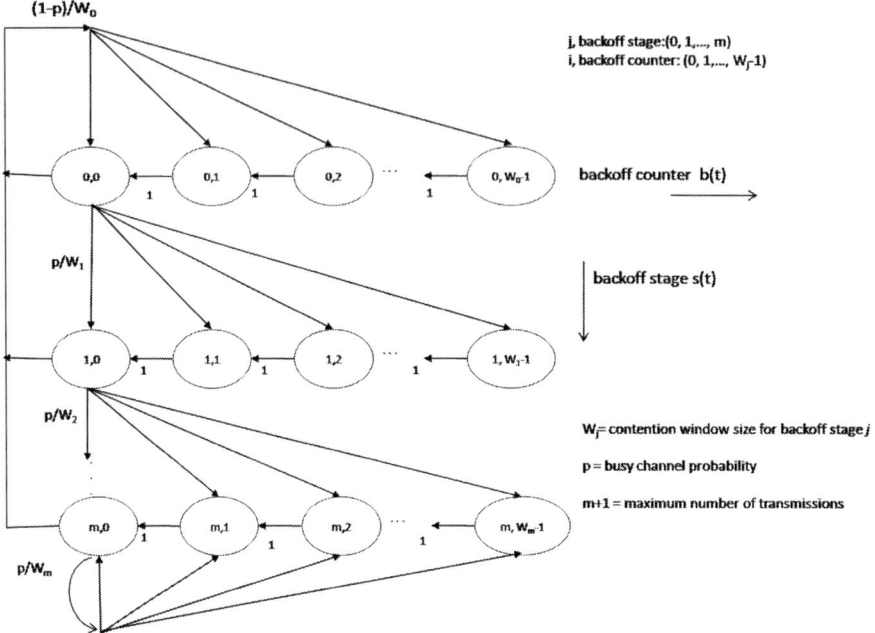

Fig. 4.18 Markov chain model for the backoff window size. Adapted from [4]

decrement of the backoff time counter when the channel is detected as idle; the fact that a new packet following a successful transmission starts with a backoff stage 0; and the fact that after an unsuccessful transmission at backoff stage, the backoff follows an exponential increase.

Several models have been proposed in the last decade to capture the dynamics of the 802.11 MAC behavior by improving the accuracy of the Bianchi's model [4], e.g., by considering non-saturation conditions [32], finite retry attempts [11], EDCA prioritization [22, 38].

Further mathematical approaches have been leveraged to understand the 802.11 dynamics. For instance, in [6] the backoff procedure of the IEEE 802.11p is modeled as a *p-persistent CSMA/CA*. Unlike the standard protocol, where the backoff interval is binary exponential, in the p-persistent CSMA/CA, the backoff interval is based on a geometric distribution with a specific probability of transmission, p. It provides a very close approximation to the IEEE 802.11.

4.6.2.2 Modeling the 802.11p MAC Protocol: Main Issues

Vehicular networks have peculiarities that make previous analytical models of generic 802.11 networks unsuitably matched as discussed in [8].

4 The MAC Layer of VANETs

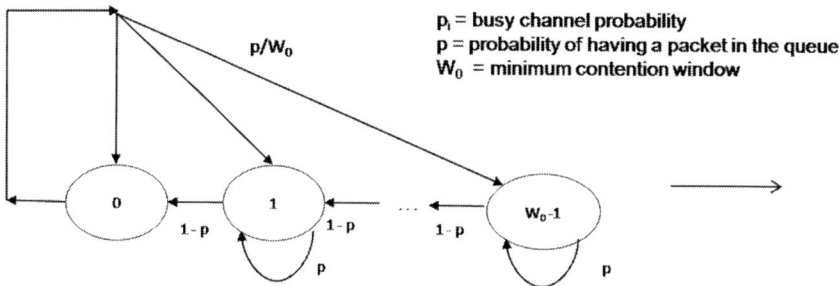

Fig. 4.19 Example of a one dimensional Markov chain for modeling broadcast traffic

Heterogeneous Applications First, a large set of heterogeneous safety and non-safety vehicular applications need to be supported on top of the 802.11p MAC protocol. Applications differ in terms of (1) the communication primitive (both unicast and broadcast communications need to be supported), (2) the traffic generation pattern (i.e., saturation, non-saturation), (3) relevant evaluation metrics and QoS demands.

Existing 802.11 models typically consider unicast data exchange. The performance of 802.11p when dealing with broadcast traffic should be modeled, instead, given its role of a basic and crucial communication mode for several road-safety and traffic-efficiency applications and service advertisement dissemination purposes. To model broadcast, a Bianchi's like model would translate in a one-dimensional Markov chain, since no retransmissions occur and the backoff stage is not modeled, as in [30], see Fig. 4.19.

Traffic generation patterns are quite different, even when focusing on safety applications only. Event-triggered transmissions need to be modeled when focusing on safety-critical alert transmissions (DENMs). Periodical transmissions, instead, characterize CAM messages. Henceforth, it is clear that traditional models focusing on saturated conditions do not well suit these kinds of traffic. Indeed, for typical arrival rates of broadcast safety data, a vehicular network operates far from saturation with no queues in the MAC layer.

To model unsaturated conditions, a quite common assumption is that data packets are generated at each station according to a Poisson process with rate λ [packets/s], e.g., in [20, 30].

As a further distinguishing feature, applications differ for the relevant metrics and the QoS demands. Some of them require high reliability and time-bounded delay (e.g., safety-critical messages), others gracefully degrade as the quality decreases.

Mean packet queuing delay is the traditional performance metric in the non-saturated case; but for the time-critical broadcast traffic, the focus must be on successful frame delivery probability as the main metric. For better dissecting the MAC dynamics, collision probability could be also analytically derived.

Mobility and Vehicle Distribution First, the high mobility of nodes could highly affect results. However, it is reasonable to neglect mobility for safety applications involving the exchange of a single packet: the one-hop neighborhood of a node does not vary during a packet transmission time (around 2 ms) [8].

Several models assume that nodes are deployed in a one dimensional network, which is a good approximation of highway scenarios, where the width of the streets can be neglected and can be associated with lines. Many authors rely on the assumption that the positions of the vehicles can be statistically modeled with a Poisson Point Process [29, 30].

Although largely neglected in the current literature relying on the simplifying assumption of a single collision domain, the presence of hidden terminals should be considered when dealing with broadcast traffic, since there is no MAC-level recovery or retransmission of frames for reliable broadcasting.

The main issues in modeling hidden terminals are: (1) how many the hidden transmitters are, (2) how their transmissions overlap, and (3) how long the transmission overlaps.

In [37] the number of hidden nodes is set to half of the overall number of nodes, which is a simplistic assumption. In [16] a Markov chain is proposed for hidden terminal. However, accurately modeling hidden terminals is still a tricky task.

In session-based applications that may require the exchange of several packets, connectivity dynamics should be carefully considered. The precise number of nodes and their residence time under the coverage of the RSU, along with their speed, should be adequately modeled. Indeed, unfairness issues when vehicles move at different speeds may arise, since 802.11p does not consider the residence time [1] and vehicles may experience different delivery performance.

Propagation Vehicular environments are characterized by harsh propagation conditions. In MAC analytical models, it is quite common to consider ideal propagation conditions with a fixed communication range for each vehicle and no channel errors, although over-estimated performance figures are achieved. As an alternative, a simple noise model may be considered to model error-prone channels and the impact of different packet lengths. Bit errors are independent and occur with a fixed bit error rate (BER); typically, the header is always assumed to be received successfully, while even a 1-bit error in the payload destroys the whole frame [8].

The Nakagami radio propagation model has been widely used in the research literature on vehicular networking and even included in simulation tools, since it shows good match with empirical data collected from mobile communications experiments [35, 36]. The effect of Nakagami fading can be included in models by using closed form expressions as done in [31].

Prioritization IEEE 802.11p heavily relies on prioritization. Differentiated channel access is mandatory in vehicular networks. More aggressively differentiated priorities are set in 802.11p/WAVE compared with generic 802.11 networks and the mutual effects of different traffic (e.g., event-triggered time-bounded safety and periodical broadcasting) should be evaluated more carefully.

In [30] a 1-D discrete-time Markov chain is proposed for packet delivery and delay performance evaluation of emergency and periodical messages under saturated and unsaturated conditions. Traffic prioritization is only based on contention window differentiation while ignoring AIFS, as instead done in [21], where a 2-D embedded Markov chain models the impact of differentiated AIFSs.

An accurate EDCA prioritization is provided in [33] and [17], whereas two ACs are considered in [8].

Multi-Channel Operations In addition, 802.11p is designed to work on multiple channels, and therefore, analytical models should take into account multi-channel operations (e.g., channel switching, adjacent channel interference, synchronization of backoff processes at the beginning of the CCH/SCH interval [7]).

In [2], the performance of the WAVE channel coordination mechanism under different traffic patterns and channel interval durations is analytically investigated by combining probabilistic analysis, M/G/1 queuing, and EDCA Markov chain. Unsaturated conditions are modeled, according to which a single priority class at each node generates Poisson-distributed packets. Attention is focused on unicast transmissions.

In [8], the authors presented a comprehensive analytical model for broadcasting over CCH of 802.11p networks that explicitly takes into account the alternating channel switching, broadcast traffic prioritization, and error-prone channel conditions. It relies on a recursive approach and unlike traditional models for 802.11 network analysis, the approach disregards the common assumption about the independent operation of an arbitrary node and constant channel access probability, which demonstrated its validity for saturation cases [23]. It precisely models underlying stochastic process covering mutual influence among nodes for a wide spectrum of input parameters, including a small number of nodes, small contention windows.

In [9] the delivery performance of short-lived event-triggered safety critical messages under the hypothesis of channel switching is modeled. Further efforts are still required to achieve accurate modeling of multi-channel operations.

All in all, we can state that the definition of analytical models for VANETs encompasses tricky assumptions. Table 4.7 summarizes the main assumptions of existing models in the literature.

Table 4.7 Main assumptions for MAC analytical models in VANETs

Application	Safety	Non-safety
Traffic pattern	Periodic short status (CAM) and/or event-based (DENM) messages	Drive-thru internet-based traffic
Communication pattern	Single-hop broadcast V2V	Single-hop unicast V2I
Metrics	Packet delivery, collision probability, delay	Throughput, delay
Mobility	Static nodes	Vehicles moving with a constant speed

4.6.3 Field Operational Tests

It may seem an obvious statement, but measurements are reality. To say it more openly, whatever is the model adopted in a simulation or in a computation:

- it will subtend some hypotheses or simplifications whose validity may hold only under certain conditions (e.g., a specific propagation model, the lack of interference, the correct working of nodes, specific—often ideal—mobility model, the different behavior of devices by heterogeneous manufacturers, etc.);
- it certainly will neglect (intentionally or not) some phenomena: we have seen that there are several phenomena occurring at physical layer (attenuation, fading, reception chain); consequently, results might change depending on which ones have been taken into account in the model. To make some examples, results could change a lot if one considered the real antenna diagrams, the effects of mounting it on the target cars and the directions of incoming rays. The obstructions by cars are often neglected as well, or modeled in a simplified way (see Chap. 13) and might strongly affect results. Modeling could also hardly capture together (hence, currently tend to neglect) diversity and/or Multiple Input Multiple Output (MIMO) and detailed propagation phenomena (scattering, diffusion, etc.);
- it does not allow to discover phenomena which have not been modeled and may just happen in reality. To say it by an example, hidden terminals had been considered negligible in VANETs, until someone included obstructions in simulations (see Chap. 15).

In other words, one cannot get along without practical experiment to validate a solution, simply because one cannot be sure that reality will/might present something which was not taken into account.

On the other hand, measurements give you the ground-truth but often have a limited scope which needs to guide their use, in that:

- you will get some output metrics but, often you will just infer, not precisely know, what led to such results. For instance, you will know that you have received only 80 % of the frames from an RSU at a given distance, but you will not know whether this is due to interference, collisions, the antenna gain, the attenuation and fading or to a bad equalization (due to coherence time or coherence bandwidth);
- measurements are extremely time-consuming and require a proper setup, possibly involving multiple nodes only when one node is transmitting and one is receiving so to check what is effectively on the air;
- it will be difficult to involve large number of nodes, at least at the current stage. Additionally, the more the involved nodes, the less comprehensible the achieved results.

Altogether, it is preferable to use tests to pursue the understanding of simple phenomena (attenuation/fading, receiver characterization, background noise, effect of the antenna), possibly to feed simulation tools with realistic data, so to study

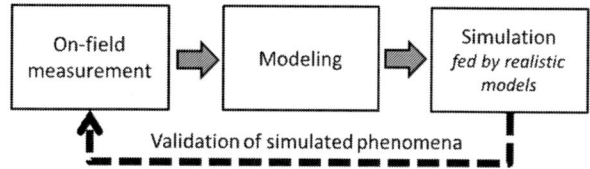

Fig. 4.20 Possible links between measurements and simulations

more complex cases relying on realistic starting points. Consequently, a nice process would foresee, as depicted in Fig. 4.20; measurement, modeling for simulation, realistic simulations and, possibly, validation.

Another possible role for measurements would be the identification (for instance, during the validation of a model) of possible novel phenomena to model: in case this is not possible, a more detailed modeling of underlying mechanisms or hypothesis might be required. This process, for instance, partially happened with hidden terminals: measurements fed new propagation models for obstructed scenarios (such as RUG [14] and Corner [18]) which permitted the study of the effect of hidden terminals in the centers of the crossroads [14]; these are still awaiting on-field validation.

To conclude, it may be worth mentioning also **emulation**. Emulation basically is adopted to challenge a transceiver with tests involving harsh propagation: typically a multipath emulator (or even a programmable *fading emulator*[6]) and programmable attenuators interconnect two small (even *portable*) anechoic chambers where one transmitter and one receiver can be, respectively, placed. More complex configuration would permit to test also MIMO and interferences. While emulators involve expensive equipment and are not meant for tests with manifolds nodes, they permit rapid and repeatable tests.

4.7 Conclusions

In this chapter we have provided the description of the ongoing MAC standardization activities for vehicular networking and an overview of the main evaluation tools for MAC analysis.

VANETs adopt the MAC layer of the IEEE 802.11 standard with proper enhancements to cope with the peculiarities of the vehicular environment.

The CSMA/CA mechanisms at the foundations of 802.11 can intrinsically work in a distributed way, hence natively supporting communications in ad hoc mode among vehicles, without requiring any centralized coordination; it is simple, well known, and several products based on its philosophy are available and proved.

---

[6]Programmable fading emulator may be configured in terms of coherence time, coherence bandwidth, and maximum delay spread (see Chap. 3)

Conversely, it unfortunately suffers from lack of deterministic QoS guarantees to any transmission either (1) in terms of access delay (CSMA/CA can unpredictably defer also high-priority transmissions) or (2) in terms of collision prevention (especially when the number of stations grows high and for broadcast unacknowledged transmissions). Moreover, this protocol cannot prevent the issue of hidden terminal.

So, despite its strengths, CSMA/CA shows relevant weaknesses which are pushing the research community to investigate: (1) its improvements and dynamic adaptations (see Chap. 15 for a survey of the related literature), (2) possible alternatives to the general idea of CSMA/CA scheme, e.g., synchronous time-slotted solutions (see Chap. 15), and (3) different technologies instead of 802.11, e.g., the Long Term Evolution (LTE) cellular system (see Chap. 16) and visible light communications (see Chap. 15).

References

1. Alasmary W, Zhuang W (2012) Mobility impact in IEEE 802.11 p infrastructureless vehicular networks. Ad Hoc Netw 10(2):222–230
2. Badawy G, Misic J, Todd T, Zhao D (2010) Performance modeling of safety message delivery in vehicular ad hoc networks. In: 2010 IEEE 6th international conference on wireless and mobile computing, networking and communications (WiMob). IEEE, New York, pp 188–195
3. Batsuuri T, Bril R, Lukkien J (2010) Application level phase adjustment for maximizing the fairness in vanet. In: 2010 IEEE 7th international conference on mobile adhoc and sensor systems (MASS), pp 697–702. doi:10.1109/MASS.2010.5663797
4. Bianchi G (2000) Performance analysis of the IEEE 802.11 distributed coordination function. IEEE J Sel Areas Commun 18(3):535–547
5. Brevi D, Cozzetti H, Scopigno R, Xu Q (2011) Effectiveness of visual analytics in supporting the reasoning on vehicular ad-hoc networks. In: 2011 IEEE vehicular networking conference (VNC), pp 138–245. doi:10.1109/VNC.2011.6117106
6. Calì F, Conti M, Gregori E (2000) Dynamic tuning of the IEEE 802.11 protocol to achieve a theoretical throughput limit. IEEE/ACM Trans Networking 8(6):785–799
7. Campolo C, Molinaro A (2013) Multichannel communications in vehicular ad hoc networks: a survey. IEEE Commun Mag 51(5):158–169
8. Campolo C, Molinaro A, Vinel A, Zhang Y (2012) Modeling prioritized broadcasting in multichannel vehicular networks. IEEE Trans Veh Technol 61(2):687–701
9. Campolo C, Molinaro A, Vinel A, Zhang Y (2013) Modeling event-driven safety messages delivery in IEEE 802.11 p/wave vehicular networks. IEEE Communications Letters, 17(12):2392–2395
10. Caragnano G, Goga K, Brevi D, Cozzetti H, Terzo O, Scopigno R (2012) A hybrid cloud infrastructure for the optimization of vanet simulations. In: Sixth international conference on complex, intelligent and software intensive systems (CISIS), 2012, pp 1007–1012. doi:10.1109/CISIS.2012.151
11. Chatzimisios P, Boucouvalas AC, Vitsas V (2003) IEEE 802.11 packet delay-a finite retry limit analysis. In: IEEE global telecommunications conference, GLOBECOM'03 2003, vol 2. IEEE, New York, pp 950–954
12. Cozzetti H, Scopigno R (2011) Scalability and qos in ms-aloha vanets: forced slot re-use versus pre-emption. In: 2011 14th International IEEE conference on intelligent transportation systems (ITSC), pp 1759–1766. doi:10.1109/ITSC.2011.6082985

13. Cozzetti H, Campolo C, Scopigno R, Molinaro (2012) A urban vanets and hidden terminals: evaluation through a realistic urban grid propagation model. In: 2012 IEEE international conference on vehicular electronics and safety (ICVES), pp 93–98. doi:10.1109/ICVES.2012.6294332
14. Cozzetti H, Campolo C, Scopigno R, Molinaro A (2012) Urban vanets and hidden terminals: evaluation through a realistic urban grid propagation model. In: 2012 IEEE international conference on vehicular electronics and safety (ICVES), pp 93–98. doi:10.1109/ICVES.2012.6294332
15. ETSI: Intelligent transport systems (ITS); on the recommended parameter settings for using stdma for cooperative its; access layer part. ETSI TR 102 861, European Telecommunication Standards Institute, Sophia Antipolis (2011)
16. Fallah YP, Huang CL, Sengupta R, Krishnan H (2011) Analysis of information dissemination in vehicular ad-hoc networks with application to cooperative vehicle safety systems. IEEE Trans Veh Technol 60(1):233–247
17. Gallardo JR, Makrakis D, Mouftah HT (2010) Mathematical analysis of edca's performance on the control channel of an IEEE 802.11 p wave vehicular network. EURASIP J Wirel Commun Netw 2010:5
18. Giordano E, Frank R, Pau G, Gerla M (2011) Corner: a radio propagation model for vanets in urban scenarios. Proc IEEE 99(7):1280–1294. doi:10.1109/JPROC.2011.2138110
19. Hafeez K, Zhao L, Ma B, Mark J (2013) Performance analysis and enhancement of the dsrc for vanet's safety applications. IEEE Trans Veh Technol 62(7):3069–3083. doi:10.1109/TVT.2013.2251374
20. Hassan MI, Vu HL, Sakurai T (2010) Performance analysis of the IEEE 802.11 mac protocol for dsrc with and without retransmissions. In: 2010 IEEE international symposium on a world of wireless mobile and multimedia networks (WoWMoM). IEEE, New York, pp 1–8
21. He J, Tang Z, O'Farrell T, Chen TM (2011) Performance analysis of dsrc priority mechanism for road safety applications in vehicular networks. Wirel Commun Mob Comput 11(7):980–990
22. Huang CL, Liao W (2007) Throughput and delay performance of IEEE 802.11 e enhanced distributed channel access (edca) under saturation condition. IEEE Trans Wireless Commun 6(1):136–145
23. Huang K, Duffy KR, Malone D (2010) On the validity of IEEE 802.11 mac modeling hypotheses. IEEE/ACM Trans Networking 18(6):1935–1948
24. IEEE Standard 802.2-2009 (ISO/IEC 8802-2:1998) - Part 2: Logical link control (1998)
25. IEEE 1609.3-2010 (2010) IEEE standard for wireless access in vehicular environments (wave) - networking services
26. IEEE 1609.4-2010 (2011) IEEE standard for wireless access in vehicular environments (wave) - multi-channel operation
27. IEEE Standard 802.11-2012 - Part 11: Wireless LAN medium access control (MAC) and physical layer (PHY) specifications (2012)
28. Intelligent Transport Systems (ITS) (2011) European profile standard for the physical and medium access control layer of intelligent transport systems operating in the 5 GHz frequency band, ETSI ES 202 663 (V1.1.0)
29. Kafsi M, Papadimitratos P, Dousse O, Alpcan T, Hubaux JP (2008) Vanet connectivity analysis. In: Proceedings of the IEEE workshop on autonet, New Orleans, LA, USA
30. Ma X, Chen X (2007) Delay and broadcast reception rates of highway safety applications in vehicular ad hoc networks. In: 2007 Mobile networking for vehicular environments. IEEE, New York, pp 85–90
31. Ma X, Yin X, Wilson M, Trivedi KS (2013) Mac and application-level broadcast reliability in vanets with channel fading. In: 2013 International conference on computing, networking and communications (ICNC). IEEE, New York, pp 756–761
32. Malone D, Duffy K, Leith D (2007) Modeling the 802.11 distributed coordination function in nonsaturated heterogeneous conditions. IEEE/ACM Trans Networking 15(1):159–172

33. Misic J, Badawy G, Misic VB (2011) Performance characterization for IEEE 802.11 p network with single channel devices. IEEE Trans Veh Technol 60(4):1775–1787
34. Schumacher H, Schack M, Kurner T (2009) Coupling of simulators for the investigation of car-to-x communication aspects. In: IEEE Asia-Pacific services computing conference, 2009. APSCC 2009, pp 58–63. doi:10.1109/APSCC.2009.5394139
35. Taliwal V, Jiang D, Mangold H, Chen C, Sengupta R (2004) Empirical determination of channel characteristics for dsrc vehicle-to-vehicle communication. In: Proceedings of the 1st ACM international workshop on vehicular ad hoc networks. ACM, New York, pp 88–88
36. Torrent-Moreno M, Mittag J, Santi P, Hartenstein H (2009) Vehicle-to-vehicle communication: fair transmit power control for safety-critical information. IEEE Trans Veh Technol 58(7):3684–3703
37. van Eenennaam M, Wolterink WK, Karagiannis G, Heijenk G (2009) Exploring the solution space of beaconing in vanets. In: 2009 IEEE vehicular networking conference (VNC). IEEE, New York, pp 1–8
38. Xiao Y (2005) Performance analysis of priority schemes for IEEE 802.11 and IEEE 802.11e wireless lans. IEEE Trans Wireless Commun 4(4):1506–1515

Chapter 5
Message Sets for Vehicular Communications

Lan Lin and James A. Misener

Abstract VANET technology includes Vehicle-to-Vehicle and Vehicle-to-Infrastructure communications. It supports the realization of a large variety of Cooperative Intelligent Transport System (C-ITS) applications and services by enabling real-time data exchanges between vehicles, between vehicles and infrastructure systems. C-ITS technologies extend the driver perception to the traffic ahead, avoid potential road hazard situations, collision risks and improve traffic efficiency. Automobile and road operator stakeholders in Europe and in North America have been jointly driving research and development of the C-ITS for more than a decade. Message sets specifications, standardization, and validation consist of one key activity for VANET technology development in world wide. Even though some differences are observed in technical features of message sets in order to satisfy regional specific requirements, commonality is often found in terms of basic features and application usages. Recently, some regions in the EU and in North America have realized large-scale field tests and entered the phase of C-ITS pilot deployment. The present chapter provides an overview of the C-ITS core message sets being standardized and planned to be deployed in the EU and in North America.

Keywords VANET • BSA • CAM • DENM • BSM • SPAT • MAP

5.1 Introduction

The Cooperative ITS (C-ITS) technologies include Vehicle-to-Vehicle (V2V) and Vehicle-to-Infrastructure (V2I) communications. They enable the real-time infor-

---

L. Lin (✉)
Hitachi Europe SAS. – Information and Communication Technologies Laboratory (ICTL),
Sophia Antipolis, 06560, France
e-mail: lan.lin@hitachi-eu.com

J.A. Misener
Qualcomm Technologies, Inc. – Qualcomm Standards and Industry
Organizations (QSIO), San Diego, CA 92121, USA
e-mail: jmisener@qti.qualcomm.com

mation exchanges between vehicles, between vehicles and infrastructure systems, in order to extend the driver perception to the traffic ahead and avoid potential road hazard situations. A variety of services may be provided to road users with C-ITS, categorized in road safety, traffic efficiency, environment friendly and infotainment services. VANET messages and the corresponding message exchange protocols enable exchange of application data in vehicular communication networks. A large set of real-time raw data or processed data are transported in message sets.

The C-ITS deployment does not only rely on the technical aspects such as message sets specifications, communication protocol design and access technology specifications, etc., but also on many non-technical aspects such as organizational, legal, or operational aspects. C-ITS deployment requires cooperation among stakeholders from different domains including automobile industries, road operators, public authorities, technology providers, telecommunication operators, etc., to reach agreement on a common deployment roadmap at vehicle and infrastructure side. The roles and responsibilities of different stakeholders need to be defined. These nontechnical aspects have strong impacts on final choice of the appropriate technical solutions. For example, car makers and road operators work together to define data sharing needs and to set requirements on data quality and information dissemination. These requirements are guiding the message content definition, message exchange protocol design as well as the applications that may be realized.

Key stakeholders in North America and the EU have been driving research and development of the C-ITS for more than a decade, with message set specification and validation as one of the core activities. In the last several years, standardization organizations are playing a key role in the development and tests of message sets and protocols standards, in order to enable an early day one deployment of the C-ITS. Actors in public and private sectors are involved actively in these standardization and test activities. In North America, United States Department of Transportations (US DOT) coordinates with standardization organizations Society of Automotive Engineers (SAE) and Institute of Electrical and Electronics Engineers (IEEE) to develop required standards for the deployment of Dedicated Short Range Communication (DSRC) Vehicular Communication Systems. In the EU, European Commission has issued a Standard Mandate 453 [14] to European Telecommunications Standards Institute (ETSI) and European Committee for Standardization (CEN) to deliver

> a minimum set of European standards... to ensure interoperability for vehicle to vehicle communications, for vehicle to infrastructure communications and for communications between infrastructure operators.

In parallel to the standards development, large-scale Field Operational Test (FOT) has been realized to validate the standard specifications and to evaluate C-ITS applications in real traffic environment. In the EU, the large-scale FOT project DRIVE C2X operates the C-ITS real site tests in seven European countries, with partners from car manufacturers, highway operators, suppliers, and research institutes.[1] DRIVE C2X results have shown that more than two-thirds of investigated customers claim that they are excited about C-ITS technology and applications [22]. In North America, automakers, state U.S. Department of Transportation (DOT), and the US DOT have been conducting research, primarily on applications of V2V and V2I safety communications. A large-scale, 1-year field operational test, the Safety Pilot Model Deployment[2] has investigated V2V safety applications. The Federal Highway Administration (FHWA) Crash Avoidance Metrics Partnership (CAMP) Light Vehicle Driver Acceptance Clinics exposed a broad array of drivers to V2V safety applications at six US locations [29]. The response was overwhelmingly positive, as 85 % of the subjects liked or felt reassurance from the warnings and 60 % found them effective. Based on these encouraging results, the EU and North America have recently started planning the phase of pilot deployment of the C-ITS. For example, on 10 June 2013 ministers representing Germany, Austria, and the Netherlands signed a Memorandum of Understanding (MOU) to initiate a C-ITS Corridor joint deployment.[3] In North America, a series of Connected Vehicle Pilot Deployments are planned.[4] Similar activities are also observed in other regions, such as Japan, Canada, and China with different progress status. In Japan, ITS Spot services are already deployed in about 1,600 places centering on expressways to provide real-time road traffic information, safety driving support, and Electronic Toll Collection (ETC) services to road users [23]. In China, standardization activities on C-ITS have started since recently. A China ITS Industry Alliance was initiated in 2013 grouping 50 industrial actors with objective to develop ITS standardization and testing activities [28].

The present chapter provides an overview of the core message sets being standardized and planned to be deployed for vehicular communication technologies in the EU and North America. It is organized as follows: Sect. 5.1 provides background information; Sect. 5.2 provides a high level introduction on application requirements that guide the message sets specifications; Sect. 5.3 gives an overview of the message sets in the EU and in North America; Sects. 5.4–5.9 present detailed specifications on core message sets as introduced in Sect. 5.3; Finally, Sect. 5.10 concludes the present chapter.

It should be noted that, like other technologies of C-ITS, message sets specification is a continuous process. The message format may need to be revised and/or new features be added in order to support new ITS applications needs and data exchange needs. Such updates may happen, even during the operational phase of an implemented product. A release mechanism has been adopted in main standardization organizations. During the standard development phase, special

---

[1] DRIVE C2X is a EU funded project, see http://www.drive-c2x.eu/project.

[2] Safety Pilot Model Deployment is a major research initiative that involves US DOT, vehicle manufacturers, public agencies, and academia. See more at: http://www.its.dot.gov.

[3] A description of C-ITS corridor joint deployment can be found at http://www.bmvi.de/SharedDocs/EN/Anlagen/VerkehrUndMobilitaet/Strasse/cooperative-its-corridor.pdf.

[4] Information on the pilot deployment is available at http://www.its.dot.gov/pilots/.

attention has been taken to ensure a certain level of backward compatibility, in order to enable the interoperability between standard releases. In the EU, the release one of message set standards are under development by ETSI and CEN. In North America, message sets are incorporated in the SAE J2735 (2009) standard [25]. At this writing, this standard is undergoing revision.

5.2 Application Requirements

A wide variety of applications may be developed using C-ITS technologies. There exist in the research community many methods to classify the application categories. One popular method is the classification based on Time-To-Collision (TTC) parameter. TTC is a parameter that measures the traffic conflict probability as the time required for two vehicles to collide if they continue at their present speed and on the same path. It has been proposed by Hayward [15].

At its establishment in 2008, ETSI TC ITS has defined a Basic Set of Application (BSA) [4], that may be enabled by C-ITS technologies and deployable

within three years time frame after the standards have been completed.

A high level classification of the BSA applications according to TTC is presented in Fig. 5.1. The TTC values are illustrated as examples.

- **Road traffic info and telematics applications** are telematics services provided by public operators or by private service providers. Upon reception of road traffic

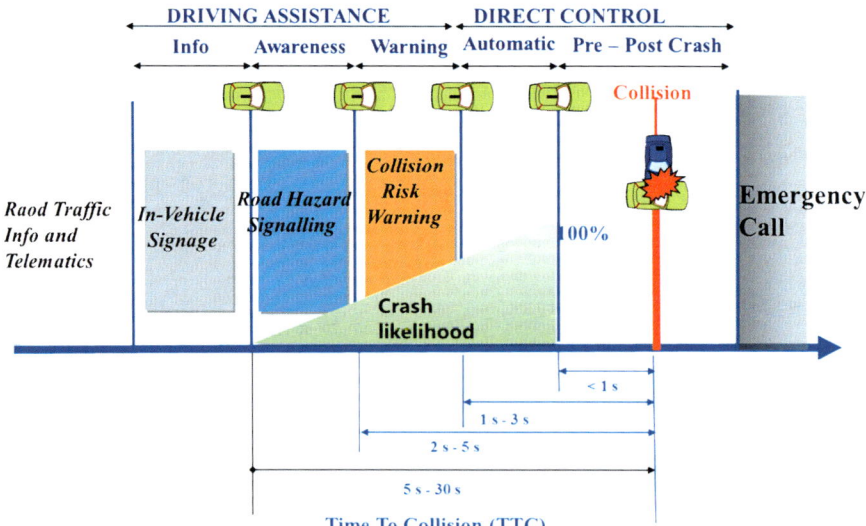

Fig. 5.1 High level classification of applications based on TTC. (source: ETSI TC ITS)

information, vehicles check the relevance of the received information and show to vehicle users as information, when no imminent danger is detected. This information may help users to adjust the navigation plan and be informed of road traffic status, to improve the driving comfort. At large scale, telematics services may also play a role in improving traffic efficiency and reducing road traffic pollution. An example of road traffic information application is Transport Protocol Experts Group (TPEG) application [24].

- **Driving assistance applications** are proposed by automobile Original Equipment Manufacturer (OEM)s to assist driver in dense traffic or in dangerous situations where attention should be paid by driver to overcome potential safety risks. The information provided to driver may be a remind information for traffic rules (e.g., speed limit) or road conditions (traffic jam), an awareness information of detected road hazards (e.g., stationary vehicle at road side, hard break vehicle ahead), or even a warning message that requires immediate actions of driver to avoid potential collision. At this stage, there is no automatic control to the in-vehicle systems. Driver keeps the control on the vehicle maneuvering. An example of awareness application is Road Hazard Signalling (RHS) application [11]. Examples for warning application are Longitudinal Collision Risk Warning (LCRW) application [12] and Electronic Emergency Brake Light (EEBL) application [6].

- **Direct control application** takes over the control of vehicle by automatic control systems to avoid potential collision in case the TTC is further reduced. If the collision cannot be avoided, the pre-crash application may be launched to mitigate the collision impact and reduce the damage to vehicle driver or passengers. Automated driving applications are within this category.

- **Post crash application** consists of cooperating with rescue organizations to reduce the accident rescue latency time and improve the rescue efficiency. In Europe, eCall is a pan-European in-vehicle emergency call system for this purpose. It aims to deploy an in-vehicle device that will automatically dial 112 in case of road accident and establish voice connection with back-end office. In addition, a set of in-vehicle sensor data such as airbag deployment, impact sensor information, and Global Positioning System (GPS) coordinates will be transmitted to local emergency agencies via the wireless communication infrastructure such as cellular networks. The emergency agencies can then organize the rescue plan properly. In April 2013, the European Commission adopted a proposal for a regulation of the eCall in-vehicle system [30].

At the time of writing the present chapter, the C-ITS standardization and validation activities in the EU are mainly focused on the driving assistance applications. During the standard development process, technology agnostic approach is used for the C-ITS applications and message sets specifications. It is assumed that any communication technology may be used to realize a C-ITS application, as long as it is suitable to satisfy the application functional and operational requirements. For example, ITS G5 technology at 5.9 GHz is considered by automobile stakeholders in the EU as main candidate technology to initiate the deployment of V2V-based road safety applications and a set of V2I applications. A MOU [3] on deployment strategy for C-ITS in Europe has been signed by OEMs in Car-to-Car Communication Consortium (C2C-CC)[5] to promote the ITS G5 enabled C-ITS and a set of day one applications. This is in order to ensure the interoperability between different implemented systems. This initiative has been followed up by main road operator association Amsterdam Group.[6] The group has published a deployment roadmap [1] announcing that

> It is generally agreed to follow a phased deployment approach with an initial deployment of simple non-complex Day One services where user benefits are achieved even with limited penetration of ITS in vehicles and equipped road side units in hot spot areas and corridors.

It has been demonstrated in various FOT projects that V2V and V2I technologies can address a large majority of road safety issues and as well improve the traffic efficiency.[7] Table 5.1 summarizes a non-exhaustive list of C-ITS applications that are selected by different European initiatives, e.g., ETSI, DRIVE C2X, C2C-CC, etc.

In North America, US DOT National Highway Traffic Safety Administration (NHTSA) has announced in early 2014 that it will begin taking steps to enable DSRC V2V communication technology for light vehicles [33]. Subsequently, in September 2014, NHTSA announced an Advanced Notice of Public Rulemaking (ANPRM)[8] and published an accompanying research report on readiness of V2V technology for application [32]. The US objective is to progress toward a Notice of proposed rulemaking (NPRM) in 2016, which formally solicits public commentary that will shape a mandate for V2V equipment, the suite of DSRC transceivers and positioning equipment, to be on new production vehicles by a few years. Data analysis has been conducted based on data collected from 3,000 DSRC-equipped vehicles participating to one-year duration Safety Pilot Model Deployment in Ann Arbor, Michigan, USA. The results show significant reduction in crashes, as a consequence lives saved and economic benefit increased. This motivation, when coupled with emerging minimum performance standards and a proposed standardized solution for system security form the basis of a safety-of-life argument for a mandate to initiate C-ITS deployment in North America. Using the TTC classification in Fig. 5.1 as a reference, the North American model for initial deployment thusly focuses almost exclusively on in the collision risk warning (2–5 s) window. To that end, the Safety Pilot Model Deployment was implemented and mined as significant data source to critically examine the technical performance of standardized DSRC messages and to project widespread crash avoidance or societal benefit for following V2V applications:

---

[5]C2C-CC (www.car-to-car.org) is a European Automobile Industrial consortium.

[6]Amsterdam Group (https://amsterdamgroup.mett.nl) is a Strategic European Road Operators consortium.

[7]See Sect. 5.1 for more details about field operational tests.

[8]See docket ID NHTSA-2014-0022 at http://www.regulations.gov.

Table 5.1 Examples of C-ITS application

Application	Short description
Longitudinal collision risk warning	Warns the driver when a longitudinal collision risk with neighbor vehicles is detected by processing the received messages from these vehicles.
Intersection collision risk warning	Warns the driver of a potential collision risk with other vehicles at an intersection area by processing the received messages from these vehicles.
Traffic light violation warning	Warns the driver of a potential traffic light violation if speed is not reduced, by processing the traffic light status information received from road side.
Lane change warning	Warns the driver who plans to change lanes if there is a vehicle in the blind spot or an overtaking vehicle.
Cooperative awareness	Vehicles are made aware of other vehicles' position, speed, and basic sensor status in its vicinity by processing the received messages from them.
Emergency vehicle approaching	Emergency vehicles transmit a message to other vehicles in its path for them to take appropriate actions to give priority.
Slow vehicle approaching	Vehicle driving at low speed on highway, e.g., a roadworks vehicle or a road surface cleaning vehicle transmits a message to announce its presence.
Emergency electronic brake light	A hard brake vehicle transmits a warning message to vehicles behind in order to avoid rear end collision.
Stationary vehicle	Stationary vehicle on the road surface due to accident, breakdown or other reasons transmits a warning message to oncoming vehicles from upstream traffic.
Hazardous location	Vehicle detecting hazardous road conditions e.g., obstacles on the road using its on-board sensors transmits a message to oncoming vehicles.
Roadworks warning	Roadwork vehicles or road side units transmit roadwork information as well as relevant speed limit information to oncoming vehicles.
Green light optimal speed advisory	Upon reception of traffic light phase and timing information from road side, vehicles may calculate optimal speed to cross the road intersection and avoid traffic light violation or reduce fuel consumption.
In vehicle signage	Road side infrastructure transmits static or dynamic road sign information to vehicles for in-vehicle presentation.
Speed limit information	Road side infrastructure transmits static or dynamic speed limit information to vehicles for in-vehicle presentation.

- **EEBL**: Onset of braking from the forward vehicle is passed to the next vehicle, and a Human Machine Interface (HMI) is actuated to provide EEBL warning to the driver.
- **Forward Collision Warning (FCW)**: An algorithm to warn the driver and as the TTC continues to diminish, graduated intensity of HMI warning, followed by brake control is provided.
- **Blind Spot Warning/Lane Change Warning (BSW/LCW)**: Vehicles in the rear-view mirror blind spot or fast-approaching vehicles elicit a warning through the HMI.
- **Do Not Pass Warning (DNPW)**: A warning drivers in the presence of hazardous oncoming vehicles in a passing situation is provided through the HMI.
- **Intersection Movement Assist (IMA)**: As drivers move directly through or conduct a turning movement within an intersection, HMI alert with warnings of crossing-path hazards posed by vehicles approaching on other intersection legs is provided.
- **Left Turn Assist (LTA)**: At unprotected left turns (in countries where vehicles are driven on the right), warnings of vehicles approaching from the opposite direction are provided when TTC thresholds are exceeded.

The aforementioned ANPRM focused on just IMA and LTA as justification for a potential NHTSA mandate, most likely because the other crash types listed here are also being addressed through other NHTSA crash avoidance rulemakings and because these intersections crash categories represent significant risk.

At the time of writing the present chapter, the first release of standards are close to finalization in the EU and in North America, in preparation of a day one deployment. At this first phase deployment, the system penetration rate in the overall car park and at road side may be limited and requires some time to progress. In addition, the communication reliability and quality of some key data e.g., position accuracy, may not be sufficient to develop direct control applications. On the other hand, new research initiatives have already been launched to study new features to support more advanced applications such as automated driving based on vehicular communication technologies. These research inputs will be used for future standard development.

5.3 Message Sets Overview

This section presents an overview of core message sets in the EU and North America. A VANET message refers to an *Application* or *Facilities layer* entity in the standardized ITS communication reference architecture [5], as illustrated in Fig. 5.2. A system that implements the ITS protocol stacks and ITS applications is denoted as an ITS station (ITS-S). An ITS-S may be integrated in a vehicle (vehicle ITS-S) or at road side (road side ITS-S). A vehicle ITS-S is also named as On Board Unit (OBU), a road side ITS-S is also named as Road Side Unit (RSU).

5 Message Sets for Vehicular Communications

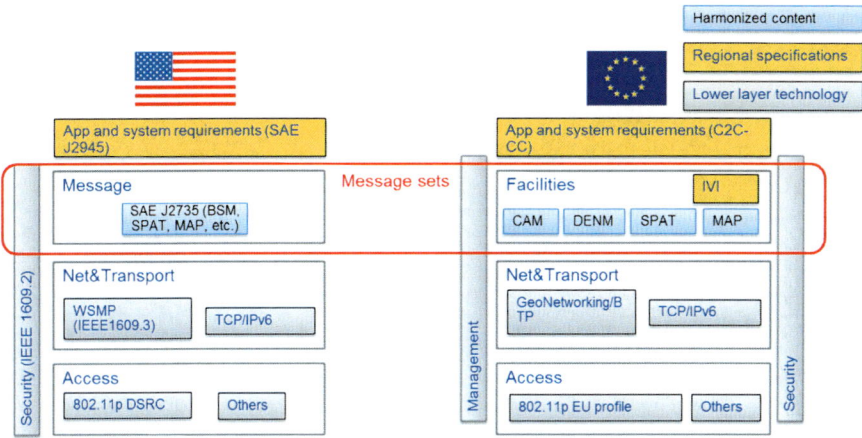

Fig. 5.2 Message sets in reference ITS communication architecture

In general, specifications of a message include a data dictionary definition (or message format definition) that defines the syntax and semantics of the message, and a related communication protocol for the message exchange. A message entity relies on the lower layer protocol stacks for dissemination. The Service Access Point (SAP)s enable data exchange between layers. The exchanged data via SAP for message transmission includes the message payload as well as communication requirement parameters. Figure 5.2 illustrates the set of messages developed in the EU and North America as well as lower layer standards applied for the message dissemination.

Stakeholders both at vehicle side and infrastructure side are cooperating together for message set specifications. The targeted applications include road safety, traffic efficiency, and added value applications. Among these applications, road safety applications represent the most stringent requirements in terms of time latency, data quality, communication reliability, and system reliability. A set of V2V and V2I messages are specified, in order to support the selected set of applications for deployment. Standardization of message format and communication protocols plays an essential role to ensure the communication interoperability between vehicles as well as between vehicles and infrastructure systems. This standardization work needs to take into account:

- **Application requirements** that define data exchange needs and communication requirements between ITS-Ss;
- **Message format and data dictionary** that defines unambiguous syntax and semantics of data being included in a message, in order to avoid any potential misinterpretation at receiving ITS-S;
- **Communication capacities** in order not to bring unnecessary overload traffic to communication channel; and

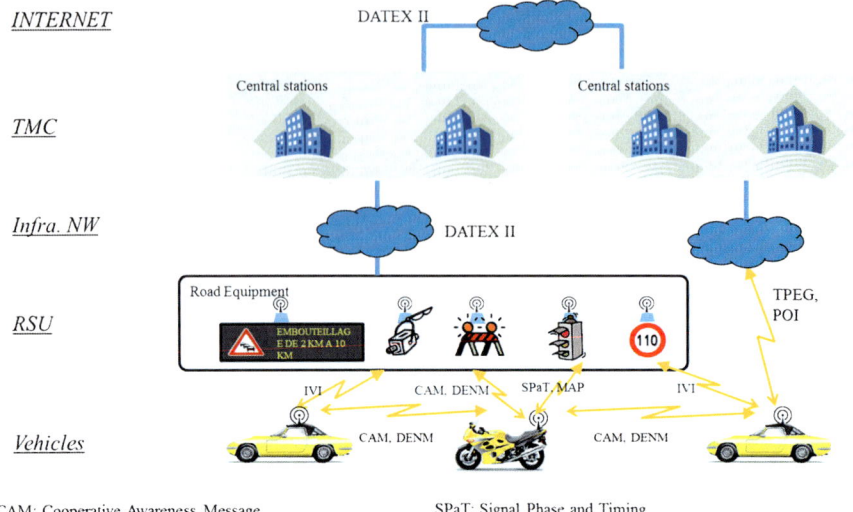

Fig. 5.3 High level overview of ITS protocols used in the EU

- **Harmonization of message sets**. The harmonization mainly includes two aspects: On the one hand, harmonization of message sets standards at International level enables cost reduction for industrial implementation at different markets. On the other hand, harmonization with existing relevant standards in the same region, in particular the existing road infrastructure message sets standards, would facilitate the interoperability between vehicles and road side infrastructures.

For illustration purpose, Fig. 5.3 presents a high level overview of ITS message sets and application layer protocols that are used in ITS domain in the EU. In North America, similar architecture may be observed, with corresponding message sets and protocols applied locally.

In the present chapter, we will provide detailed description for message sets that are used for VANET networks, i.e., Cooperative Awareness Message (CAM), Decentralized Environmental Notification Message (DENM), Basic Safety Message (BSM), Signal Phase and Timing (SPAT), Map data (MAP), and In Vehicle Information (IVI) message. For other message sets and application protocols included in Fig. 5.3, they are not used in VANET, but the harmonization with these message sets are taken care during the specifications phase. The message sets is summarized as follows:

- **CAM** is specified by ETSI TC ITS for European deployment. It is a *heartbeat* message that is transmitted periodically from OBU and RSU to announce the

position, movement, and basic attributes of the transmitting ITS-S. Detailed description is provided in Sect. 5.4.
- **DENM** is specified by ETSI TC ITS for European deployment. It is an *event-driven* message that is transmitted from OBU or RSU at the detection of a traffic event or road hazard. Detailed description is provided in Sect. 5.5.
- **BSM** is specified by SAE TC DSRC for North America deployment. It provides functions equivalent to CAM and DENM adapted to targeted applications selected in North America. Detailed description is provided in Sect. 5.6.
- **SPAT** is specified by SAE TC DSRC, it has been extended for deployment in North America, EU, and Japan. It is transmitted from an RSU to provide *phase and timing* information of one or a set of traffic lights. Detailed description is provided in Sect. 5.8.
- **MAP** is specified by SAE TC DSRC, it has been extended for deployment in North America, EU, and Japan. It is transmitted from an RSU to provide *road topology and geometry* information of a road segment or an intersection area. It is used by a receiving ITS-S to map the SPAT data to the intersection topology. Detailed description is provided in Sect. 5.9.
- **IVI** is initiated by CEN and specified by CEN/ISO as world side standard. It is transmitted from an RSU to provide static or dynamic *road signage* information for in-vehicle presentation. Detailed description is provided in Sect. 5.7.
- **DATEX 2** is a web service based application protocol already deployed in the EU for the communication between Traffic Management Center (TMC)s to exchange *road traffic or traffic management* information.[9] DATEX 2 specification includes a set of data dictionary for the description and definition of traffic situation, traffic rules, and road topology information. This data dictionary is considered during the I2V message specifications, e.g. the IVI message.
- **TPEG** specifications provide *traffic and travel information* services to road users. TPEG messages and protocols are developed by stakeholders in Traveller Information Services Association (TISA) and adopted as International Organization for Standardization (ISO) standards for Road Traffic and Traveller Information (RTTI) Service.[10] TPEG also includes a data dictionary for road traffic and traveller information. In particular, the traffic event data dictionary is considered during the specifications of event-driven message DENM.
- **POI** refers to a set of messages and application protocols that provide *Point of Interest* Information to road users. A POI message set has been specified in ETSI TC ITS. For example, the POI message for electric vehicle charging spot information provides up-to-date availability and characteristics of charging stations to electric vehicle users [7]; The POI message for tyre pressure pump and

---

[9]DATEX 2 specifications are available at www.datex2.eu.

[10]TISA (www.tisa.org/technologies/tpeg/) is a market-driven membership association established as a non-profit company.

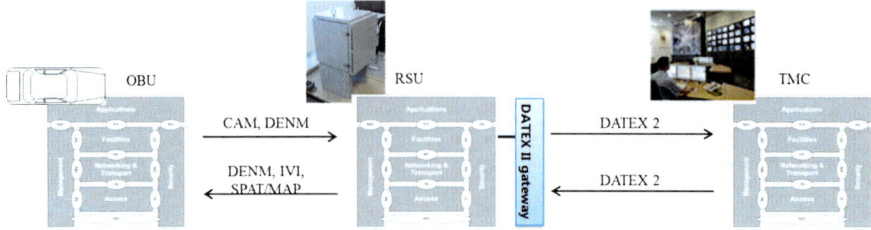

Fig. 5.4 RSU gateway function for the inter-networking between VANET and road infrastructure network

gauge station provides availability of the tyre pressure gauge and pump station information to road users.[11]

In the EU, road operators are testing functionalities to interconnect VANET and road infrastructure networks. In one representative application, VANET messages transmitted from vehicles are collected by RSU for aggregation then provided to TMCs as probe data. For example, in French FoT project SCORE@F,[12] each RSU collects CAM, DENM messages transmitted from vehicles in its vicinity, then realizes a simple processing of the collected data to derive local traffic status such as average driving speed, average travel time, etc. The aggregated traffic status data is then transmitted up-link to TMC using DATEX 2 protocol. In down-link, a TMC may send a DATEX 2 message to an RSU, providing traffic management, speed limit, and traffic status information that is relevant to the local traffic close to the RSU communication coverage range. After processing the received DATEX 2 message, the RSU generates a VANET message accordingly and transmits it to vehicles located nearby. A gateway function is implemented in RSU to interface the VANET and road infrastructure network, as illustrated in Fig. 5.4. Such gateway function further facilitates the vehicle and infrastructure integration.

Table 5.2 presents a match between ITS applications as listed in Sect. 5.2 and VANET message sets that may be used to support the ITS application is question.

The transmission and dissemination of a VANET message rely on the lower layer protocol stacks. In the EU, the GeoNetworking and Basic Transport Protocol (BTP) stacks are specified, enabling a multi-hop, geographical position-based addressing scheme for message dissemination. This allows the dissemination of data packets to a geographical area according to the ITS application needs.[13] In North America, IEEE 1609.3 [19] standard specifies a network layer protocol for DSRC message dissemination in one hop dissemination mode. For access technology, IEEE 802.11p protocol at 5.9 GHz spectrum provides an ad-hoc and low latency access to the

---

[11] At the time of writing this chapter, this standard is still under development in ETSI TC ITS.

[12] SCORE@F is a French Government funded FOT project for C-ITS. More detailed information can be found at https://project.inria.fr/scoref/en/.

[13] See Chap. 8 for details on GeoNetworking.

Table 5.2 Matching of ITS applications and VANET messages

Application	CAM	DENM	BSM	SPAT	MAP	IVI
Longitudinal collision risk warning	X	X	X	–	–	–
Intersection collision risk warning	X	X	X	X	X	X
Traffic light violation warning	X	X	X	X	X	–
Lane change warning	X	–	X	–	–	–
Cooperative awareness	X	–	X	–	–	–
Emergency vehicle approaching	X	X	X	–	–	–
Slow vehicle approaching	X	X	X	–	–	–
Emergency electronic brake light	X	X	X	–	–	–
Stationary vehicle	X	X	X	–	–	–
Hazardous location	–	X	X	–	–	–
Roadworks warning	X	X	–	–	–	X
Green light optimal speed advisory	–	–	–	X	X	–
In-vehicle signage	–	–	–	–	–	X
Speed limit information	–	–	–	–	–	X
FCW	X	X	X	–	–	–
BSW/LCW	X	–	X	–	–	–
DNPW	X	X	X	–	–	–
IMA	X	X	X	X	X	X

radio channel, enabling broadcast and unicast of data packets without the needs of communication infrastructure. In addition, other protocol stacks such as legacy IPv6 protocol and cellular technologies may also be used.[14] This is made possible thanks to the technology agnostic approach adopted for message sets standard development.

5.4 Cooperative Awareness Message

The generation, transmission, and management of CAM is realized by the Cooperative Awareness (CA) basic service component. The CA basic service specifications have been initially introduced by European C2C-CC as a European Automobile stakeholders joint effort [2]. The CA basic service has been prototyped, validated, and tested in multiple EU FOT projects. Technical findings and lessons learned of these initiatives are fed back to ETSI, which has developed a European Norm (EN) [9]. At the time of writing the present chapter, this EN document has been submitted to National Vote in the EU. CAM is a core message that is required to support the day one deployment of VANET system in Europe.

---

[14] See Chap. 9 for the usage of IPv6 for vehicular communications.

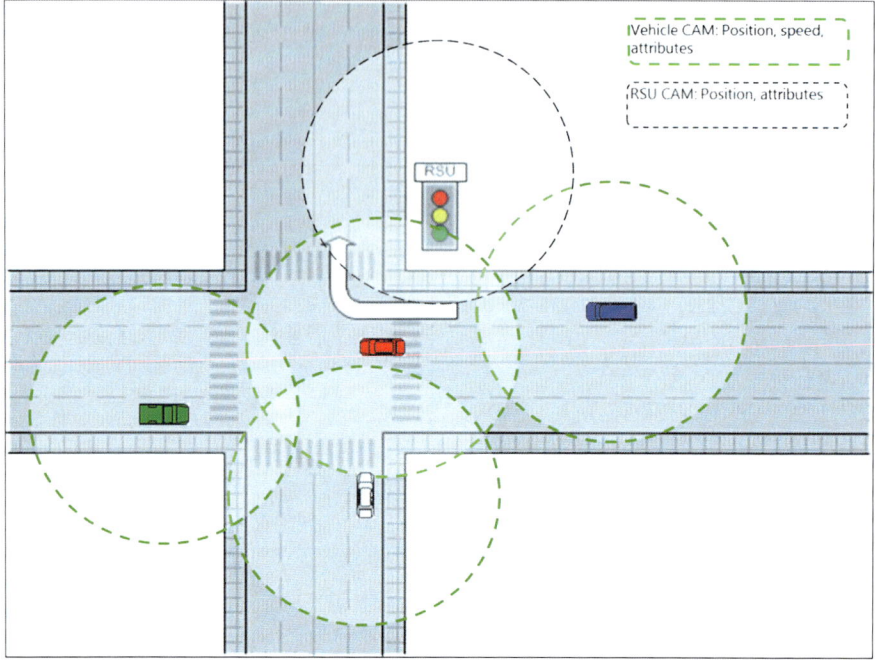

Fig. 5.5 Example scenario of CAM transmission

5.4.1 The CA Basic Service Overview

CAM contains real-time vehicle data or RSU data. It is transmitted with high frequency from an OBU or RSU to other OBUs or RSUs located in the close vicinity, in order to

> create and maintain awareness of each other and to support cooperative performance of vehicles using in the road network [9].

Figure 5.5 provides an overview of the CAM transmission in a road network. Currently in ETSI standard [9], vehicle CAM and road side CAM are specified. A vehicle CAM contains information of vehicle position, vehicle movement, vehicle basic attributes, and sensor data. An RSU CAM announces the basic attributes of the road side equipment. For example, [9] specifies an RSU CAM that announces one or more CEN DSRC road tolling gate positions, as well as a coverage range around the CEN DSRC station that requires being protected against any interference of ITS G5 radio equipment. Upon reception of this RSU CAM, a vehicle is made aware of the position of the CEN DSRC station and performs appropriate mitigation techniques to avoid potential interference to the CEN DSRC transaction.[15]

---

[15] In some European countries, CEN DSRC tolling service operating at 5.8 GHz has been deployed. Simulation studies have shown that interference may exist between ITS-S operating ITS G5 and

5 Message Sets for Vehicular Communications   137

CAM is mainly designed for road safety applications. In one example use case where ITS G5 technology is used, an ego vehicle receiving CAMs from the surrounding vehicles has awareness of the movement and basic sensor status of these vehicles. The ego vehicle can therefore detect in a short time latency an abnormal maneuvering of the surrounding vehicles such as a hard brake situation. The transmitting vehicle also provides its historical path and path prediction information in CAM, allowing the receiving ego vehicle to estimate the probability of path crossing with the transmitting vehicle in order to estimate the collision risk. Once a safety risk is detected, a warning is delivered to the driver of the ego vehicle, who may take appropriate actions to reduce the risk and improve the driving safety.

CAM may also be used in non-safety applications. In particular, CAMs transmitted by vehicles may be collected by RSUs and forwarded to TMC. TMC may further process the received CAM data for traffic monitoring and traffic management applications. A short description of this use case is provided in Sect. 5.3.

5.4.2 CAM Dissemination and Transmission Protocol

In Europe, ITS G5 technology is considered as main technology for CAM dissemination in support of road safety applications. However, CAM dissemination is not limited only to the ITS G5 technology. For example, for CAM collection by TMC application as mentioned in Sect. 5.3, cellular network may be used to transmit CAM and DENM directly to the TMC as road traffic floating car data.

5.4.2.1 CAM Dissemination

In case ITS G5 is used, CAM is broadcasted over Control Channel (CCH). The Single-Hop Broadcast (SHB) protocol of the *GeoNetworking/BTP* [8] is used for CAM dissemination. As consequence, CAM is broadcasted to vehicles located in the direct communication range of the transmitting node. Given the high update and transmission frequency, the CAM packet life time set for the SHB protocol is set to a small value (1 s) to avoid unnecessary queuing at network layer. The priority (denoted as *Traffic Class* [8]) for CAM is set to a high value. This is to guarantee a prioritized access to the radio resource for CAM transmission. The CAM transmission is independent to any specific applications. Its transmission is activated as long as the vehicle is located in the public road and the OBU is activated.

CEN DSRC tolling station when two stations are within the vicinity of each other. Mitigation techniques are currently under specifications in ETSI TC TC to avoid such potential interference.

5.4.2.2 CAM Transmission Protocol

According to [9], the CAM transmission frequency varies between 1 and 10 Hz. This requires the OBU to collect up-to-date information at least 10 Hz rate for CAM construction. Typically, the construction of a vehicle CAM requires the CA basic service to access to geographic positioning systems such as GPS and to the in-vehicle network, e.g. Controller Area Network (CAN).

A protocol is defined in [9] to dynamically adjust the CAM transmission interval between the upper limit (1,000 ms) and lower limit (100 ms), according to the vehicle dynamics and the ITS G5 *channel congestion* status. This protocol defines a set of conditions under which a new CAM shall be generated, as follows:

1. If the absolute difference between the current heading of vehicle and the heading included in the previous generated CAM exceeds 4 deg, or
2. If the distance between the current position of the vehicle and the position included in the previous generated CAM exceeds 4 m, or
3. If the absolute difference between the current speed of the vehicle and the speed included in the previously generated CAM exceeds 0.5 m/s, and
4. The transmission interval is equal or greater than the allowed transmission interval set by the Decentralized Congestion Control (DCC) functionality. DCC includes a set of functionalities including transmission interval control in order to limit the channel congestion level for ITS G5 radio channels under a target threshold. Its standardization work is ongoing in ETSI TC ITS.
5. In addition, if the CAM transmission interval is reduced, i.e. the vehicle dynamics is reduced, the same transmission interval value should be maintained for at least a predefined number of consecutive CAM generations. This is to improve the CAM reception probability even at vehicle low dynamics situation, in particular when the packet lost may be increased in some radio propagation conditions, e.g. in non-Line-of-Sight (LoS) conditions at road intersection area.

As for RSU CAM, the default transmission interval is set to 1 Hz. However, this may be adjusted depending on the application needs. For RSU CAM that provides CEN DSRC position as specified in [9], 1 Hz transmission rate is set to allow vehicles entering the direct communication range of the RSU have opportunity to receive at least one CAM during the passing-by time period.

5.4.3 CAM Format and Data Requirements

Depending on the type of transmitting node (e.g., light vehicle, public transport vehicle, or RSU, etc.), a CAM provides:

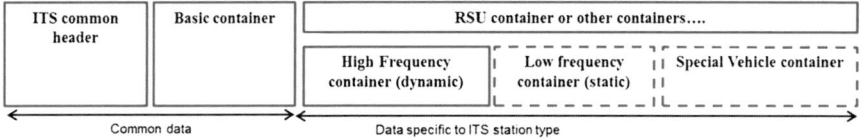

Fig. 5.6 Overview of CAM structure

- **ITS-S attributes** such as vehicle type, vehicle role (e.g. emergency vehicle),[16] vehicle size, road side equipment type information.
- **ITS-S movement status** such as position and time, moving speed, heading and the path history of the vehicle.
- **vehicle basic sensor data** including acceleration status, exterior light status, steering wheel angle, yaw rate information, vehicle moving path curvature, etc.
- **special vehicle information** including additional data for special vehicles such as light bar and siren status for emergency vehicle, roadwork type information for roadwork trailer cars, etc.

A CAM is constructed with a set of data containers, either *mandatory* or *optional*. Mandatory container mainly contains highly dynamic data, i.e. vehicle movement status and vehicle basic sensor data. Mandatory container is present in all CAMs transmitted by an ITS-S. Optional containers contain low dynamic data of the vehicle or data that is available under specific conditions when vehicle is operating in a specific role like roadwork vehicle, emergency vehicle, public transport vehicle, etc. Figure 5.6 provides an overview of the CAM format as defined in [9]. The main purpose of container-based CAM format design is to enable the flexibility of the message structure for future extensions of the message content, when new application needs are identified. Optional containers reduce unnecessary transmission of the low dynamic data at high frequency, in order to reduce the message size for ITS G5 communications.

The Abstract Syntax Notation One (ASN.1) unaligned Packed Encoding Rules (PER) are used for CAM encoding and decoding, in order to optimize the message size for ITS G5 technology.

One main performance indicator for CAM data quality is sensor data freshness. Data freshness indicates up-to-date level of the provided data, essential for the estimation of collision risk. Each CAM is timestamped according to the position data. Other sensor data included in CAM should ensure that the data age with regard to the timestamp is less than a certain predefined threshold. In V2V road safety application requirement standards defined in ETSI TC ITS [11], the data age of high dynamic sensor data is used to classify the OBU performance into two categories (class A and class B system). A threshold data age of 150 ms is used

---

[16]The same vehicle type may play different roles in different situations, depending on local regulations and the assigned authorization.

for this classification, denoted as the time interval between time at which a sensor data is available and the time at which a CAM is timestamped. This threshold is derived from the assumption that an overall end-to-end time latency of 300 ms is required for CAM dissemination, so that a receiving vehicle is able to perceive the vehicle dynamics or any potential sudden maneuvering in time for the realization of a collision risk warning application. Otherwise, an awareness information should be provided to driver, when applicable.

In addition to the data age, another important performance indicator is the position and time accuracy. Position and time accuracy determines to which level the ego vehicle may correctly estimate its relative position with regard to the transmitting vehicle (lane alignment or road alignment). Depending on the application requirements, the position accuracy requirement may vary. Typically, for an application that relies on CAM to estimate the longitudinal collision risk, the position accuracy requirement is set to 1 m, so that the ego vehicle can judge if it is located in the same lane as the transmitting vehicle [12]. Otherwise, for applications that provide awareness information to driver, the position accuracy requirement can be relaxed to 10–15 m.

5.4.4 CAM Security

The security mechanism for CAM dissemination considers the authentication of messages transferred between ITS-Ss with certificates. The CAM signing and verification are processed at lower layer *GeoNeworking/BTP* protocol stack. The objective of this mechanism is to provide authentication service not only to the CAM message payload, but also to the GeoNetworking packet headers, which contain also security sensitive data such as node position information. A CAM security profile is standardized in [13].

Figure 5.7 presents an overall message frame for a secure CAM, when the *GeoNeworking/BTP* protocol stack is used for CAM dissemination. A CAM is delivered to or received at the *GeoNeworking/BTP* stack, which will send a request to the security functionalities for signing and verification, together with the GeoNetworking headers and CAM data. Format and content of the GeoNetworking headers are specified in [8].

A CAM certificate indicates the permission of the transmitting ITS-S using the parameter Service Specific Permissions (SSP). CAM SSP is defined based on CAM content, in particular the vehicle role and vehicle permission relevant data. More specifically, a CAM certificate indicates if an ITS-S is entitled to transmit a CAM

MAC header	LLC header	GeoNetworking Basic header	Secure packet (Geonetworking common header, extended header and message)

Fig. 5.7 Message frame for a secure CAM

5 Message Sets for Vehicular Communications   141

with a specific role setting such as emergency vehicle, or with a specific permission to override some traffic regulations such as requiring the traffic light preemption in an emergency situation. An incoming signed CAM is accepted by the receiving ITS-S if the CAM content is consistent with the SSP in its certificate.

The SSP indicates if the transmitting ITS-S is compliant to certain rules defined by a deployment stakeholder such as industrial consortium, to a compliance assessment standard, or to the local law or regulations. The SSP assignment procedure will be ensured by Public Key Infrastructure (PKI) in a real deployment.

5.5 Decentralized Environmental Notification Message

The generation, transmission and management of DENM is realized by the Decentralized Environmental Notification (DEN) basic service component. The DEN basic service has also been initially introduced by European C2C-CC [2]. In complementary to CAM, the DEN basic service is considered as another core message required for the vehicular communication system deployment since day one in the EU. At the time of writing the present chapter, an EN document developed by ETSI TC ITS [10] has been submitted to National Vote.

5.5.1 The DEN Basic Service Overview

The DEN basic service is an entity that supports the exchange of event-driven DENM in vehicular communication networks. A DENM contains information related to a road traffic event, e.g., traffic jam, break down vehicle, roadworks, etc. The DENM transmission is triggered by an application, upon the detection of an event. DENM may be transmitted with high frequency (1–10 Hz) to other vehicles or road users within a predefined geographical area. In [10], DENM is defined as follows:

> A DENM contains information related to an event that has potential impact on road safety or traffic condition. An event is characterized by an event type, an event position, a detection time and a time duration. These attributes may change over space and over time... The DENM protocol is designed to manage the event detection, event evolution and event termination

Figure 5.8 provides an example scenario of the DENM transmission to announce a roadwork event. DENMs may be transmitted by a road work vehicle equipped with a vehicle ITS-S or by an RSU upon the request of a road operator. Currently in [10], a variety of event types are specified, ranging from events detected by vehicles with in-vehicle sensors such as electronic brake lights warning, collision risk warning, to events related to the traffic and driving environment such as roadworks, extreme weather conditions, etc.

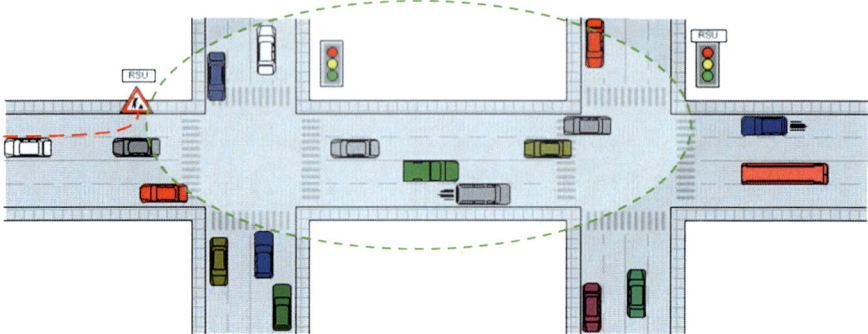

Fig. 5.8 Example of DENM transmission

Compared to CAM which is an application independent protocol, the DENM transmission is controlled by applications. The standardization of the DEN basic service is therefore scoped by applications selected for initial deployment, while keeping in mind the potential extension needs in the future.

The DEN basic service is mainly used for road safety applications. Its dissemination area may range from several hundreds of meters to several kilometers, depending on the application requirements. A node receiving a DENM may forward it to further distance, in order that vehicles located within the dissemination area may also receive the event information. If the received event information is considered relevant, a warning or information may be provided to driver, who takes appropriate actions to bypass the event position safely.

5.5.2 DENM Dissemination and DENM Protocol

In Europe, ITS G5 technology is considered as one of the candidate technologies for DENM dissemination in support of road safety applications. In example illustrated in Fig. 5.8, RSU roadworks warning application requires DENMs to be broadcasted to all vehicles located within an area of relevance (or area of destination) located in the upstream traffic. In complement, other communication technologies such as cellular network may also be used.

5.5.2.1 DENM Dissemination

The *GeoBroadcast* protocol of the *GeoNetworking* functionality [8] may be used for DENM dissemination. It supports multi-hop packet forwarding functionalities, in order to route a DENM packet from the source to the defined geographical destination area. In case ITS G5 is used, DENM is broadcasted over CCH for the

1st hop. Service Channel (SCH) may be used for broadcast from the following hops until reaching the destination area. In case the detected event requires immediate actions to avoid potential collisions, the *Traffic Class* for DENM may be set to the highest value, in order to guarantee a prioritized access to the radio resource.

Alternatively, the DEN basic service also includes a forwarding mechanism at the facilities layer, in complementary to the network layer routing protocols. The main objective is to enable forwarding of the most updated DENM among multiple received ones of the same event. This forwarding mechanism may be useful in situation where the event is highly dynamic and requires continuous update of the DENM content (e.g., a moving emergency vehicle event). Another useful situation is when an event covers an area and/or persists during some time, more than one passing vehicle ITS-Ss may detect this event at different positions and times (e.g. an extreme weather condition event).

It should be noted that the DEN basic service standard does not specify requirements on the conditions under which the DENM transmission is triggered or terminated. These conditions are defined per application as application requirements. For example, automobile stakeholders at C2C-CC and road operators in Amsterdam Group define a set of *triggering condition* documents that specify DENM transmission triggering and termination conditions for different events.

5.5.2.2 DENM Protocol

In [10], the main technical features of the DENM protocol are related to the management of DENM transmission during different phases of the event evolution. For this purpose, multiple types of DENMs are defined:

- **new DENM** refers to a DENM generated by an ITS-S that detects an event for the first time. A new DENM contains event information, such as event position, event type, and optionally an estimated (or pre-set) validity duration of the event.
- **update DENM** refers to a DENM generated when an ITS-S detects an update of the event, such as the event position change.
- **cancellation DENM** refers to a DENM generated by the same ITS-S that has generated the corresponding new DENM, when this ITS-S detects that the event has ended before the originally set validity duration.
- **negation DENM** is generated by an ITS-S to announce the event termination. This ITS-S did not generate new DENM for this event but has received one from another ITS-S some time ago before arriving to the event position. When this receiving ITS-S arrives the event position and detects that the event has terminated before the expiration of the received event validity duration, it may generate a negation DENM according to a set of triggering conditions. A negation DENM differs from a cancellation DENM, in the sense that the negation DENM provides a possibility for the event termination from an ITS-S other than the one that has originally detected the event for the first time, referred as third-part termination. In situation where the ITS-S that has generated the new DENM has

lost the capability to transmit a cancellation DENM (e.g., OBU fail in accident) or has moved away from the event position therefore can not detect the event termination by itself, a negation DENM may be useful. Even though the feature is enabled by the standard, there remains research topics for third-part event termination, in particular how to ensure the liability of the information and how to avoid the misuse of the negation DENM.

The type of DENM to be generated is determined according to the type of application request when the event is newly detected, updated, or terminated. The DENM protocol operation is realized using several parameters:

- *actionID* is composed of the station ID of the detecting ITS-S and a sequence number. The concept of the *actionID* is introduced as the event identifier. An *actionID* enables a receiving ITS-S to distinguish an event detected by different ITS-Ss, or different events detected by the same ITS-S.
- *referenceTime* is the parameter that enables the distinction of different DENM updates about one event.
- *termination* allows the receiving ITS-S to derive the DENM type. If present in DENM, it includes two values i.e., cancellation DENM or negation DENM.
- *validityDuration* parameter indicates the end of a DENM validity. It may be used to indicate an estimated or preset duration of the event persistence, in case such duration is known in advance. This parameter may not be present in a DENM, in case the detecting ITS-S is not able to provide the event duration information. In this case, a default value is set by ITS-S for internal protocol operation.
- *repetitionDuration* and *repetitionInterval* are parameters to control the DENM repetition. In case DENM includes event information that is static, a DENM repetition may be triggered, to transmit DENMs to oncoming vehicles entering the destination area. These parameters are used for protocol operation at transmitting ITS-S, therefore not included in a DENM.
- *transmissionInterval* is present in DENM when facilities layer forwarding is activated. It indicates the time interval of DENM transmission at the originating ITS-S.

The DEN basic service processes received DENMs using these parameters, then it provides up-to-date event information to applications or redelivers the DENM to the ITS networking and transport layer for forwarding. In one possible forwarding protocol as introduced in [10], a receiving ITS-S may forward the most up-to-date DENM of a specific *actionID*, if it does not receive any repeated or updated DENM from other ITS-Ss within a time period (e.g., three times of *transmissionInterval*) and is still located inside the DENM destination area when this time period is expired.

5 Message Sets for Vehicular Communications

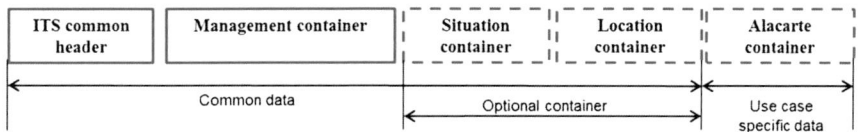

Fig. 5.9 Overview of DENM structure

5.5.3 DENM Format and Data Requirements

Figure 5.9 provides an overview of the DENM format as defined in [10]. Similar to CAM, DENM is structured with data containers. Each container is extensible to support potential extension of the DENM content for future ITS application needs.

- **ITS PDU header** in a common header for all VANET message types. It includes the protocol version, message ID and station ID.
- **Management container** contains information for the DENM protocol operation. This container is mandatory and shall be present in all transmitted DENM. The receiving ITS-S relies on the content of this container for protocol operation.
- **Situation container** contains event type information and an indicator of the event detection performance. Each event type is identified with an integer type event code. This list of event codes is extensible.
- **Location container** contains information that describes the *location referencing* information at the event position. The *location referencing* information for DENM is a list of *traces*. Each *trace* is composed of a list of *waypoints* that construct a path approaching to the event position. This *location referencing* information enables receivers to estimate its relevance to the event, by comparing its own itinerary path to each *trace* contained in the received DENM. In addition, the location container may also include information that represents the detection history of a plain event (e.g., an extreme weather condition event), if the same event was detected by a moving vehicle along its travel path in the past.
- **Alacarte container** contains optional data specific to an ITS application. For example, in a pre-crash application, an *alacarte* container is defined to enable exchange of detailed vehicle size, shape, and passenger presence information for collision mitigation purpose.

The ASN.1 unaligned PER encoding rules are used for DENM encoding and decoding, in order to optimize the message size.

5.5.4 DENM Security

Similar to CAM, the security mechanism for DENM considers the authentication of messages transferred between ITS-Ss with certificates. The DENM signing and verification are realized by lower layer, i.e. *GeoNeworking/BTP* protocol stack. A DENM security profile is standardized in [13].

DENM SSP is defined based on event type code included in the DENM situation container. It indicates if an ITS-S is entitled to transmit a DENM with a specific event code. This permission implies that the ITS-S is compliant to certain event detection and data quality requirements defined by Industrial Consortium such as C2C-CC, to a compliance assessment standard, or to the local laws or regulations.

5.6 Basic Safety Message

An overview of the full set of standards underway to enable DSRC deployment in the USA can be found in [21]. A brief description of the protocols germane to generation message sets are those developed, and in some cases still under development, and described in that thorough overview. However, in keeping with the safety focus of the initial C-ITS deployment in the USA, the primary discussion of this section will detail the definition and use of the BSM. The BSM is a broadcast message with data elements that provide kinematic and other state information sufficient to develop V2V safety-of-life applications.

Despite this dominant set of applications, it is important to first conceptually understand the standardization work in defining the middle layer to enable the host of eventual C-ITS applications, as the initial implementation of V2V safety services portends those other mobility and environmental services that will be delivered with the V2V and V2I market penetration. This middle layer is developed by standards produced by the IEEE P1609 Working Group.[17] Specifically, IEEE 1609.12 [16] defines the Provider Service Identification (PSID), IEEE 1609.3 [19] defines networking services, and IEEE 1609.4 [18] defines multichannel operations. For completeness, note also that IEEE 1609.0 defines the WAVE architecture enabling the use of IEEE 802.11p WAVE, and IEEE 1609.2 [17] defines security services.

This short operationally oriented description of the so-called middle layer begins by considering that IEEE 1609.3 defines the WAVE Short Message (WSM) to contain three extensions: channel number, data rate, and transmit power. These parameters enable higher layer to indicate the communication requirements for a message dissemination. It also contains a PSID, WSM length, followed by a higher layer message payload, e.g. BSM. The Wave Short Message Protocol (WSMP) is a networking protocol that delivers a WSM, when requested by higher layer, to required destination nodes. In case for BSM dissemination, the WSMP transmits the corresponding WSM to direct network neighbors (one hop broadcast).

---

[17]IEEE 1609 Working group develops DSRC standards for Wireless Access for Vehicular Environments (WAVE), including V2V and V2I communications. More details can be found at http://standards.ieee.org/.

At receiving side, the parameter EtherType is used to distinguish WSMP stack from IP stack, and PSID is used to deliver the received WSM to corresponding higher layer entities. A PSID value is assigned to each message, e.g. BSM, based on a well-defined registration procedure as specified in IEEE 1609.12 [16]. In addition, the WSMP protocol specified in [19] includes a management plane protocol WAVE Service Advertisement (WSA). A WSA is sent in CCH channel to indicate whether a delivered C-ITS service (i.e., denoted as PSID and other service context information) is through IPv6 or the WSMP stack. A WSA contains well-defined fields that indicates repeat rate, transmit power, location and confidence, among other values. These parameters enable an ITS-S receiving WSM to properly configure the system to access to the announced service, if interested. It is again worth noting that the associated standards are comprehensively described in [21]. At this writing there is considerable activity within IEEE P1609 to refine the WSMP and the security protocol, IEEE 1609.2, so the community should anticipate some changes in the detail of the standard. The current work is very focused on maturing the IEEE 1609 set of standards to accommodate a potential V2V rulemaking or mandate.

Given the impending mandate and associated V2V deployment, the set of BSM content and performance standards drive the key C-ITS standardization activities in the USA. The content or data dictionary is given in SAE J2735 [25] and described in Tables 5.4 and 5.5; the performance standards work is in progress at SAE J2945 [26]. In light of the impending V2V mandate in the United States, both are presently undergoing scrutiny and revision within the SAE DSRC Technical Committee. To set the context, consider that the SAE J2735 defines nearly 150 data elements, organized into data frames. These are fifteen explicit standard message sets in the standard, as illustrated in Table 5.3.

Table 5.3 VANET message sets specified in SAE J2735

No.	Standard message name
1	A La Carte (ACM)
2	Basic Safety Message (BSM)
3	Common Safety Request (CSR)
4	Emergency Vehicle Alert (EVA)
5	Intersection Collision Avoidance (ICA)
6	MAP
7	NMEA (GPS) Corrections (NMEA)
8	Probe Data Management (PDM)
9	Probe Vehicle Data (PVD)
10	Road Side Alert (RSA)
11	RTCM Corrections (RTCM)
12	SPAT
13	Signal Request Message (SRM)
14	Signal Status Message (SSM)
15	Traveler Information Message (TIM)

While the ACM (Standard Message 1) is purposely designed to be flexible, there is by design significant flexibility within SAE J2735. The re-use of modular data elements and data frames is anticipated in the standard. Message sets can be user-defined to address an assortment of C-ITS applications from a wide variety of data elements. The rather extensive SAE J2735 (2009) [25] is therefore designed to be comprehensive. It is analogous to the combination of CAM, DENM, and IVI messages used in Europe. Since the original creation of SAE J2735, US DOT commissioned a contracting team to undertake a requirements-driven system-engineered process wherein a set of primarily public sector stakeholders was engaged to derive user needs, and where existing and prospective concepts of operations were extracted from the ongoing Dynamic Mobility Applications Program.[18] This work was documented in what has become the SAE J3067 Information Report [27]. From the list of 15 SAE J2735 standard messages, the SPAT (Standard Message 12) and MAP (Standard Message 6) are close matches to the EU SPAT and MAP messages. In order to achieve this harmonization, the SAE versions of the SPAT and MAP are under revisions, with a final SAE J2735 step in final ballot stages at this writing. This activity will enable SPAT and MAP from Europe (and Japan) to be additionally described within SAE J2735 with regional extensions. This harmonization is fostered under ISO/TC204 Working Group 18 (Cooperative ITS) development of Technical Standard (TS) 19091, which addresses the dynamic messages to enable harmonized mobility and safety applications, particularly for public sector transit and traffic operations. With the SPAT and MAP revisions within TS 19091 essentially complete, the SRM (Standard Message 13) and SSM (Standard Message 14) are the next targets for potential ISO harmonization. Similar to the EU messages, all the messages specified in J2735 except the BSM use ASN.1. Outside the separate DSRC Message ID, the BSM has only one encoding tag and the message, a BSM blob is fixed in length and order. This saves message size and for good reason: the most pertinent message for V2V or initial deployment in North America is the BSM, as the BSM Part 1 is a mandatory representation of the vehicle state, consisting of its kinematic state and other pertinent information. It is defined by the data elements enumerated in Table 5.4. In North America Safety Pilot Model Deployment tests, BSM transmission rate is fixed to 10 Hz. This transmission rate may be reduced in case channel load is too high.

The BSM Part II is optional, need and broadcast frequency in V2V safety will likely be lower. BSM Part II data elements and broadcast frequencies are listed in Table 5.5. If no frequency is specified for a certain data element, it will be transmitted when status changes are detected.

The BSM Part II data elements necessary to implement V2V safety are at this writing not normative. Event flags (consisting of application designer-selected values from Element Numbers 1 to 13 above), path history (Element Number 14), path prediction (Element 14), and RTCM correction (Element 16) have been

---

[18]Dynamic Mobility Applications Program is a US DOT initiative, see http://www.its.dot.gov/dma/.

Table 5.4 BSM Part I content

No.	BSM Part I data element
1	DSRC message ID
2	Message count
3	Temporary ID
4	Current time, 1 ms resolution
5	Lat/Long, resolution 0.1 μdeg
6	Elevation from sea level, 0.1 m
7	Position accuracy, 1 standard deviation per major and minor axes
8	Transmission (gear) and Vehicle speed, 1 cm s^{-1}
9	Heading, 1/80 deg
10	Steering wheel angle, 1.5 deg
11	Acceleration (three axes and yaw rate)
12	Braking state (control, boost, auxiliary) for each of four wheels
13	Vehicle size (length and width), 1 cm
14	Path history (sequence of position vectors for recent past)
15	Path prediction (radius of curvature)
16	Differential GPS corrections
17	Lights status (headlights, running lights, hazard lights, turn signals)
18	Light bar status (for emergency responders, school buses, special vehicles)
19	Front wiper status (on, off, intermittent)
20	Front wiper rate (sweeps per minute)
21	Rear wiper status (on, off, intermittent)
22	Rear wiper rate (sweeps per minute)
23	Braking status (brake applied, ABS, stability, traction control auxiliary, boost systems active)

used in an optional vehicle safety extension frame field tested within the Safety Pilot Model Deployment. Changes to the above may be considered in future revisions of the applicable SAE standards in support of the potential NHTSA V2V rulemaking. As an example, the safety extension frame might be transformed to a mandatory element. Moreover, the earlier-referenced SAE J2945.1 DSRC Vehicle BSM Communication Minimum Performance Requirements may rigorously specify the BSM Part I and Part II data elements based on a flowdown of the overall V2V safety performance and the data from the Safety Pilot Model Deployment. Sensor accuracy will be considered, along with the necessary BSM sending rate and transmit power, the latter two contingent on channel congestion issue currently under study by NHTSA and their CAMP partners.

Table 5.5 BSM Part II content

No.	BSM Part II data element	Frequency
1	Hazard lights active	N/A
2	Vehicle expected to violate stop bar	N/A
3	Antilock brake system active over 100 ms	Sec.
4	Traction control system active over 100 ms	Sec.
5	Stability control system active over 100 ms	Sec.
6	Vehicle placard as HazMat carrier	N/A
7	Public safety vehicle responding to an emergency	N/A
8	Recent or current hard braking (> 0.4g)	0.1 s
9	Light status changed	N/A
10	Wiper status changed	All
11	Flat tire	N/A
12	Vehicle is disabled	N/A
13	Airbag has deployed	N/A
14	Path history (sequence of position vectors for recent past)	N/A
15	Path prediction (radius of curvature)	0.1 s.
16	Differential GPS corrections (RTCM)	N/A
17	Lights status (headlights, running lights, hazard lights, turn signals)	Sec.
18	Light bar status (for emergency responders, school buses, special vehicles)	N/A
19	Front wiper status (on, off, intermittent)	Min.
20	Front wiper rate (sweeps per minute)	Min.
21	Rear wiper status (on, off, intermittent)	Min.
22	Rear wiper rate (sweeps per minute)	Min.
23	Braking status (brake applied, ABS, stability, traction control auxiliary, boost systems active)	Sec.
24	Level of brake application	Sec.
25	Road coefficient of friction	Sec.
26	Sunlight level	Min.
27	Rain type	Min.
28	Ambient air temperature	Min.
29	Ambient air barometric pressure	Min.
30	Confidence- steering wheel angle	N/A
31	Confidence- steering wheel rate of change	N/A
32	Front wheel angle	0.1 s
33	Vertical acceleration over threshold	N/A
34	Confidence- yaw rate	N/A
35	Confidence- acceleration	N/A
36	Confidence- set of values	N/A
37	Distance to obstacle on the road	N/A
38	Azimuth to obstacle on the road	N/A
39	Date/Time of obstacle detection	N/A
40	Confidence- time	N/A

5 Message Sets for Vehicular Communications   151

Table 5.5 (continued)

No.	BSM Part II data element	Frequency
41	Confidence- position	N/A
42	Confidence- speed/heading/throttle	N/A
43	Throttle position (percent)	0.1 s
44	Vehicle height	N/A
45	Bumper heights	N/A
46	Vehicle mass	Trip start
47	Trailer weight	Trip start
48	Vehicle type	Trip start
49	Descriptive vehicle identifier	Trip start
50	Vehicle identification number (VIN)	Trip start
51	Fleet owner code	Trip start
52	Vehicle group	N/A
53	Responder group	N/A
54	Incident responder equipment group	N/A
55	J1939-71 Tire conditions	N/A
56	J1939-71 Axle location/weight	N/A
57	J1939-71 Cargo weight	Trip start
58	J1939-71 Steering/drive axle condition Info	N/A
59	Daily solar radiation (solar energy detected over previous 24 h)	Min.
60	GPS status/quality	Sec.
–	*Note*: J1939-71 is a SAE standard for conventions and notations of vehicle application layer	–

5.7 In-Vehicle Information

IVI is a message that enables the transmission of road side sign information to road users for in-vehicle presentation. It provides static sign or dynamic sign (e.g., Variable Message Sign (VMS)) data from infrastructure to vehicles or to mobile devices. The receiving ITS-S processes the received IVI data and estimates the relevance of the information to the driver. When appropriate, the information is delivered to driver as warning or as information. If required by the local regulation, presentation format of a signage, e.g. layout, font size, color, etc. may also be included in IVI for the in-vehicle presentation.

IVI is one of the infrastructure messages that is selected by stakeholders in the EU for day one deployment. The standardization of the IVI is undertaken by ISO TC 204, taking into account the feedbacks from Industrial Consortium such as C2C-CC, Amsterdam Group and FOT projects. Multiple technologies can be appropriate candidates for IVI dissemination to satisfy requirements like covered area size, availability of the communication infrastructure, communication cost, etc.

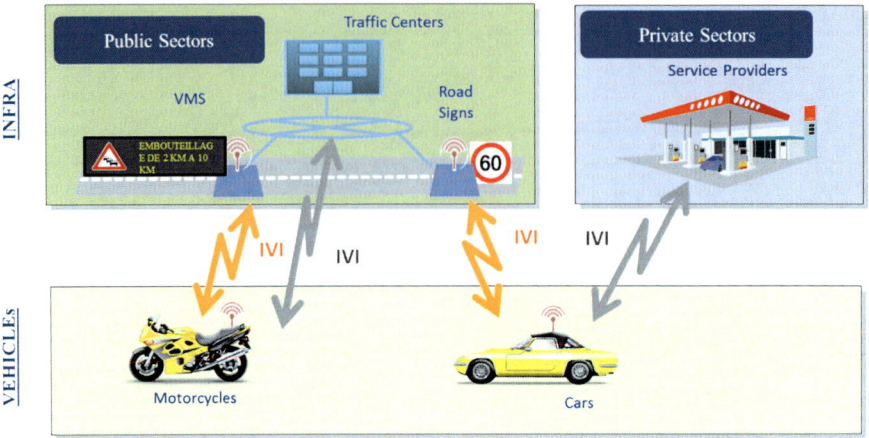

Fig. 5.10 Example scenario of IVI application

5.7.1 IVI Application

The IVI application consists of providing authenticated road sign information from road side to vehicles and to road users. However, the information chain may go beyond the transmission of messages itself. A series of back-end actions in road ITS infrastructure are required before hand, from the data collection, data processing, to data generation and authorization/authentication procedures. The IVI message communicates the results of previous back-end actions in a specific message format with corresponding syntax and semantics definitions, enabling the receiving ITS-S to present the information in a proper manner and timing. The IVI application is one representative C-ITS application to further facilitate the vehicle and infrastructure integration.

Figure 5.10 provides an example scenario of the IVI application. The road signage information, e.g. VMS, speed limit sign, fuel station sign, etc., may be transmitted by RSU, by TMCs, or by private service providers. Standardization of IVI application is ongoing in ISO TC 204 ISO TC 204 Technical Specification 17425.

5.7.2 IVI Message Overview

The structure of an IVI message is derived from DENM. This is motivated by the similarity between the two, in terms of data exchange needs and content requirements. For example, a dynamic road side sign information is also characterized by a sign type, duration, position, and a relevance area, comparable to an event as indicated by a DENM. Generally speaking, an IVI contains information

5 Message Sets for Vehicular Communications

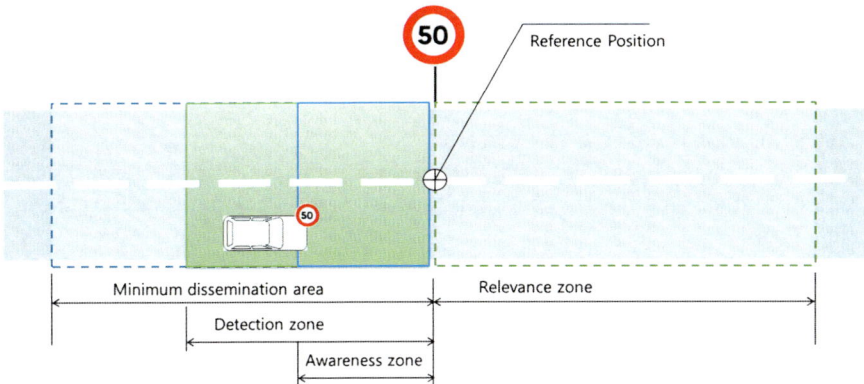

Fig. 5.11 Overview of IVI geographical validity

authorized and authenticated by public authorities or by road operators. Compared to a distributed ITS applications where the relevance check is completely at the shoulder of the receiving side, road operators or public authorities may have the control on the sign systems on the road and set requirements on the relevant road sections that a sign information should be informed to road users. This requirement is reflected in the IVI message structure, where the location container is extended with enriched content. Nevertheless, the receiving ITS-S remains the decision maker on the final presentation of a sign to driver or to road user, based on the overall in-vehicle information processing load and the HMI system design.

Typically, an IVI is transmitted from an RSU which is in connection with a data center (e.g., a TMC) providing the IVI content. The IVI message is under standardization in ISO TC 204 Technical Specification 19321.

Figure 5.11 provides an overview of the area concepts being relevant to the IVI application, described as follows:

- **Minimum Dissemination Area** defines the area of the IVI message dissemination. It is provided to lower layer stacks, for example, the *BTP/GeoNetworking* stack for routing.
- **Detection Zone** gives indication of an area in which the received IVI message should be processed by an ITS-S. Vehicles entering this zone would probably passing by the road sign in the short future and the information of IVI may be relevant for in-vehicle presentation.
- **Driver Awareness Zone** is typically determined by receiving ITS-S, taking into account a minimum time at which an IVI data is shown to driver before entering the IVI relevance zone, based on its motion status, e.g. driving speed, driving itinerary. Alternatively, it may also be included in an IVI message and transmitted from road side. In this case, it can be seen as an indication from IVI information providers (e.g., a TMC operator) when the IVI should be informed to drivers. However, the final decision on whether and when to show in vehicle resides on

vehicle side. For example, in case when vehicle encounters dangerous situations near the sign, the OBU application may ignore or delay the IVI presentation in order not to avoid further increasing driver's work load. Typically, a Driver Awareness Zone is a sub-set of the detection zone.

- **Relevance Zone** defines the zone of relevance for a sign, for example as shown in Fig. 5.11 for Speed Limit Sign, the relevance zone is located in the downstream traffic from the sign position during a certain distance, e.g. until the next speed limit sign.
- **Reference Position** is the starting point for the definition of all different zones. By definition, the reference position point is covered by all zones of the IVI. In the example shown in Fig. 5.11, the reference position is in the middle of the road surface of speed limit sign position. However, the reference position does not necessarily always correspond to the position where a sign is physically installed. Actually, road signs are often installed a certain distance prior to the position where the sign information becomes effective. For these signs, the reference position should therefore refer to the sign effective position.

Each IVI message includes at least one reference position and one area. One IVI message may be linked to other IVI messages, in case relevance (or overriding) of multiple signs occurs. For example, for a heavy vehicle, a speed limit sign for heavy vehicles should override the speed limit sign in case of bad weather, the lowest speed limit should apply for in-vehicle presentation.

Similar to CAM and DENM, IVI standard does not specify the receiving side protocol of the message processing. This is left at the discretion of implementors.

5.7.3 IVI Format

The container-based structure of a IVI message is illustrated in Fig. 5.12. Each IVI contains at least one management container and optionally one or more location containers or one or more application containers.

The management container contains management information for an IVI message. The IVI protocol also supports similar life cycle management functions like the DENM protocol, including IVI trigger, IVI update, IVI cancellation, and IVI

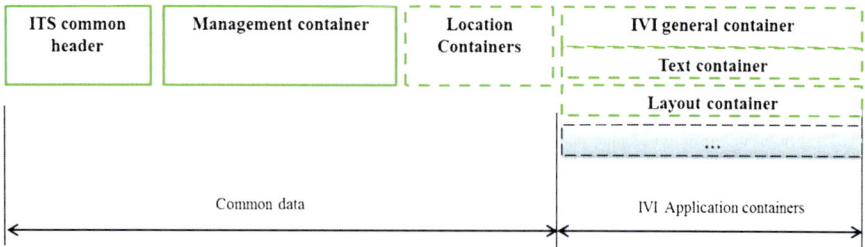

Fig. 5.12 Overview of IVI structure

negation. This life cycle management is adopted to mainly support the dynamic sign information management. In addition, the management container may include an identifier of the provider authorities, or identifier of other relevant IVI messages.

The location containers contain one or more reference position and description information for one or more zones. A zone may be described using the combination of geographical points and distance attributes starting from the reference position, forming a circular (combination of reference position and a radius distance) or a polygon shape. Alternatively, the location referencing may also include a road segment ID of a map database or other relevant road topology information.

The application containers contain IVI application data. Currently three types of application containers are defined, namely the IVI general container, IVI text container, and IVI layout container:

- **IVI general container** provides information on the content and type of a sign, denoted as road sign code. Each sign is linked with one or more zones described in the location containers, either as detection zone, driver awareness zone or relevance zone. The road sign codes are inherited from a road sign catalog such as defined in Vienna Convention for EU road signs [31]. Other catalogs may apply such as ISO14823 codes [20], for usage in other regions or according to local regulations. In Vienna Convention catalog, a road side code includes typically a class, an integer code, a value with unit, some text and additional information. For example, the speed limit sign is assigned as code C.2, in addition to a value that indicates the speed limit value with a unit, e.g. 70 km/h. A primary sign may be associated with secondary sign, which limits the application conditions of the primary sign. Optionally, an IVI general container may also be linked to a specific layout container, in case a certain layout is required to be respected for in-vehicle presentation.
- **Text container** provides possibility to include free text information to road users as it may be the case for VMS text in some European Regions. Like IVI general container, it should be linked to one or more zones, be the relevance zone, driver awareness zone or a detection zone.
- **Layout container** defines a set of layout information for road sign. A layout container may be linked by an IVI general container or a text container. At receiving the IVI, the in-vehicle system may present the information to driver as defined in layout container. This is to enable an identical road sign layout in vehicle as at road side.

5.7.4 IVI Security

The current draft IVI standard defines the IVI SSP based on the type of applications that it may support, including road signage information, contextual speed information, roadwork warning, restriction information, rerouting, and traveller information, etc. The SSP indicates if the transmitting ITS-S is entitled to provide the corresponding IVI content in support of the ITS application in question.

5.8 Signal Phase and Timing Message

SPAT is a message that provides the traffic light phase and timing information from road side ITS-S to vehicles or to mobile devices. One SPAT message may include traffic light status information of one or multiple intersections. The receiving ITS-S should process the received SPAT data together with the intersection topology data, in order to match each individual traffic light status data to the corresponding road segment in the intersection to which the traffic light status information is relevant. The intersection topology information may be made available by an RSU, by transmitting a road topology MAP message. Detailed description of the MAP message is provided in Sect. 5.9.

The SPAT and MAP message are standardized by SAE DSRC Technical Committee in the standard SAE J2735 [25], currently under revision to take into account regional requirements of USA, Europe, and Japan.

SPAT and MAP may be used by different ITS applications at intersection area. By knowing the light status and status switch timing before approaching to an intersection, OBU may provide speed advice and warning to driver to avoid traffic light violation or to smooth the intersection crossing.

5.8.1 SPAT Overview

Typically, SPAT is broadcasted from an RSU to vehicles located near the intersection area in question. In order to obtain the real-time traffic light phase and timing information, the RSU should interface with traffic light controller systems of the traffic light system. Depending on the deployment of the traffic light control system, a traffic light controller may be equipped locally in the intersection or in a specific traffic light control network. In urban environment, one traffic light controller may coordinately control a series of traffic lights within a road segment or within an area. Therefore, the availability and accuracy of the content in a SPAT depends directly on the data made available by the traffic light controller.

Figure 5.13 provides an example scenario of the Green Light Optimal Speed Advisory (GLOSA) application realized by SPAT and MAP messages. This application is developed and validated in European FOT project DRIVE C2X. This application provides speed advice to pass the next traffic lights during a green phase. In case it is not possible to provide a speed advice, the remaining time to green is displayed. This helps to reduce stop time and unnecessary acceleration/brake in urban traffic situations to reduce fuel consumption and pollution emissions.

The SPAT message sends the current movement state of each active phase of the traffic light, including values of what lights are active and values of phase duration if available. Movements are mapped to specific lanes and approaches by use of the lane numbers included in the message. These lane numbers correspond to the specific lanes described in the MAP message for that intersection.

5 Message Sets for Vehicular Communications   157

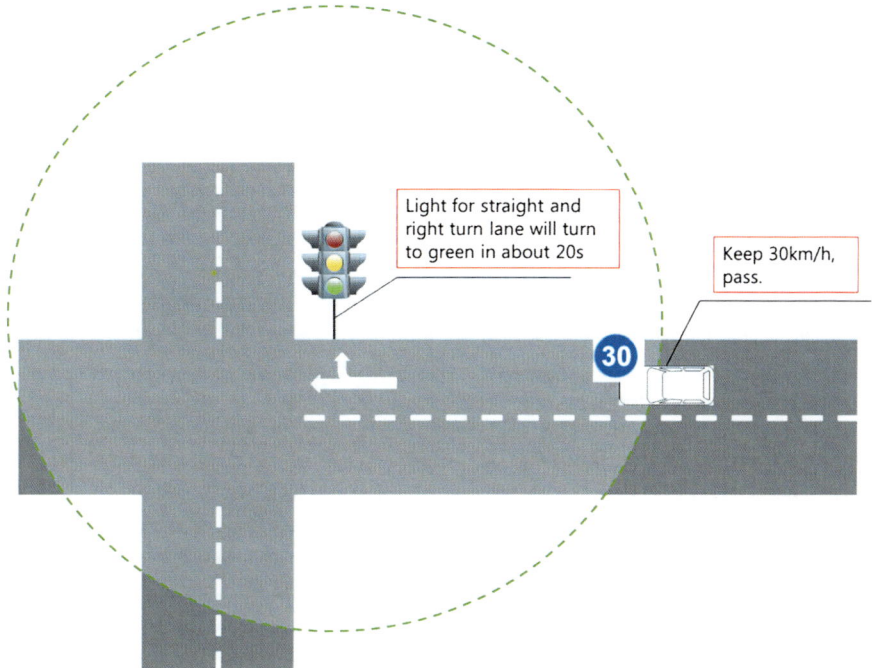

Fig. 5.13 Example scenario for GLOSA application

Fig. 5.14 SPAT message format

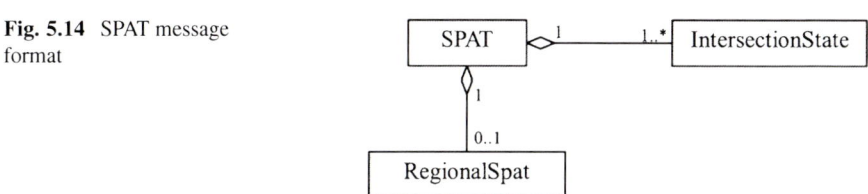

5.8.2 SPAT Format

The SPAT message format can be illustrated in UML class diagram in Fig. 5.14. A SPAT message contains traffic light status of one or more intersections, each denoted as IntersectionState data frame. Optionally, SPAT message may contain regional SPAT content extensions, to provide regional additional content to describe the traffic light status, e.g. traffic light priority or preemption information.

Each of the InstersectionState data frame may be composed of the following categories of information to describe the traffic light status of one intersection:

- **Message management data** includes general information of the message, including a message count and a message generation time.

- **Intersection data** includes general information of the intersection, including an intersection ID, an enabled lane list, and general status information. Optionally, the enabled lane list may be used to indicate the list of lanes that are active (or open to traffic) within the intersection. This list may be changed over time, e.g. a right turn lane may be enabled even when the straight traffic light is in red, or a lane is open to traffic only during some period of the day. The intersection status data indicates the traffic light controller state of the intersection in question. Depending on the local configuration of the traffic light control system, a traffic light controller may be operating in active, stand by, off or failure mode, the operation time interval maybe fixed or dynamic, the priority mode may be activated or deactivated. In addition, the intersection status may also indicate the status of the SPAT and MAP transmitting system itself e.g. if an MAP update is expected to process the SPAT, if the active lanes are updated etc.
- **Traffic light status data** provides in turn all traffic light status of the intersection, denoted as movement state, each applying to a set of lanes inside the intersection. An ID is assigned to each movement state, which will be used to match to lane descriptions of the corresponding MAP message. This ID should be made unique, at least within the intersection. In one movement event, the movement state may provide the phase type, preset or estimated phase change time, and optionally the advisory speed information.
- **Maneuvering assistance data** provides additional information to assist receiving vehicles to exit the intersection. For example, the SPAT message may include estimated queue length or presence of pedestrian information, if such information is made available by, e.g., equipped sensors.
- **Other data** for regional extensions.

5.9 Map Data

MAP is a message that provides road topology and geometry information from road side ITS-S to vehicles or to mobile devices. One MAP message may include road topology and geometry information of one or multiple intersections or one or more road segments. The receiving ITS-S may process the received MAP data together with the SPAT message of the corresponding intersection, in order to match each individual traffic light status data to the corresponding road segment in the intersection to which the traffic light status information is relevant.

Based on the standard work in SAE TC DSRC, CEN TC 278 has established a standard work item TS 19091 to profile the European SPAT/MAP message. This standardization work is ongoing.

MAP may be used by different ITS applications. Generally speaking, MAP is jointly processed with other road infrastructure message, in order to enable the matching of road infrastructure information to the local road topology. For example, MAP message may be used to describe road curve geometry, in order to support curve speed warning application.

5 Message Sets for Vehicular Communications 159

Fig. 5.15 Map message format

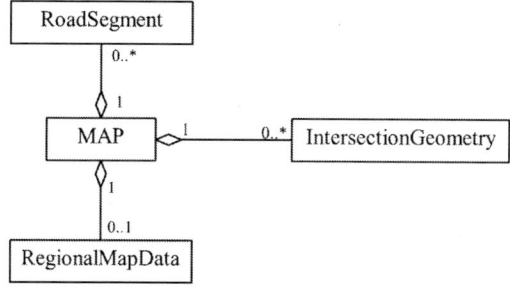

5.9.1 MAP Overview

Typically, MAP is broadcasted from an RSU to vehicles located near the road segment or intersection in question. Detailed road topology and geometry information may be obtained from different sources. Even though one may expect that MAP contains mainly static information, the content of MAP message may be updated, or re-organized, in order to support application needs.

The MAP message format can be illustrated in UML class diagram in Fig. 5.15. A MAP message may be used to describe intersection or road segment topology and geometry information. Optionally, MAP message may contain regional MAP content extensions, to provide regional additional content. Other type of road topology may be added in the future.

The content of a MAP message includes:

- **Message management data** includes general information of the message, including a message count, message ID, etc.
- **Map meta data** provides meta data of the MAP content, including road topology type, e.g. intersection, curve, parking area, etc., data source, e.g. data provision agency, version date, etc., as well as restriction information in case the road is restricted to specific users.
- **Intersection geometry** provides detailed description of one or more intersections. For one specific intersection, lanes are described one by one. Each lane shape is described with a list of attributes, including attributes of the lane itself such as an ID, lane type, lane width, lane geographic description (described using list of node coordinates), etc., as well as attributes of the traffic supported by the lane for intersection traffic clearance such as allowed traffic, allowed maneuver, egress/ingress direction type, applied speed limit, etc. In addition, MAP message also includes information on how lanes are connected (or over-cross) with each other, as well as the allowed maneuver enabled by such connection. If MAP is used to support SPAT message processing, a lane may be assigned to a group of lanes, each group should correspond to one traffic light movement state. This group ID is used in SPAT message, in order to enable the matching between SPAT data and intersection topology. Optionally, lanes may also be grouped and

indexed, in order to be used in SPAT for description of detected traffic presence, as included in *Maneuvering assistance data* in SPAT message.
- **Road segment geometry** provides detailed description of one or multiple road segments. Similar to intersection, information is provided by lanes inside the road segment in question.
- **Other data** for regional extensions.

5.9.2 Example

Figure 5.16 gives an intersection example for SPAT and MAP implementation. This example illustrates an intersection at downtown of city of Paris in France. This intersection includes motor vehicle lanes, bicycle lanes, and pedestrian lanes, with possibility to exit to an underground parking area. For the presentation purpose, not all lanes of the intersection are illustrated. In this example, red lines indicate lanes, shaped with list of node sets marked as red circles. Each lane may be described with attributes such as lane type, lane direction (ingress, egress lane), etc. It should be noted that an egress lane may be the ingress lane of the neighbor intersection, as it is the case for three egress lanes in the left bottom of Fig. 5.16. Therefore, for optimization purpose, one may include the description of only ingress lanes for one intersection, in case multiple neighbor intersections are described. In this case, a lane is identified by the lane ID together with an ID of the intersection, these IDs should be assigned in a way that a lane can be uniquely identified for the application usage. In Fig. 5.16, blue lines indicate the connection between lanes. In connections that are controlled by signals, the signal group as well as allowed maneuvers are provided. In this example, a right turn light is attached, allowing vehicles to perform right turn maneuver from est to north. Therefore, this connection belongs to a signal group different from the one for straight and left turn maneuvers, as illustrated in the figure. Not all lanes include connections in one intersection, for example in the case of median lanes.

5.10 Conclusion

The VANET message sets enable exchanges of application data between vehicles as well as between vehicles and infrastructures. Multiple road safety, traffic efficiency, or added value applications may be realized. In order to enable the C-ITS deployment, standardization of the message sets is essential to ensure the communication interoperability. Key stakeholders in public and private sectors are collaborating with each other during the standardization, validation, and promotion of the C-ITS technologies in Europe and in North America.

The present chapter describes the core VANET messages standardized in Europe and in North America. Among many applications that may be enabled by VANET,

5 Message Sets for Vehicular Communications

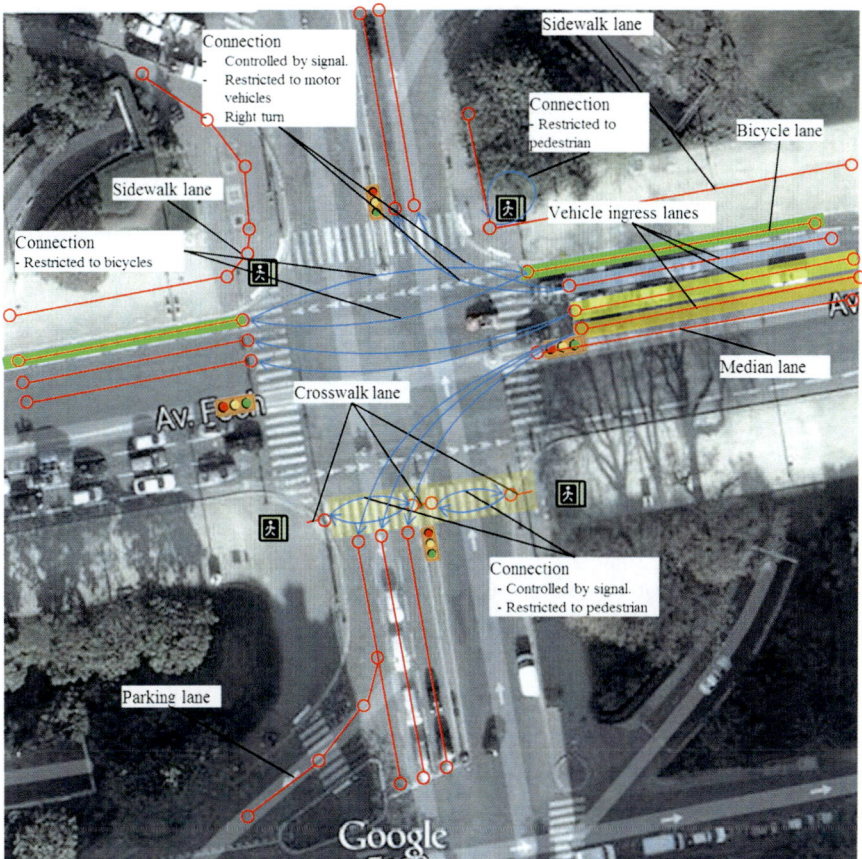

Fig. 5.16 SpatMapExample

the road safety applications are the most demanding ones in terms of communication latency, data quality, and reliability requirements. These requirements guide the standard development process in terms of message format definition and communication protocol design. In addition, message sets specifications should also take into account requirements such as message size optimization, future extensibility, and technology agnostic requirements. ITS G5 and other technologies like cellular ones are candidate technologies to realize C-ITS applications. In the EU, industrial consortium like C2C-CC together with road operator associations has defined deployment roadmap for C-ITS underlining VANET message sets transmission over ITS G5 technology for the day one deployment, ensuring the communication interoperability. In North America, an ANPRM for DSRC device in light vehicles is initiated by US DOT, with active participation of industrial partners. In addition, usage of other technologies as complementary solution may be at the discretion of implementation choice.

In parallel to the standardization, the core message set has been tested and validated within multiple FOT projects and standards testing events, which provide valuable feedbacks to Standard Development Organizations to correct the specifications errors and improve the standard quality. At international level, collaboration efforts are established, promoted by industrial consortium and by government agencies, to enable message sets harmonization. At the time of writing the present chapter, the standard development of the first message sets is close to its finalization. Future releases of the message set specifications are under planning to support new technical features.

References

1. Amsterdam Group (2013) Roadmap between automotive industry and infrastructure organisations on initial deployment of cooperative ITS in Europe
2. C2C-CC (2007) Car2Car communication consortium manifesto
3. C2C-CC (2011) Memorandum of understanding for OEMs within the CAR 2 CAR communication consortium on deployment strategy for cooperative ITS in Europe
4. ETSI (2009) Intelligent Transport Systems (ITS); Vehicular Communications; Basic Set of Applications; Definitions
5. ETSI (2010) ETSI EN 302 665 (V1.1.1) - Intelligent Transport Systems (ITS); Communications Architecture
6. ETSI (2011) Vehicle Safety Communications Applications (VSC-A), Final Report
7. ETSI (2012) ETSI TS 101 556 -1 (V1.1.1) - Intelligent Transport Systems (ITS); Infrastructure to Vehicle Communication; Electric Vehicle Charging Spot Notification Specification
8. ETSI (2013) ETSI EN 302 636-4-1 (V1.2.0): "Intelligent Transport Systems (ITS); Vehicular Communications; GeoNetworking; Part 4: Geographical addressing and forwarding for point-to-point and point-to-multipoint communications; Sub-part 1: Media-Independent Functionality"
9. ETSI (2013) ETSI EN 302 637-2 (V1.3.0) - Intelligent Transport Systems (ITS); Vehicular Communications; Basic Set of Applications; Part 2: Specification of Cooperative Awareness Basic Service
10. ETSI (2013) ETSI EN 302 637-3 V1.2.0 - Intelligent Transport Systems (ITS); Vehicular Communications; Basic Set of Applications; Part 3: Specification of Decentralized Environmental Notification Basic Service
11. ETSI (2013) ETSI TS 101 539-1 (V1.1.1): "Intelligent Transport Systems (ITS); Vehicular Communications; Part 1: Road Hazard Signalling (RHS) application requirements specification"
12. ETSI (2013) ETSI TS 101 539-3 (V1.1.1): "Intelligent Transport Systems (ITS); Vehicular Communications; Part 3: Longitudinal Collision Risk Warning (LCRW) application requirements specification"
13. ETSI (2013) ETSI TS 103 097 V1.1.1 - Intelligent Transport Systems (ITS); Security; Security Header and Certificate Format
14. European Commission (2009) Standardisation mandate addressed to CEN, CENELEC and ETSI in the field of information and communication technologies to support the interoperability of co-operative systems for intelligent transport in the European Community, p 5
15. Hayward JC (1972) Near-miss determination through use of a scale of danger. Highway Research Record, vol 384
16. IEEE: 1609.12 - IEEE Standard for Wireless Access in Vehicular Environments (WAVE) - Identifier Allocations

17. IEEE: 1609.2 - IEEE Standard for Wireless Access in Vehicular Environments - Security Services for Applications and Management Messages
18. IEEE: 1609.4 - IEEE Standard for Wireless Access in Vehicular Environments (WAVE) - Multi-channel operation
19. IEEE: IEEE 1609.3 - IEEE Standard for Wireless Access in Vehicular Environments (WAVE) - Networking Services
20. ISO (2008) ISO/TS 14823: Traffic and travel information—messages via media independent stationary dissemination systems—graphic data dictionary for pre-trip and in-trip information dissemination systems
21. Kenney JB (2011) Dedicated short-range communications (DSRC) standards in the United States. Proc IEEE 99(7):1162–1182
22. Malone K, Rech J (2013) User related results from DRIVE C2X test sites. http://www.drive-c2x.eu/tl_files/publications/3rd%20Test%20Site%20Event%20TSS/6%20DRIVE%20C2X%203rd%20Test%20site%20event_Kerry%20Malone_Users_20130715.pdf
23. Ministry of Land, Infrastructure, Transport and Tourism: Deployment of ITS Spots (2012). http://www.mlit.go.jp/road/ITS/topindex/ITSSpot.pdf
24. MobileInfo Automotive WG (2006) TPEG TEC application specification
25. SAE (2009) SAE J2735: Dedicated short range communications (DSRC) message set dictionary
26. SAE DSRC Technical Committee (2011) SAE Draft Std. J2945.1, Revision 2.2.: Draft DSRC message communication minimum performance requirements: basic safety message for vehicle safety applications
27. SAE DSRC Technical Committee (2014) SAE J3067: Information report on candidate improvements to dedicated short range communications (DSRC) message set dictionary [SAE J2735] using systems engineering methods
28. Song X (2014) Cooperative vehicle-infrastructure system - activities in China. In: 6th ETSI TC ITS workshop, Berlin, February 2014. http://docbox.etsi.org/Workshop/2014/201402_ITSWORKSHOP/S02_ITS_SomeBitsFromtheWorld/CHINA_NATIONALCENTREofITS_SONG.pdf
29. Stevens S (2013) HS63A3 Project memorandum: safety pilot—preliminary analysis of the driver subjective data for integrated light vehicles
30. Union E (2013) Commission delegated regulation (EU) No 305/2013 of 26 November 2012 supplementing Directive 2010/40/EU of the European Parliament and of the Council with regard to the harmonised provision for an interoperable EU-wide eCall
31. United Nations: Convention on road signs and signals done at Vienna on 8 Nov 1968
32. U.S. Department of Transportation, National Highway Traffic Safety Administration (2014) Vehicle-to-vehicle communications: readiness of V2V technology for application
33. US DOT (2014) RU.S. Department of transportation announces decision to move forward with vehicle-to-vehicle communication technology for light vehicles. http://www.nhtsa.gov/About+NHTSA/Press+Releases/2014/USDOT+to+Move+Forward+with+Vehicle-to-Vehicle+Communication+Technology+for+Light+Vehicles

Chapter 6
Decentralized Congestion Control Techniques for VANETs

Dieter Smely, Stefan Rührup, Robert K. Schmidt, John Kenney, and Katrin Sjöberg

Abstract Direct vehicle-to-vehicle (V2V) communication must operate reliably and with low latency under all circumstances to allow for imminent collision prevention. However, the amount of data to be exchanged is limited by the physical limitations of the communication channel. In vehicular ad hoc networks (VANETs), this limitation is particularly relevant when vehicle density rises and thus lots of information needs to be exchanged, bringing the communication system into a congested state. Rapidly changing topology, channel characteristics, distributed medium access, and challenging active safety application requirements cause many challenges to the resource management. This chapter reviews these issues to motivate suitable Decentralized Congestion Control (DCC) algorithms and cross-layer coordination mechanisms. In particular, it is shown that performance degradations like packet losses, the reduction of the effective communication range, and packet transmission delays are correlated with the channel load. The strategy of the DCC is to avoid these degradations by limiting the load offered by each vehicle to the radio channel, such that a certain channel load threshold is not significantly exceeded. This mitigates the communication range degradation and keeps the packet latencies moderate. The DCC is already part of specifications in the European Telecommunications Standards Institute (ETSI), e.g. ETSI TS 102 687.

D. Smely (✉)
Kapsch TrafficCom AG, Vienna, Austria
e-mail: Dieter.Smely@kapsch.net

S. Rührup
Telecommunications Research Center Vienna (FTW), Vienna, Austria
e-mail: ruehrup@ftw.at

R.K. Schmidt
DENSO AUTOMOTIVE Dtld. GmbH, Eching, Germany
e-mail: r.schmidt@denso-auto.de

J. Kenney
Toyota InfoTechnology Center, Mountain View, CA, USA
e-mail: jkenney@us.toyota-itc.com

K. Sjöberg
Volvo GTT, Göteborg, Sweden
e-mail: katrin.sjoberg@volvo.com

Keywords VANET • DCC • Congestion • Channel busy ratio • Cross-layer DCC • Transmit power • Transmit rate • LIMERIC • AIMD

6.1 Introduction

The vehicular ad hoc network (VANET) in support of road traffic safety and road traffic efficiency applications, also known as cooperative intelligent transport systems (C-ITS), provides a direct information exchange between vehicles. The majority of the intended applications are using broadcasts to convey messages to other vehicles. For the channel access as well as for the physical layer, the IEEE 802.11 [9] standard for Wireless LANs has been extended for the use in vehicular communication and adopted in ITS standardization in the USA and Europe [3]. Broadcast transmissions are performed using the same channel, which is a limited resource. The resource allocation in vehicular environments, however, is not managed centrally, and the channel access mechanism of IEEE 802.11 cannot prevent channel congestion, especially when broadcast communication is used.

Broadcasts are not acknowledged, and no re-transmissions are made. Furthermore, the so-called contention window, which describes the time period in which the next channel access attempt should be made, is not enlarged. Thus, the mechanism to distribute the load by transmission attempts over a longer period is not applied in broadcast communication. This behavior is beneficial for the MAC transmission delay, but the packet collision probability is higher compared to acknowledged packets. However, both the MAC transmission delay and the number of packet collisions rapidly grow when the channel load is increased above 40% of the theoretical maximum channel capacity. The MAC transmission delay will cause safety messages to arrive late, and a high number of packet collisions lead to a lower reception probability and hence an effective radio range reduction.

For a safety-related system excessive transmission delays and a poor radio range are not acceptable, therefore, overloading the radio channel has to be avoided by an additional algorithm called Decentralized Congestion Control (DCC).

The goal of the DCC is to minimize packet collisions and provide similar channel access opportunities to all stations under the same channel load conditions, even for rapid network topology changes that can appear in the vehicular environment. A commonly agreed input to the congestion control algorithm is the Channel Busy Ratio (CBR), which is a measure for channel load (Sect. 6.3.1.1). The CBR is defined as ratio between the time the channel is sensed as busy and the total observation time (which is usually set to 100 ms). Based on the CBR, the transmission opportunity for each station is determined by the congestion control algorithm in use. There are several possibilities to realize DCC, and a selection of these algorithms is presented in Sect. 6.3 of this chapter.

6 Decentralized Congestion Control Techniques for VANETs

Before going into the detailed algorithm descriptions, Sect. 6.2 reviews the basic communication mechanisms in VANETs based on IEEE 802.11 and the problem of congestion of the communication channel. Section 6.3 introduces possible congestion mitigation strategies, as a foundation for congestion control. Selected DCC algorithms are presented and discussed. The use of DCC in the overall protocol stack and the evolution of DCC specifications by international standardization bodies is outlined in Sect. 6.4.

6.1.1 Relevant Road Traffic Scenarios and Vehicle Densities

DCC becomes relevant if the load offered by many vehicles reaches the channel capacity. Such scenarios are imaginable in dense traffic, but vehicle density is not the only criterion to describe relevant scenarios.

In urban traffic, vehicles generally move at low to medium speeds with the exceptions of intra-city expressways. Only parking lot scenarios with densely parked or waiting vehicles pose a risk of high channel load if a static high message generation rate would be applied. However, it is already obvious that a simple velocity-based rate adaptation would solve this problem. Thus, the more challenging scenarios may be found on highways, as characterized in the following.

From a higher perspective, the movement of the vehicles on a highway is rather one-dimensional. The traffic is floating along the highway on multiple parallel lanes into the same direction. Oncoming traffic is usually separated by a central barrier and represents no threat. Most dangerous are situations in which a vehicle approaches or passes by with high relative speed, especially at the end of traffic jams.

On a typical highway, the vehicle density does not change rapidly but rather slightly and continuously. Kerner characterizes the state transitions by his three-phase traffic flow theory in [13]. Usually, vehicles float in free flow. Due to perturbations at junctions or on-ramps, the state changes to synchronized flow where there is nearly the same speed of vehicles on all lanes of the same direction. Due to further slight distortions, severe road traffic congestion occurs, also referred to as wide moving jams. These transitions of traffic flow are most relevant for DCC as the traffic density is high while at the same time the velocity is also considerably high. Traffic perturbations may cause emergency breakings and accidents, which should be mitigated with the aid of inter-vehicle communication.

6.1.2 Offered Load and Impact on Position Data Accuracy

The channel load that a vehicle perceives depends on the number of vehicles in range and their individual message generation rates. As an example, 300 vehicles in communication range and a message rate of 1 Hz result in 40 % offered load at a

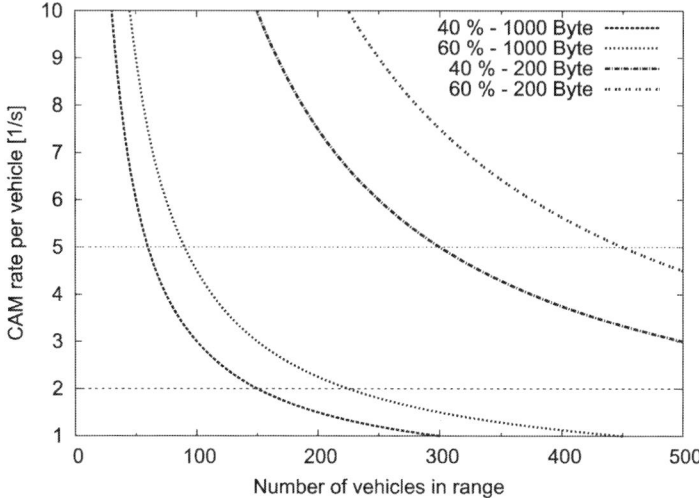

Fig. 6.1 Offered load given by a message rate (CAM rate [6]) at a data rate of 6 MBit/s for channel loads of 40 % and 60 %, and message lengths of 200 and 1,000 bytes, depending on the number of vehicles in communication range

data rate of 6 MBit/s, if the size of a message is 1,000 bytes. The same channel load is reached, if 150 vehicles sending 2 messages per second. Figure 6.1 shows two levels of offered load (40 % and 60 %) with a common static message rate, using medium-size packets with 200 bytes and large packets with 1,000 bytes per packet. The example shows that high message rates cannot be supported for all vehicles in case of high vehicle densities or larger packets.

To limit the offered load the message rate must be adapted. But with the resulting reduced update rate, the per-vehicle position data accuracy that a vehicle can achieve from the received messages gets worse.

Basically, one can derive the offered load for a given average speed (of all vehicles) and a particular message rate. From this the per-vehicle position data accuracy, and the achieved (average) position data accuracy[1] can be estimated.

However, the heterogeneity of velocities should be taken into account in the discussion of the message rate adaptation. Imagine a traffic jam with vehicles driving slowly at the same velocity. At a constant message rate, the position of these vehicles can be determined with the same accuracy. But, oncoming traffic may not be jammed and travels at high speeds. For the same message rate the accuracy would be much lower. For these vehicles a much higher message rate is required to meet the same accuracy requirements. This could result in situations where the offered load exceeds the channel capacity. Summarizing, the message rate should be adapted based on a vehicle's current context.

---

[1]The complete definition of VANET position accuracy can be found in [19].

6.2 The Congestion Problem in CSMA Networks

Starting with a quick recap of the MAC layer, and showing that the channel access mechanism of IEEE 802.11 [9] cannot avoid an overload of the radio channel, this section will present details of the performance reductions caused by high channel loads.

6.2.1 CSMA Medium Access

Typically, MAC protocols schedule the access to the medium to avoid concurrent access in the presence of multiple transmitters. For this, Carrier Sense Multiple Access with Collision Avoidance (CSMA/CA) has been selected for VANETs, in particular the algorithm specified in the IEEE 802.11 standard (see also Chap. 4). Its basic principle and the QoS extensions for broadcast communication are described in the following, to understand the root cause of channel congestion.

6.2.1.1 MAC Backoff Procedure

Basically, in CSMA/CA a station observes the channel status prior to a transmission. If the so-called carrier sensing indicates an idle channel for a predetermined time, the station is allowed to transmit directly. If the channel indication is busy during the period, the station performs a backoff procedure, i.e., the station has to defer its access according to a randomized time period. In IEEE 802.11, the predetermined period is called arbitration interframe space (AIFS).

Simplified, the backoff value to avoid concurrent access is determined as follows:

1. Draw an integer from a uniform distribution [0,CW], where CW refers to the size of the contention window;
2. Decrease the backoff value only when the channel is free, one decrement per slot time; after carrier sensing indicates a clear channel, all stations must apply the AIFS waiting time before the backoff value can resume;
3. Upon reaching a backoff value of 0, transmit. In broadcast operation the station will only invoke the backoff procedure once during the initial listening period.

In enhanced distributed channel access (EDCA), every station maintains queues, a.k.a. access categories (AC), with different AIFS values and CW sizes for the purpose of increasing the probability that data traffic with higher priority can access the channel before data traffic with lower priority.

6.2.1.2 Carrier Sensing

In IEEE 802.11, carrier sensing (CS) is done by two mechanisms, Virtual CS implemented in the MAC layer and Physical CS implemented in the PHY layer. Both basically define when a station is *not* allowed to access the medium:

1. Physical CS: Do not transmit while the own receiver is busy, determined by the Receiver Sensitivity (−85 dBm).
2. Physical CS: Do not transmit while sensing high energy on the channel, determined by the CCA threshold (−65 dBm).
3. Virtual CS: When the receiver lost the signal, wait until the calculated end of transmission based on the frame header information.[2]

These principles support a high amount of spatial medium reuse to optimize throughput, which was the original target for short-range unicast communication. If the DCC limits the channel load to moderate values, these principles also work well for broadcast messages in VANETs.

The CCA threshold is only considered by the CS if two or more concurrent transmissions overlap and thereby the signal-to-interference ratio gets too low to decode the OFDM signals. Thus, controlling the CCA threshold is directly related to the spatial reuse of the communication channel (see also [18]).

The carrier sensing mechanism can also be used to determine the channel load as input to the DCC algorithm. Therefore the receiver sensitivity and the CCA threshold can influence the DCC behavior [8].

6.2.1.3 Broadcasting in IEEE 802.11 VANETs

In broadcast communication, the EDCA parameter CW_{min} for the minimum contention window size always denotes the upper limit for the backoff counter. This constraint increases the likeliness of packet collisions. In case there are two (or more) stations where the backoff counter times out simultaneously, they will start the transmission at the same time resulting in a packet collision. This can happen if two stations try to get a transmit opportunity while the channel is busy and select the same backoff counter at the same time. When the channel is highly loaded, the probability that this happens when the counters are different is much higher (see also Fig. 6.2). In this case several stations are in defer state and wait for a transmit opportunity. After the channel was busy they all sense the channel for the AIFS time and count down their backoff counter if the channel was idle. All stations that reached the count zero start to transmit immediately. Since the radio waves need some 100 ns to propagate from one station to the other, these packet collisions by simultaneous start of transmissions cannot be prevented.

---

[2]How to calculate the end of a transmission in broadcast mode is described in [20].

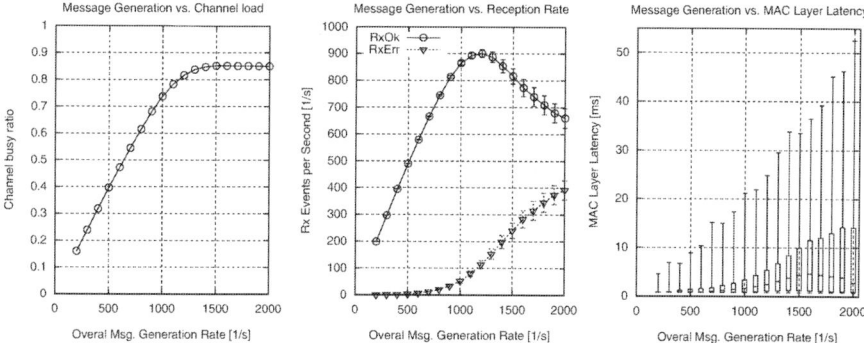

Fig. 6.2 Channel load and reception rate under increasing transmit events for plain IEEE 802.11 broadcasts with 800 μs frame duration. Transmit events are generated in a network of 200 stations and message generation rates between 1 and 10 Hz. The curves are averaged over 200 stations; error bars in the left two plots show the standard deviation; the right box-and-whisker plot shows quartiles (*box*) as well as minimum and maximum (*whiskers*). All stations are within radio range

6.2.2 Performance Degradation due to High Channel Load

6.2.2.1 Impact of Channel Load on Packet Loss and MAC Layer Delays

To study the impact of high channel loads on the MAC performance for broadcast messages, the influence of radio channel issues are neglected in this section. We assume that all stations can sense each other, and that several concurrent packets always lead to *one* receive error (RxErr). Hence, no reception of concurrent messages is possible in this simplified setup. In practice, up to moderate channel loads this assumption is close to a scenario where the VANET is limited by the vehicle cluster size and not by the radio range.

Under increasing message generation rates, the CBR increases until the capacity of the channel is reached, as shown in Fig. 6.2 (left) for a 200 station example. Close to the capacity limit, stations experience a busy channel and have to defer their intended transmission by the contention mechanism ($CW_{min} = 15$, $AIFSN = 6$). Additionally, the number of packet collisions increases.

The center plot in Fig. 6.2 shows over the message rate how the receive errors (RxErr) increase, while the successful receptions (RxOk) decrease after reaching an optimum. Operating the medium access at this optimum or close to the optimum has already a drawback of a significantly increased latency, which is shown in the right figure. Up to a CBR of 40 % the maximum channel access delay stays below 10 ms and the number of packet collisions can be neglected.

In practice this means that up to a CBR of 40 % the influence of MAC and radio channel issues can be safely neglected for vehicle cluster sizes that are small compared to the radio range. For higher CBR values or scenarios that also contain vehicles close or outside the radio range, additional effects as described in the following sections must be considered.

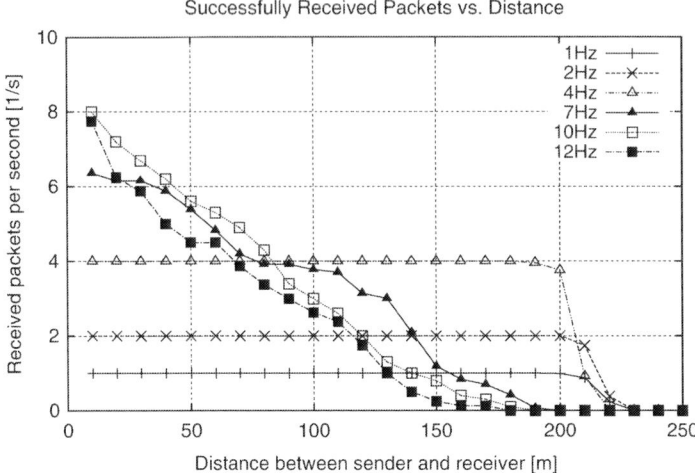

Fig. 6.3 Impact of channel load on receive rates: Successfully received packets per second for a traffic jam scenario with 1,200 vehicles placed in distances of 10 m on four lanes. Messages with 500 bytes payload and a frame duration of 800 μs (at 6 MBit/s) are generated at rates between 1 and 12 Hz. The CSMA parameters are $CW_{min} = 15$, $AIFSN = 6$, and $CCA = -85$ dBm. Propagation is modeled with a dual-slope lognormal shadowing model without fading. No packet capturing is assumed

6.2.2.2 Impact of Channel Load on Effective Transmission Ranges

In contrast to the aforementioned model (Sect. 6.2.2.1), here, a vast traffic jam scenario and a simple dual-slope lognormal radio channel model are used to study the impact of the channel load on the radio range.

Under high channel load conditions, the likelihood of simultaneous transmissions and thereby radio interference rises. Frames can only be decoded, if the signal-to-interference-plus-noise ratio (SINR) is sufficient—this is usually the case when the transmitter is close and the interferer far (the higher the data rate the higher the SINR must be). Figure 6.3 shows the successfully received transmissions from a designated transmitter to receivers in different distances. In this selected traffic jam example the transmissions can be received up to the full radio range, where the noise limitation becomes dominant, when all transmitters generate their messages with not more than 4 Hz. For a message generation rate of 7 Hz and more the number of received packets is significantly lower than the number of transmitted packets. For data rates of 6 MBit/s as used in this example, the message reception rate can be improved by an increased message generation rate only for distances up to 25 % of the full radio range, for longer distances the interference due to overlapping frames becomes dominant and higher message rates are detrimental.

6.2.2.3 Communication Range Under Interference

Interference due to hidden stations or simultaneous sending has an impact on the effective communication range. This can be modeled as follows: Assuming two transmitting stations, a transmitter T and an interferer I, and a receiving station R. The receiving station can decode a packet by T, if the SINR, i.e. the ratio of the signal strength over interference and noise, exceeds a certain threshold.

To calculate the signal strengths of transmitter and interferer at the receiver position, the log-distance path loss model is used. According to this model, the received power level $P_r(d)$ at distance d is $P_r(d) = \frac{P_0}{d^\rho}$, where P_0, which is proportional to the transmit power level, is the received power level at a unit distance of 1 m, and ρ is the path loss coefficient which depends on the environmental conditions for RF wave propagation. For free space environments ρ is equal to 2. Recent measurements show that, depending on scenario, ρ may be slightly lower, i.e. $1.6 \leq \rho \leq 1.8$ in urban and suburban areas [11].

Using the log-distance path loss model, the SINR can be determined as follows

$$\text{SINR} = \frac{P_{rT}(d_{TR})}{P_{rI}(d_{IR}) + P_N} = \frac{\frac{P_{0T}}{d_{TR}^\rho}}{\frac{P_{0I}}{d_{TI}^\rho} + P_N} \quad (6.1)$$

where d_{TR} is the distance between transmitter and receiver and d_{IR} the distance between interferer and receiver. P_{0T} and P_{0I} are the received power levels at unit distance for the transmitter and the interferer.

Figure 6.4 shows how the effective range of the transmitter is reduced by a concurrent transmission of an interfering hidden station, assuming a necessary SINR of 6 dB, $\rho = 2$, equal transmit power levels of 4 dBm, and a noise figure of −95 dBm. The full radio range is a circular area shown as dashed line, where the radius of this circle is given by the noise limit when the interferer is not transmitting and P_{0I} in Eq. (6.1) is zero. It is important to understand that the radio range is

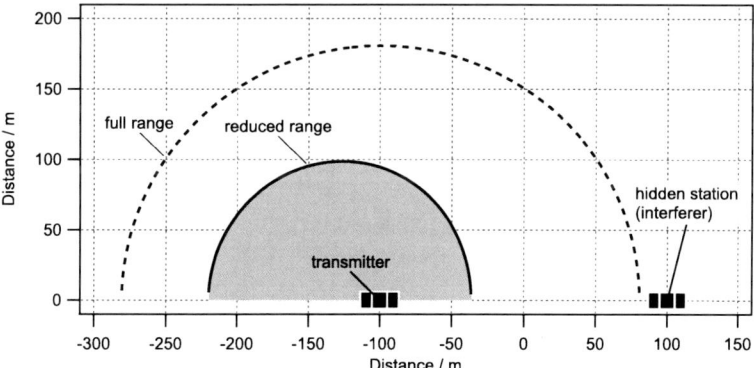

Fig. 6.4 Spatial communication range under interference

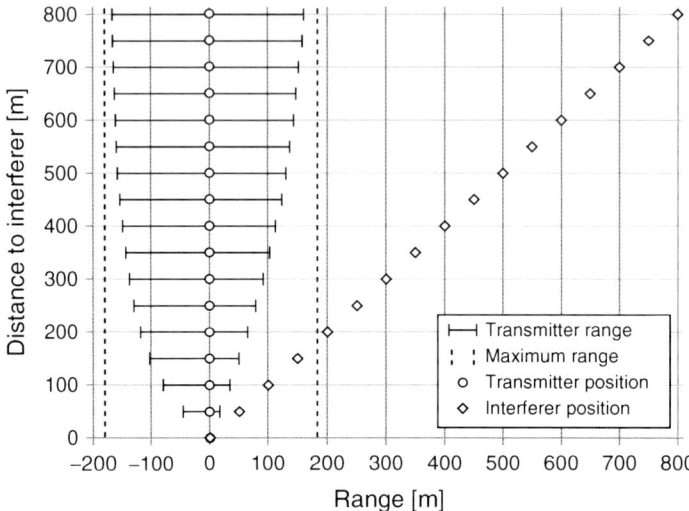

Fig. 6.5 Communication range of a transmitter along the street for different interferer distances

reduced in all directions. Under interference only receivers in the shaded region can decode the messages from the transmitter.

For the same radio parameters, Fig. 6.5 shows the radio range along a straight street as function of the distance between the transmitter and the interferer. The range reduction is stronger on the interferer side, but cannot be neglected in opposite direction. When the interferer is further away than the maximum range (dot-dashed line) it cannot sense the transmitter—the station is hidden. As outlined in the previous subsections, concurrent transmissions can also occur within sensing range, in this case the range reduction is dramatic. But even when the interferer is in a distance of three times the radio range, it will degrade the effective range by around 25 %. On the other hand, the probability that two hidden stations transmit simultaneously is proportional to the channel load, which is much higher than the packet collision probability for transmitters within sensing range. The results of these effects for a complex road traffic scenario were already shown in Fig. 6.3.

6.2.2.4 Classification of Packet Loss

The following list summarizes the reasons for packet loss and hence degradation of the effective communication range under high load:

- *Simultaneous Sending*: Regardless of the distance between two transmitters, a collision can occur if at least two stations have the currently lowest backoff slot. Hence, they start the transmission at the same time. This also allows for two colliding transmissions within receiving range of each other which is in contrast to the classic hidden station (see Sect. 6.2.1.3).

- *Single/Multiple Hidden Station(s)*: During the reception, the SINR deteriorates due to colliding medium access by a station that is out of carrier sensing range (hidden station). Packet losses occur, if the minimum SINR of the receiver is underrun during reception, or the receiver is not even able to sense the packet from the beginning (see Sects. 6.2.2.2 and 6.2.2.3). Moreover, many simultaneous transmitting stations can together cause sufficiently high accumulated interference even if they are far away.
- *Exposed Station*: Packet loss implicitly occurs due to local message congestion if the medium cannot be accessed in time. Many messages overcrowd the local message queue, so that messages have to be dropped before transmission.
- *Near Adjacent Station*: Interference caused due to adjacent channel interference from near-by transmitting stations can also reduce the SINR below the receiver requirement.

Concluding, with increasing channel load, these effects decrease the number of received messages. When designing DCC mechanisms, these effects should be well understood to engage appropriate mitigation strategies.

6.3 Fundamental Congestion Control Mechanisms

The aforementioned considerations show that an increasing channel load leads to frame collisions and packet loss, increasing channel access delays, and a reduction of the effective transmission range. Since broadcast channel access in 802.11-based wireless ad hoc networks is performed locally without the coordination of a central scheduler, a DCC is needed. DCC can be designed in various ways, but the general idea is that every station of a network measures the channel load and limits its own channel utilization in order to avoid an overall channel congestion. The basic DCC functionality consists of measuring the overall resource utilization, which is usually the channel load, and set an individual limit to resource allocation (e.g., transmit rate, output power). The resource allocation of the individual stations accumulates to the overall resource utilization, which is again the input to the DCC; this is sometimes called the DCC control loop for its circular dependency (see Fig. 6.6). At the heart of the control loop a DCC control function or algorithm decides about the individual resource limits. It is important to note that the kind of DCC that we discuss here only sets limits to resource allocation, but does not allocate resources itself; without having packets generated by the application layer, DCC will not use or reserve resources on the channel.

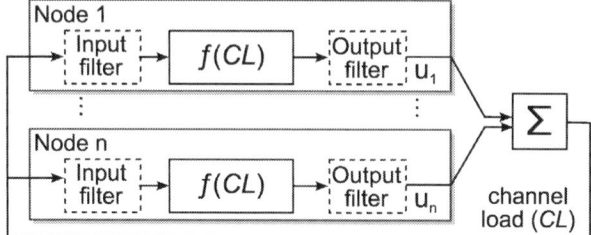

Fig. 6.6 DCC control loop model. A station i takes the channel load CL as input and applies a control function $f(CL)$ that leads to an individual channel utilization u_i. The channel utilizations of all stations K give in turn the new overall channel load: $CL = \sum_{i=1}^{K} u_i$

6.3.1 Input and Output Parameters

DCC algorithms typically take input parameters that characterize channel load, which is the factor to be controlled, e.g. the CBR. The output is a parameter or a set of parameters that determines how the individual station limits its utilization of the channel resource, e.g. a transmit rate limit or a output power limit. A detailed description of the input and output parameters is given in the following.

6.3.1.1 Input Parameters: Channel Busy Ratio and Number of Neighbors

Channel load can be characterized by the CBR, which is the ratio of the time the channel is perceived as busy and the overall observation time. A high channel load can be the result of a high number of neighbors (i.e., stations within range) or of high transmit rates or both. Thus, the CBR and the number of neighbors are important inputs to the DCC. The latter can be made available via the neighbor table associated with the networking layer (cf. Fig. 6.10).

Since the impact of interference by neighbors depends on their distance, only neighbors within a certain distance should be taken into account. Similarly, the CBR threshold should be set to a reasonable value. For the CBR determination this threshold is the RX power level that determines whether the channel is perceived as busy. With a low threshold, low received signal levels from distant transmitters add to the CBR, resulting in a higher value.

Figure 6.7 depicts a snapshot of simulation results of a multi-lane highway intersection of two major German highways with dense synchronized traffic. The comparison considers two CBR thresholds -80 and -95 dBm as basis for calculating the CBR, as well the number of neighbors in the complete set of neighbors $N^{\text{inf.}}$ and in the filtered set $N^{300\,\text{m}}$, including only vehicles within 300 m range. Each data point represents the result of the respective vehicle versus its current distance to the highway crossing middle point.

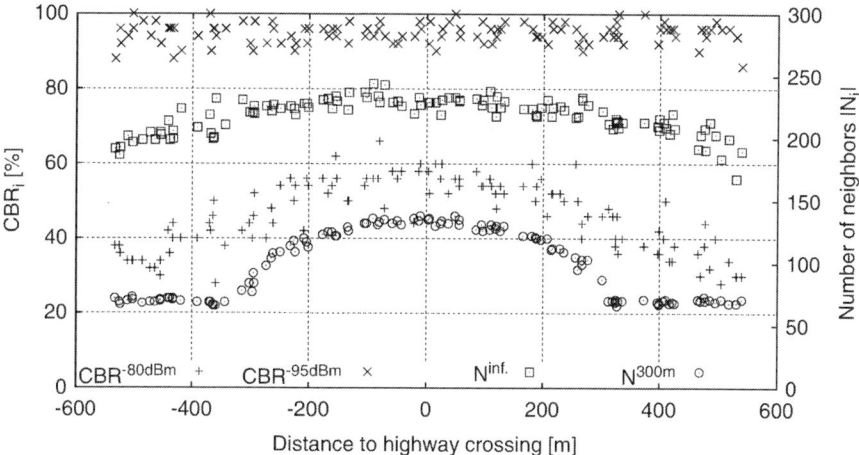

Fig. 6.7 Comparison of the input metrics CBR (*left* y-axis) and the number of neighbors (*right* y-axis) in a typical highway crossing scenario, with a message rate of 5 Hz. For the CBR, two different thresholds are applied. The number of neighbors is either limited by the distance or unlimited (total number of neighbors)

$CBR^{-95\,dBm}$ always determines a highly loaded channel, reaching even full channel load. There is no statistically significant difference between vehicles located exactly at the crossing and vehicles located farther away. By contrast, for $CBR^{-80\,dBm}$ the distribution of the results is closer to the desired classification, i.e. the high load detection at the crossing. Thus, to classify high load with satisfying resolution, a CBR with the higher threshold is more suitable. A lower threshold, however, provides a higher resolution of low channel loads. In order to be able to detect changes at high and low channel load, it seems advisable to observe two different CBR thresholds.

The comparison of the results of the number of neighbors is quite similar. For the total number $N^{inf.}$, only a slight increase of the vehicle density can be detected, from 180 to 230 vehicles. The filtered set $N^{300\,m}$ shows a clearer increase of the density, from 70 to 140. Thus, without a street map and within a few meters of movement, a vehicle is able to detect that the traffic situation has significantly changed.

Summarizing, the presented metrics are suitable to classify channel load, especially in a combination of the two basic metrics *CBR with two thresholds* and the *filtered number of neighbors*.

6.3.1.2 Output Parameters

The output parameters or return parameters of a DCC algorithm are the factors that influence transmit behavior. This is mainly the transmit rate or transmit interval. The transmit rate determined by DCC can directly be used to trigger message transmissions, or it can be used as an upper limit for the message rate. In the latter case, the message rate is at least given by the actually generated messages or at most by the DCC message rate limit.

A second important output parameter is the transmit power level or output power level. A reduction of the transmit power level reduces the spatial utilization of the channel. This power level adaption, however, has to be used carefully. If two stations contend for transmission on the channel without equal transmit power levels, a fair distribution of the channel resource cannot be guaranteed, and severe hidden station problems can occur. Therefore, a power control should go hand in hand with a multi-hop information exchange among stations.

Other output parameters are the receiver sensitivity and the transmit data rate (i.e., the modulation and coding scheme). The first tunes the perception of the channel, the latter influences the transmit durations of frames and the SINR. These parameters were proposed in the initial ETSI standard on DCC [2], but for simplification not further used in DCC standardization.

Since the 802.11 MAC layer has four access categories, output parameters of DCC can be made specific for the four access categories.

6.3.2 Performance Metrics

Several performance metrics have been proposed in the literature for an assessment of DCC algorithms. They can be classified into MAC layer metrics, application and facilities layer metrics, as well as algorithm quality metrics. While MAC layer metrics directly assess the impact on the usage of the wireless medium, the application and facilities layer metrics quantify the impact of a DCC algorithm on how well other vehicles are perceived. Finally, the quality metrics focus on how fast a DCC algorithm reaches its goal and how well it can distribute the channel resource to the participating stations. Table 6.1 gives an overview, a summary can be found in [7].

Most *MAC layer metrics* are not to be seen as individual minimization or maximization goals. Instead, a group of metrics can express a trade-off. As an example, the average CBR as a network-wide metric should stay below a certain level, while the transmit rates of individual stations should be high enough to successfully update their neighbors. Receivers should be able to receive a high number of messages from other nearby vehicles. It is also desirable to receive messages from other vehicles in regular intervals, which is expressed by the update delay [14]. It depends on the application, how the update delays from nearby vehicles and distant vehicles should be balanced. The metrics however allow to rate DCC algorithms according to several desired properties.

Application layer metrics are defined in the context of the usage of data that is especially transported by Cooperative Awareness Messages (CAMs) [6] or Basic Safety Messages (BSMs) [16]. CAMs carry position information, therefore the deviation between the last known position of a neighboring vehicle and the actual

Table 6.1 Performance metrics for DCC

Metric	Definition
MAC layer metrics	
Average channel busy ratio (CBR)	Average of CBR values of all stations in the network for a given time interval.
Frame collision rate	Rate of receiver errors due to synchronous and overlapping frame transmissions. Can be given for a single receiver.
MAC layer delay	Duration between sending a message down to the MAC layer until receiving it at the MAC layer of the receiver. This includes the channel access and queuing delays. Can be given for any pair of sender and receiver.
Reception rate	Number of packets received from a single station within 1 s. It can be seen as inverse of the packet inter-reception time averaged over a second. It can be given for a single receiver or for a transmitter-receiver-distance.
Packet inter-arrival time or update delay [14, 22]	Time between subsequent Cooperative Awareness Message (CAM) receptions from the same sender.
Transmit rate	Number or transmit opportunities within a given time interval of a given station.
Application and facilities layer metrics	
End-to-end latency	Time duration from message generation at application layer (or facility layer) to reception. It can be given for any pair of sender and receiver.
Awareness quality [17]	Percentage of vehicles within a given distance from which CAMs were received and have not expired yet. This is given for a certain vehicle.
Position error [19]	Average deviation of the position data from the last received transmission of a vehicle and its actual position. It can be given for any pair of sender and receiver.
Quality metrics for a DCC algorithm	
Convergence	Time until the resource allocation of each station stays within a certain interval, given a specific start condition.
(Max-min-)Fairness	Ratio between maximum and minimum of the number of channel access opportunities or the transmit durations among all stations within a given time interval.

position is a metric that is influenced by the ability of neighboring vehicles to access the channel and broadcast their position information often enough.

The awareness quality metric [17] goes one step further. Awareness in the road traffic context refers to the relation between knowledge of vehicles that *are* stored in a vehicle's neighbor table and the knowledge of vehicles that *should be* stored. Awareness Quality is then defined as the level of awareness averaged over all stations for a given time interval. This metric is suitable to take into account application requirements and reflect the limitations of communication. It needs global knowledge to be evaluated, just like the (network-wide) average CBR.

6.3.3 Reactive Dynamic DCC Algorithms

Reactive Dynamic DCC Algorithms are characterized by the way their output depends on the input parameter(s). Reactive algorithms use the input, which is usually the CBR, *directly* to determine the output value. The output value can be the result of a formula or a table lookup.

The first ETSI standard on DCC [2] describes a reactive dynamic DCC algorithm. It is based on a state machine definition with basically three states named RELAXED, ACTIVE, and RESTRICTIVE. While in RELAXED state the DCC imposes no limitations on transmit rate or output power, in ACTIVE state, the transmit rate is limited moderately, while in RESTRICTIVE state only the lowest defined transmit power and the lowest transmit rate is allowed. For each of the states there is a channel load threshold defined. Once this threshold is exceeded for a certain duration Δ, the state machine is switched to the respective state and the corresponding transmit rate or power limitations are activated. The ACTIVE state may contain several sub-states (see Fig. 6.8) that allow a fine-grained assignment between a CBR and a transmit rate.

Since there is a transition from each state into each other state according to [2] and since the transition only depends on a threshold for the target state, the definition of a state machine is actually obsolete. Instead, the choice of a transmit rate $r(t)$ from a set of pre-defined rates depending on a set of CBR thresholds can be specified as follows:

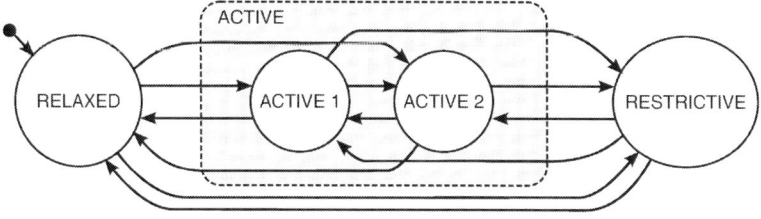

Fig. 6.8 DCC State machine of [2]. Basically, there are direct transitions between all states. Note that the small *black circle* indicates the start state

$$r(t) = \begin{cases} r_{\text{RESTR.}} & \text{if} \qquad\qquad\qquad \text{CBR}(t') > \text{CBR}_{\text{RESTRICT.}} \quad \text{for all } t' \in [t-\Delta, t] \\ r_{\text{ACTIVE 2}} & \text{if } \text{CBR}_{\text{RESTRICT.}} \geq \text{CBR}(t') > \text{CBR}_{\text{ACTIVE 2}} \quad \text{for all } t' \in [t-\Delta, t] \\ r_{\text{ACTIVE 1}} & \text{if } \text{CBR}_{\text{ACTIVE 2}} \geq \text{CBR}(t') > \text{CBR}_{\text{ACTIVE 1}} \quad \text{for all } t' \in [t-\Delta, t] \\ r_{\text{RELAXED}} & \text{otherwise.} \end{cases}$$

Note that this definition only includes the transmit rate r. The standard [2] also defines parameters for output power level, the modulation and coding scheme as well as receiver sensitivity depending on state and access category. Furthermore, for exceeding or falling below a threshold, two different durations are defined instead of a single duration Δ in the definition above.

The standard allows to extend the state-machine by further active sub-states. This way, the relation between perceived channel load and a transmit limitation can be modeled in a more fine-grained way; e.g., several active sub-states can be used to define a step-wise function that sets the transmit interval proportional to the CBR. However, it is not possible to define a direct relation between channel load and a transmit rate of a single station, that is optimal for all vehicle densities. In sub-optimal cases, the assignment leads to oscillations between the states. The state-machine approach has been evaluated in simulations in [21], which gives some insights into its advantages and disadvantages.

6.3.4 Adaptive DCC Algorithms

In the previous subsections we introduced approaches to DCC that react directly to the measured CBR. The transmit behavior of the DCC algorithm, i.e. the message rate in these examples, is pre-determined for any given CBR value. The function between CBR and rate determines the behavior and convergence of the control loop when the number of stations in range or their message rate changes. This freedom makes them very flexible, but analytically difficult to handle and understand.

By contrast, the algorithms presented in this subsection act on the difference between the measured CBR and a CBR target value, adapting the message rate in order to drive the CBR toward the target. By this simple measure, adaptive control algorithms can ensure that the channel load does not exceed a given limit (the CBR target value).

Two adaptive approaches are presented here, differing primarily in the way the CBR error (i.e., difference between CBR target and measured CBR) is used to adapt the message rate. In the first approach, the entire error signal is used so that the adaptation is a linear function of the error. In the second, only the sign of the error signal is used, as an indicator of whether the message rate should be increased or decreased according to a formula.

6.3.4.1 Linear Adaptive Control: LIMERIC

The linear adaptive control approach is represented by the LIMERIC (LInear MEssage Rate Integrated Control) algorithm [12]. LIMERIC is a distributed algorithm, executed independently by each station in a neighborhood. If $r_j(t)$ is the message rate of the jth station at time t, and CBR(t) is the CBR value measured at time t, then at each update interval LIMERIC adapts the message rate according to:

$$r_j(t+1) = (1-\alpha)r_j(t) + \beta \left(\text{CBR}_{\text{target}} - \text{CBR}(t)\right). \tag{6.2}$$

In this equation, α and β are algorithm parameters that influence stability and adaptation speed, among other features. β is related to a certain default frame duration d_{def}. If a frame duration d different to this is used, β must be scaled by d_{def}/d to achieve the same control behavior. With K, which denotes the number of LIMERIC stations sensing and adapting to CBR in a given region, and assuming no frame collisions, CBR can be modeled according to

$$\text{CBR}(t) = d \sum_{j=1}^{K} r_j(t). \tag{6.3}$$

For equal rates r_j this simplifies to $\text{CBR}(t) = dK r_j(t)$. From this the CBR converges in steady state under the assumption of $r_j(t+1) = r_j(t) = r_j$ to

$$\text{CBR} = \frac{dK\beta}{(\alpha + dK\beta)} \text{CBR}_{\text{target}}. \tag{6.4}$$

The rates converge to the same value r_j, since LIMERIC allocates the channel resource in a fair manner. Satisfying the stability condition for a desired operating range of K is one of the key considerations in the choice of α and β. Even if K is unexpectedly large such that the inequality is not satisfied, LIMERIC employs a gain saturation feature that guarantees stability by limiting the magnitude of the $\beta(\text{CBR}_{\text{target}} - \text{CBR}(t))$ term at each update. The convergence and fairness properties of LIMERIC are illustrated in Fig. 6.9 (left), which plots the message rate over time for two stations among a set of 180 stations. In this simulation, the 180 stations are divided into four groups, with initial message rates for each group set to 0, 2, 7, and 10 messages/s, respectively. Stations 1 and 2, whose message rate trajectory is plotted in the figure, belong to the initial-condition groups with the highest and lowest rate (this is done purely to illustrate fairness, not to suggest that such a distribution of initial rates is likely). Notice that not only do the message rates converge to the same value independent of the initial message rate, but they do so with no steady state variation, i.e. the convergence is perfect in this idealized example. In a more practical implementation of LIMERIC, inaccuracies in the measurement of CBR will create small random variations in message rate.

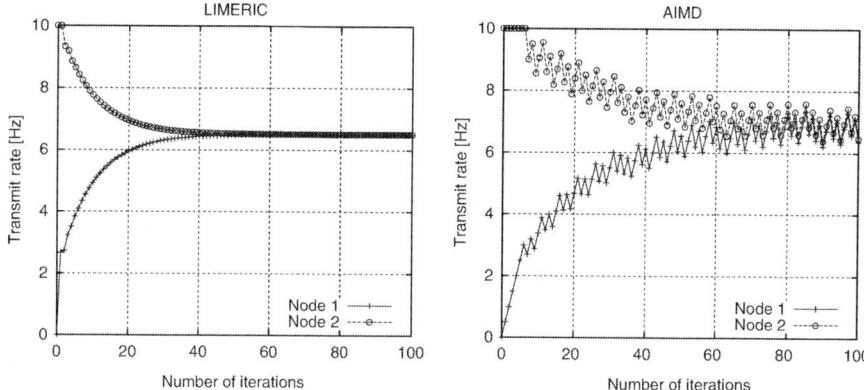

Fig. 6.9 LIMERIC and AIMD convergence and fairness for 180 station example, where two stations were selected from groups initialized with transmit rates of 0 and 10 messages/s

The convergence of the LIMERIC algorithm for synchronized updates can be shown as follows. The sequential case can be found in [12]. Let $u_i = r_i \cdot d$ be the individual channel utilization of station i, consisting of a message rate r_i and a frame duration d. Let $\mathrm{CBR}(t)$ be the overall channel utilization of all stations. From Eqs. (6.2) and (6.3) the overall load results to

$$\mathrm{CBR}(t) = d \sum_{i=1}^{K} \left((1-\alpha) r_i(t-1) + \beta(\mathrm{CBR}_{\mathrm{target}} - \mathrm{CBR}(t-1))\right) \quad (6.5)$$

$$= (1-\alpha)\mathrm{CBR}(t-1) + dK\beta \cdot \mathrm{CBR}_{\mathrm{target}} - dK\beta \cdot \mathrm{CBR}(t-1) \quad (6.6)$$

$$= (1-\alpha - dK\beta)\mathrm{CBR}(t-1) + dK\beta \cdot \mathrm{CBR}_{\mathrm{target}} \quad (6.7)$$

$$= (1-\alpha - dK\beta)^2 \mathrm{CBR}(t-2) + dK\beta \cdot \mathrm{CBR}_{\mathrm{target}}((1-\alpha - dK\beta) + 1) \quad (6.8)$$

$$= (1-\alpha - dK\beta)^t \mathrm{CBR}(0) + dK\beta \cdot \mathrm{CBR}_{\mathrm{target}} \sum_{i=0}^{t-1} (1-\alpha - dK\beta)^i \quad (6.9)$$

The convergence from an initial load $\mathrm{CBR}(0)$ can be expressed by $\lim_{t \to \infty} \mathrm{CBR}(t)$. From the expression above, we can see that a limit exists if $|1 - \alpha - dK\beta| < 1$. Hence, LIMERIC is stable if

$$\alpha + dK\beta < 2. \quad (6.10)$$

Under this condition the first part of the expression $(1 - \alpha - dK\beta)^t \mathrm{CBR}(0)$ tends to zero, the remaining part becomes an infinite geometric series.

$$\lim_{t\to\infty} \text{CBR}(t) = 0 + dK\beta \cdot \text{CBR}_{\text{target}} \frac{1}{1-(1-\alpha-dK\beta)} = \text{CBR}_{\text{target}} \frac{dK\beta}{\alpha + dK\beta} \tag{6.11}$$

Thus, LIMERIC converges to $\text{CBR}_{\text{target}} \cdot dK\beta/(\alpha + dK\beta)$, required that $0 < \alpha + dK\beta < 2$. For $\alpha = 0.1$, $\beta = 1/(150\,\text{ms})$, and $d = 1\,\text{ms}$, it converges to $1/(1 + \frac{15}{K})$ for $K < 285$ vehicles.

To summarize this subsection, LIMERIC is a linear distributed adaptive control algorithm that adapts message rates to ensure that CBR does not exceed a desired limit. It has provable stability, convergence, and fairness properties [12].

6.3.4.2 Additive Increase Multiplicative Decrease

One of the most well-known approaches to congestion control is the algorithm built into the TCP protocol, the so-called *Additive Increase Multiplicative Decrease* (AIMD) algorithm [10]. This algorithm controls TCP's congestion window, which limits the number of sent TCP segments with pending acknowledgements. As long as transmitted segments are successfully acknowledged, it increases the congestion window additively (i.e., slowly); once a transmission fails, the congestion window is decreased by a factor. This rule can be applied to controlling congestion on the wireless medium as well using the excess of a given channel load threshold $\text{CBR}_{\text{target}}$ as a binary decision criterion for the decrease step: As long as the perceived CBR is below this threshold, the transmit rate is increased by an increment α, otherwise decreased with a factor β, where $0 < \beta < 1$:

$$r(t) = \begin{cases} r(t-1) + \alpha & \text{if } \text{CBR}(t) < \text{CBR}_{\text{target}} \\ \beta \cdot r(t-1) & \text{otherwise.} \end{cases}$$

Due to the binary decision, AIMD does not converge to a single value, but to a corridor where it oscillates around the target value $\text{CBR}_{\text{target}}$. In order to ensure that these oscillations are bounded, we require that the overall load does not increase when the threshold is exceeded:

$$\text{CBR}(t) > \text{CBR}_{\text{target}} \Rightarrow \text{CBR}(t+1) < \text{CBR}(t) \tag{6.12}$$

$$\Leftrightarrow \sum_{i=1}^{K} r_i(t+1) \cdot d < \sum_{i=1}^{K} r_i(t) \cdot d \tag{6.13}$$

$$\Leftrightarrow \sum_{i=1}^{K} \beta \cdot r_i(t) \cdot d < \sum_{i=1}^{K} r_i(t) \cdot d \Leftrightarrow \beta < 1 \tag{6.14}$$

Similarly, we require that the overall load does not decrease when the load is below threshold:

$$\text{CBR}(t) < \text{CBR}_{\text{target}} \Rightarrow \text{CBR}(t+1) > \text{CBR}(t) \tag{6.15}$$

$$\Leftrightarrow \sum_{i=1}^{K} r_i(t+1) \cdot d > \sum_{i=1}^{K} r_i(t) \cdot d \tag{6.16}$$

$$\Leftrightarrow \sum_{i=1}^{K} r_i(t) \cdot d + \alpha > \sum_{i=1}^{K} r_i(t) \cdot d \quad \Leftrightarrow \alpha > 0 \tag{6.17}$$

Thus, AIMD converges to bounded oscillations as long as $\alpha > 0$ and $\beta < 1$. A full analysis can be found in [1].

To summarize, AIMD is a non-linear, discrete-event algorithm. The additive increase step provides utilization of available resources, while the multiplicative decrease step ensures fairness, since all participants decrease their resource utilization proportional to their current allocation. The repeated increase and decrease steps lead to oscillations around the target threshold. Detailed simulative results can be found in [15].

6.3.5 Discussion

In the previous subsections, several DCC algorithms have been described. A table based or state machine approach was outlined in Sect. 6.3.3 for a reactive DCC algorithm. An adaptive algorithm was introduced in detail in Sect. 6.3.4, where two control strategies were presented: A time discrete linear approach (LIMERIC) and a time discrete differential approach (AIMD). This already shows that there is no ultimate solution how to build a working DCC. But some fundamental building blocks have to stay the same for all solutions.

This fundamental outline is shown in Fig. 6.6. DCC can be regarded as a control loop with a primarily unknown number of distributed controller stations, where the feedback in the control loop is based on the overall channel load (CL). All aforementioned algorithms use CBR as input, which is considered as a measure for this channel load. For the following considerations, we assume that channel load is given by the sum of the channel utilizations each of all K stations is producing. If all stations are identical and message collisions can be neglected (normal DCC operation), the total channel load is given by $CL = \sum_{i=1}^{K} u_i = K \cdot u$, where $u = u_1 = \cdots = u_K$ is the identical channel utilization of each station.

From this, it already gets obvious that when using only the CL as DCC input, the steady state controller equilibrium must be a function of the unknown number of stations K. This is in contradiction to the DCC requirement, that only the CL is the reason for the performance degradations, and that the optimum channel load CL_{opt} is a fixed value independent of K. In our simplified example the optimum channel utilization of each station is then $u_{opt} = CL_{opt}/K$.

There is another reason why, when using the CL as only input to the DCC algorithm, the CL must be a function of K. The value of K is needed to divide the total available channel resources equally between all stations, this is called fairness requirement. If K cannot be at least implicitly derived from the DCC input, one station could take all resources for its own. All presented DCC algorithms exhibit a continuous raise of the CL over K, and therefore can fulfill the fairness criterion.

From these thoughts, it can be seen that the DCC algorithm could benefit from additional inputs like the number of reachable stations K. It is possible to estimate K from the CL value when the DCC algorithm is known, but this suffers from the inaccuracies of the CL determination. Just counting all stations from which messages were received could be outdated in a dynamic vehicular scenario and does not take hidden stations into account that also add to the CL.

To address the hidden station problem a multi-hop dissemination of DCC relevant parameters like CL, number of neighbors K, or TX power levels could be advantageous for coming DCC algorithms. But all these measures add complexity to the distributed control loop that makes it very hard to prove its system-wide stability and convergence behavior.

Concerning convergence and stability, DCC algorithms that are based on the simple structure outlined in Fig. 6.6, can be analyzed in a common way. For a first order low pass input filtering and assuming $CL = u \cdot K$ the control equation for the time discrete control loop shown in Fig. 6.6 can be written as follows:

$$u(t+1) = (1-A) \cdot u(t) + A \cdot f(u(t) \cdot K), \qquad (6.18)$$

where A is a filter parameter and $f(CL)$ is the control function. From this follows that the steady state solution, where $u(t+1) = u(t) = \tilde{u}$, is given by:

$$\tilde{u} = f(\tilde{u} \cdot K). \qquad (6.19)$$

From this the control equilibrium for a certain number of stations K can be calculated. An example for this calculation was shown in Sect. 6.3.4.1. It can also be used to find an adequate control function $f(CL)$ for a given relation between K and CL.

Hence, the filter parameter A is only influencing the dynamic control behavior, but not the control equilibrium. If A is a constant in the range of $0 < A < 1$, then the dynamic behavior is that of a first order time discrete low pass filter at the controller input. If A is a function of $u(t)$ and/or K, a nonlinear filter of first order can be realized with this concept.

The dynamic behavior of $u(t)$ can be analyzed by mapping the discrete time realization to a continuous time description and solve following differential equation, which can be derived from solving Eq. (6.18) for $u(t+1) - u(t)$:

$$u'(t) = \frac{A}{\Delta t} \cdot (f(u(t) \cdot K) - u(t)), \qquad (6.20)$$

where Δt is the time step size of the discrete time realization.

When the continuous time solution converges for all stations to the same stable value $\tilde{u}$, the discrete time DCC algorithm can still be unstable when $u(t)$ changes in one time step by at least twice the distance to $\tilde{u}$ (overshoot). Because of the continuous time mapping, the $u(t)$ step size equals $u'(t)\Delta t$ and the discrete time realization is stable when $|u'(t)|\Delta t < 2|\tilde{u} - u(t)|$ for all $t \geq 0$ and all possible boundary conditions. It is also stable if there is no overshoot ($|u'(t)|\Delta t < |\tilde{u} - u(t)|$) for all positive or negative differences $\tilde{u} - u(t)$.

6.4 DCC in the ITS Protocol Stack

6.4.1 Cross-Layer Aspects

Channel load can be directly controlled on the medium access layer, where the actual decision about accessing and using the channel is made. DCC has to effectively limit the input of messages into the MAC layer queues. This control of the MAC layer queues needs to be placed in or directly above the MAC layer.

However, in the overall protocol stack DCC should not only be focused on the MAC layer. First, it needs the information about the channel load from the physical layer. Second, the messages to be transmitted are generated on the higher layers, which are the application layer and the facilities layer in the ETSI architecture. Using information about the current road traffic scenario and the available channel utilization on the upper layers can already reduce the load on the internal message queues by adapting the data traffic to safety needs. If the usage of the full available channel resources do not significantly increase road safety, the message rate should be reduced to utilize only the necessary share of the channel (see also Sect. 6.1.2). The DCC algorithm will then automatically distribute the remaining channel resources between all other stations. As an example the CAM rate is standardized in ETSI EN 302 637-2 [6] to be dependent on the vehicle dynamics (e.g., speed).

On the network layer, further information is available to support DCC: a neighbor table contains information about neighboring vehicles that also contribute to channel load and possibly channel congestion. This number can be a valuable input to DCC, because a first estimate on the available share of the channel resource can be derived from it. Figure 6.10 shows the ETSI ITS station architecture, which gives an example on which layers the DCC collects information or controls the message flow. According to the ETSI architecture, the parts of DCC, which span across the layers, are coordinated in a management plane.

6.4.2 Evolution of DCC Standards

In Europe, DCC is required by regulation in EN 302 571 [4] by referencing TS 102 687 [2]. TS 102 687 was approved and released by ETSI TC ITS in July 2011 and it is the first standard addressing congestion control for VANET operation. TS 102 687 describes the different approaches available for congestion control such

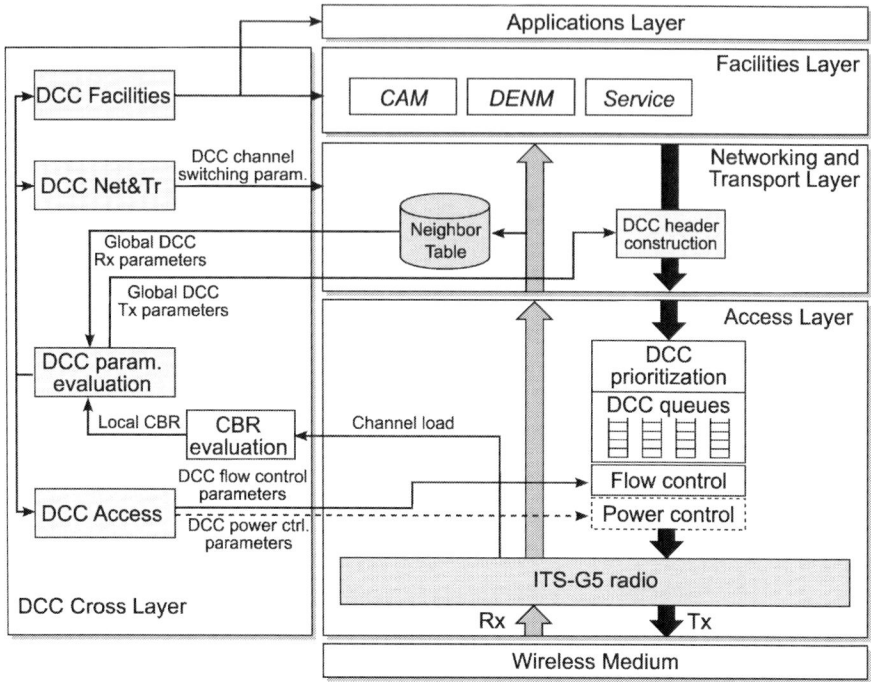

Fig. 6.10 DCC in the ETSI ITS station architecture (see also [7])

as transmit power control (TPC), transmit rate control (TRC), transmit data rate control, (TDC), DCC sensitivity control (DSC), and transmit access control (TAC). Further, it proposes a state machine for controlling the network load. In Fig. 6.10, the DCC architecture for the European C-ITS protocol stack is depicted and the DCC components are spread throughout the stack. TS 102 687 belongs to the access layer. TS 102 636-4-2 [5], which was approved in October 2013, addresses DCC at the network and transport layer and its main contribution is the possibility to disseminate local CBR and the highest received CBR information to neighboring stations through a field in the GeoNetworking header. In TS 102 636-4-2, the highest CBR (received or local) is used as input to the congestion control algorithm. The dissemination of the CBR information has the advantage that all co-located stations will adjust their congestion control algorithm based on the same input, even when their local CBR results differ because of measurement uncertainties. Further, a station that perceives a low local CBR but receives high CBR values from neighbors reduces its load contribution to hidden congested areas.

In Table 6.2, a summary of the different technical specifications addressing DCC at the different layers of the ETSI TC ITS protocol stack is shown. At the end of 2015, there will be a holistic view on DCC for the European C-ITS protocol stack.

In the USA, DCC for DSRC safety communication has been an active research topic for several years, both in the academic and industrial communities.

Table 6.2 Overview of ETSI standards related to DCC

Layer	DCC component	Technical specification	Status	Description
Facilities	DCC Facility	TS 103 141	Under development	Planned final draft for approval October 2015.
Networking and transport	DCC Net&Tr	TS 102 636-4-2	V1.1.1	Approved in October 2013.
Access	DCC Access	TS 102 687	V1.1.1 currently being revised	Approved in July 2011. Currently being revised for accommodating the latest findings within congestion control.
Management	DCC Mgt.	TS 103 175	Under development	Planned final draft for approval October 2015.

A consortium of automakers, in cooperation with the US Department of Transportation, has implemented and tested two main DCC approaches. One approach is based on the LIMERIC adaptive message rate control algorithm described above. The other is based primarily on a reactive transmit power control. Both methods also consider vehicle dynamics and send messages more frequently when necessary to keep tracking error low. In the future, US DSRC standards are expected to reflect the industry consensus that emerges from this research and testing.

6.5 Conclusion

DCC is an important mechanism to mitigate channel overload in VANETs that are based on the IEEE 802.11 standard. It becomes necessary when broadcast communication with high rates or a high number of participants dominates. As discussed in Sect. 6.2 of this chapter, high channel load results directly in degradation of reception rates and effective reception ranges. This, in turn, reduces the information exchange among vehicles and limits cooperative awareness.

DCC algorithms reduce the individual contribution to channel load by limiting transmit rate and transmit power depending on channel load measurements. Typical approaches to DCC have been discussed in Sect. 6.3: reactive algorithms use a measured channel load to directly set a fixed transmit rate. This mapping between channel load and transmit rate allows a high flexibility in parameterization, but cannot ensure optimal utilization when the number of vehicles changes. Adaptive algorithms use a channel load target and adapt the transmit rate based on the difference between actual and desired load. For such algorithms it is important that the channel load converges to a state where all vehicles obtain a fair share of the channel resource. We have seen two examples for adaptive algorithms (LIMERIC and AIMD) that achieve this goal, required that channel load is perceived equally.

The control of transmit rates is ideally done on the medium access layer. This is an essential function of DCC, but DCC is not limited to this layer. In a larger scope, information from the neighbor table on the network layer can add valuable input, and congestion can already be controlled at message generation on the application and/or facilities layer. Furthermore, DCC can benefit from an exchange of information among vehicles including transmit rate, the number of neighbors, or the perceived channel load, but this also adds complexity.

DCC is already part of specifications in the ETSI and under consideration in the USA. It can be expected that DCC becomes an essential part of protocol stacks for vehicular networks.

References

1. Chiu DM, Jain R (1989) Analysis of the increase and decrease algorithms for congestion avoidance in computer networks. Comput Netw ISDN Syst 17(1):1–14
2. ETSI TS 102 687 V1.1.1 ITS; Decentralized congestion control mechanisms for intelligent transport systems operating in the 5 GHz range; Access layer part (2011)
3. ETSI EN 302 663 V1.2.1 (2013-07) ITS; Access layer specification for intelligent transport systems operating in the 5 GHz frequency band (2013)
4. ETSI EN 302 571 V1.2.0 (2013-05) ITS; Radiocommunications equipment operating in the 5 855 MHz to 5 925 MHz frequency band; harmonized EN covering the essential requirements of article 3.2 of the R&TTE directive (2013)
5. ETSI TS 102 636-4-2 V1.1.1 (2013-10) ITS; Vehicular communications; GeoNetworking; Part 4: Geographical addressing and forwarding for point-to-point and point-to-multipoint communications; Sub-part 2: Media-dependent functionalities for ITS-G5 (2013)
6. ETSI EN 302 637-2 V1.3.1 (2014-09) ITS; Vehicular communications; basic set of applications; Part 2: Specification of cooperative awareness basic service (2014)
7. ETSI TR 101 612 V1.1.1 (2014-09) ITS; Cross layer DCC management entity for operation in the ITS G5A and ITS G5B medium; Report on cross layer DCC algorithms and performance evaluation (2014)
8. Fuxjager P, Ruehrup S, Smely D (2013) Impact of CCA threshold, contention window, and transmit rate on VANET simulations. In: 39th Annual conference of the IEEE industrial electronics society (IECON 2013), pp 6885–6890
9. IEEE Std 802.11-2012. Part 11: Wireless LAN Medium Access Control (MAC) and Physical Layer (PHY) Specifications (2012)
10. Jacobson V (1988) Congestion avoidance and control. In: Symposium on communications architectures and protocols, SIGCOMM '88, pp 314–329
11. Karedal J, Czink N, Paier A, Tufvesson F, Molisch A (2011) Path loss modeling for vehicle-to-vehicle communications. IEEE Trans Veh Technol 60(1):323–328
12. Kenney JB, Bansal G, Rohrs CE (2011) Limeric a linear message rate control algorithm for vehicular DSRC systems. In: 8th ACM international workshop on vehicular inter-networking, VANET'11, pp 21–30
13. Kerner BS (2004) The physics of traffic, empirical freeway pattern features, engineering applications, and theory. Springer, Berlin. ISBN 978-3-642-05850-9

14. Kloiber B, Strang T, de Ponte-Müller F, Rico Garcia C, Röckl M (2010) An approach for performance analysis of ETSI ITS-G5A MAC for safety applications. In: 10th ITST
15. Ruehrup S, Fuxjaeger P, Smely D (2014) TCP-like congestion control for broadcast channel access in VANETs. In: 3rd International conference on connected vehicles and expo (ICCVE)
16. SAE J 2735; Dedicated short range communications (DSRC) message set dictionary (2014)
17. Schmidt R, Lasowski R, Leinmüller T, Linnhoff-Popien C, Schäfer G (2010) An approach for selective beacon forwarding to improve cooperative awareness. In: Vehicular networking conference (VNC), pp 182–188
18. Schmidt RK, Leinmüller T, Schäfer G (2010) Adapting the wireless carrier sensing for VANETs. In: 6th International workshop on intelligent transportation (WIT), Hamburg
19. Schmidt RK, Leinmüller T, Schoch E, Kargl F, Schäfer G (2010) Exploration of adaptive beaconing for efficient intervehicle safety communication. IEEE Netw Mag 24:14–19. Special Issue on Advances in Vehicular Communications Networks
20. Schmidt RK, Brakemeier A, Leinmüller T, Kargl F, Schäfer G (2011) Advanced carrier sensing to resolve local channel congestion. In: 8th ACM international workshop on vehicular inter-networking, VANET'11, pp 11–20
21. Subramanian S, Werner M, Liu S, Jose J, Lupoaie R, Wu X (2012) Congestion control for vehicular safety: synchronous and asynchronous MAC algorithms. In: 9th ACM international workshop on vehicular inter-networking, systems, and applications (VANET'12), pp 63–72
22. Tielert T, Jiang D, Chen Q, Delgrossi L, Hartenstein H (2011) Design methodology and evaluation of rate adaptation based congestion control for vehicle safety communications. In: Vehicular networking conference (VNC), pp 116–123

Chapter 7
Multi-Channel Operations, Coexistence and Spectrum Sharing for Vehicular Communications

Jérôme Härri and John Kenney

Abstract DSRC has been allocated (three in EU, seven in US) dedicated channels at 5.9 GHz for vehicular communications. Although resource allocations on the common control channel (CCH) reserved for safety-related applications have been well investigated, efficient usage of the other SCHs is less developed. With new ITS safety-related applications appearing, such as autonomous driving or truck platooning, as well as the expected coexistence between ITS and non-ITS technologies for smart mobility applications, operating on multiple channels and efficiently sharing the ITS spectrum become critical. First, multi-channel operations aim at mitigating the communication load on specific channels by offloading part of traffic to alternate channels. Second, multi-channel operations aim at providing mechanisms to dynamically change channels and fit to the service requirements as function of external interferences or to varying traffic conditions. In this chapter, we describe the regulations and mechanisms for ITS multi-channel operation and coexistence in the US and in the EU. We first provide an overview of the frequency allocations and access restrictions for ITS, and then describe the protocols available in standards and R&D for multi-channel operations.

Keywords VANET • WAVE • ETSI • Multi-channel • Spectrum allocation • Multi-channel congestion control • Offloading • Service management • Coexistence • WAVE 1609.3 • WAVE 1609.4 • ETSI EN 302 571 • ETSI TS 102 724 • ETSI TS 102 636-4-2 • ETSI TS 103 165

J. Härri (✉)
Mobile Communications EURECOM, Sophia Antipolis, France
e-mail: haerri@eurecom.fr; jerome.haerri@eurecom.fr

J. Kenney
Toyota InfoTechnology Center, Mountain View, CA, USA
e-mail: jkenney@us.toyota-itc.com

© Springer International Publishing Switzerland 2015
C. Campolo et al. (eds.), *Vehicular ad hoc Networks*,
DOI 10.1007/978-3-319-15497-8_7

7.1 Introduction

Since the pioneer developments of DSRC technologies for Cooperative Intelligent Transportation Systems (C-ITS), road safety applications had the favors of car industry as a major factor of industrial growth and revolutionary driving experience. At that time, all traffic safety applications involved the transmission of safety-related messages—CAMs/DENMs in the EU or BSMs in the US—on a single common well-known safety channel to detect and anticipate road hazard (see Chap. 5 for a detailed description of these messages). Considering the limited capacity of this channel, most of the scientific, standardization and industrial Research & Development (R&D) aimed at developing smart cooperative communication and network strategies to mitigate congestion on this channel.

Yet, more than one channel are available for C-ITS applications. In 1999, the Federal Communications Commission (FCC) in the US allocated seven 10 MHz channels for C-ITS in the 5.9 GHz band, while in 2008 the Electronic Communication Committee (ECC) in the EU allocated three 10 MHz channels, including four extra 10MHz channels to be allocated in the future. And despite their early availability and the potential innovations from C-ITS applications, the available C-ITS spectrum has not been well used.

C-ITS applications considered for *Day One* deployments, such as *Road Hazard Warning* or *Intersection Collision Warning* have been specified to only use a single of these channels, the channel called CCH in the EU, and Channel 172 in the US. Yet, with the appearance of C-ITS applications considered for *Day Two* deployments, such as *autonomous driving* or *platooning*, as well as the future coexistence of DSRC with non-ITS technologies, such as WiFi-Giga or LTE-Direct, smart and dynamic multi-channel mechanisms for a fair and efficient usage of the overall C-ITS spectrum at 5.9 GHz are expected to become critical.

Multi-channel mechanisms have three objectives. First, they allow to efficiently use the resources of all available channels by offloading some type of traffic to adjacent channels. Second, they specify mechanisms for Service Providers (SPs) to dynamically offer services on various channels and Service Consumers (SCs) to switch to the corresponding channel to consume the offered service. Finally, they provide the opportunity for various technologies to coexist in same spectrum bands, by detecting potential harmful interferences and dynamically move to other channels.

Standards describing these mechanisms showed different maturity evolutions between the US and the EU, and although globally sharing same objectives, they differ in some aspects. The 70 MHz spectrum in the US motivated the early development of multi-channel switching mechanisms, including network-layer primitives. Accordingly, the IEEE 1609.4 [21] describing multi-channel switching mechanisms, and the IEEE 1609.3 [22] describing multi-channel service management primitives have been completed as early as 2010. In the EU, the smaller 30 MHz ITS spectrum instead motivated the development of smart traffic offloading or relaying mechanisms on alternate channels. Accordingly, the ETSI provided TS 102 636-4-2 [15] describing the DSRC network-level support for multi-channel

7 Multi-Channel and Coexistence   195

operations, and the TS 103 165 [17] describing multi-channel congestion control mechanisms. However, these ETSI standards were still under development at the time of writing of this chapter.

Using C-ITS channels depends on specific per-channel access restrictions, both in terms of radiation and applications. Channels may be a control channel, a channel open to any C-ITS services, or a channel restricted to only one type of C-ITS application. In Sect. 7.2, we survey the different channel allocation plans in the US and in the EU, emphasizing their differences and similarities, as well as describing the use of the different channels for C-ITS applications.

Not all DSRC devices are born equal. Different types exist, from those that are only static or mobile, to those that are capable of switching and those that are not, and those that include multiple transceivers and those that do not. In Sect. 7.3, we review the different ITS stations and capabilities in the US and in the EU. We will describe the differences between On-Board Units (OBUs) and Road Side Units (RSUs), as well as their capabilities to access different types of channels.

Efficiently using all available channels for DSRC devices supporting one or more transceivers requires mechanisms to switch between channels either synchronously or asynchronously, monitor the load on various channels and offload traffic to available channels, and provide mechanisms for C-ITS SPs and SCs to *rendezvous* on common channels to consume services. In Sect. 7.4, we take a holistic view and describe the multi-channel mechanisms first in the US standards and in the EU standards, emphasizing the multi-channel switching mechanisms and their applications for multi-channel congestion control and multi-channel service management.

Finally, available spectra is usually very scarce in the telecommunication domain. When 70 MHz is reserved and not efficiently used, it attracts the attentions of other technologies. The DSRC technology is therefore expected to have to co-exist, but the sensitivity of some C-ITS applications, and the resource greediness of non-ITS applications are expected to make such coexistence very challenging. In Sect. 7.5, we introduce the reasoning behind such coexistence, from the resource requirements of the new very high speed IEEE 802.11ac [26] standard, to potential strategies for efficient and fair coexistence between ITS and non-ITS devices in the 5 GHz band.

We conclude this chapter in Sect. 7.6, and emphasize future challenges in multi-channel operations and coexistence for DSRC and C-ITS applications.

7.2 Frequency Allocation

One of the foundations of the Cooperative ITS system architecture (also referred to in the US as DSRC) is the spectrum in which the Vehicle-to-Vehicle (V2V) and Vehicle-to/from-Infrastructure (V2I) communication takes place. In both the US and Europe, authorities have allocated licensed spectrum for this purpose. In this section we present details of these allocations, including frequency ranges, channelization, power limits, and prescribed uses. Fortunately, many of these details are the same or similar between the US and Europe. The harmonization of spectral allocation facilitates common hardware platforms for deployment in both regions.

7.2.1 Allocation in the US

In the US, the FCC is responsible for spectrum regulation. In 1999, the FCC allocated 75 MHz of spectrum in the range 5.850–5.925 GHz (commonly called the 5.9 GHz band) for ITS, and specifically for the DSRC Service. In 2003, based on input from the ITS community and US Department of Transportation (DOT), the FCC issued licensing and service rules for the spectrum [28]. These rules include a division of the spectrum into seven non-overlapping 10 MHz channels, and a 5 MHz unused band at the low end. As shown in Fig. 7.1, these 10 MHz channels are numbered with even numbers 172 through 184. It is also permitted to operate in 20 MHz channels 175 and 181, each of which overlaps with two 10 MHz channels. Each channel includes a maximum conducted power and Effective Isotropic Radiated Power (EIRP) limit; in some cases, there are separate limits for public (i.e., government operated) devices and for private devices. While these power limits typically permit 33 dBm EIPR, or higher, key applications will more often use transmit power in the range of 10–20 dBm. These lower powers are chosen to achieve a desired transmission range without causing excess interference at longer distances.

Each channel is further classified as either a CCH or an SCH. Channel 178, in the middle of the seven 10 MHz channels, is the CCH. The other six 10 MHz channels are classified as SCHs, as are the two 20 MHz overlapping channels. The roles of the CCH and SCHs are not specified in detail in Fig. 7.1, but as explained below these distinctions are key to multi-channel operation in the US. The 2003

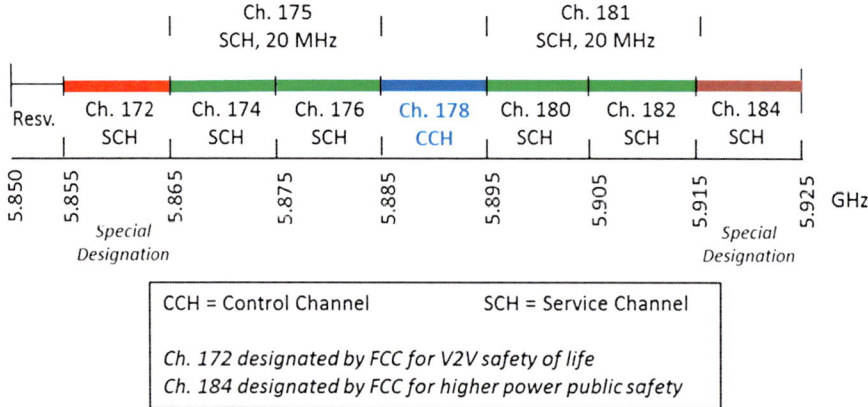

Fig. 7.1 Channels allocated in the US by FCC for ITS

7 Multi-Channel and Coexistence

rules also require equipment to conform to the Physical (PHY) layer and Medium Access Control (MAC) sublayer protocols defined in ASTM standard E2213-03. This standard was replaced in 2010 by the IEEE 802.11p-2010 Wireless Access in Vehicular Environments (WAVE) amendment [23] to the popular IEEE 802.11 standard. However, the FCC regulations have not yet been updated to require conformance to the new standard. The 802.11p amendment was subsequently incorporated into the integrated IEEE 802.11-2012 standard, which continues to be amended and revised. Conformance to the IEEE 802.11 standard in the 5.9 GHz band in both the US and Europe requires use of *communication outside the concept of a Basic Service Set (BSS)*, abbreviated as OCB, which is the principal novelty of the IEEE 802.11p amendment (see Chaps. 3 and 4, respectively, for details of the PHY and MAC standards).

7.2.2 Allocation in the EU

In Europe, the ECC is responsible for spectrum regulation, and the European Commission (EC) is responsible to enforce that the allocated spectra are made available in all states of the EU. In 2008, an ECC *Decision* [8] made 30 MHz of spectrum in the range 5.875–5.905 GHz (commonly called *ITS-G5A*) available for ITS restricted to safety-related communications, as well as an extra 20 MHz of spectrum in the range 5.905–5.925 GHz (commonly called *ITS-G5D*) for future ITS extensions. Also in 2008, an ECC *Recommendation* [9] made 20 MHz of spectrum in the range 5.855–5.875 GHz (commonly called *ITS-G5B*) available for ITS non-safety communications. Figure 7.2 illustrates the FCC frequency allocation plan and their different classes. As for the FCC spectrum allocation, the ECC allocation comprises of six SCHs and one CCH. Yet their EU-wide availabilities as well as their usage slightly differ from the FCC allocation.

As specified in [10], the ITS-G5A frequency band contains channels CCH, SCH1, and SCH2, which are restricted to ITS road safety-related communications. SCH3 and SCH4 are contained in the frequency band ITS-G5B and are intended for

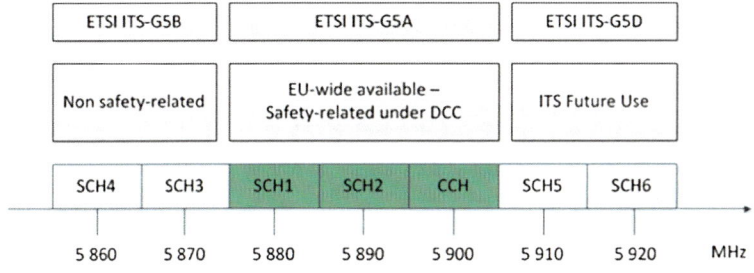

Fig. 7.2 ITS carrier frequencies in the EU [8, 9]; ITS-G5A is the only frequency band currently available EU-wide [1]

ITS non-safety communications. Finally, SCH5 and SCH6 are part of the frequency band ITS-G5D and reserved for future ITS extensions. At the time of writing, the ITS-G5A band is the only ITS spectrum currently usable European-wide following a 2008 EC Decision [1]. The other ITS bands (ITS-G5B, ITS-G5D) have been allocated but not enforced to be made available by this EC decision. Depending on the states in the EU, these bands may or may not be available and usable.

One important difference between the US and the EU is that BSMs in the US are not sent on the CCH but rather sent on Channel 172, an SCH specially designated for this purpose by the FCC, while the analogous CAMs and DENMs in the EU are sent on the CCH.

Another difference between the FCC and the ECC is that although the ITS-G5A spectrum is restricted to safety-related communications, it is not restricted to a specific technology [1, 8]. In principle, if other technologies than ITS-G5 (e.g., 3GPP LTE, WiFi-Giga) could operate in 10 MHz OFDM channel for traffic safety, they could operate in the ITS-G5A band. Complementary to the ECC allocation, the ETSI [10, 16] yet requires technologies operating on the ITS-G5A to operate under the control of the ETSI DCC [11, 17] mechanisms, which are responsible for per-packet transmit power and rate restrictions (see Chap. 6).

As illustrated in Fig. 7.3, each ITS channel in the ITS-G5A, ITS-G5B, and ITS-G5D bands has spectral power restrictions [16], which in turn restrict the ITS applications that can be operated in them. The CCH and SCH channels have a 23 dBm/MHz transmit power restriction (33 dBm EIRP on 10 MHz channel). Due to adjacent channel interferences, the SCH2 has a stronger spectrum power restrictions to 13 dBm/MHz (23 dBm EIRP on 10 MHz channel), which restrict any Inter-Vehicle Communication (IVC) to short range transmissions. The ITS-G5C band also follows the spectral power restrictions of the Radio Local Area Network (RLAN) bands (e.g., WiFi-5) as described in [13].

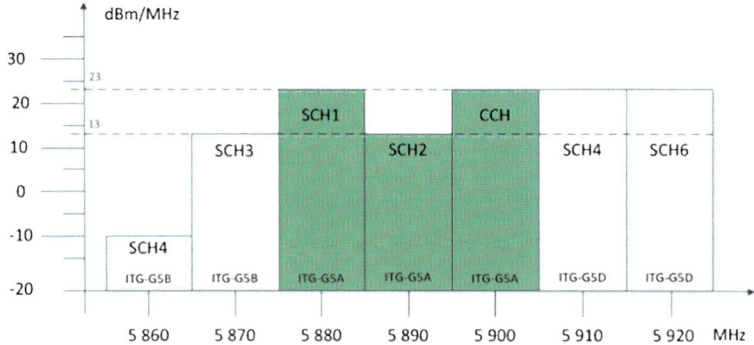

Fig. 7.3 Power spectral limits on each ITS channel [13], adapted from [15]

7.3 ITS Station Types and Restrictions

ITS stations have different functionalities depending if they are mobile (on-board of vehicles) or static (as road-side units). ITS stations may also be equipped with more than one transceiver. As such, ITS stations may access different type of channels, implement different multi-channel mechanisms, and have different transmit restrictions. We describe in this section the access restrictions of DSRC devices in the US, as well as ITS Station (ITS-S) in the EU.

7.3.1 DSRC in the US

DSRC devices must be licensed to operate in 5.9 GHz band. The FCC recognizes two types of device, each with its own licensing status. OBUs are permitted to operate while in motion, and can operate anywhere vehicles or pedestrians are allowed. OBUs are licensed implicitly by rule according to Part 95 of the FCC regulations. By contrast, RSUs are required to be stationary when operating, and are licensed explicitly for operation at a site or in a region. RSU operations are specified in Part 90 of the FCC regulations. OBUs may use the 5.9 GHz spectrum to communicate with other OBUs or with RSUs. RSUs are only permitted to communicate with OBUs in the 5.9 GHz spectrum.

In February 2013 the FCC issued a Notice of Proposed Rulemaking (NPRM) [30] concerning operation of unlicensed devices in various portions of the 5 GHz band. In this NPRM, the FCC solicited comments about the possibility of unlicensed devices (e.g., IEEE 802.11 or other devices) sharing the 5.9 GHz DSRC spectrum, with a condition that no unlicensed device would be allowed to harmfully interfere with a DSRC device.

In 2006, again at the request of the ITS community, the FCC updated the 2003 rules by adding special designations to Channels 172 and 184 [29]. Channel 172 is designated exclusively for vehicle-to-vehicle safety communication for accident avoidance and mitigation, and safety of life and property applications. Thus, Channel 172 is where vehicles will exchange BSM, as well as send and receive other messages integrally related to this mission (e.g., Intersection MAP (MAP) or SPAT messages[1] are likely to be sent by RSUs on Channel 172). As a historical note, in early phases of testing the BSM was sent on the CCH (Ch. 178), but following the 2006 FCC decision there was a change in concept of operation so that in the US the BSM is now sent on Channel 172.

Channel 184 is designated exclusively for high-power longer-distance communications to be used for public safety applications involving safety of life and property, including road intersection collision mitigation. One key application for Channel

---

[1] See Chap. 5 for a detailed description of such messages.

184 is signal preemption by emergency vehicles. In this application, an emergency vehicle can send a message to a signal controller requesting that the signal state be set so that the emergency vehicle does not contend with other traffic. The option to use power up to 40 dBm allows this preemption request to reach a kilometer or more in many settings, which gives the signal controller time to safely clear the intersection. Note that Channels 172 and 184 remain classified as SCHs, along with these special designations.

7.3.2 ITS-S in the EU

The ETSI is responsible for the ITS-S plans in Europe, whether it is for static RSUs or mobile OBUs. This section describes the ITS-S architecture for operation in ITS-G5 band. An ITS-S station may contain one or more ITS G5 transceivers, each of them camping to one or more ITS channels. ITS-Ss maybe classified into three types depending on their functions: safety-related, traffic efficiency, commercial applications. The supported configurations (operating channels and channel switching) for ITS transceivers (number and type of supported ITS G5 transceivers) for ITS Stations are described in the ETSI specification [12]. At the time of writing, [12] was being re-opened to allocate channels for the ETSI ITS Day two applications. Without lost of generality, we provide one approach currently considered.

IVC communication on ITS-G5 should be capable of operating on single channels or on multiple channels according to the requirements of the ITS applications. As a scan mode does not exist in ITS-G5 transceivers, a base channel is specified for each transceiver configuration. This base channel corresponds to where an ITS-G5 transceiver may expect to receive unsolicited traffic.

An ITS-S using one or more ITS-G5 transceivers shall operate each transceiver in one of the following configurations, as depicted in Fig. 7.4:

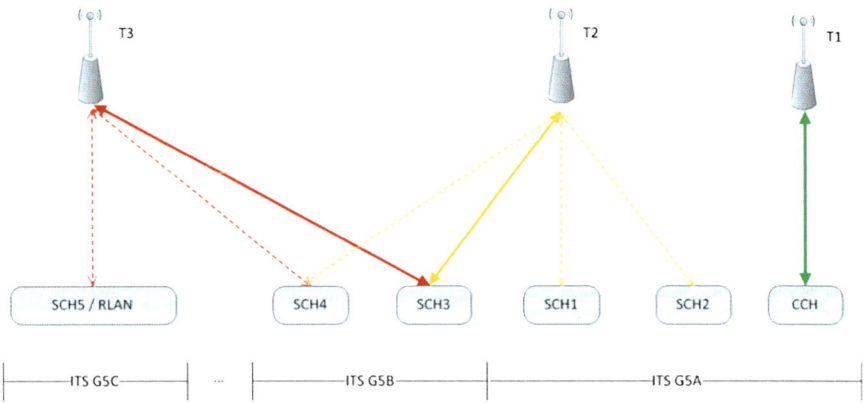

Fig. 7.4 ITS-S types and restrictions as defined at ETSI [12]

- *Transceiver Configuration 1 (T1)*: This configuration corresponds to a transceiver strictly operating for safety-related ITS applications. A T1 station never switches channels and must strictly remain on its assigned channel.
- *Transceiver Configuration 2 (T2)*: The transceiver can be tuned to arbitrary ITS-G5A or ITS-G5B channels. T2 transceivers may support synchronous or asynchronous channel switching mechanisms, described later in this chapter.
- *Transceiver Configuration 3 (T3)*: The transceiver can be tuned to arbitrary ITS-G5B or ITS-G5C channels. It is the only transceiver type than can share spectrum with WiFi on the RLAN bands.

The ETSI regulation mandates that at least one T1 transceiver of any ITS-S be constantly tuned to the CCH to send and receive CAM and DENM when the vehicle is considered *in traffic*. This clause is subject to interpretations. In Europe, being *in traffic* corresponds to having the engine *turned on*, regardless of moving or being static, being on the street or not.

Yet, T1 transceivers are not restricted to the CCH, and may be tuned to other ITS-G5A channels (SCH1 and SCH2) (not shown in Fig. 7.4) for other safety-related applications such as platooning or autonomous driving. For non-safety related and commercial traffic, T2/T3 transceivers are used. Service Announcement Messages (SAMs) [18] are required to let SCs become aware of the services offered by SCs, and which must be sent on a well-known channel to all SPs and SPs. Accordingly, T2/T3 transceivers have a base channel *SCH3* corresponding to where SAMs can be received, and which may also be used to coordinate ITS-G5 and any other type of networks (WiFi, LTE, etc.) if necessary.

Also, both T1 and T2 transceivers must support the ETSI DCC specification, while T3 transceivers do not, as T3 are primarily operating in the ITS-G5C band. Finally, depending on the context, an ITS-S may reconfigure transceivers. For example, a single transceiver ITS-S not considered *in traffic* may reconfigure its T1 transceiver to a T2 or T3 transceiver.

An ITS-S consists of one or more ITS-G5 transceivers, and vehicles or infrastructures may have one or more ITS-Ss. This leads to the following requirements:

- *Single ITS-S, single ITS-G5 transceiver*—the ITS-G5 transceiver must be T1 while *in traffic*. It can be reconfigured to T2 or T3 transceivers when not *in traffic*.
- *Single ITS-S, multiple ITS-G5 transceivers*—at least one ITS-G5 transceiver must be T1 while *in traffic*, while other transceivers may be of different types.
- *Multiple ITS-Ss, multiple ITS-G5 transceivers*—at least one ITS-G5 transceiver of at least one ITS-S must be T1 while *in traffic*, while other transceiver and the other ITS-Ss may operate in other bands.

This classification enforces that the first category of ITS-Ss configurations can only support safety-related applications operating on the CCH. If other type of safety-related applications, such as platooning or highly autonomous driving, are required, at least another T1 transceiver must be available.

In the rest of this chapter, we will describe the mechanisms available for T1, T2, and T3 transceivers for multi-channel operations.

7.4 Multi-Channel Operations

Multi-channel operation means to tune a given transceiver to different wireless channels at different times. It is sometimes also referred to as channel switching. Given the large number of channels available for DSRC (US) or C-ITS (EU) communication, multi-channel operation is a desirable alternative to statically assigning one device to each channel, or to foregoing operation on some channels to which no device is statically assigned. This section describes the protocols and processes designed to facilitate multi-channel operation in the 5.9 GHz band.

7.4.1 US Regulations: WAVE 1609.3 and 1609.4

As shown in Fig. 7.1, there are seven 10 MHz and two 20 MHz channels identified in the US DSRC spectrum. Some of them have special designations, while others might be used to support a wide variety of services. The large number of channels provides not only a high aggregate data communication capacity, but also flexibility in the assignment of applications and services to specific channels. Road geometry, traffic movement, the location of RSUs and of various sources of interference, and the set of applications supported in a region may dictate a certain set of channel assignments at a given time and place, but changes to those variables might make an alternate assignment desirable at another time or place. A given vehicle or infrastructure device might find it useful to utilize all of those channels at one time or another, but over any short period of time (hundreds of milliseconds) the device likely will not need to utilize more than two or three of those channels. While one could theoretically build the device with nine radios, one tuned permanently to each of the channels, it would be an inefficient use of resources. An intelligent and efficient alternative is provided in the IEEE 1609 suite of standards. In particular, the IEEE 1609.4-2010 Multi-Channel Operation [21] and IEEE 1609.3-2010 Networking Services [22] standards define a flexible approach based on time division and channel switching.

7.4.1.1 Channel Switching Principles

As the DSRC technology does not support the WiFi *scanning* phase, any DSRC device must somehow become aware on which channel it can meet other DSRC devices. For BSM, the channel is well known (Channel 172). But for non-safety services, any of the other channel could theoretically be used. Accordingly, IEEE 1609 channel switching conveys service information using announcements.

At a high level, IEEE 1609 channel switching uses the following paradigm. Time is divided into alternating intervals, and all devices that wish to participate are synchronized to a common clock so they know what interval is active at any given time. One type of interval is the *Control Channel Interval (CCH Interval)*. The other type of interval is the *Service Channel Interval (SCH Interval)*. During the CCH Interval, devices wishing to offer services (i.e., Provider devices) and devices wishing to utilize provided services (i.e., User devices) all tune a radio to the CCH (see Fig. 7.1)so that Providers can advertise services via theWSA and Users can listen to the advertisements. This can be referred to as a rendezvous operation, i.e. it is a way for devices to find each other without prior arrangement. A given WSA specifies on which SCH the Provider offers the service. If a User wishes to participate in an advertised service, it tunes a radio to the indicated SCH during the following SCH Interval, perhaps by switching the radio that was previously tuned to the CCH. During the next (or a subsequent) CCH Interval, the device switches a radio back to the CCH to again listen for advertised services.

At this high level one can see that this paradigm supports a single radio switching among any or all of the channels to access desired services, one channel at a time. Note that channel switching is optional within the IEEE 1609 standards. A single-radio device might use channel switching or not. A device might also have multiple radios, some of which switch and some do not. One configuration expected to be common for OBUs will be to have two radios, one of which is statically tuned to Channel 172 for safety communication and one of which follows the switching paradigm to access other services.

The IEEE 1609 standards describe an optional internal mechanism that a User device can utilize to manage its participation in services. The management function of the device maintains a list of service requests made by higher layers in the device. If a service in the service request list appears in a received WSA, the management function initiates a channel switching operation to the SCH indicated in the WSA. Services in the request list can also register a priority, which the management function uses to arbitrate requests in the case that more than one is available at the same time. The standards specify primitives for the maintenance of this service request list. The service request list is an optional mechanism, not required for over-the-air interoperability.

Services are identified in the WSAs, and in the service request table, using the Provider Service Identifier (PSID) value. Each PSID value is associated with an application area. Examples of application areas for which PSID values have been allocated are: *vehicle to vehicle safety and awareness*, *traveler information and roadside signage*, and *electronic fee-collection*. At the time of writing, the allocation of PSID values to application areas is documented in the IEEE 1609.12 Identifier Allocations standard [24]. In the future, this registration function may be performed by another organization, for example the *IEEE Registration Authority*. The IEEE 1609 WG coordinates the allocation of PSIDs from the same number space that ISO and ETSI allocate ITS Application Identifier (ITS-AIDs). The PSID is a variable

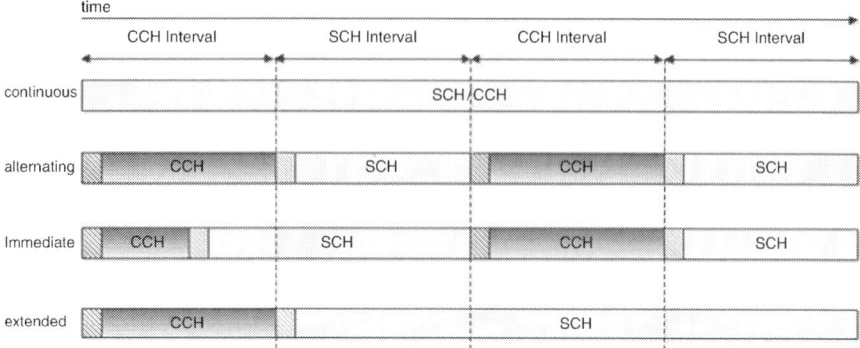

Fig. 7.5 Channel switching principles as defined in IEEE 1609.4 [21]. Adapted from [21]

length value, occupying one to four bytes. The format is specified in IEEE 1609.3. The PSID value is also used in the WAVE Short Message Protocol (WSMP) (IEEE 1609.3) [22] and in WAVE Security Services (IEEE 1609.2) [25].

Application areas are by design somewhat general. In order to include more specific information about an advertised service, the WSA may also include a Provider Service Context (PSC) field for each advertised PSID. The PSC is a variable length field, up to 31 bytes. The format of the PSC field is specific to the PSID value and is specified by the organization to which the PSID value is allocated, for example ISO or SAE.

The high level switching operation described above, in which a device tunes to the CCH during each *CCH interval* and to an SCH during each *SCH interval*, is described in the IEEE 1609 standards as *alternating channel access*. As noted, a device might also utilize *continuous channel access* to tune to one channel indefinitely. The standards provide for two additional channel access modes, *immediate channel access* and *extended channel access*. These latter two can be used in combination if desired. Figure 7.5 illustrates the four channel switching principles.

In the case of *immediate channel access*, when a User device wishes to access a particular advertised service, it switches to the indicated SCH immediately, not waiting for the start of the next *SCH interval*. In the case of *extended channel access*, the User device remains tuned to the indicated SCH through one or more subsequent *CCH intervals*, until a specified time interval has expired. When extended access is completed the User device switches back to the CCH. These additional modes are designed to reduce the latency associated with accessing a service. An SP can indicate in the WSA whether it is capable of providing the service only during the *SCH interval*, or during *both* CCH and SCH intervals. The User device can use this information when deciding whether to initiate either immediate or extended access. A given device can utilize any of these channel access modes: continuous, alternating, immediate, or extended, and some devices will use different modes at different times. In this way the IEEE 1609 channel switching mechanism provides a User device with a high degree of flexibility in accessing DSRC services.

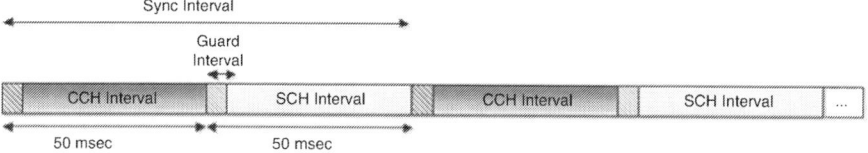

Fig. 7.6 *Alternating Channel Access* switching principle as defined in IEEE 1609.4 [21]

Figure 7.6 illustrates the time division associated with IEEE 1609.4 channel switching. Every 100 msec period, synchronized with the GPS second boundary, constitutes a *Sync Interval*. The first 5 msec of each sync interval is the CCH Interval, while the latter 50 msec is the SCH Interval. These durations are default values in the standard, and have been used in most testing. The standard permits other duration values, with the constraint that the ratio of 1 s to the Sync Interval be an integer.

Each CCH interval and SCH interval begins with a short *Guard Interval*. This is to accommodate the switching of radio resources from one channel to another. Devices participating in channel switching are encouraged to abstain from transmitting during the Guard Interval, since the receiver(s) might not be able to receive yet. The default Guard Interval is 2 msec.

Another potential complexity associated with channel switching is that without special care there could be a large amount of traffic waiting to be transmitted at the start of a given SCH (or CCH) Interval. The channel access procedures of the IEEE 802.11 MAC protocol would arbitrate that access in a way that could result in an artificially high level of frame collisions, compared to a case in which that traffic was spread out over the Interval. To mitigate this, a device that enqueues a packet for a specific SCH (or CCH) Interval that has not yet begun declares the channel busy when the Interval begins. This forces the channel access to use the IEEE 802.11 backoff procedures to reduce the number of collisions. Furthermore, transmitters are encouraged to further spread out their channel access during the entire Interval.

Various studies investigated the performance of the IEEE 1609.4 switching mechanism. For example, in [20], Hong et al. investigated the performance of safety-related applications and proposed slight modifications to the IEEE 1609.4 standard to improve it. A similar study has been conducted by Di Felice et al. in [7], which proposed a modification of the MAC procedures to mitigate the influence of the switching mechanism. In [31] Wang and Hassan instead evaluated the 1609.4 capacity to support non-safety related DSRC traffic. Various other approaches were also proposed to enhance the switching mechanisms with different synchronous or asynchronous *rendezvous* mechanisms. They are surveyed in [3] and described in more detail in [2, 4, 19].

7.4.2 EU Regulations: ETSI ITS

Multi-channel operations proposed by the ETSI bear many similarities with that of the IEEE 1609.4 and IEEE 1609.3. They yet fundamentally differ with the station architecture previously described, which enforces that at least one ITS-S T1 transceiver be always tuned to the CCH. Accordingly, this T1 transceiver cannot have multi-channel operations. The major advantage of this proposal is to maximize the safety-related communication capacity on the CCH. The synchronous channel switching mechanism specified in 1609.4 notably reduces the CCH channel capacity during the time the ITS-S is switched to an SCH. Considering that even a full capacity of the CCH might not even be sufficient to support safety-related applications (see Chap. 6), an ITS-S must always be capable of receiving on the CCH. The major disadvantage of this approach is that if an ITS-S also need to offer ITS services, it then requires to have at least two transceivers.

Compared to the situation in the US, the EU can only effectively use three 10 MHz channels at the time of writing. The benefit of having a second ITS-G5 supporting channel switching for operating on the other two SCHs is therefore less straightforward compared to having that ITS-G5A transceiver constantly tuned on one of the SCHs. Accordingly, the ETSI did not primarily focus on ITS service managements. Yet, considering the scarce channel resources on CCH from ITS *Day One* applications, and the availability of a second ITS-G5 transceiver constantly tuned on another ITS-G5A channel, the ETSI focused on providing mechanisms to mitigate channel congestions by benefiting from traffic offloading on two other ITS-G5A channels. Accordingly, we will first describe multi-channel congestion control, and then briefly introduce the general directions envisioned by the ETSI for multi-channel service management.

7.4.2.1 Multi-Channel Congestion Control

At the time of writing this chapter, standards describing multi-channel congestion control at the ETSI were still being finalized. We provide here the current trends, but some details might evolve in the future. We suggest interested readers to refer to the corresponding standard [15, 17]. Also, at the time of writing this chapter, very few studies could be found, which investigated the performance of safety-related traffic offloading on adjacent channels. One of them is provided by De Martini and Härri in [5].

As described in Chap. 6 dedicated to congestion control, the load on the CCH must be regulated to keep a communication quality for safety-related applications. Regularly found mechanisms include regulating the transmit power or the transmit rate. Yet, when the channel load reaches a limit when packets either must be dropped by the ITS-G5 T1 transceiver, or cannot be generated according to the ITS application requirements, an alternative approach is to transmit these packets on alternate channels.

7 Multi-Channel and Coexistence

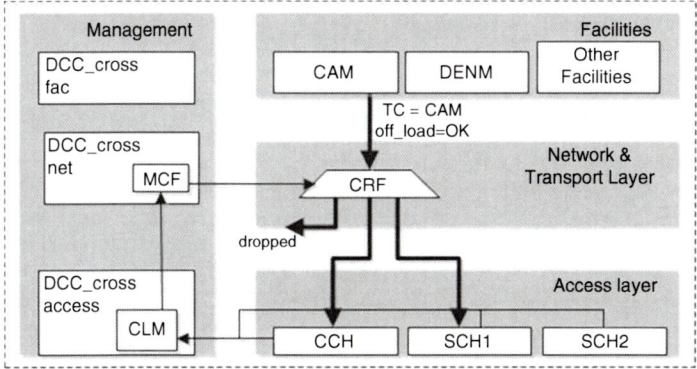

Fig. 7.7 Multi-Channel Congestion Control architecture based on the ETSI DCC architecture [17]

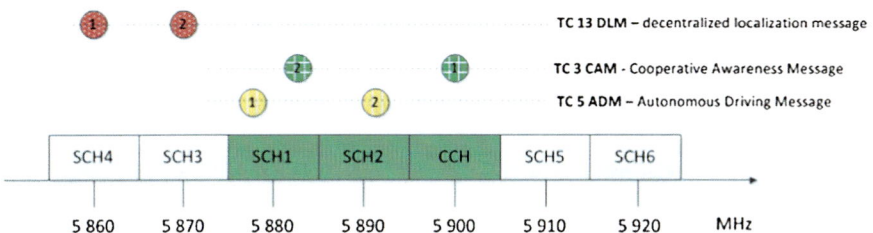

Fig. 7.8 Exemplary primary and secondary assignment concept illustrated with CAM and two fictional messages ADM and DLM

As illustrated in Fig. 7.7 depicting the ETSI DCC management architecture, the DCC-net block includes a Multi-Channel Function (MCF) on the management plane, and a Channel Routing Function (CRF) in the data plane. The MCF aims at providing the CRF with the load and remaining capacities for the CRF to be able to offload a particular Traffic Class (TC) to an adjacent channel. The primary and alternate channels for a particular TC is known to all ITS-Ss and depicted in Fig. 7.8, where two fictional messages (Autonomous Driving Message (ADM) and Decentralized Localization Message (DLM)) have been used as examples. This figure shows that a typical TC, for instance $TC3$ corresponding here to a high priority CAM message, has the CCH as primary channel and Service Channel 1 (SCH1) as secondary channel. In order not to perturb traffic on the secondary channels, a TC being offloaded may also lose priority. For example, the TC corresponding to a high priority CAM may become the lowest ITS-G5A Access Category (AC) on SCH1 to help mitigate congestion with a TC $TC5$ corresponding to an ADM. The selection of the primary and secondary channels, as well as their level of ITS-G5 AC priorities is out of the scope of the multi-channel mechanisms.

The CRF therefore takes as input from the MCF the load and remaining capacities on all channels under multi-channel operations, as well as the TC of a packet. As specified in the ETSI technical specification [15] and depicted in Fig. 7.9,

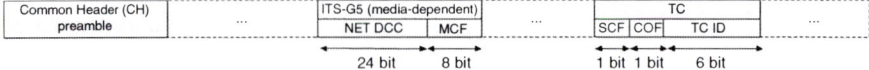

Fig. 7.9 ETSI GeoNetworking Common Header, and the multi-channel related fields: COB and MCF as defined in [14, 15]

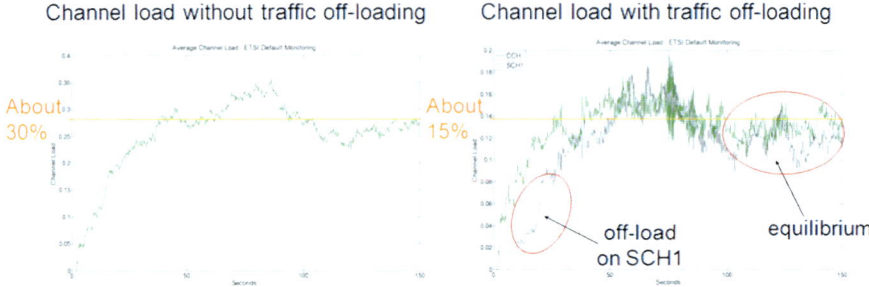

Fig. 7.10 Illustration, appeared in [5], of the potential reduction of channel load on the CCH from offloading traffic on a secondary SCH

the TC includes a 1-bit field *Channel Offloading Bit (COB)*, which indicates if the ITS application generating the packet tolerates multi-channel offloading. In the negative, the CRF operates as a single channel DCC as described in Chap. 4. In the positive, depending on the status of the DCC function on the primary channel, the packet may be sent on the primary channel, offloaded on a secondary channel or dropped.

Figure 7.10 illustrates the multi-channel congestion control benefits to regulate the load on safety-related channels. To obtain these results, the multi-channel congestion control mechanism has been implemented on the iTETRIS ITS simulation platform [27], configured with the channel load monitoring and offloading parameters indicated on Table 7.1. More detailed results are available in [5].

Each vehicle transmits a CAM at the intended 10 Hz rate, where CAM primary and secondary channels are the CCH and SCH1, respectively. When the MCF indicates 50% channel load on CCH and less than 50% channel load on SCH1, the CRF offloads all CAM to SCH1. In order to avoid oscillating behaviors of the CRF offloading on the SCH1 when the channel load is slightly higher or lower than 50%, a configurable hysteresis is added. In the results shown in Fig. 7.10, the hysteresis corresponds to 10% of the target channel load. In that case, we can see that the channel load first increases on CCH, before a gradual offloading on SCH1 starts. If multi-channel offloading is not supported, we can see that the channel load reaches 30%.[2] But when offloading is triggered, the load generated by the CAM messages is shared between CCH and SCH1 and reaches only 15%.

---

[2]The IEEE 802.11-2012 default value of the 802.11 Clear Channel Assessment (CCA) threshold has been considered for the measure of the ITS-G5A channel load.

Table 7.1 iTETRIS ITS multi-channel simulation parameters

Parameter	Value
Simulator	iTETRIS (ns-3 + ETSI ITS stack)
PHY/MAC	ETSI ITS-G5
Channels	CCH, SCH1 (ITSG5A)
Switch delay	1 msec
Fading	Log distance
Attenuation	2.3
Mobility	Highway, three lanes, two directions
Generation^{vehicle}	Erlang $\lambda = 2$ s
Speed^{vehicle}	[20–40 m/s]
T_{Sync}	1,000 msec
T_{SAM}	200 msec
T_{offset}	$Uniform[0, T_{Synch} - T_{SAM}]$
$CCA_{Threshold}$	-85 dBm
T_{mon}	100 msec
$CL^{offload}_{threshold}$	20 %
$CL^{offload}_{hist}$	5 %
SAM^{TX}	5 Hz
CAM^{TX}	10 Hz
SAM size	500 Byte
CAM size	500 Byte

7.4.2.2 Channel Switching Principles

At the time of writing this chapter, standards describing multi-channel switching at the ETSI were still being finalized. We provide here the current trends, but some details might evolve in the future. We suggest interested readers to refer to the corresponding standard [15, 17]. Also, at the time of writing this chapter, very few studies could be found, which investigated the performance of the ETSI channel switching standard. One of them is provided by De Martini and Härri in [6].

According to the channel plans depicted in Fig. 7.2 services may be offered on four SCHs. The ECC regulations enforce that only safety-related services be transmitted on ITS-G5A channels (SCH1 and SCH2), but ITS-G5B channels (SCH3 and SCH4) are also open to non-safety-related traffic. Service management (safety-related or not) again bears similarities with IEEE 1609.3. A *Service Provider* announces the presence of an ITS service with a SAM [18] indicating the service and on which SCHs it can be found. *Service Consumers* interested in ITS services tune to the channel, where SAM are sent, and then go to the indicated SCH to consume the service. Service providers and service consumers must therefore first *rendezvous* for service announcement, and then *rendezvous* again where the service is actually being offered. While the SAM message indicates the channel on which the ITS service will be provided, service providers and service consumers must still

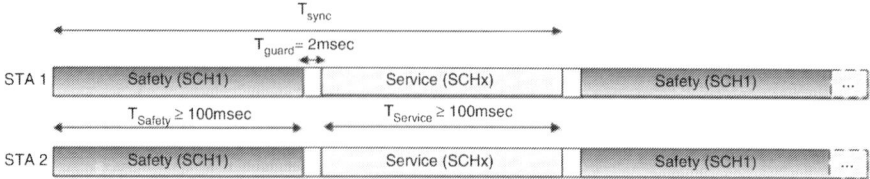

Fig. 7.11 Synchronous Channel Switching for ITS-S Services, adapted from [15]

agree on a common *rendezvous* channel for SAM. In EU, the channel on which SAM are sent corresponds to the base channel of the respective ITS transceiver (see Fig. 7.4) and further referred to as Service Announcement Channel (SACH) in the rest of this chapter.

As indicated in Fig. 7.8 and described in Sect. 7.4.2.1, road safety-related messages, such as CAM or DENM, may be offloaded or relayed on secondary SCH. ITS-Ss operating on the ITS-G5A spectrum and supporting multi-channel operations (i.e., the T2 transceivers as described in Fig. 7.4) are therefore enforced to monitor these channels. Considering the safety-of-life content of safety-related messages, even when offloaded, channel switching must be synchronous between ITS-Ss to guarantee that all T2 ITS-Ss operating on ITS-G5A are on the same secondary SCH when safety-related messages are offloaded or relayed. The synchronous channel switching mechanism is very similar to IEEE 1609.4, but restricted on SCHs.

The synchronous channel switching mechanism for the ITS-G5A spectrum has two switching phases as depicted on Fig. 7.11:

- *Safety*: the ITS-S is tuned to the SCH1 during a *Safety* interval corresponding to T_{Safety}.
- *Service*: the ITS-S is tuned to any other SCHs during a *Service* interval corresponding to T_{Service}.

Similarly to IEEE 1609.4, ITS-S transceivers alternate between a safety phase and a service phase following two intervals T_{Safety} and T_{Service}, both configurable but widely known to all ITS-Ss. DCC requires a minimum channel load monitoring interval $T_{\text{mon}} = 100$ ms. Accordingly, both T_{Safety} and T_{Service} must be integer values of T_{mon}. A guard interval, T_{gard} will also be added for time synchronization reasons, during which both channels are unavailable. Finally, T_{Safety}, T_{Service}, and T_{gard} sum up to a T_{Synch}.

For ITS-S transceivers not operating on ITS-G5A, the strict and synchronous *rendezvous* on an SCH may be relaxed. The principle is similar to the synchronous mechanism previously described and is illustrated in Fig. 7.12. The *Safety* phase is replaced by a SAM phase, and ITS-S transceivers are only required to be on a SAM phase *globally* at the same time. This has two advantages: first, ITS-Ss do not need to be synchronized to operate in this mode. Second, the SAM phase may be adjusted

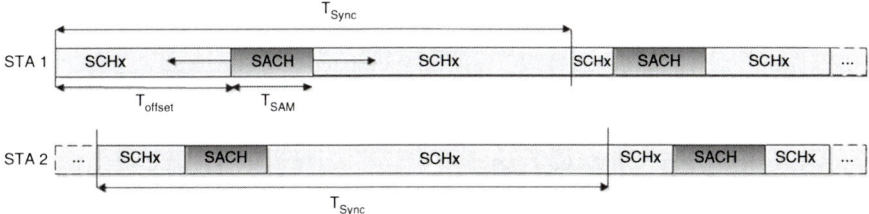

Fig. 7.12 Asynchronous Channel Switching mechanism, adapted from [15]

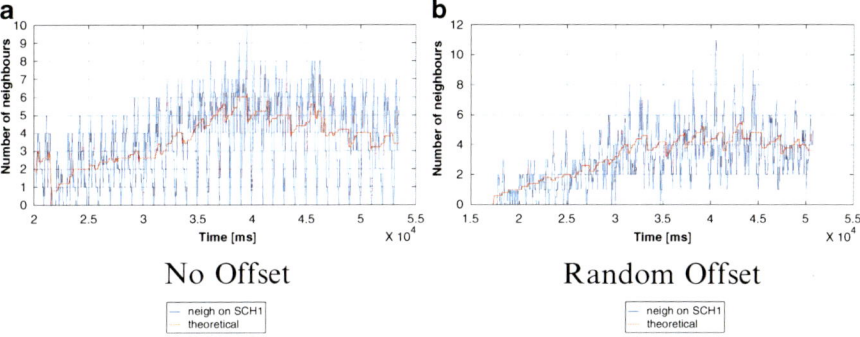

Fig. 7.13 Illustration, appeared in [5], of the impact of the offset parameter in increasing the chances of a Rendezvous point. (**a**) No offset. (**b**) Random offset

at any time during the *synch interval* T_{Synch}. The length of a SAM phase is also not strict, but must be sufficient for channel load monitoring. So, $T_{SAM} > T_{mon}$.

The *rendezvous* of ITS-Ss during a SAM phase is yet statistically enforced. The start of a SAM phase is triggered according to a *time offset* T_{offset} randomly assigned in an interval $[0, (T_{Synch} - T_{SAM})]$. When an ITS service provider announces a service, it cannot expect to have all ITS-Ss tuned on the SACH, but may statistically estimate the number of neighbors that will be present at any time instant on the SAM phase as: $N_{Nb}^{SAM} = N_{Nb} * \frac{T_{SAM}}{T_{Synch}}$.

The channel switching principles described here have been implemented on the iTETRIS ITS simulation platform [27], configured according to the parameters indicated in Table 7.1. An illustration of the impact of the T_{offset} on the probability of finding a neighbor on the reference channel is depicted in Fig. 7.13. Considering that ITS-Ss are not synchronous, enforcing a SAM phase at the beginning of the T_{Synch} interval shows a high variance in the number of ITS-Ss jointly being on the SACH during a SAM phase. When a time offset is applied, we can see that the variance is significantly reduced, which is a stability indicator for ITS service providers to offer services on a SACH during a SAM phase. More detailed results are available in [5].

7.4.2.3 Multi-Channel Service Management

Multi-channel service management is the process for SPs or SCs to offer and consume ITS services, following the synchronous or asynchronous channel switching mechanisms previously described. In this section, we provide an example of how these mechanisms interact for SPs and SCs to operate ITS services inspired from the ETSI multi-channel operation mechanisms, even though ITS service management is not standardized by the ETSI at the time of writing.

As previously described, most of the ITS services need to be a priori advertised on a SACH on which SPs and SCs must *rendezvous*. The challenge is therefore to evaluate the time before which SPs and SCs manage to rendezvous, and the time required to start consuming an ITS service. Figure 7.14 depicts this process, where one SP offers a service to three SC. As SPs are not synchronized on the SAM phase, it takes multiple *synch intervals* before all SC are on the SACH at the same time as the SP. On the first *synch intervals*, only SC3 receives a SAM, while SC1 and SC2 are on different channels. During the second *synch intervals*, SC2 is able to receive a SAM, but still not SC1. In this figure, we also illustrate the probability (even though highly unlikely) that an SC, here SC1, never receives a SAM. Once SC have received a SAM, they switch to the SCH on which the service provided by SP1 is offered.

The time required for this process therefore depends on various criteria (e.g. T_{Synch}, T_{SAM}, periodicity of SAM, etc.), one important one being whether the SP and SCs are already on the same SACH when the SAM message is sent or if they

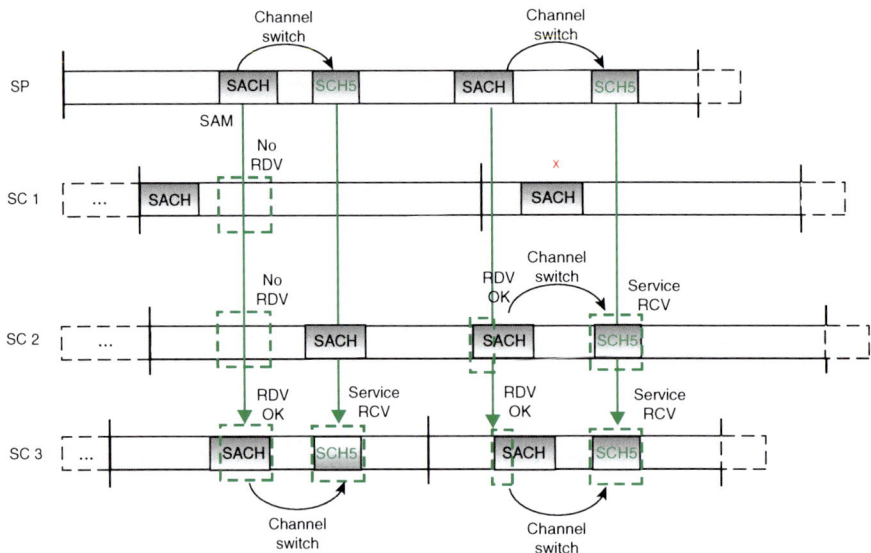

Fig. 7.14 Illustration, appeared from [15], of ITS service management between an SP and SCs considering a asynchronous channel switching mechanism

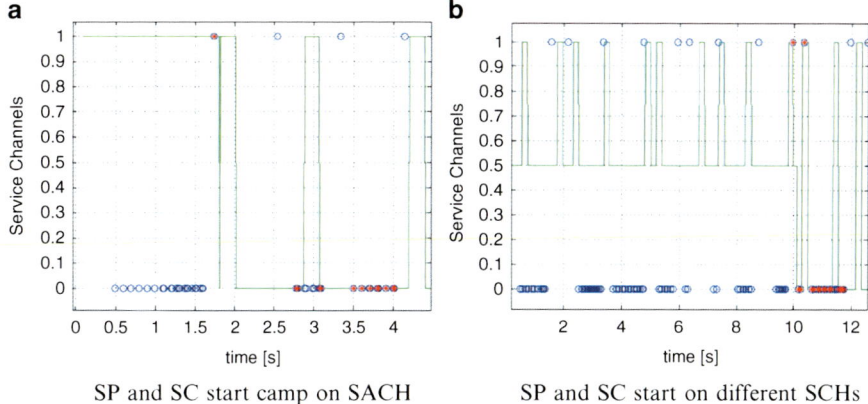

Fig. 7.15 Illustration of the required time for SP and SC to rendezvous as function of their initial channel, where blue and red circles respectively represent a missed or a met rendezvous. (**a**) SP and SC start camp on SACH. (**b**) SP and SC start on different SCHs

are on different channels. We can note that the case where SPs and SCs are on the same SACH is conceptually similar to a synchronous channel switching mechanism, whereas when they are on different SCHs, it corresponds to an asynchronous switching mechanism.

Services are identified in SAMs using an ITS-AID. At the time of writing, there is no clear process or responsible entity to administrate ITS-AID. This duty is shared by the various standard organizations (IEEE, ISO, SAE, ETSI). A list of existing ITS-AIDs are maintained by the ISO.[3] At the time of writing, the ITS-AID is a fixed length value occupying three bytes, although there are ongoing proposals to make ITS-AID a variable length value occupying one to three bytes in order to allocate them on ETSI ITS packet headers.

We illustrate the performance of this ITS service management in Fig. 7.15. The service management has been implemented on the iTETRIS ITS simulation platform [27], configured with the service management parameters indicated in Table 7.1. The x-axis corresponds to the simulation time, the green lines represent the channel switch operated by the ITS-S, and the y-axis corresponds to channels, on which the ITS-S is currently tuned: the y-label 1 corresponds to the SACH, the y-label 0 corresponds to the Service Provider Service Channel (SP-SCH), while the y-label 0.5 corresponds to any other SCH. Also, the blue, respectively red circles, on the SACH (Channel "1") are SAM being sent, respectively received. The blue, respectively red circles, on the SP-SCH (Channel "0") are SP packets being sent, respectively received. Accordingly, by observing transitions of the blue circles to red circles, we can see the service management mechanisms in operation.

---

[3]ISO 17419: Available ITS-AIDs—http://standards.iso.org/iso/ts/17419/TS17419%20Assigned%20Numbers/TS17419_ITS-AID_AssignedNumbers.pdf.

On the left side (Fig. 7.15a), we can see that as both SP and SC are camping on the SACH at the same time, it only takes one SAM transmit interval to switch to the target SCH (SP-SCH) and consume the service. On the right side (Fig. 7.15a), we can see that, as SP and SC are away on different SCHs, it requires multiple iterations of the asynchronous channel switching mechanism before both SC and SP *rendezvous* on the SACH and then switch to the SP-SCH. From a time aspect, the difference between asynchronous and synchronous channel switching mechanisms bring an order of magnitude five to the service management convergence time. It should yet be noted first that this time penalty strongly depends on the SAM phase time T_{SAM} and the synch interval T_{Synch}. In this simulation, $T_{SAM} = 200$ ms and $T_{Synch} = 1$ s. Shorter synch intervals automatically gives a faster *rendezvous* convergence. Also, when an SC or even SPs are away on a different channels, it also means they are either consuming or producing already other services, so this does not correspond to wasted time. More detailed results are available in [5] and in [6].

7.5 Coexistence Issues and Future Challenges

In 2013 the US FCC issued an NPRM concerning the use of the 5 GHz spectral band by *Unlicensed National Information Infrastructure (U-NII)* devices. Devices that implement the IEEE 802.11 protocols, frequently referred to as Wi-Fi, are the most common type of U-NII device. The need for more spectrum in which Wi-Fi devices can operate was the primary motivation behind the FCC NPRM. It divides the 5 GHz band into several sub-bands, as shown in Fig. 7.16. U-NII operation has been permitted in the following sub-bands by previous FCC decisions: U-NII-1, U-NII-2a, U-NII-2c, and U-NII-3. The NPRM proposed a variety of changes to U-NII operation in those bands. It also asked whether U-NII operation should be permitted, and if so on what basis, in the U-NII-2b and U-NII-4 sub-bands, where U-NII operation was not previously permitted. U-NII operation in the 5 GHz band is sometimes referred to as *spectrum sharing*, because the U-NII devices use spectrum that is allocated on a primary basis to licensed devices in these sub-bands. U-NII operation, according to Part 15 of the FCC rules, must not lead to harmful interference of any licensed communication.

The 5.9 GHz DSRC band shown in Fig. 7.1 is designated by the FCC as U-NII-4 for purposes of the NPRM. This band is allocated on a primary, licensed basis to DSRC services as well as to some radar and satellite services. The IEEE 802.11ac-2013 [26] amendment defines high bit rate Wi-Fi that utilizes 80 and 160 MHz channel bandwidths. The Wi-Fi community was especially interested in gaining access to the U-NII-4 band. The combination of the U-NII-3 and U-NII-4 sub-bands would permit one additional 80 MHz channel and one additional 160 MHz channel, both with upper frequency 5.895 GHz. The potential for spectrum sharing in the DSRC band created significant concerns in the DSRC community. Many DSRC stakeholders worried that U-NII devices using the band would interfere with the DSRCs safety-of-life mission.

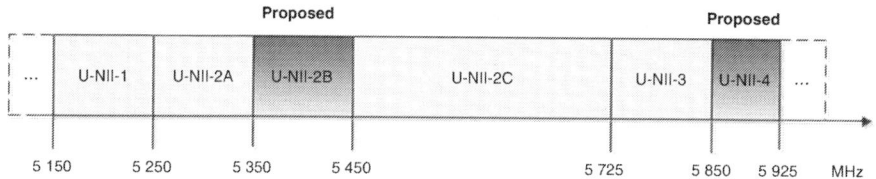

Fig. 7.16 The US U-NII bands, and coexistence with RLANs and DSRC spectra

A similar proposal was made to allow Wi-Fi RLANs to operate in the ITS-G5 spectrum (Fig. 7.2) in Europe. In September 2013, the European Commission mandated the European Conference of Postal and Telecommunications Administrations (CEPT) to investigate the issue. The task was assigned to CEPT Spectrum Engineering Group 24 (*SE-24: Short Range Devices*). A major limitation of this proposal in Europe is related to the existence of the Comité Européen de Standardization (CEN) DSRC allocated spectrum for toll collection, which would not be able to coexist with other technologies sharing its dedicated spectrum. At the time of writing, SE-24 continues to study the issue of sharing ITS-G5 spectrum between C-ITS and RLAN devices.

In early 2013 key DSRC stakeholders in the US reached out to the Wi-Fi community to initiate discussions about whether sharing is possible and how it might be regulated. This led to DSRC experts providing educational tutorials at IEEE 802.11 Standards Working Group (WG) meetings, as well as several face-to-face meetings between the DSRC and Wi-Fi stakeholders. In August 2013 the IEEE 802.11 WG formed a technical *Tiger Team* to investigate coexistence between DSRC and Wi-Fi devices.

Wi-Fi members proposed two types of sharing solutions in the Tiger Team. One called for DSRC devices operating in the U-NII-4 band to detect the presence of DSRC transmissions, and to stop using the band for a period of time when DSRC devices were detected. Since DSRC is based on the IEEE 802.11 protocol, detection of DSRC can be done using the IEEE 802.11 *listen before talk* Carrier Sense Multiple Access/Collision Avoidance (CSMA/CA) MAC protocol (see Chap. 4). The CSMA function for detection is called Clear Channel Assessment (CCA), so this sharing proposal is known as the CCA proposal. The second proposal, called the *Rechannelization* proposal, calls for the DSRC community to move BSM communication from Channel 172 into one of the channels above 5.895 GHz, and to use two 20 MHz channels between 5.855 and 5.895 GHz, the portion of the spectrum that IEEE 802.11ac devices also want to use. It also calls for the FCC to draw the upper edge of the U-NII-4 band at 5.895 GHz instead of 5.925 GHz, so that the upper 30 MHz of the DSRC band does not overlap with U-NII-4.

The DSRC community has indicated to the Tiger Team, to the FCC, and to other US Government decision makers that it is opposed to the *Rechannelization* proposal for a number of reasons, but that it thinks the CCA proposal has potential to enable sharing without harmful interference. The DSRC community has encouraged the Wi-Fi community to further develop the CCA proposal into a complete sharing solution, so that it can be tested with DSRC devices. All parties agree that any potential sharing solution must be rigorously tested before sharing could be allowed. The FCC has not completed its consideration of U-NII device sharing of the DSRC band. At the time of writing, the Tiger Team is discussing these proposals according to the feedbacks from the DSRC community.

7.6 Conclusion

We reviewed in this chapter the basic multi-channel mechanisms for vehicular communications, as standardized in the US and in Europe. Although sharing similar objectives and bearing resemblance to many aspects, the source of their difference is to be found in the different spectra allocated in the US and in the EU: the US has seven allocated DSRC channels, which makes it critical to have multi-channel switching mechanisms, as it is economically impossible to operate all seven with different DSRC transceivers. In Europe, only three ITS-G5 channels have been allocated since 2008, and due to the reduced spectrum, multi-channel switching appeared less critical. The ETSI therefore rather focused on mitigating channel congestion by offloading traffic on the other available channels. We first introduced the channel allocations and different ITS stations (i.e., RSUs and OBUs), we then described the channel switching principles proposed by IEEE 1609.4 and IEEE 1609.3 in the US and ETSI TS 724-4-2 and ETSI TS 103 165 in Europe. We illustrated the performance of potential multi-channel congestion control and showed how offloading or relaying part of the safety-critical traffic from the ETSI CCH on secondary SCH is another viable mechanism to regulate the load on CCH. We also proposed and tested an asynchronous multi-channel service management compliant with the current standards and emphasized its flexibility for service providers and consumers to rendezvous and consume services.

Although mildly followed and supported, efficient multi-channel mechanisms are expected to become critical at the eve of the ITS Day Two applications, and mostly with the expected future requirement for DSRC to coexist and share the ITS spectrum with alternate technologies (i.e., WiFi-Giga or LTE-A). It is expected to be highly unlikely that all ITS bands be strictly assigned to a particular type of traffic. Early proposals already suggest to rely on cognitive principles to dynamically move traffic between channels as function of the coexistence with other type of traffic. It is therefore expected that the very mechanisms described in this chapter to evolve and expand to support a larger category of ITS services, spanning from safety-related to fully commercial traffic. If the initial years of the DSRC/ITS-G5 technology have been focused primarily on single channels and on congestion *control*, the future years are expected to be focused on multi-channels dynamic spectrum sharing and on congestion *avoidance*.

Acknowledgements EURECOM acknowledges the support of its industrial members: SFR, Orange, ST Microelectronics, BMW Group, SAP, Monaco Telecom, Symantec, IABG.

The authors would also like to thank Laura De Martini for her support in the simulation results of the ETSI channel switching mechanisms, as well as the members of the ETSI Specialist Task Forces (STF) 420, 447, and 469, and the Editors of this Book for their fruitful suggestions and remarks.

References

1. 2008/671/EC European Communication (EU) (2008) EU Decision of 5 August 2008 on the harmonized use of radio spectrum in the 5 875–5 905 MHz frequency band for safety-related applications of Intelligent Transport Systems (ITS). Decition 2008/671/EC, EC
2. Bi Y, Liu KH, Cai L, Shen X, Zhao H (2009) A multi-channel token ring protocol for qos provisioning in inter-vehicle communications. IEEE Trans Wirel Commun 8(11):5621–5631
3. Campolo C, Molinaro A (2013) Multichannel communications in vehicular ad hoc networks: a survey. IEEE Commun Mag 51(5):158–169
4. Chu JH, Feng KT, Chuah CN, Liu CF (2010) Cognitive radio enabled multi-channel access for vehicular communications. In: IEEE 72nd Vehicular technology conference fall (VTC 2010-Fall), pp 1–5
5. De Martini L, Härri J (2013) Design and evaluation of multi-channel mechanisms for efficient ITS spectrum usage. Technical Report. RR_13_289, EURECOM
6. De Martini L, Härri J (2013) Short paper: design and evaluation of a multi-channel mechanism for vehicular service management at 5.9 GHz. In: 2013 IEEE vehicular networking conference (VNC), pp 178–181
7. Di Felice M, Ghandour A, Artail H, Bononi L (2012) On the impact of multi-channel technology on safety-message delivery in IEEE 802.11p/1609.4 vehicular networks. In: IEEE 21st International conference on computer communications and networks (ICCCN), pp 1–8
8. ECC/DEC/(08)01 Electronic Communications Committee (ECC) (2008) ECC Decision (08)/01 on the harmonized use of the 5875–5925 MHz frequency band for Intelligent Transport Systems (ITS). Decision ECC/DEC/(08)01, ECC
9. Electronic Communications Committee (ECC) (2008) ECC Recommendation (08)01. use of band 5855–5875 MHz for Intelligent Transport Systems (ITS). Recommendation ECC/REC/(08)01, ECC
10. European Telecommunications Standards Institute (2009) Intelligent Transport Systems (ITS); European profile standard for the physical and medium access control layer of intelligent transport systems operating in the 5 GHz frequency band. ES 202 663 V1.1.0, ETSI
11. European Telecommunications Standards Institute (2011) Intelligent Transport Systems (ITS); decentralized congestion control mechanisms for intelligent transport systems operating in the 5 GHz range; Access layer part. TS 102 687 V1.1.1, ETSI
12. European Telecommunications Standards Institute (2012) Intelligent Transport Systems (ITS); harmonized channel specifications for intelligent transport systems operating in the 5 GHz frequency band. TS 102 724 V1.1.1, ETSI
13. European Telecommunications Standards Institute (2012) Broadband Radio Access Networks (BRAN); 5 GHz high performance RLAN; harmonized EN covering the essential requirements of article 3.2 of the R&TTE Directive. EN 301 893 V1.7.1, ETSI
14. European Telecommunications Standards Institute (2013) Intelligent Transport Systems (ITS); vehicular communications; GeoNetworking; Part 4: Geographical addressing and forwarding for point-to-point and point-to-multipoint communications; Sub-part 1: Media-Independent Functionality. TS 102 636-4-1 V1.1.s2, ETSI

15. European Telecommunications Standards Institute (2013) Intelligent Transport Systems (ITS); vehicular communications; GeoNetworking; Part 4: Geographical addressing and forwarding for point-to-point and point-to-multipoint communications; Sub-part 2: Media-dependent functionalities for ITS-G5. TS 102 636-4-2 V1.0.0, ETSI
16. European Telecommunications Standards Institute (2013) Intelligent Transport Systems (ITS); radiocommunications equipment operating in the 5 855–5 925 MHz frequency band; harmonized EN covering the essential requirements of article 3.2 of the R&TTE Directive. EN 302 571 V1.2.0, ETSI
17. European Telecommunications Standards Institute (2015) Intelligent Transport Systems (ITS); cross layer DCC management entity for operation in the ITS G5A and ITS G5B medium; cross layer DCC control entity. TS 103 165 V1.0.0, ETSI
18. European Telecommunications Standards Institute (in progress) Intelligent Transport Systems (ITS); facilities layer function; Part 2: Services announcement specification. TS 102 890-2 V 0.0.2, ETSI
19. Han C, Dianati M, Tafazolli R, Liu X, Shen X (2012) A novel distributed asynchronous multichannel mac scheme for large-scale vehicular ad hoc networks. IEEE Trans Veh Technol 61(7):3125–3138
20. Hong K, Kenney J, Rai V, Laberteaux K (2010) Evaluation of multi-channel schemes for vehicular safety communications. In: IEEE 71st Vehicular technology conference (VTC 2010-Spring), pp 1–5
21. IEEE Standard (2010) IEEE Standard for Wireless Access in Vehicular Environments (WAVE)—multi-channel operation. Technical Report. 1609.4-2010, IEEE
22. IEEE Standard (2010) IEEE Standard for Wireless Access in Vehicular Environments (WAVE) Networking services. Technical Report. 1609.3-2010, IEEE
23. IEEE Standard (2010) IEEE Wireless LAN MAC and PHY Specifications Amendment 6: Wireless Access in Vehicular Environments. Technical Report. IEEE 802.11p-2010
24. IEEE Standard (2012) IEEE Standard for Wireless Access in Vehicular Environments (WAVE)—identifier Allocations standard. Technical Report. 1609.12-2012, IEEE
25. IEEE Standard (2013) IEEE Standard for Wireless Access in Vehicular Environments (WAVE)—security services for applications and management messages. Technical Report. 1609.2-2013, IEEE
26. IEEE Standard (2013) IEEE Wireless LAN MAC and PHY Specifications Amendment 4: Enhancements for very high throughput for operation in bands below 6 GHz. Technical Report. IEEE 802.11ac-2013, IEEE
27. Rondinone M, Maneros J, Krajzewicz D, Bauza R, Cataldi P, Hrizi F, Gozalvez J, Kumar V, Röckl M, Lin L, Lazaro O, Leguay J, Härri J, Vaz S, Lopez Y, Sepulcre M, Wetterwald M, Blokpoel R, Cartolano F (2013) iTETRIS: a modular simulation platform for the large scale evaluation of cooperative ITS applications. Simul Model Pract Theory 34:99–125. doi:10.1016/j.simpat.2013.01.007
28. US Federal Communications Commission (2003) DSRC Licensing and Service Rules. Technical Report. FCC-03-324A1, FCC
29. US Federal Communications Commission (2006) Memorandum report and order. Technical Report. FCC-06-110, IEEE
30. US Federal Communications Commission (2013) Notice of proposed rulemaking. Technical Report. FCC-13-22, IEEE
31. Wang Z, Hassan M (2008) How much of dsrc is available for non-safety use? In: Proceedings of the fifth ACM international workshop on VehiculAr inter-NETworking (VANET '08), pp 23–29

Part III
Vehicular Networking and Security

Chapter 8
Forwarding in VANETs: GeoNetworking

Andrea Tomatis, Hamid Menouar, and Karsten Roscher

Abstract Nowadays connectivity in the vehicles is becoming the base to enable safety, traffic efficiency, and infotainment applications. These applications require to exchange information with specific geographical locations. After a decade of research and standardization activities, a dedicated technology to support forwarding in Vehicular Ad-hoc Network (VANET) called GeoNetworking has been completed in Europe. This chapter will provide an overview of GeoNetworking features and functionalities describing geographical addressing and geographical forwarding in an easy and understandable way for both technical and non-technical readers.

Keywords VANET • GeoNetworking • Geocast • Geounicast • Geobroadcast • Media-dependent • Media-independent

8.1 Introduction

Vehicular applications often require communication among vehicles located within a specific geographical area or at specific geographical locations. For instance, if there is a traffic jam or a collision, all vehicles heading towards this traffic jam should be warned. On the other hand, for the vehicles moving away from this area, this information is not relevant anymore. This is where GeoNetworking could offer great support to these applications.

A. Tomatis (✉)
Hitachi Europe - Information and Communication Technologies Laboratory (ICTL),
Ecolucioles B2, 955, route des Lucioles, 06560, Valbonne - Sophia Antipolis, France
e-mail: andrea.tomatis@hitachi-eu.com

H. Menouar
Qatar Mobility Innovations Center - QMIC, Qatar Science & Technology Park, Doha, Qatar
e-mail: hamidm@qmic.com

K. Roscher
Fraunhofer Institute for Embedded Systems and Communication Technologies ESK,
Hansastraße 32, 80686 Munich, Germany
e-mail: karsten.roscher@esk.fraunhofer.de

© Springer International Publishing Switzerland 2015
C. Campolo et al. (eds.), *Vehicular ad hoc Networks*,
DOI 10.1007/978-3-319-15497-8_8

A simple definition of GeoNetworking is: *a set of Network Layer functionalities that uses geographical information of the involved vehicles to enable ad-hoc communication without the support of fixed infrastructure.* In other words it provides wireless ad-hoc communication among vehicles and to roadside units. GeoNetworking is part of the ITS Station reference architecture [1] and supports ITS applications for safety, traffic efficiency, and infotainment with network functionalities. In particular, it offers broadcasting of safety related messages in a certain geographical destination region, using multi-hop communication if needed and unicast communication for Internet applications [3].

GeoNetworking uses the geographical position of vehicles and roadside units to disseminate the information. GeoNetworking has two main functionalities, geographical addressing and geographical forwarding which will be depicted in this chapter.

GeoNetworking has been standardized by ETSI TC ITS in a multi-part standard series which describes the specific details of the network layer. In particular, two major documents have been published:

- Media Independent [4]: This document specifies the principal functionalities of the network layer which are independent from the media used by GeoNetworking.
- ITSG5 Media Dependent [5]: This document specifies the functionalities which are related to ITSG5 media (e.g., DCC, etc.).

The following sections introduce the GeoNetworking functionalities as defined in the ETSI Standards in an easy to understand way for both technical and non-technical readers.

At first, the base of routing is described in Sect. 8.2. In Sect. 8.3 the different addressing techniques used in GeoNetworking are described. Section 8.4 follows describing the principles of the position-based forwarding. The protocol functionalities of GeoNetworking are deeply described with examples in Sect. 8.5. Finally the security aspects, the duplicate packet detection, and the GeoNetworking special features are, respectively, described in Sects. 8.6–8.8.

8.2 Routing Based on Positions

Routing or *forwarding* in a communication network is the process of transporting information from a source towards its destination. Forwarding usually involves intermediate nodes that relay data packets on behalf of the source. It is the task of a routing protocol to find the correct path to reach the destination.

Most routing protocols leverage information about the topology of the network to calculate a path between a source and a destination. The topology is defined by direct communication links between peers as well as the properties of such links. However, the topology of vehicular ad-hoc networks is highly dynamic. Thus, links often exist only for a brief moment and can undergo significant fluctuations during their lifetime. Using link state information to determine end-to-end paths

8 Forwarding in VANETs: GeoNetworking

under these conditions requires a massive overhead to distribute link state updates. Furthermore, discovery of an entire path might not be feasible at all if the process takes longer than the average lifetime of the communication links.

Shifting forwarding decisions for each communication hop (intermediate forwarder) to the individual nodes rather than calculating an entire path at once can avoid such problems. Nonetheless, a globally available metric defining how to *get closer* to the destination is required for the routing process to converge. Absolute geographic positions are an ideal basis for such a metric. This leads to a simple requirement: Each node must have the ability to determine its current geographical position. However, since modern vehicles are usually equipped with navigation devices it can be assumed that they have the ability to locate themselves using GPS, GLObalnaya NAvigatsionnaya Sputnikovaya Sistema (GLONASS) or similar technologies with reasonable accuracy.

Based on position information several routing algorithms have been proposed. Since they offer distinct properties and trade-offs between reliability and efficiency, ETSI GeoNetworking offers the choice between different algorithms. The basic principles of these algorithms are described in Sect. 8.4. An additional benefit of a position-based networking approach is a simple and efficient implementation of packet delivery based on the current location of potential receivers. This leads to new features in terms of addressing which are the topic of Sect. 8.3.

8.3 Addressing

Addressing specifies the intended recipients of a data packet. Four different addressing schemes can be identified in most modern networks:

- **Unicast** packets are delivered to a single, specific node uniquely identified by a destination address.
- **Broadcast** packets are delivered to all hosts in a certain network.
- **Anycast** addressing is used to send data to at least one member of a distinct group sharing common criteria.
- **Multicast** packets, on the other hand, are sent to all members of a group.

However, GeoNetworking uses slightly modified approaches to cope with the special requirements of information dissemination in vehicular safety and traffic efficiency applications.

8.3.1 GeoUnicast

GeoUnicast addresses a single and specific node. This is similar to the common unicast, e.g. in IP-based networks. However, in addition to a unique node identifier the position of the node is used to route the packet towards its destination. Figure 8.1 provides an example. If the current location of the destination is not available at the source node, a *Location Service* is used to gather this information.

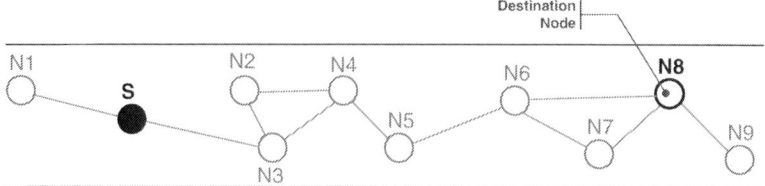

Fig. 8.1 GeoUnicast example

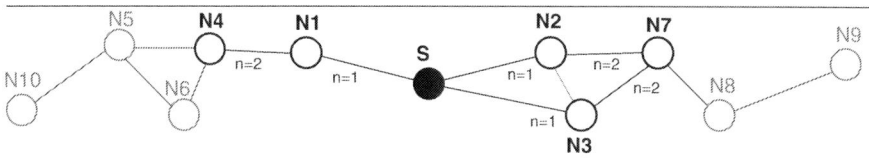

Fig. 8.2 Topological Broadcast example with $n = 2$

8.3.2 Topological Broadcast

Topological Broadcast addresses all nodes that can be reached with at most n forwarding operations (hops). This process is similar to the common broadcast with added control over the dissemination distance (in terms of hops). Figure 8.2 illustrates the process for $n = 2$. Packets that are forwarded only once ($n = 1$) are referred to as SHB.

8.3.3 GeoBroadcast/GeoAnycast

GeoBroadcast and GeoAnycast address, respectively, at least one (anycast) and all (broadcast) nodes within a specific destination area. This destination area is defined and provided by the upper layer generating the request. The source of the packet may or may not be within the destination area. Figure 8.3 illustrates communication targeting all nodes within a circular destination GeoArea.

There are similarities between GeoBroadcast/GeoAnycast and common multicast/anycast operations if all nodes in the destination area are considered to be part of a multicast or anycast group. However, groups in IP-based networks are either predefined (with a predefined address) or explicitly formed using some kind of subscription model because they are intended to be used for multiple transmissions over a longer period of time. Groups in GeoNetworking are formed implicitly based on the current location of involved nodes, which is often very dynamic. In most cases they are valid only for a single packet.

8 Forwarding in VANETs: GeoNetworking

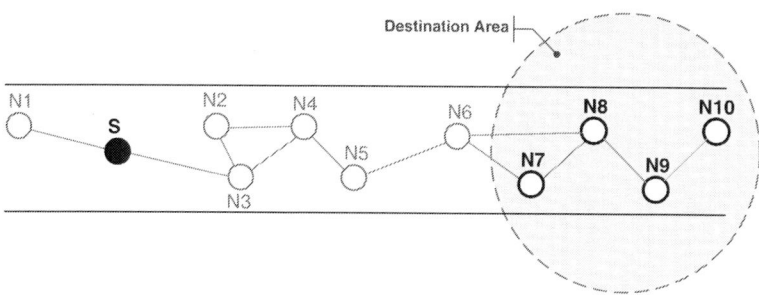

Fig. 8.3 GeoBroadcast example

8.4 Basic Principles of Position-Based Forwarding

Depending on the addressing scheme there are different phases of the forwarding process with distinct characteristics and specific algorithms. Topological Broadcast packets are forwarded with a limited *flooding* approach using simple re-broadcasts as long as the maximum number of hops is not reached. GeoUnicast requires forwarding of a packet towards a specific location, more precisely the location of the destination. This process is called *line forwarding*. GeoAnycast and GeoBroadcast use line forwarding as well to route a packet towards a node in the destination area if the sender is not located within that area. In case of a GeoBroadcast different flooding algorithms are used to efficiently distribute the information once the packet is in the destination area. Table 8.1 summarizes the different phases for each addressing type.

For line forwarding there are two basic approaches: explicit selection of a next hop by the current forwarder (sender-based) or implicit selection of a forwarder among all candidates through a decentralized coordination function (receiver-based). The former is used by Greedy Forwarding, the latter is the basis for Contention-Based Forwarding (CBF).

CBF—or variations of it—are also used to improve the flooding efficiency for geo-broadcast packets within the destination area. More information can be found in [11].

In addition to the forwarding from node to node, *store-and-forward* mechanisms can be used to temporarily buffer a packet for which forwarding is not possible at the moment. The buffer contents are reevaluated whenever the context of the node changes, e.g. a new neighbor was detected or an existing neighbor moved into a better position.

Table 8.1 Forwarding phases for different addressing schemes

Addressing	To destination	Within area
Topological Broadcast	–	Flooding (simple)
GeoUnicast	Line forwarding	–
GeoAnycast	Line forwarding	Broadcast once if source is within area
GeoBroadcast	Line forwarding	Flooding (simple or advanced)

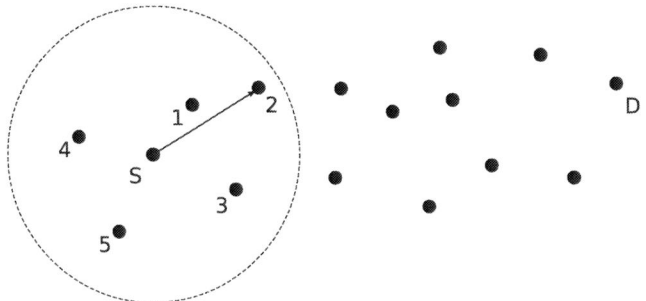

Fig. 8.4 Greedy forwarding example

8.4.1 Greedy Forwarding

Greedy forwarding was introduced as part of Greedy Perimeter Stateless Routing (GPSR) [10]. In greedy forwarding each node that needs to forward a packet explicitly selects the next hop among all its known neighbors. That requires up-to-date location information from surrounding nodes. This information can be distributed via periodic heart beats (beacons) or added to sent data packets. Applying the *most forward within radius* rule the peer closest to the destination is selected. This approach maximizes the distance progress per hop and thus results in fewer forwarding steps and efficient usage of the wireless channel.

Figure 8.4 illustrates the process with an example: The source node S wants to send a packet to the destination D. Within its transmission range there are five other nodes 1–5 with node 2 being the closest to the destination. Thus node 2 is selected and the packet sent to it. On reception of the packet the same process for selecting the next forwarder is repeated until the destination is reached.

The appeal of greedy forwarding lies in its simplicity and efficient use of the communication channel. Nevertheless, the selection of a specific node can have negative consequences for reliability. If the designated next hop moved out of transmission range or disappeared completely, data packets might get lost. This can partly be avoided with additional packet buffering and recovery strategies for transmission failures on the network layer but retransmissions increase packet latency in any case.

8.4.2 Contention-Based Forwarding

CBF [8] aims at improving reliability by leveraging multiple forwarding candidates for each step. It belongs to the opportunistic routing approaches where the receiving node decides whether it should continue forwarding or not.

Packets are sent using the broadcast mechanism of the wireless medium. Every node that received the packet checks whether it is closer to the destination than the previous forwarder in which case it is considered to be a forwarding candidate. The location of the previous transmitter is required for this process. It can either be piggy-backed on the data packet itself or derived from periodic information about neighbor positions similar to greedy forwarding.

For each step multiple forwarding candidates need to be coordinated to avoid concurrent access to the wireless channel as well as unnecessary repeated transmissions of all candidates. This can be done without additional communication overhead by applying a distinct forwarding delay at each node before sending the packet again. This delay is also called *contention timer*. If a forwarding candidate receives the packet a second time while the timer is still running it refrains from sending the packet itself since it has already been forwarded by someone else.

The delay time for each node is inversely proportional to the distance progress over the previous forwarder [8]. Thus, the node with the greatest progress has the shortest delay and tries to forward the packet first. This leads to an efficient channel usage since the distance covered with each forwarding step is maximized.

Figure 8.5 illustrates the process with an example: Source node S wants to send a packet to node D. In transmission range are five nodes that will receive the packet broadcast. Nodes 4 and 5 are farther from D than S and are therefore no candidates for forwarding. Nodes 1–3 provide progress over S. Each of them calculates its contention timer value t, buffers the packet, and waits for the determined time. Since node 2 is closest to D it has the shortest delay and will rebroadcast the packet on expiry of its timer after 10 ms. Nodes 1 and 3—still waiting—also receive the transmission of node 2, cancel their timers, and stop forwarding the packet.

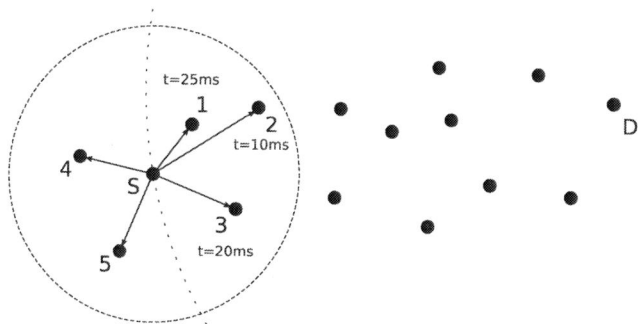

Fig. 8.5 Contention-based forwarding example

CBF offers increased reliability over the greedy approach through multiple forwarders at the cost of higher latency and increased channel usage. The individual delays of each step accumulate over the entire path. An effect that obviously gets more severe with increasing distance between source and destination. Algorithms with explicit selection of a next hop are guaranteed to use only a single transmission for each forwarding step.[1] CBF, on the other hand, can cause multiple transmissions on the wireless channel if at least two candidates are not within each other's communication range. In such cases the suppression of additional forwarding attempts fails and the packet is sent a number of times.

8.5 Protocol Functionalities

This section presents the main GeoNetworking functionalities of an ITS Station as defined in the ETSI TC ITS Specifications [4, 6] in an abstracted and easy way to understand.

8.5.1 GeoNetworking Beacon

In GeoNetworking each mobile node has to periodically inform neighbors about its presence by periodically broadcasting a specific packet called GeoNetworking Beacon.

Based on GeoNetworking Beacons received from the neighbors, a node builds a Location Table which can be consulted at any time to know the neighboring nodes and their locations. The Location Table is populated not only with information about direct neighbors, i.e. those that are located within one hop communication range and from which beacons are received, but also with farther neighbors, i.e. those located at two-hops distance and more. The farther neighbors can be added to the Location Table if a GeoNetworking Packet is received from them through multi-hop communication.

8.5.1.1 Scenario

Figure 8.6 shows a representative example of a GeoNetworking Beacon exchange and Location Table update. In the shown example, a node is surrounded by direct neighbors (N1–N3) as well as non-direct neighbors (N4–N8). Direct neighbors are inserted into the Location Table of S as soon as a beacon is received from them, while non-direct neighbors are added to the Location Table of S only if a GeoNetworking Packet that contains the location information of that neighbor is received.

---

[1]Retransmissions not taken into account.

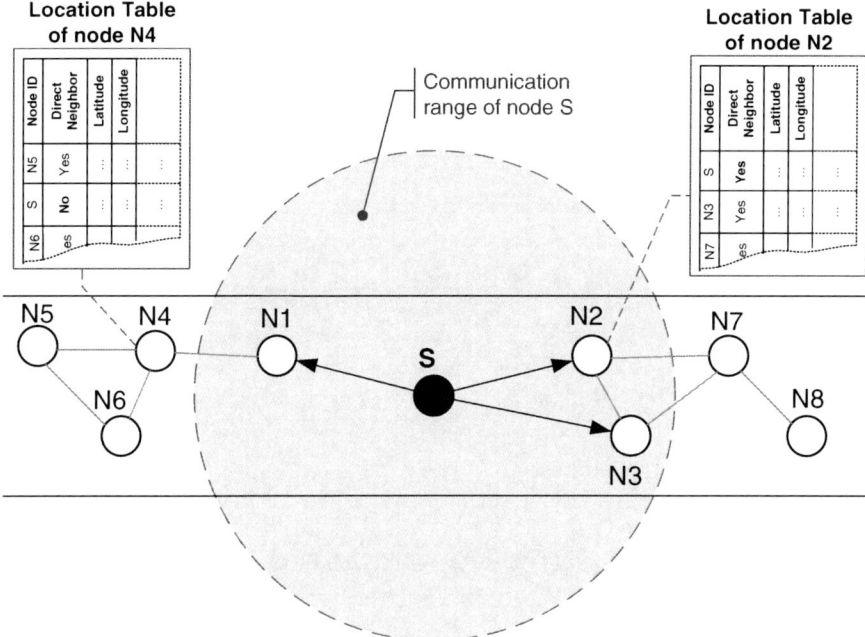

Fig. 8.6 GeoNetworking Beacon exchange and Location Table update

8.5.1.2 Protocol Operations

Before digging into the details of the GeoNetworking Beacon protocol operations, it is important to understand the data structure of the GeoNetworking packet. A GeoNetworking Packet is composed of three parts as shown in Fig. 8.7, where the first two parts are always included, while the last part is added only when needed. The fields included inside the basic header can be changed by the forwarding nodes, while the fields in the common header are set at the beginning by the originator node and then remain the same till the packet reaches its destination. This is important to make the cost of the security as low as possible, where the common header needs to be signed only once at the originator node, and then remains unchanged throughout the forwarding path.

The Basic Header is a set of information fields which is inserted in every GeoNetworking packet, and it expresses the basic information about the packet as follows:

- **Version**: expresses the version of the utilized GeoNetworking protocol.
- **NH**: stands for Next Header and it expresses the type of the header which comes after the basic header. Here in case of Network Beacon, NH indicates Common Header as next header.

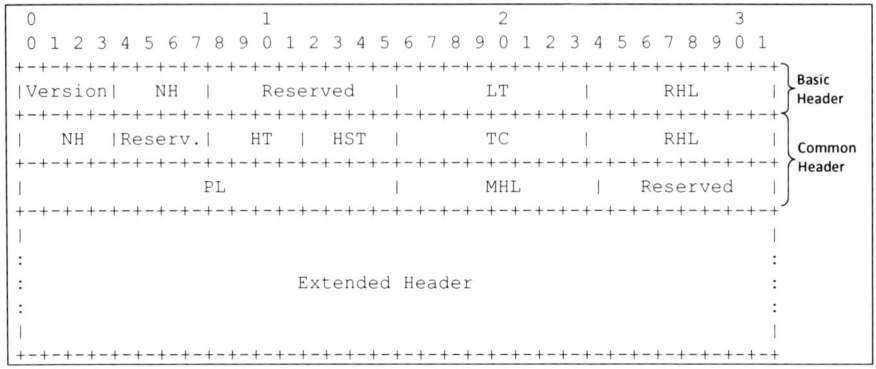

Fig. 8.7 The structure of a GeoNetworking packet

- **LT**: stands for Life Time and it expresses the time a packet is maintained alive till it reaches its destination.
- **RHL**: stands for Remaining Hop Limit and it expresses the remaining number forwarding hops. It is decreased by one at each forwarder.

The Common Header is another set of basic information fields which is also inserted in every GeoNetworking packet to express common information about the packet. These information fields are set initially by the originator node and remain unchanged till reaching the destination. The common header information fields are presented briefly as follows:

- **NH**: stands for Next Header and it expresses the type of the header coming next i.e. the type of protocol at the upper layer (Transport Layer).
- **HT**: stands for Header Type and it expresses the type of the GeoNetworking header as well as protocol, e.g. Network Beacon, GeoBroadcast, GeoUnicast, etc.
- **HST**: stands for Header Sub-Type and it is used to express the sub-type of the GeoNetworking header as well as protocol, e.g. to indicate Single-Hop or Multi-Hop Communication protocol under Topology Scooped Broadcast communication.
- **TC**: stands for Traffic Classes and it expresses upper layer requirements, e.g. in terms of priority.
- **Flags**: is partially used to express whether the source node is mobile or not.
- **PL**: stands for Payload Length and it expresses the length of the upper layer payload in bytes. Here in case of GeoNetworking Beacon it is set to zero as there is no payload attached.
- **MHL**: stands for Maximum Hop Limit and it expresses the maximum distance in terms of hops that the packet can be forwarded away from the source node. In the case of a GeoNetworking Beacon packet, the field MHL is always set to 1 as the beacon packet must not be forwarded.

8 Forwarding in VANETs: GeoNetworking

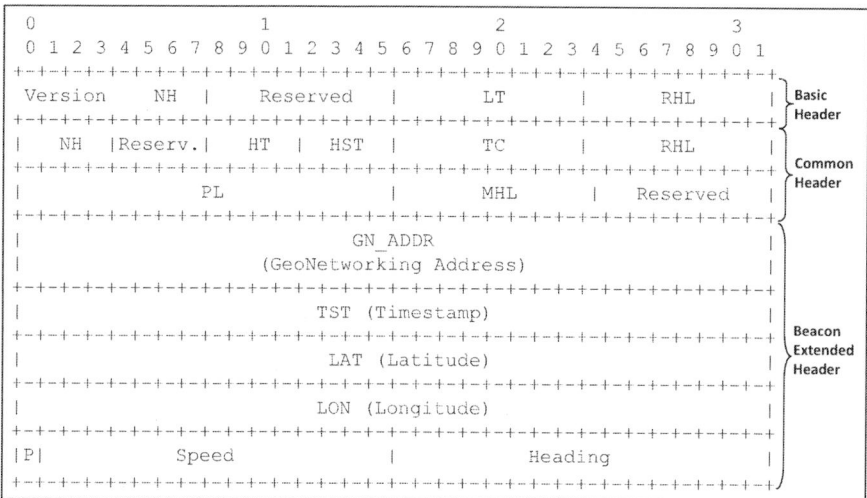

Fig. 8.8 Packet structure and headers of the GeoNetworking Beacon packet

The GeoNetworking Beacon includes a Basic Header as well as a Common Header, followed by an extended header as shown in Fig. 8.8, and which provides the unique identifier of the sender node and information about its most recent location.

Each node in the network has to inform other nodes in its surrounding about its presence in the network. To achieve this, each node creates a GeoNetworking Beacon packet as shown in Fig. 8.8, and fills the necessary information before transmitting it out. Among the necessary information, the node has to fill the GN_ADDR field with its own GeoNetworking Address. This address represents the principal identifier of the node and it contains the following information:

- **M**: this field allows to distinguish between a manually configured station and a station which is automatically configuring its address. For example, in case the pseudonyms are used see Sect. 8.6 for more details.
- **ST**: this field identifies the ITS Station type as described in [7].
- **SCC**: stands for ITS Station Country Code. Currently this field is not used.
- **MID**: this field represents the Link Layer address of the principal radio device or alternatively it contains the pseudonym (see Sect. 8.6 for more details).

Additionally the node has to fill the remaining fields in the extended header with the most updated information about its location and the Timestamp (TST) field is set with the Timestamp representing the time when the location information has been calculated.

The Beacon is sent with updated location information every few seconds. As the same information, i.e. GN_ADDR and node location, is also included in the extended header of other GeoNetworking packets, e.g. Single-Hop Broadcast, there is no need to transmit the Beacon packet if another packet containing the GN_ADDR and location of the node has been transmitted before the expiry of the Beacon timer.

All nodes located within the communication range of the sender should receive its GeoNetworking beacons. And when receiving a GeoNetworking beacon packet, or any other GeoNetworking packet which contains the GN_ADDR and location information of the node from which the packet has been received, a receiver node decodes that packet and extracts the information about the sender (i.e., GN_ADDR and Location). This information is inserted in a local location table, which lists the GeoNetworking nodes present in the surrounding. Each node listed in the local location table is flagged either as a direct or an indirect neighbor. When receiving a GeoNetworking Beacon from a node, that means this node is a direct neighbor, and therefore when inserting it in the location table it is labeled as a direct neighbor.

8.5.2 Single-Hop Broadcast

SHB is a simple and basic broadcast protocol, and it is a sub-set of the Topological Broadcast protocol as defined in Sect. 8.5.3. In SHB, the transmitted packet is never forwarded by the receivers, and therefore it reaches only neighbor nodes within the communication range of the sender. Such a broadcast protocol can be used by safety applications that require fast information dissemination in the surrounding space.

8.5.2.1 Scenario

Figure 8.9 shows an example where a node S uses SHB communication to disseminate information to its neighbors. The SHB dissemination reaches only those nodes that are located within the communication range of node S, i.e. nodes N1–N3.

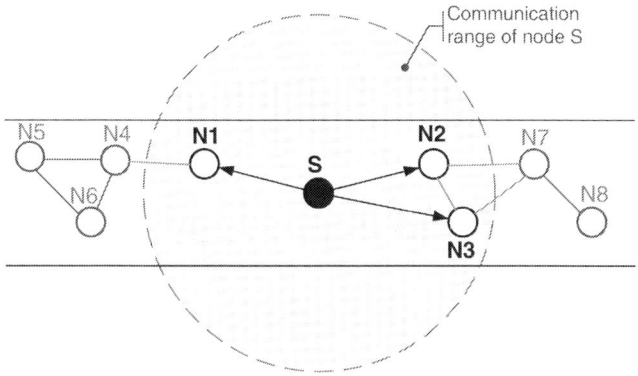

Fig. 8.9 Single-Hop communication scenario

8.5.2.2 Protocol Operations

The size as well as the content of the SHB packet has been reduced as much as possible as shown in Fig. 8.10. The SHB packet includes a basic header and a common header, followed by an extended header which includes only information about the sender, i.e. GeoNetworking Address and Location of the sender.

When the sender node receives a payload from upper layer to transmit throughout the network using an SHB scheme, an SHB packet (as shown in Fig. 8.10) is created and filled with the necessary information as follows:

- The GN_ADDR field is set with the GeoNetworking address of the sender node,
- the LAT/LON fields are filled with the most updated location coordinates of the sender node,
- the TST field is set with the Time Stamp (in milliseconds) by when the most updated location coordinates of the sender node have been calculated, and finally,
- the Speed and Heading fields are filled with the sender movement Speed and Heading, respectively. The field P is used to express whether the location coordinates of the node are accurate enough or not.

Four bytes have been reserved at the end of the SHB extended header, to be used in the future by media dependent functionalities (see Sect. 8.8.3).

Whenever a node receives an SHB packet, the node decodes it first and then checks if the packet has not been already processed previously. This duplicated

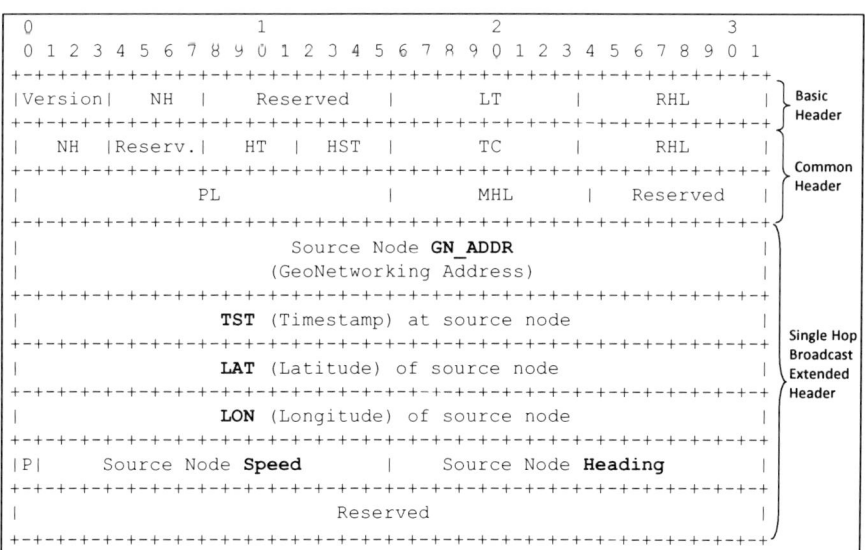

Fig. 8.10 Packet structure and headers of the GeoNetworking SHB packet

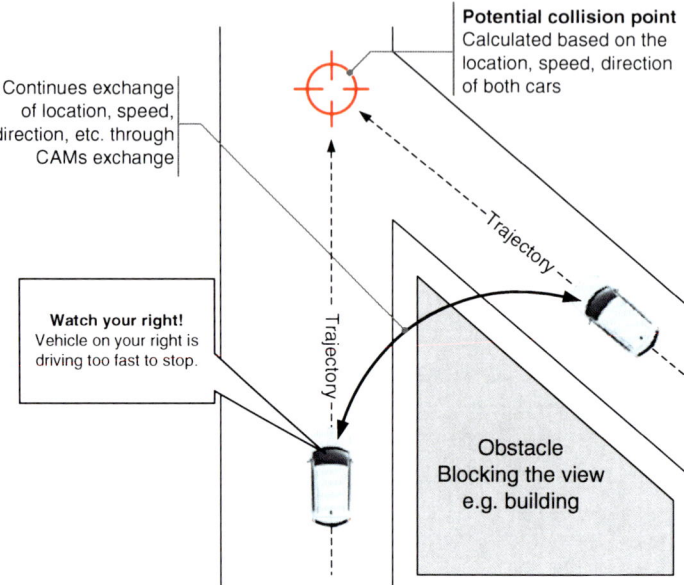

Fig. 8.11 Collision avoidance use-case enabled by continuous exchange of CAMs through GeoNetworking SHB. communications

packet check is achieved by comparing the GN_ADDR and acTST fields in the received packet against local information about packets that have been received and processed in the past (see Sect. 8.7).

8.5.2.3 Example of Usage

Figure 8.11 shows a typical use-case example where SHB communication is used to avoid collisions at an intersection. Such use-case is enabled based on the continuous exchange of the location, speed and direction of the cars through the cooperative awareness message (CAM).[2] The CAM is continuously and periodically broadcasted by each car using SHB.

8.5.3 Topological Scoped Broadcast

When tackling short range wireless communication systems and dealing with a limited communication range, any packet that needs to be transmitted beyond the communication range has to fly throughout multi-hop communications, where intermediate nodes (called also relays) forward the packet till reaching its destination.

---

[2]See Chap. 5 for more details about CAM messages.

8 Forwarding in VANETs: GeoNetworking

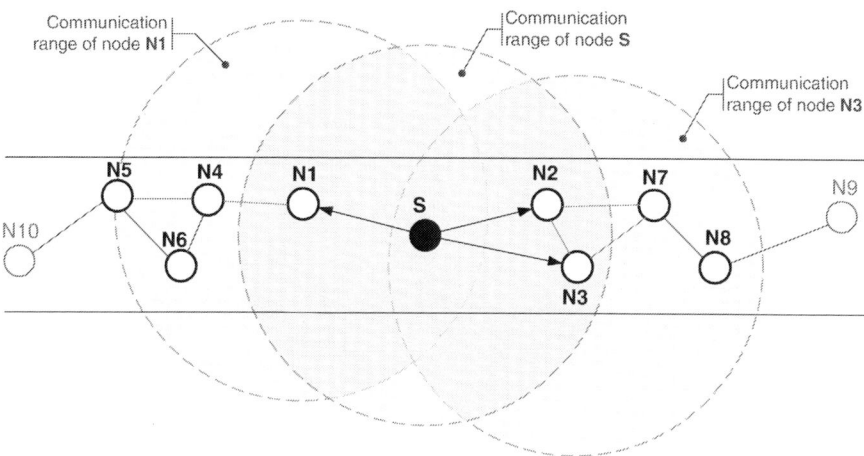

Fig. 8.12 Topological Broadcast (two hops) scenario

Topology Broadcast is well known not only in the vehicular communication field, but also in the ad-hoc communication field in general as it has been one of the first and basic forwarding schemes adopted in the community.

8.5.3.1 Scenario

Figure 8.12 presents the Topological Broadcast scenario through a 2-hop Topological-Broadcast example. In the shown example, a node S initiates a 2-hop Topological Broadcast dissemination, by transmitting a Topological Broadcast packet to all nodes within its communication range, i.e. nodes N1–N3. As the communication is intended for 2-hops distance, the packet is forwarded once by selected forwarders among the direct neighbors, to reach the second hop neighbors, i.e. nodes N4–N8.

8.5.3.2 Protocol Operations

Whenever a node receives a payload from upper layer to disseminate within a defined number of hops, a Topological Broadcast scheme is triggered and a Topological Broadcast packet is created (as shown in Fig. 8.13). The necessary information is then filled in the Topological Scoped Broadcast (TSB) packet similar to SHB as explained previously, i.e. GeoNetworking address and location fields in the Extended header with the sender information. As packets forwarding is allowed in Topological Broadcast, a same packet might reach a destination or an intermediate node through different neighbors (i.e., different forwarders). In such a case, only the very first packet is processed, and all the other duplicates are just dropped. To enable

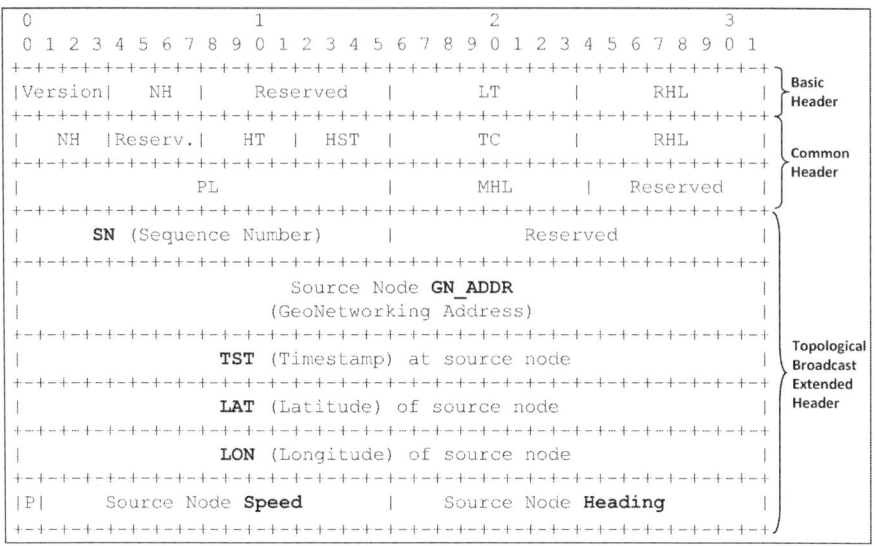

Fig. 8.13 Packet structure and headers of the GeoNetworking Topological Broadcast packet

such a duplicate packet detection mechanism, a Sequence Number (SN) has been added to the extended header of the Topological Broadcast packet, as well as in the other forwarding broadcast protocols (e.g., GeoBroadcast, etc.). See Sect. 8.7 for more details.

Differently from SHB, in Topological Broadcast the fields RHL and MHL in the Basic and the Common headers, respectively, are set with the number of hops requested by the user.

When receiving a new Topological Broadcast packet, the receiver node decodes it and checks if a similar packet has not been already received and processed in the past. If the packet has not been processed before, the contained payload is extracted and passed to the upper layer. Then, the value in the RHL field is reduced by one. If the new value of the RHL in the basic header is greater than one, the packet is rebroadcasted.

8.5.3.3 Example of Usage

Figure 8.14 shows an example of a vehicle using topological broadcast communication to search for a wanted service or information within the surrounding environment. A request for the specific service or information is then disseminated to not only those nodes within the communication range, but also to far-away neighbors through multi-hop forwarding. Any node (vehicle or road-side unit) that provides the requested service or information, can reply to the requester through a defined communication mechanism.

8 Forwarding in VANETs: GeoNetworking 237

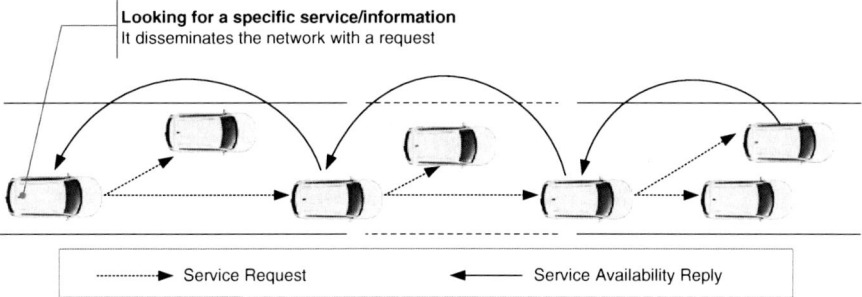

Fig. 8.14 A vehicle disseminates the network with a service request through a topological-broadcast communication to search for any node that can provide the desired service/information

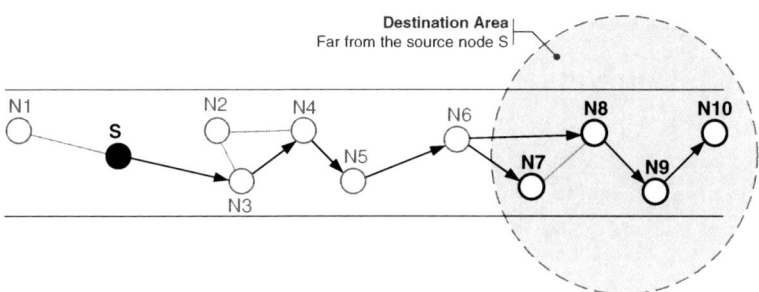

Fig. 8.15 GeoBroadcast communication scenario—sender outside the destination area

8.5.4 GeoBroadcast/GeoAnycast

GeoBroadcast and GeoAnycast protocols are very similar, especially from the view point of the data structure. This is the main reason why they are presented here under the same section. For the same reason, in other materials such as [4] the two protocols are presented together. Both protocols target a geographical area as destination. In GeoBroadcast the transmitted information is intended for all nodes located within the destination area, while in GeoAnycast the target is only the first node located within the destination area.

8.5.4.1 Scenario

GeoBroadcast intends to distribute information to a geographical area. Depending if the sender is inside the destination area or not, two types of GeoBroadcast scenarios can be identified as shown in Figs. 8.15 and 8.16. In Fig. 8.15 the sender is outside of and away from the dissemination area. In such a scenario, the GeoBroadcast packet first has to reach the destination area via a line forwarding (i.e., GeoUnicast fashion) and, once inside the destination area the packet is broadcasted to all nodes.

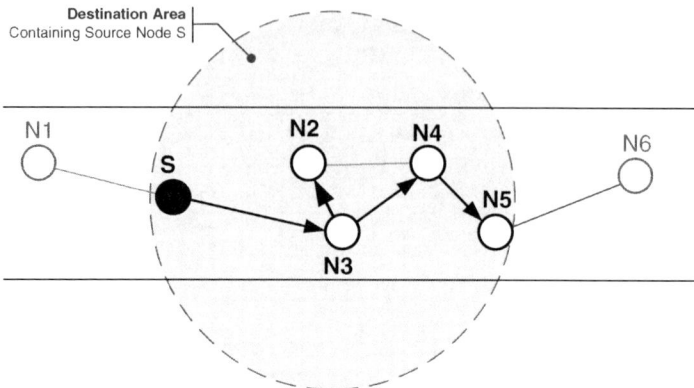

Fig. 8.16 GeoBroadcast communication scenario—sender inside the destination area

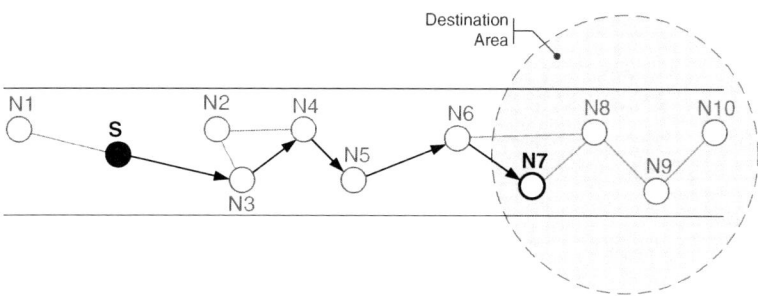

Fig. 8.17 GeoAnycast communication scenario

In Fig. 8.16, the sender is already inside the destination area, and the GeoBroadcast is immediately broadcasted to all nodes.

GeoAnycast can be looked at as a sub-set of GeoBroadcast, as the packet is always forwarded towards a geographical area. While GeoBroadcast targets all nodes within the destination area, the GeoAnycast target only one node within the destination area. Figure 8.17 shows a GeoAnycast communication scenario, where the packet is line forwarded till it reaches a first node inside the destination area (so far similar to GeoBroadcast), and then stops at that first node as it is considered as the destination node. In the case of GeoBroadcast, it would not have stopped at this node, but broadcasted to all other nodes within the destination area.

8.5.4.2 Protocol Operations

Whenever a node receives a payload from upper layer to disseminate to any node or all nodes within a specific geographical area, a GeoNetworking GeoBroadcast/GeoAnycast communication scheme is triggered and a GeoBroadcast/GeoAnycast packet is created (as shown in Fig. 8.18). The HST field in the

8 Forwarding in VANETs: GeoNetworking

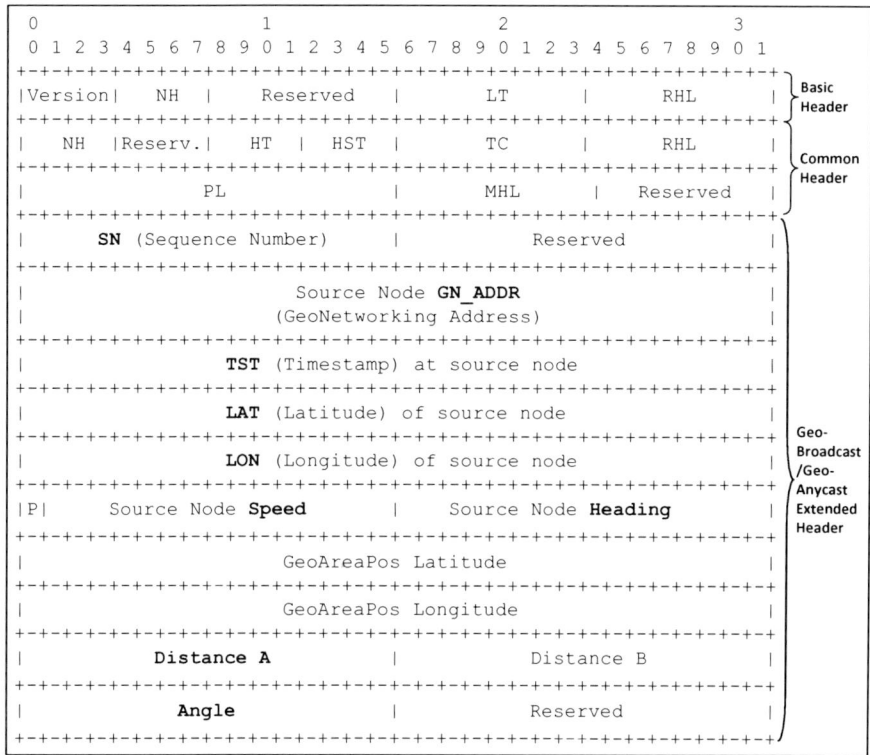

Fig. 8.18 Packet structure and headers of the GeoNetworking GeoBroadcast and GeoAnycast packets

common header is filled in a way to reflect whether the destination is any node (GeoAnycast) or all nodes (GeoBroadcast) in the destination area. The Extended header is composed of two parts, a first part dedicated to carry the information (GN_ADDR, timestamps, and location) of the sender (source node), and a second part is dedicated to carry the information of the destination area. Once the packet is prepared and filled in, the sender checks whether it is inside the destination area or not. If the sender is inside the destination area, the packet is broadcasted to all nodes around, otherwise the packet is line forwarded towards the destination area (i.e., towards the geographical point represented by GeoAreaPos Latitude and GeoAreaPos Logitude information in the extended header).

If a forwarder node receives a new GeoBroadcast or a new GeoAnycast packet intended for a destination area which does not surround that receiver node, and this packet has been received from a sender node which does not belong to the destination area either, then it forwards the packet to continue the line forwarding towards the destination. In case the packet has been received from a sender node which is located inside the destination area, that packet is ignored and dropped immediately.

If a node receives a GeoBroadcast or a GeoAnycast packet for which it is within the intended destination area, the packet is consumed immediately (decoded and payload moved up to upper layer). If the received packet is a GeoBroadcast packet, and the received node has been selected for forwarding the packet, then the packet is re-broadcasted to all nodes around to continue the dissemination of the packet within the destination area.

A GeoNetworking packet might be buffered for a defined duration of time due to the unavailability of the forwarding resources or for other reasons as explained in Sect. 8.4.

8.5.4.3 Example of Usage

Figure 8.19 shows an example to help understanding the usage of line forwarding and GeoBroadcast mechanisms, which are the key features in both GeoBroadcast and GeoAnycast protocols. In the example an emergency vehicle wants to inform all relevant vehicles located around the next road intersection about its arrival to allow a smooth and safe intersection crossing. The information about the approaching emergency vehicle is disseminated to all nodes within the destination area, including both vehicles and traffic lights. As the destination area is far-away, the packet is first line-forwarded through intermediate forwarders till it reaches a vehicle or any communicating node inside the targeted area, then broadcast within that area.

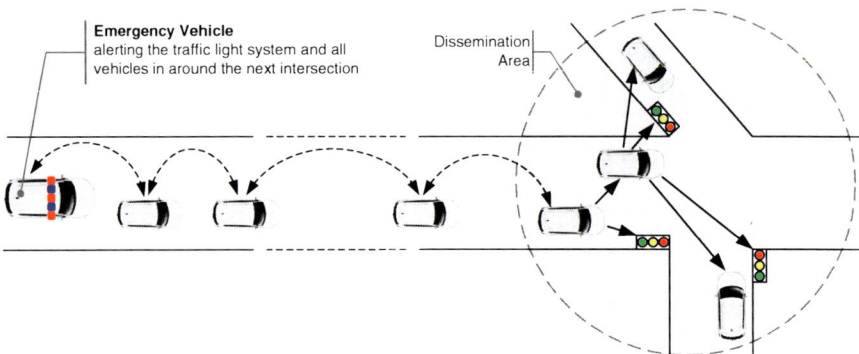

Fig. 8.19 Approaching emergency vehicle use case enabled by dissemination of Decentralized Environmental Notification Message (DENM) (See Chap. 5 for more details about DEN Messages.) through GeoNetworking GeoBroadcast communications

8 Forwarding in VANETs: GeoNetworking

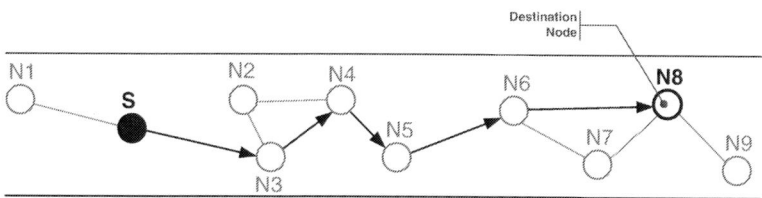

Fig. 8.20 GeoUnicast (Peer-to-Peer) communication scenario

8.5.5 GeoUnicast

GeoUnicast is usually intended for disseminating non-critical safety applications, as the number of beneficiaries is limited to two peers: the sender and the receiver. GeoUnicast is used to transmit a packet from one node (sender node) to another node (destination node) either through a direct communication link (One hop) if the two peers are within the communication range, or through multiple forwarding steps utilizing intermediate nodes.

8.5.5.1 Scenario

Figure 8.20 presents the GeoUnicast scenario through a peer-to-peer communication example. In the shown example, a node S initiates a peer-to-peer communication with a node N8 which is located beyond its communication range. A GeoUnicast packet is transmitted by the source node S, then forwarded by a selected neighbor N3, then N4 and so on till it reaches the destination node N8 through N6.

8.5.5.2 Protocol Operations

Whenever a node receives a payload from upper layer to transmit to a specific GeoNetworking node, a GeoNetworking GeoUnicast communication scheme is triggered and a GeoUnicast packet is created (as shown in Fig. 8.21). The GeoUnicast extended header is composed of two parts: a first part dedicated to carry the information (GN_ADDR, timestamps, and location) of the sender (source node), and a second part is dedicated to carry the information of the destination node. The upper layer provides the GeoNetworking address (GN_ADDR) of the destination node, without its location. Therefore, the source node first checks locally in its location table if there is any fresh entry that can provide the location of the destination. If the location of the destination node is missing, then a special mechanism (Location Service) as described in Sect. 8.5.5.3 is triggered to recover it. Such mechanism is necessary because the GeoUnicast packet can be transmitted only is the location information of the destination node is known.

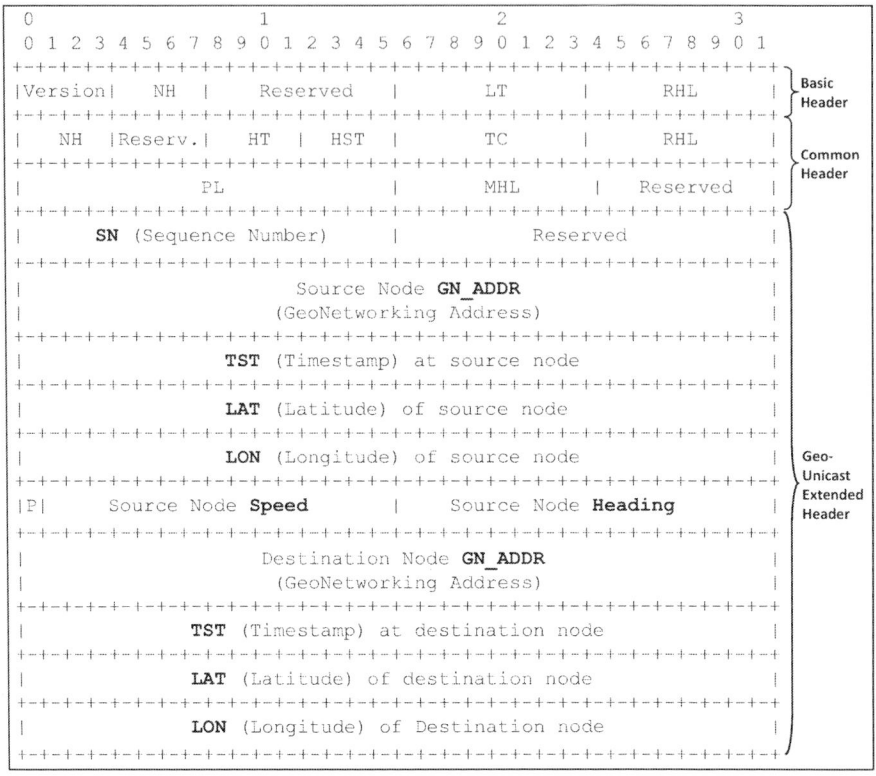

Fig. 8.21 Packet structure and headers of the GeoNetworking GeoUnicast packet

Once the GeoUnicast packet is prepared and filled in with all required information, it is transmitted either in broadcast mode or unicast mode depending on either sender-based or receiver-based forwarding mechanism is used (see Sect. 8.4 for more details). If sender-based forwarding is used, then the source node has to select the best forwarder among all its neighbors and then transmit the packet to that selected node through a unicast transmission. If a receiver-based forwarding mechanism is used, then the source node does not care about the selection of the next forwarder, and immediately broadcast the packet. Such packet is received by all neighbors, which enter in sort of competition to elect the next forwarder. When receiving a new GeoUnicast packet, a forwarder node has to check if the packet is not intended to itself (i.e., if it is not the destination). If it is the case, the packet is consumed immediately, otherwise the forwarding mechanism continues through selected neighbors, i.e. next forwarders.

8 Forwarding in VANETs: GeoNetworking

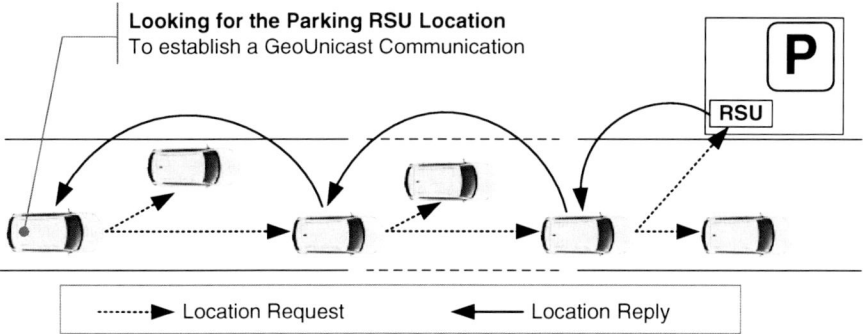

Fig. 8.22 Location Service Request and Reply packets exchange to recover the location of GeoUnicast destination node

8.5.5.3 Location Service

It might happen that the local location table does not contain fresh location information of the destination node. In such a case, the GeoUnicast cannot be performed. To overcome this, a location service is used to recover the location information of the destination node. Such a service is simple and basic, where the sender floods the network (surrounding nodes) with a Location Request packet by means similar to the Topological Broadcast scheme. Whenever it reaches either the requested node (i.e., the destination node of the GeoUnicast communication), or any other node in the network which has up-to-date information about the location of the destination node, a Location Reply is immediately transmitted towards the Source node (i.e., the node which initiated the GeoUnicast communication) using a mechanism similar to GeoUnicast forwarding. Figure 8.22 shows a typical example of Location Request and Reply packets exchange.

Once the Location Request packet reaches the requesting node with fresh information about the location of the designated destination node, all the GeoUnicast packets that have been buffered and that are destined to that particular destination node are extracted, updated (filling the destination node information in the GeoUnicast extended header), and then transmitted.

8.5.5.4 Example of Usage

Figure 8.23 shows a peer-to-peer communication based on GeoNetworking GeoUnicast exchange of packets between a vehicle and a parking lot. In the shown example, a vehicle sends a GeoUnicast packet which contains a request for parking reservation. The parking RSU replies with a GeoUnicast packet to confirm the parking availability along with a reservation number.

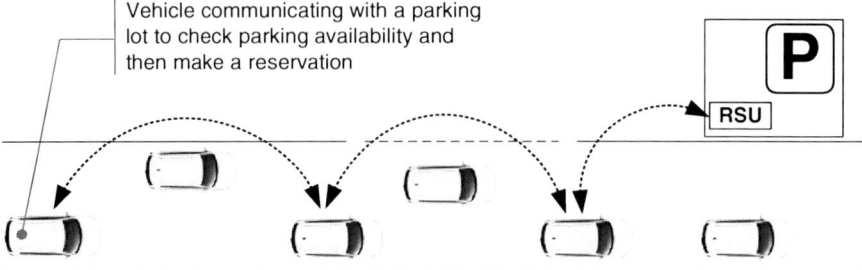

Fig. 8.23 GeoUnicast communication between a vehicle and a parking RSU to check parking availability and make a reservation

8.6 Security

This section is intended to give a brief overview on how security, as part of the ITS Station reference architecture, is integrated into GeoNetworking. For more details on security in VANETs, see Chap. 10.

The GeoNetworking layer may use the Security entity of the local node to provide secure communication services. This includes signing of messages and signature validation as well as end-to-end encryption. The selection of the mechanisms for a packet is based on a security profile supplied by the application or facility layer.

Security can be enabled via the `itsGnSecurity` protocol option. In secure mode GeoNetworking uses the following security services:

- **SN-ENCAP** is used to encapsulate each packet generated by the local node in a security envelope. Depending on the security profile, packets are signed, encrypted or both. The security entity also manages the inclusion of certificates on a regular basis or on demand.
- **SN-DECAP** is used to decapsulate a secured packet and remove the security envelope. Depending on the received packet this process includes verification of the signature, decryption of the packet or both. The result of the decapsulation and the packet without the security envelope are returned back to GeoNetworking for further processing.
- **SN-IDCHANGE-SUBSCRIBE**, **SN-IDCHANGE-UNSUBSCRIBE**, and **SN-IDCHANGE-EVENT** are used by GeoNetworking to be notified about changes of the node's pseudonym. Since some parts of the local GeoNetworking address are derived from the pseudonym the address is also updated with each pseudonym change.

Figure 8.24 illustrates the basic flow of protocol operation for an outgoing packet. First, all headers (basic, common and the packet type specific extended header) are created. In case security is enabled, the packet is handed to the Security

8 Forwarding in VANETs: GeoNetworking

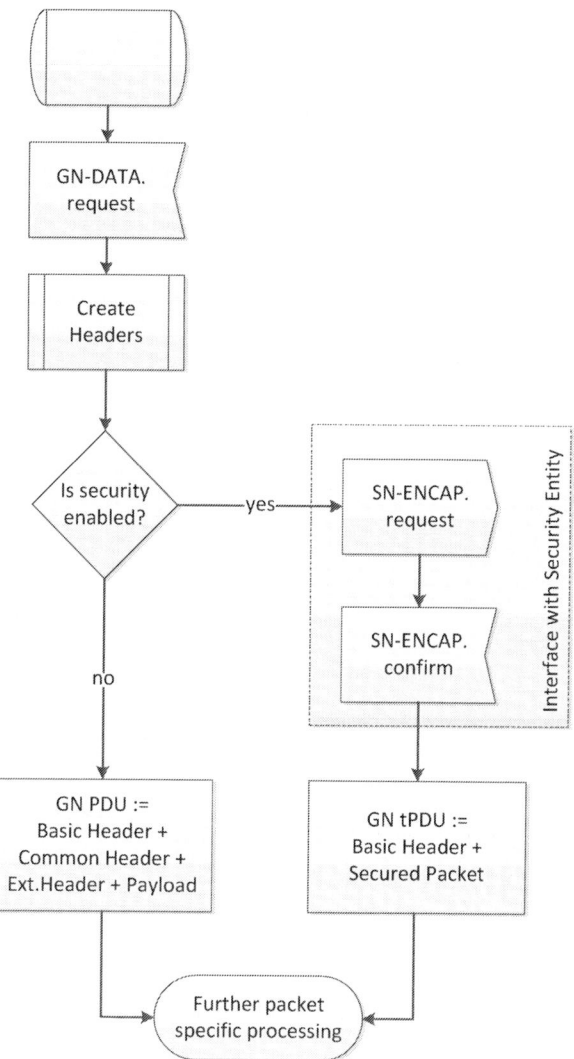

Fig. 8.24 Protocol operation for outgoing packets

Entity (SN-ENCAP.request). A *Secured Packet* is returned (SN-ENCAP.confirm) that contains all the information from the common and extended header as well as the packet payload in a signed and/or encrypted format. The specific security operations applied to create the Secured Packet are defined by the *Security Profile* provided with the encapsulation request. The Secured Packet is appended to the basic header to create the GeoNetworking PDU.

Figure 8.25 shows the protocol operation for an incoming packet. First the basic header is processed. If the field for the next header indicates a Secured Packet, the data following the basic header is handed to the Security Entity for decapsulation (SN-DECAP.request). The result (SN-DECAP.confirm) consists of a decapsulation

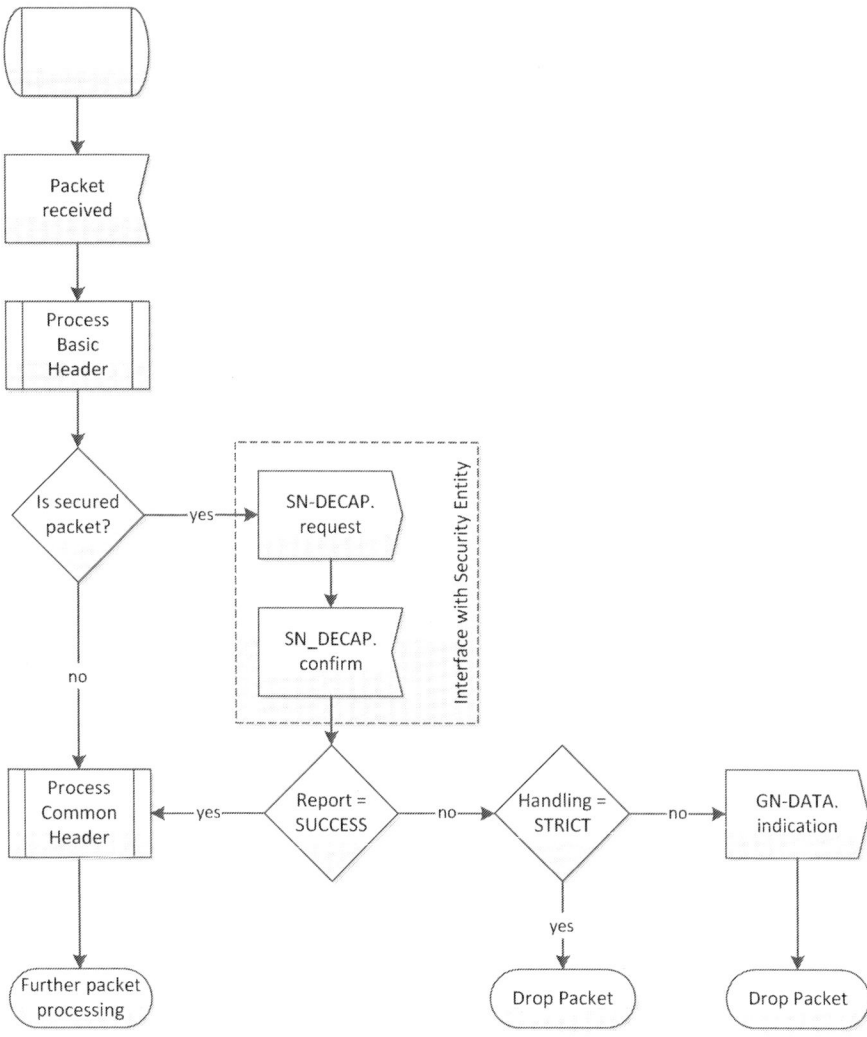

Fig. 8.25 Protocol operation for incoming packets

report, the plain packet data, and additional security related parameters. If the report indicates a success, the remaining headers and payload are processed similar to unsecured packets. In case of a failed security check packets are dropped without further processing. However, the GeoNetworking standard [4] provides a *non-strict* operation mode in which the payload of packets with failed verification is still handed to upper layers to allow for additional security evaluation on facility or application levels. Nevertheless, secured packets are only forwarded if the security check was successful.

8.7 Duplicate Packet Detection

GeoNetworking protocol is built around the concept of maximizing the spreading of information in a geographical area and for this reason each packet might be retransmitted several times within an area. For example, in the case of GeoBroadcast, each receiving host will retransmit the packet if either the hop count is greater than zero or the host is inside a defined geographical area. But, similar behavior might happen in case of routing loops, misconfiguration or replay of packets from misbehaving nodes.

For this reason, in order to control (e.g., prevent) the forwarding of duplicate packets, the GeoNetworking protocol uses mechanisms for duplicate packet detection:

- **Sequence number and time stamp-based:** This technique, applicable for Topological Broadcast, Geobroadcast, GeoAnycast, and GeoUnicast protocol operations (see Sects. 8.5.3–8.5.5), makes use of both sequence number and time stamp included in the transmitted packet to evaluate if a packet is duplicated or not.
- **Time stamp-based:** This technique makes use only of the time stamp and it is suitable for SHB protocol operation (see Sect. 8.5.2).

It is important to remind that GeoNetworking is able to detect packet duplication, however it does not provide any packet re-ordering and due to the simple duplicate packet detection, out-of-sequence packets are discarded.

8.8 Special Features

In order to provide differentiated and reliable support to vehicular applications, GeoNetworking contains special features which address Quality of Service (QoS):

- Traffic Classes
- Packet Data Rate Control
- Decentralized Congestion Control

In the case of *Decentralized Congestion Control*, the functionality might be applicable only to a particular media (e.g., ETSI ITS-G5). The other special features are instead available for any media on top of which GeoNetworking is relying.

8.8.1 Traffic Classes

QoS is especially important for safety where different applications should have a corresponding priority level. For example, a safety service should have a higher priority than a service for advertisements. For this reason, GeoNetworking uses classes to prioritize the network traffic. Each packet contains a dedicated field *Traffic Class* (TC) which contains the following information:

- **SCF** indicates whether the packet shall be buffered when no neighbor exists.
- **Channel Offload**[3] indicates whether the packet can be offloaded to another channel than specified in the TC ID.
- **TC ID** identifies which traffic class should be used.

SCF and *Channel Offload* are used within GeoNetworking to perform forwarding operations. In particular, the former is used by GeoNetworking to decide what to do once a packet should be transmitted but the location table is empty. Figure 8.26 provides an example of how GeoNetworking behaves depending on SCF value as provided by the upper layer. The latter allows (or not) GeoNetworking to transmit a packet on a different TC from the one specified inside the packet. With ETSI ITS-G5 the *TC ID* is used to decide which 802.11e [9] queue shall be used.

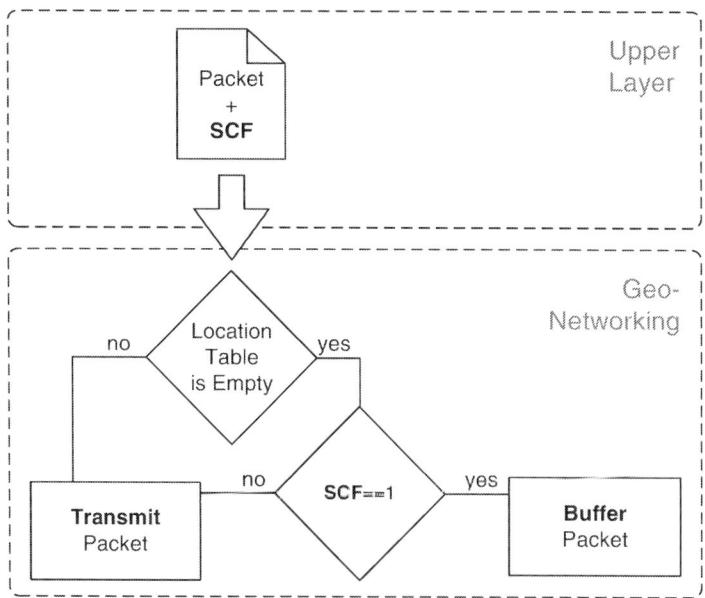

Fig. 8.26 Example of GeoNetworking behavior depending on SCF

---

[3] See Chap. 6 for more details about offloading between channels.

8.8.2 Packet Data Rate Control

Forwarding a packet is one of the main functionalities of GeoNetworking. But, in the case a misbehaving node is generating too many packets the network could be easily flooded locally and damages could be potentially created far away from the originator due to forwarding.

GeoNetworking addresses this problem by providing a *Packet data rate control* functionality which is evaluating the quantity of generated packets by each neighbor and limiting the forwarding of packets from a misbehaving node. In order to realize this, each node maintains an Exponential Moving Average (EMA) of the Packet Data Rate (PDR) for each entry in the location table. The calculation of the EMA is available in Eq. (8.1):

$$\text{PDR} = \beta \times \text{PDR}_{t-1} + (1 - \beta) \times x_t \qquad (8.1)$$

where:

x_t is the measured instantaneous value of the packet data rate upon reception of the GeoNetworking packet.
PDR is the average value of the packet data rate at time t;
PDR_{t-1} is the previous value at time $t - 1$ maintained in the location table.
β is the weighting factor ($0 < \beta < 1$)

In the case the EMA is above a predefined threshold, the packets of the particular node are not forwarded.

8.8.3 Decentralized Congestion Control at Network Layer

In case safety vehicular applications run on top of IEEE 802.11p, the so-called DCC [2][4] feature is required. In fact, it is well known that the CSMA/CA mechanism used by IEEE 802.11p provides a fair channel access to the contending nodes, i.e., on average, it lets nodes access the channel the same number of times in a given time period. However, in the short run, it is inherently unpredictable and due to the random exponential back-off procedure the interference between concurrent transmissions may lead to transmission failures. DCC makes use of the local knowledge about the channel status to adjust the transmission parameters and thus reducing channel congestion.

GeoNetworking extends this concept by providing the capabilities of disseminating the information about the Channel Busy Ratio (CBR) measured locally to surrounding nodes. By disseminating CBR values, the neighbors increase their awareness of possible channel congestion that the ego node can contribute to, even

---

[4]See Chap. 7 for more details about DCC.

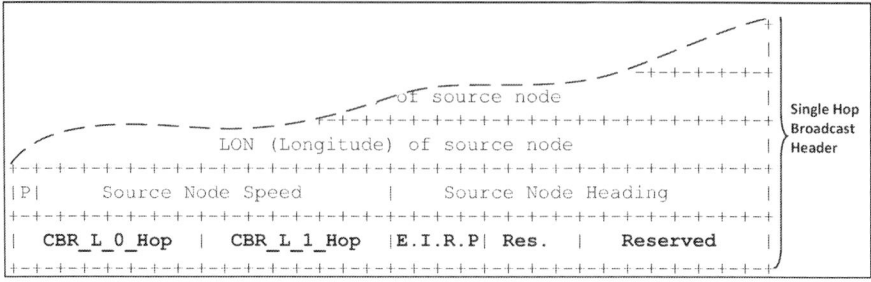

Fig. 8.27 DCC fields included in the SHB packet

though the ego node does not perceive a local congested channel status. In general the dissemination of CBR values allows to react quickly to congestion and maintains globally fair the usage of the channel. More information could be found in [12].

GeoNetworking supports the transmission of CBR values directly in the header of every transmitted SHB packet. Figure 8.27 shows the DCC related fields which are included in the SHB packet:

- CBR_L_0_Hop represent the CBR value based on neighbor information,
- CBR_L_1_Hop represent the maximum CBR_L_0_Hop value received from a neighboring node in a given interval,
- output power of the current packet—EIRP,
- reserved for future Multi Channel Operations.

8.9 Conclusion

In this chapter the base of routing has been discussed, the differences between multiple addressing techniques used in GeoNetworking have been identified and the principles in the position-based forwarding have been clarified. Such features are the core techniques on top of which Geonetworking functionalities are built. Such functionalities have been standardized in ETSI TC ITS as a multi-series standards [4, 5] which is going to be used in Europe for the initial deployment of vehicular communication. Multiple examples of the usage of Geonetworking have been provided in the chapter to simplify the comprehension of this technology for both technical and non-technical readers.

References

1. ETSI EN 302 665 (V1.1.1): Intelligent transport systems (ITS); communications architecture (2010)
2. ETSI TS 102 687 (V1.1.1): Intelligent transport systems (ITS); decentralized congestion control mechanisms for intelligent transport systems operating in the 5 GHz ranger; access layer part (2011)

3. ETSI EN 302 636-1 (V. 1.2.1): Intelligent transport systems (ITS); vehicular communications; GeoNetworking; Part 1: Requirements (2013)
4. ETSI EN 302 636-4-1 (V1.2.0): Intelligent transport systems (ITS); vehicular communications; GeoNetworking; Part 4: Geographical addressing and forwarding for point-to-point and point-to-multipoint communications; Sub-part 1: Media-independent functionality (2013)
5. ETSI TS 102 636-4-2 (V 1.1.1): Intelligent transport systems (ITS); vehicular communications; GeoNetworking; Part 4: Geographical addressing and forwarding for point-to-point and point-to-multipoint communications; Sub-part 2: Media-dependent functionalities for ITS-G5 (2013)
6. ETSI EN 302 636-3 (V1.1.2): Intelligent transport systems (ITS); vehicular communications; GeoNetworking; Part 3: Network architecture (2014)
7. ETSI TS 102 894-2 (V1.2.1): Intelligent transport systems (ITS); users and applications requirements; Part 2: Applications and facilities layer common data dictionary (2014)
8. Füßler H, Widmer J, Käsemann M, Mauve M, Hartenstein H (2003) Contention-based forwarding for mobile ad-hoc networks. Elsevier's Ad Hoc Netw 1(4):351–369
9. IEEE Computer Society. IEEE Std 802.11eT, Part 11: Wireless medium access control (MAC) and physical layer (PHY) specifications: medium access control (MAC) enhancements for quality of service (2005)
10. Karp B, Kung HT (2000) GPSR: greedy perimeter stateless routing for wireless networks. In: Proceedings of the 6th annual international conference on mobile computing and networking, MobiCom '00. ACM, New York, pp 243–254
11. Mariyasagayam M, Osafune T, Lenardi M (2007) Enhanced multi-hop vehicular broadcast (mhvb) for active safety applications. In: IEEE 7th international conference telecommunications, ITST'07, pp 1–6
12. Tielert T, Jiang D, Chen Q, Delgrossi L, Hartenstein H, (2011) Design methodology and evaluation of rate adaptation based congestion control for vehicle safety communications. In: IEEE vehicular networking conference (VNC), pp 116–123. doi:10.1109/VNC.2011.6117132

Chapter 9
The Use of IPv6 in Cooperative ITS: Standardization Viewpoint

Fernando Pereñiguez, José Santa, Pedro J. Fernández, Fernando Bernal, Antonio F. Skarmeta, and Thierry Ernst

Abstract *Internet Protocol version 6* (IPv6) is envisaged to be the cornerstone for the future development of ITS services. Both research communities and standardization forums are considering the use of IPv6 as a media-agnostic carrier for non-time critical safety applications as well as for traffic management. This interest is based on the numerous benefits brought by IPv6: large addressing space able to cope with ambitious deployment scenarios such as the vehicular one, easier support of management operations related to node auto-configuration and native support of security mechanisms. This chapter surveys the current European standard specifications in the field of IPv6-based communications applied to cooperative ITS systems. The convenience of using IPv6 technology is demonstrated by identifying a wide set of relevant ITS specifications where the application of IPv6 networking protocols outperforms other protocol solutions. Furthermore, this study is complemented with a performance assessment of IPv6 communications over a cooperative ITS scenario using a real IPv6-enabled communications stack compliant with the standardized ITS station reference architecture specified by *International Standards Organization* (ISO) and *European Telecommunications Standards Institute* (ETSI).

F. Pereñiguez (✉)
University of Murcia, Murcia, Spain

Catholic University of Murcia, Murcia, Spain
e-mail: pereniguez@um.es

J. Santa
University of Murcia, Murcia, Spain

University Centre of Defence at the Spanish Air Force Academy, Murcia, Spain
e-mail: jose.santa@cud.upct.es

P.J. Fernández • F. Bernal • A.F. Skarmeta
University of Murcia, Murcia, Spain
e-mail: pedroj@um.es;fbernal@um.es; skarmeta@um.es

T. Ernst
INRIA - Mines Paris Tech (LaRA), Paris, France
e-mail: thierry.ernst@mines-paristech.fr

Keywords VANET • ISO • ETSI • C-ITS • ITS-Station • Vehicle ITS-S • Roadside ITS-S • Central ITS-S • Networking • Mobility • IPv6 • Address autoconfiguration • ICMPv6 • NEMO • MCoA • IPsec • IKEv2

9.1 Introduction

Cooperative Intelligent Transportation Systems (C-ITS) are ITS where participating entities (e.g., cars, charging stations, traffic lights, etc.) continuously communicate and exchange information among them with the objective of improving safety, sustainability, efficiency, and comfort beyond the scope of standalone ITS. In other words, C-ITS magnifies the benefits offered by autonomous ITS, brings major social and economic benefits, and leads to greater transport efficiency and increased safety. C-ITS supports decreasing road fatalities, improving the capacity of roads, diminishing the carbon footprint of road transport, and enhancing the user experience during travels. Although there are many vehicular services envisioned for the short, medium, and long term, these are usually categorized in the next groups [1, 2]:

- *Safety.* These services are intended to reduce accidents and safeguard vehicle occupants and pedestrians lives. Some examples are collision avoidance, accident notification, or emergency vehicle approaching.
- *Traffic efficiency.* In this group there are services that improve the road network capacity and reduce the travel time. Some examples are variable speed limit, dynamic management of road intersections, or congestion detection and mitigation.
- *Infotainment.* Mainly oriented to provide value-added comfort services, Internet access, and multimedia. Some examples are context-aware touristic guidance, video under demand, and video conferencing.

For supporting this diversity of services it is clear that a generic network architecture is needed that also assures the future compatibility among different providers. Due to this, during the last year there has been an intense work on standardization activities regarding cooperative ITS. First, the ISO Technical Committee 204 released the Communications Access for Land Mobiles (CALM) concept [3], but the later created European Telecommunications Standards Institute (ETSI) Technical Committee ITS improved CALM based on the results of the COMcSafety European project through the European ITS communication architecture [4]. The last update of this common ISO/ETSI effort has been recently provided by ISO [3]. The proposed architecture by the European standardization bodies should be instantiated totally or partially on vehicles, nomadic devices, roadside units, and central points.

Despite the general architecture has been defined, there exist currently several ongoing efforts to define the operation of the different modules integrating the communications stack. In particular, a controversy remains about the protocols to be used mainly in the networking layer. So far, we can identify two main families of protocols being adopted by standardization bodies (and the academia): specific ITS protocols based on GeoNetworking [5],[1] which is a multi-hop routing protocol oriented to the geo-dissemination of information in vehicular environments; and Internet Protocol version 6 (IPv6) technologies [6], based on the evolution of well-known Internet protocols defined within the Internet Engineering Task Force (IETF).

Although GeoNetworking offers more adapted functionalities for supporting vehicular communications, such as native geographical distribution or low packet overhead, IPv6 offers a more interoperable solution with the rest of the (Future) Internet, supporting in a better way concepts such as Internet of Things (IoT) or Smart Cities, among others. Moreover, a number of well-known IETF protocols could be added for providing extra security, multicast, multi-homing, etc.

This chapter motivates the usage of IPv6 in this context, providing an overview of the current standardization efforts to develop IPv6-based communications architectures for C-ITS. The discussion is complemented with several use cases presenting the usage of IPv6-based technologies in order to develop valuable ITS application services.

This chapter is organized as follows. First, Sect. 9.2 provides a brief overview of the ISO/ETSI station reference architecture, specially attending to the standardized extensions to support IPv6-based communications. Next, Sect. 9.3 presents relevant IPv6 technologies, emphasizing their application to C-ITS. Once both the architecture and related IPv6 protocols have been described, Sect. 9.4 delves into the details of the IPv6 networking of the communications stack, attending to the concrete modules and functionalities that are expected to provide. After that, Sect. 9.5 demonstrates the usefulness of applying IPv6 technology in C-ITS by describing a set of use cases. Finally, the chapter is concluded in Sect. 9.7 by enumerating the different advantages derived from applying IPv6-based technologies in the ITS segment.

9.2 ISO/ETSI Station Reference Architecture

In an effort towards harmonization, the international ITS community agreed on the definition of a common ITS communication architecture suitable for a variety of communication scenarios (vehicle-based, roadside-based, and Internet-based) through a diversity of access technologies (802.11p, infra-red, 2G/3G, satellite, ...) and for a variety of application types (road safety, traffic efficiency, and comfort/infotainment) deployed in various continents or countries ruled by distinct policies.

---

[1] See Chap. 8 for more details about GeoNetworking.

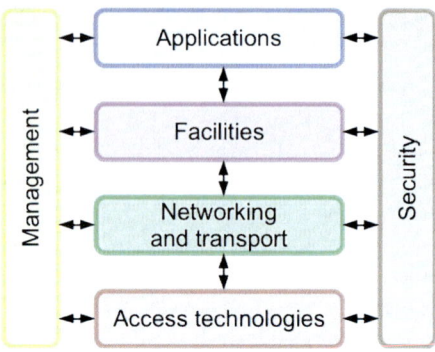

Fig. 9.1 ISO/ETSI reference communication stack

This common communication architecture is known as the *ITS station reference architecture* and is specified by ISO in [3] and by ETSI in [4]. As depicted in Fig. 9.1, the ITS station architecture follows an *Open Systems Interconnection* (OSI) like layered design: *access, networking & transport, facilities* and *applications*. Additionally, two cross-layer entities are defined: *ITS Station Management Entity* (SME) and *ITS station Security Entity* (SSE). While the former is in charge of managing the internal processes carried out within the stack at the different layers, the latter is intended to provide security services to modules requiring such need. These functional blocks are interconnected through *Service Access Points* (SAP) that allow the exchange of information between layers and entities.

One distinguishable feature of this architecture is the ability to use a variety of networking protocols in order to meet opposite design requirements i.e., fast time-critical communications for traffic safety versus more relax communication requirements for road efficiency and comfort/infotainment.

9.2.1 ITS Stations

There exist different types of ITS stations (ITS-S): *personal ITS-S* (e.g., smartphones), *vehicle ITS-S* (e.g., cars), *roadside ITS-S* (e.g., electric charging station), or *central ITS-S* (e.g., road operator control center). Each ITS station type implements a subset of the functionalities of the general ITS station reference architecture according to the played role, connectivity requirements, supported applications, etc. For example, while a personal ITS could only require 802.11g to connect with the vehicle's mobile router and gain Internet access, vehicle ITS stations may use both 802.11p and 3G wireless technologies for short-range and long-range communications, respectively. Conversely, central ITS most probably will be only accessed through wired communication media (e.g., 802.3).

The ITS station is defined as a *bounded secured managed domain*. In the most general case, the functions of an ITS-S are split into a router (ITS-S router) and hosts (ITS-S host). The ITS-S router is a node comprised of routing functionalities that is

9 The Use of IPv6 in Cooperative ITS: Standardization Viewpoint 257

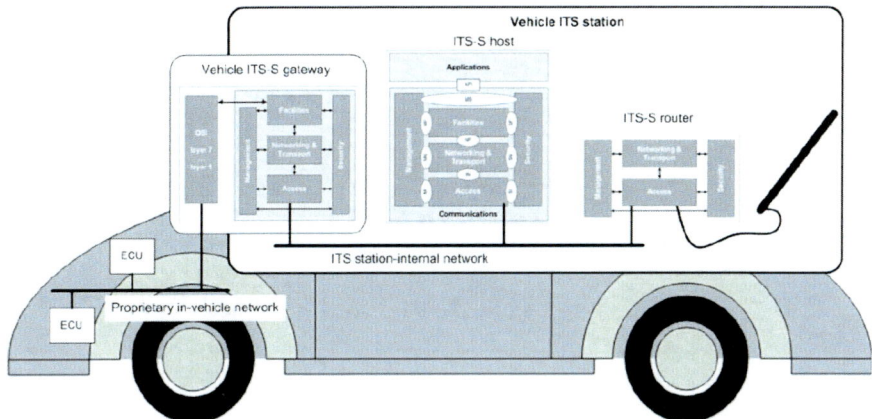

Fig. 9.2 Vehicle ITS-S [4]

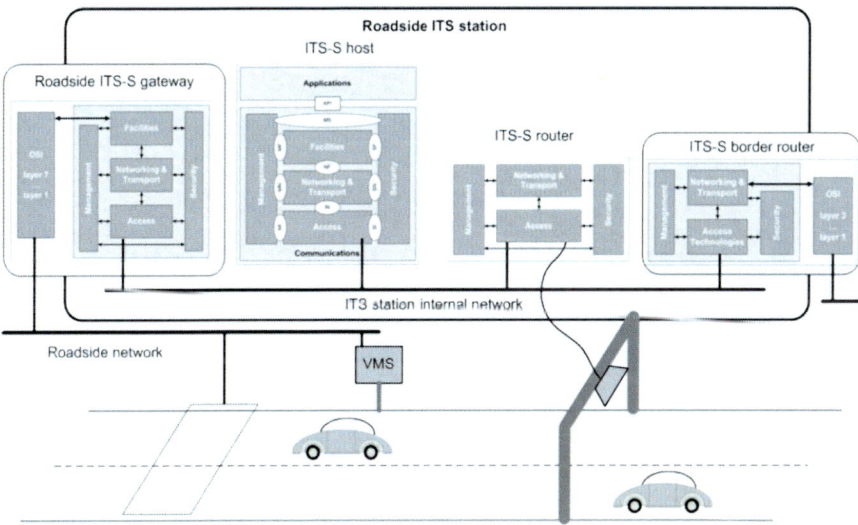

Fig. 9.3 Roadside ITS-S [4]

used to connect two networks and forward packets not explicitly addressed to itself. In contrast, ITS-S host is a final node executing some specific ITS application, for example, a collision avoidance service to the driver. ITS-S hosts are attached to the ITS-S router via some ITS station internal network. In some cases, the router and hosts functions may also be merged into a single entity. This general ITS station architecture is common to the different types of stations: vehicle ITS station (see Fig. 9.2), roadside ITS station (see Fig. 9.3), and central ITS station (see Fig. 9.4).

In this chapter, the terms *On-Board Unit* (OBU) and *Road Side Unit* (RSU) are meant for mobile router (MR) and access router (AR), respectively.

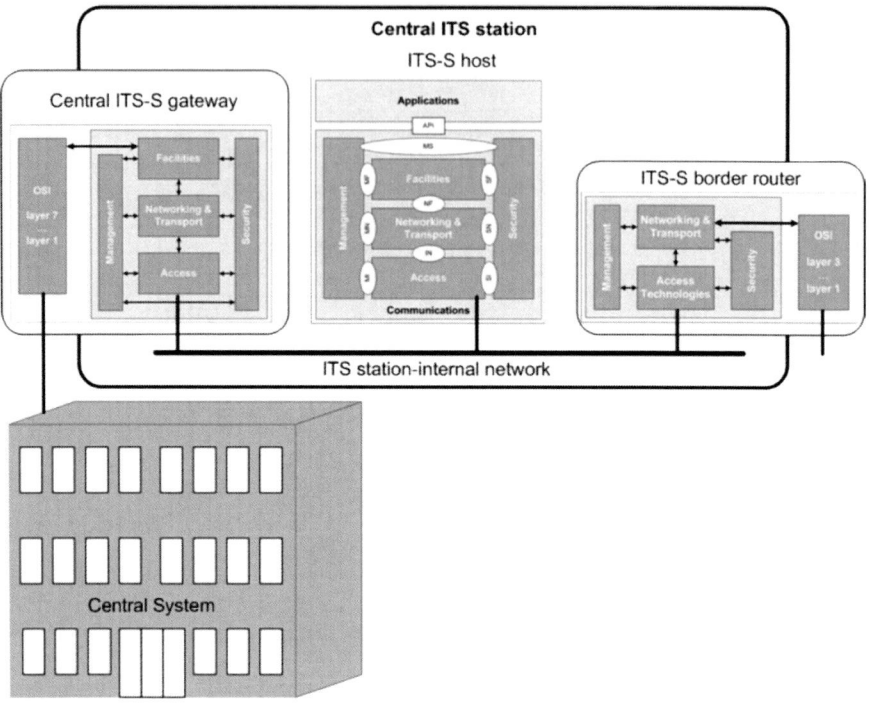

Fig. 9.4 Central ITS-S [4]

9.2.2 Stack Layers

In the following we provide a brief overview of the different layers conforming the ISO/ETSI communications stack. Note that we emphasize on specific details of this stack that allow us to explain the integration of IPv6 technologies in vehicular cooperative systems.

The *Access Layer* includes a variety of access technologies that can be used for the communication with other ITS stations, with ITS station internal network nodes, and other legacy ITS nodes. These access technologies can be used simultaneously, and vary according to the type of ITS station and its purpose. Currently, the support of several wireless access technologies (infrared [7], microwave [8], millimeter wave, 2G/3G [9, 10]) and wired access technologies (Ethernet) is already specified, in either ISO or ETSI standards. More access technologies could be supported in the future, without any impact on the other layers of the ITS station reference architecture.

The horizontal *networking & transport* layer is the core of the communications stack. This layer aggregates all the networking protocols in charge of transmitting application messages over the ITS network to reach the destination. Currently, the standardization community has regulated different networking technologies to operate in this layer. On the one hand, the FNTP protocol (Fast Networking & Transport layer Protocol, previously known as FAST) has been specified by ISO TC204 in ISO 21210 [11, 12]. On the other hand, the well-known GeoNetworking protocol (media-dependent [5] and media-independent [13]) offers a geographical routing of information.

Nevertheless, the architecture is flexible enough so that more networking protocols could be added in the future if needed. In fact, a new standardization work has recently started in order to integrate IPv6 networking technologies into the ITS-S communications stack. Section 9.4 surveys the current European standard specifications related to the application of IPv6 technologies to C-ITS communications.

The ITS station *Facilities Layer* is an intermediate layer between the *networking & transport* layer and the *applications* layer, offering them access to information collected by other ITS stations (vehicles, roadside) and freeing them from the necessary message signaling to transmit and process data exchanged between ITS stations in a broadcast fashion. The immediate benefit is the sharing of data between various applications located in different nodes by using standardized messaging. The current facilities include *Cooperative Awareness Message* (CAM) [14], which is broadcasted in a single-hop fashion by an ITS-S to inform about the ego status, and *Decentralized Environmental Notification Message* (DENM) [15], which is a multi-hop broadcast message transmitted in a given geographical area. For a detailed description of CAM and DENM, please refer to Chap. 5.

The *Applications Layer* concentrates all the ITS applications (e.g., road safety or traffic efficiency) being executed within a particular ITS station.

These applications rely partly on services offered by the ITS station facilities layer. As a result of the messages received from the ITS station facilities, applications may, for example, display messages on the navigation system of the vehicle. These applications may also issue messages to other ITS stations without subsequently involving the ITS station facilities. For example, as a result of receiving a service notification by the ITS station facilities, of a charging spot for electric vehicles, the application may directly contact a server to enquire about the availability of the charging spot and book it.

Finally, the aforementioned layers are surrounded by two vertical layers: the *ITS Station Management Entity* (SME) and the *ITS station Security Entity* (SSE). The SME is responsible for the selection of the best communication path (communication interface and end-node), according to the flow requirements expressed by the ITS station applications, the access technologies characteristics, and the current network conditions. Conversely, SSE provides common security functionalities to all the horizontal layers and maintains security credentials used by these [16]. The ITS station security entity does not perform any particular security mechanism or protocol. Protocol dedicated security mechanisms and security protocols are implemented at each layer. Instead, the SSE manages credentials (e.g., cryptographic keys or certificates) and performs atomic security operations (e.g., hashing, encryption, etc.).

9.3 Relevant IPv6 Technologies for C-ITS

Internet Protocol version 6 (IPv6) [6] is the new version of the Internet Protocol (IP) standardized by the IETF. Apart from solving the IPv4 addresses depletion, IPv6 provides extra advantages that cover important needs in cooperative vehicular communications. First, IPv6 defines addresses of a fixed 128-bit length. This allows a very large address space that is considered sufficient for most ambitious deployment scenarios such as the vehicular one, where a number of vehicles and on-board devices should be addressed. Second, IPv6 makes easier the integration of mobile IP technologies, thus allowing the support of network continuity upon the change of point of attachment. Third, IPv6 also provides node auto-configuration, which is useful for nomadic devices entering the vehicle. Finally, another advantage of IPv6 relies on its security capabilities. Unlike its predecessor, IPv4, IPv6 natively supports secure communications to assure information confidentiality, integrity, and authentication.

The next parts of this section give an introduction to some of the most important IPv6-based technologies that are being considered by current ISO/ETSI standards in the field of cooperative systems. The next sections combine both protocol description and application to the C-ITS scenario, so that the reader can better perceive how these technologies contribute to the objective of achieving efficient communications among cooperative vehicular systems.

9.3.1 Address Auto-configuration

Similarly to IPv4, in IPv6 networks hosts should be plug and play in such a manner that access to the network must be acquired in a transparent manner. For this reason, two different processes have been defined for IPv6 address auto-configuration: stateless and stateful.

Stateful auto-configuration is performed through the *Dynamic Host Configuration Protocol version 6* (DHCPv6) [17]. The operation is based closely on the DHCP protocol version designed for IPv4 networks, where a server is in charge of assigning IP addresses. The operation of DHCPv6 is called stateful due to the DHCP server and the client (i.e., IPv6 host) must both maintain state information to keep addresses from conflicting, to handle leases, and to renew addresses over the time.

The real novelty of IPv6 relies on the *stateless auto-configuration protocol* [18], a novel solution where nodes can cooperate to assign an IPv6 address by themselves without the intervention of a server. The stateless auto-configuration procedure works as follows. Upon the reception of a global prefix to be used in the IPv6 network, which is usually provided by the access router to which a node is connected, it can auto generate an IPv6 address using its *Media Access Control* (MAC) address. By using the *Duplicate Address Detection (DAD)* procedure, the uniqueness of this address is checked with the rest of nodes present in the local

9 The Use of IPv6 in Cooperative ITS: Standardization Viewpoint

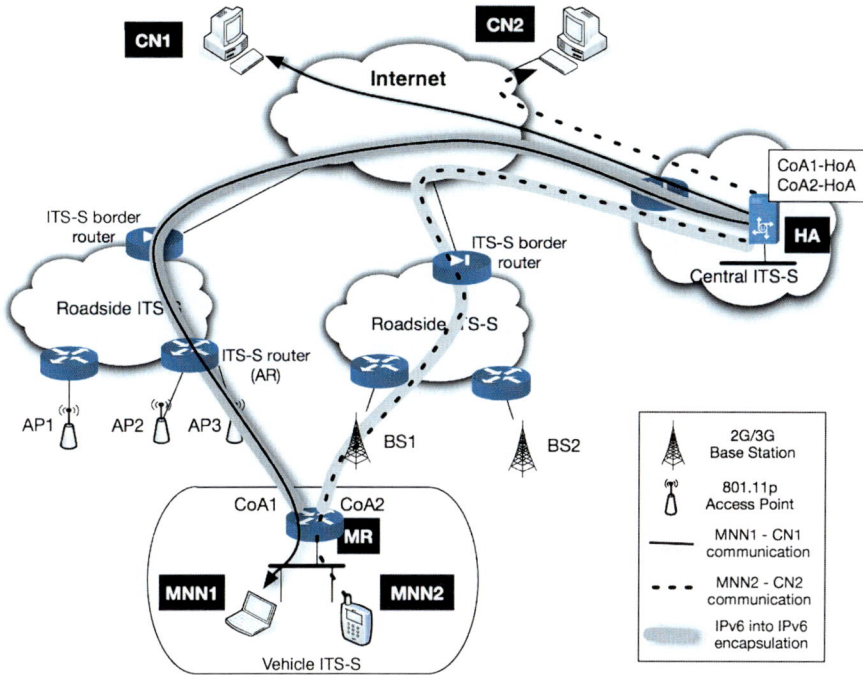

Fig. 9.5 Application of IPv6 technologies to C-ITS

network. According to this protocol, the node sends an ICMPv6 (Internet Control Message Protocol version 6) Neighbor Solicitation message using as target address the created one. If no response is received, the address is assumed to be unique and can be assigned to the node.

As observed, because no server has to approve the use of the address, stateless auto-configuration is simpler than the stateful process. For this reason, this is the default mode of operation adopted by most IPv6 systems. In any case, the IPv6 address auto-configuration mechanisms (either stateless or stateful) offer a simple procedure for ITS nodes to automatically configure valid IPv6 addresses. For example, as depicted in Fig. 9.5, in-vehicle hosts can easily acquire an IPv6 address within the vehicle to communicate with external peers in Internet through the infrastructure. The same happens with the vehicle router when it needs to configure a valid IPv6 address when visiting a roadside ITS-S.

9.3.2 NEMO

Network Mobility Basic Support (NEMO) [19] allows terminals within an IPv6 mobile network to be globally connected to the Internet. The key idea of NEMO is that the mobile network will maintain the same IPv6 addressing irrespective of the topological location of the mobile network. To accomplish this functionality, mobility capabilities are distributed between the *Mobile Router* (MR) and a server known as the *Home Agent* (HA) located in an IPv6 subnet known to the MR. The mobility operation is performed without breaking the flows under transmission, and transparently to the nodes located behind the MR and the communication peers.

The functionality provided by NEMO is useful to maintain Internet connectivity in C-ITS between all the nodes in the vehicle and the infrastructure. Thanks to NEMO, in-vehicle hosts are reachable through the infrastructure at the same IPv6 address as long as the address is not deprecated. In fact, the mobility support mechanism provided by NEMO is very easy to deploy since only the vehicle router of the vehicle ITS-S and a dedicated HA at the Central ITS-S are needed.

Figure 9.5 illustrates the operation of NEMO in C-ITS to maintain network connectivity for the vehicle network. Initially, an unchangeable IPv6 network prefix called *Mobile Network Prefix* (MNP) is delegated by the home network (Central ITS-S) to the vehicle MR. The MNP is used by the MR for assigning addresses to the *Mobile Network Nodes* (MNN). Following the NEMO model, the MR is aware of the existence of a new network when the MR attaches to a new visited network (Roadside ITS-S) and receives a *Router Advertisement* (RA) message from an *Access Router* (AR). In this case, the MR, which already has a fixed IPv6 address within the Central ITS-S (known as *Home Address* or HoA), generates a new autoconfigured IPv6 address valid within the Roadside ITS-S. This address is called *Care-of address* (CoA), and it is immediately notified to the HA.

The process of registering a new CoA at the HA is performed by the MR through a *Binding Update* message, which is acknowledged with a *Binding ACK* sent by HA. Only MR and HA are aware of the network change, since MNNs (i.e., in-vehicle hosts) continue connected with the MR using the same address configured using the MNP. Hence, when any computer outside the home network (*Correspondent Node* or CN) communicates with any of the in-vehicle hosts, it uses the MNP-generated address as destination and, hence, packets follow the route towards the home network. Then, HA redirects these IPv6 packets to the current IPv6 CoA of the MR, which finally distributes the packets within the mobile network. In the same way, when packets are sent from any in-vehicle host to a CN, they are routed by the MR towards the HA, which forwards them to the destination. This bidirectional communication between the HA and MR requires both entities to perform an IPv6 into IPv6 encapsulation in order to create a mobility tunnel.

9.3.3 MCoA

MRs can be provided with multiple communication interfaces such as IEEE 802.11b/g, IEEE 802.11p, WiMAX or UMTS, for instance. When an MR maintains these interfaces simultaneously up and has multiple paths to the Internet, it is said to be multihomed. In mobile environments, multihoming capabilities can alleviate problems suffered by MRs such as scarce bandwidth, frequent link failures, and limited coverage.

In those situations where the MR is simultaneously using multiple communication interfaces, in the basic NEMO protocol the HA is unable to keep track of the MR location. For this reason, the *Internet Engineering Task Force* (IETF) developed an extension of both Mobile IPv6 and NEMO Basic Support called *Multiple Care of Addresses Registration* (MCoA) [20]. Basically, MCoA defines extensions to allow the establishment of multiple tunnels between MR and HA. By distinguishing each tunnel with a unique *Binding Identification number* (BID), the mobility management of each interface follows the same procedure defined by NEMO.

The functionality offered by MCoA is specially useful in vehicular communications since mobile routers typically are equipped with multiple communication interfaces. For example, as depicted in Fig. 9.5, an intermittent 802.11p channel could be complemented, for instance, with a continuous 2G/3G connection.

9.3.4 IPsec and IKEv2

The *IP Security* (IPsec) [21] protocol is an enhancement to the basic IPv6 protocol that defines a set of security services for protecting IPv6 traffic. Since it is defined at IP level, the security protection is transparent to other protocols carried over IPv6. The IPsec packet protection to IPv6 packets can be applied through two security protocols: *Authentication Header* (AH) and *Encapsulating Security Payload* (ESP). While the former provides authentication and integrity protection to the IPv6 packet, the latter also provides confidentiality to the data transported within the IP packet. At the same time, these protocols can be applied in two different operation modes. In *transport mode*, the security services are applied to next layer protocols, i.e., the information carried within the IPv6 packet. Conversely, in *tunnel mode*, the protection is applied to the whole IP packet which is sent through a tunnel. Figure 9.6 describes what portion of the IPv6 packet is protected by each protocol, as well as the type of security protection provided.

The IPsec operation relies on the fundamental concept of *Security Association* (SA). An SA conceptually represents a secure connection between two entities. The establishment of an SA implies the negotiation of a set of security parameters such as cryptographic algorithms or key material that is used by the AH or ESP protocols. This protection must be performed in a secure manner, so that an attacker is unable to manipulate the negotiation process or access to the established cryptographic material (e.g. key used for encrypting).

In order to automate the IPsec SA establishment, the IETF designed the *Internet Key Exchange* (IKE) protocol. The last version of the protocol, IKEv2 [22], solves some limitations and provides a simplified protocol operation.

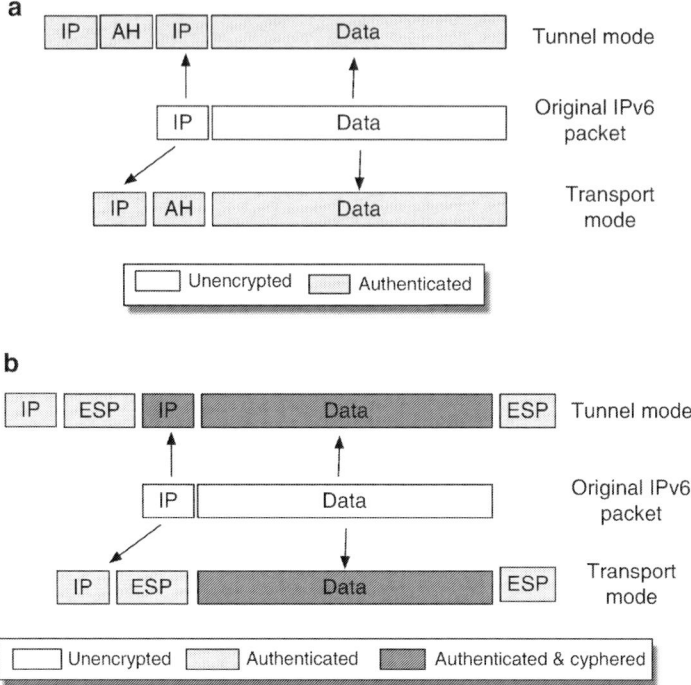

Fig. 9.6 IPsec security protection. (**a**) AH mode, (**b**) ESP mode

IKEv2 is executed between the parties willing to establish IPsec packet protection, called *initiator* and the *responder*. As their names indicate, the initiator starts the IKEv2 protocol whereas the responder acts as server during the negotiation. The protocol is composed of a well-defined set of four main exchanges (*request-response*), namely: IKE_SA_INIT, IKE_AUTH, CREATE_CHILD_SA, and INFORMATIONAL. In each exchange, IKEv2 assures reliability and confidentiality of the parameters negotiated.

The IKE_SA_INIT exchange establishes an SA at IKE level, named the IKE SA (IKE_SA), between the participant entities. This IKE_SA will protect all the following IKE exchanges. Once the IKE_SA is established, an IKE_AUTH exchange is performed in order to authenticate the parties and create the first IPsec SA (CHILD_SA) between them. These exchanges are denoted as *initial exchanges* and always must occur in this order. There are two additional exchanges used for managing the SAs. The CREATE_CHILD_SA exchange allows to create additional SAs. And the INFORMATIONAL exchange, that can be used for deleting SAs, notifies events and manages configuration issues.

The security provided by IPsec/IKEv2 is of vital importance to protect IPv6 communications in C-ITS. For example, in the scenario described in Fig. 9.5, the mobility messages exchanged between MR and HA are critical from a security viewpoint. If an attacker impersonates the MR, it can send fake registrations and modify the CoA recorded by the HA. For this reason, in this case, the application of an IPsec protection to the IPv6 communication channel established between MR and HA is highly desirable.

9.4 IPv6 Networking in the ITS Station Architecture

ISO 21210 (CALM: IPv6 Networking) [11] is the result of the work performed within ISO TC204 WG16 (CALM). It specifies IPv6 network protocols and services necessary to support global reachability of ITS stations (in the context of the ITS station architecture), continuous Internet connectivity for ITS stations, and the handover functionality required to maintain such connectivity. This functionality also allows legacy devices to effectively use an ITS station as an access router to connect to the Internet. Essentially, this specification describes how IPv6 is configured to support fixed and mobile ITS stations (vehicle, roadside, and central ITS stations) and their attached nodes. ISO 21210 does not define a new protocol, a new exchange of messages at the IPv6 layer, or new data structures. It defines how standardized IETF protocols are combined so that ITS stations can communicate with one another using the IPv6 family of protocols.

When attending to the case of IPv6 networking, the ISO 21210 [11] standard particularizes the general types of ITS stations. According to this standard, an ITS station will be linked to other ITS stations and networked entities via IPv6, either a legacy IPv6 network or a proprietary IPv6 network. In some very specific cases, for instance, direct communication between neighbor vehicle ITS stations using 802.11p, IPv6 may be replaced by some ITS-specific networking protocol like GeoNetworking or a legacy cellular network. Considering the different types of ITS stations, the following ITS-S IPv6 nodes are defined:

- In the vehicle ITS station, the nodes executing the ITS-S router functions are the *ITS-S IPv6 vehicle routers* (VRs). The ITS-S host functions may be implemented by the ITS-S IPv6 router, or by ITS-S IPv6 hosts. Vehicle ITS-S IPv6 routers and hosts are also known as mobile routers (MRs) and mobile network nodes (MNNs) when continuous Internet connectivity is supported.
- In the roadside ITS station, the nodes executing the ITS-S router functions are the *ITS-S IPv6 roadside routers* (RRs) and the *ITS-S IPv6 border routers* (BRs). Roadside ITS-S IPv6 routers are also known as the ITS-S IPv6 access router (AR) when they provide access to ITS-S IPv6 mobile router (MR). Border routers (BRs) are the ITS-S IPv6 routers connecting the ITS station to the Internet or other ITS stations. The ITS-S host functions may be implemented by an ITS-S IPv6 router, or by ITS-S IPv6 hosts.
- In the central ITS station, the nodes executing the ITS-S router functions are the *ITS-S IPv6 border routers* (BRs) connecting the ITS station to the Internet,

or other ITS stations, and the ITS-S home agents (HAs) for supporting IPv6 mobility. The ITS-S host functions may be implemented by an ITS-S IPv6 router, or by ITS-S IPv6 hosts.

Apart from establishing IPv6-enabled ITS stations, the ISO 21210 standard also defines the IPv6 networking features to be integrated in the communications stack to achieve continuous connectivity:

- **IPv6 signaling**. All the basic set of IPv6 features necessary to establish communication with peer IPv6 nodes, such as the usage of neighbor discovery [23] or ICMPv6 [24].
- **IPv6 addressing**. Each station is provided with an IPv6 prefix, allowing it to distribute its functions into multiple nodes. The prefix is allocated statically. Stateless Address Auto-Configuration (SLAAC) [25] is used to configure the IPv6 addresses of the ITS station nodes attached to the ITS station routers (ITS station internal IPv6 interface). For mobile ITS stations, transient addresses are configured as a result of handovers (on the ITS station external IPv6 interface).
- **IPv6 reachability and session continuity**. NEMO Basic Support [19] is used to maintain IPv6 reachability at a permanent address and to maintain IPv6 connectivity when a mobile ITS station is moving between IPv6 points of attachment or is changing its access technology.[2]
- **IPv6 mobile edge multihoming**. MCoA [20] is used in combination with NEMO Basic Support to maintain various IPv6 paths over multiple access technologies simultaneously. This is considered in a mobility scenario where the end point of the IPv6 communication path for the vehicle ITS-S is renewed each time it gets attached to a new access router.

The features provided by the IPv6 protocol block as specified in ISO 21210 are grouped into modules, as illustrated in Fig. 9.7. This separation into modules makes the specification of IPv6 functions for each type of ITS station much easier as not all features (thus all modules) are necessarily in all ITS station types.

The operation of the IPv6 features could be different for distinct types of IPv6 nodes, according to the role they play and the specification of the feature (e.g., the mobility feature is provided by NEMO, and NEMO behaves differently for the MR and the HA). ISO 21210 specifies which modules must be supported for each type ITS station IPv6 nodes, i.e. for MR, AR, BR, HA, and hosts, and rely on IETF RFC specifications (possibly other standards) for the definition of their operation. Those modules are:

- **IPv6 forwarding module**. This module comprises basic IPv6 features (Neighbor Discovery, Stateless Address Auto-Configuration, etc.) to acquire necessary IP parameters (IPv6 addressing) and IP next hop determination (IPv6 forwarding). It enables IPv6 to run over different lower layer technologies. It must be present in all ITS station IPv6 nodes.

---

[2]NEMO Basic Support is used instead of Mobile IPv6 to comply with the requirement that an ITS-S, in the most general case, is composed of several nodes.

9 The Use of IPv6 in Cooperative ITS: Standardization Viewpoint

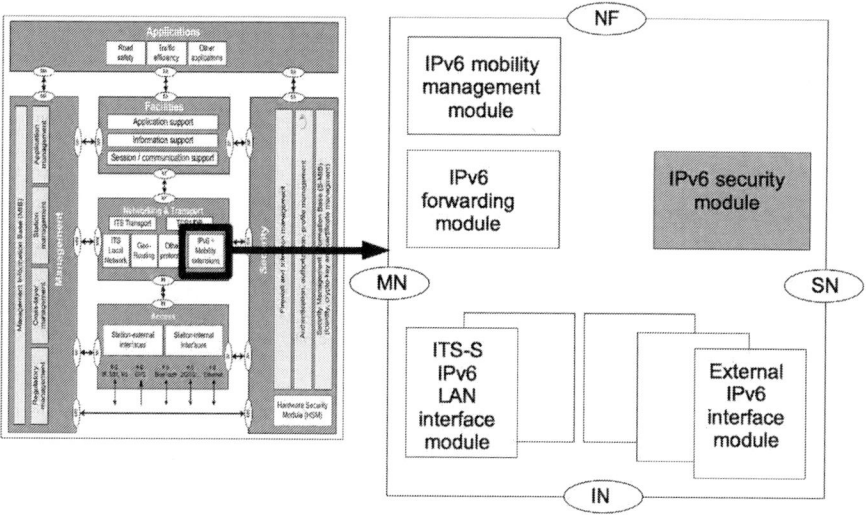

Fig. 9.7 IPv6 functional modules in ISO 21210

- **IPv6 mobility management module**. This modules comprises mechanisms for maintaining IPv6 global addressing, Internet reachability, session connectivity, and media-independent handovers (handover between different media) for in-vehicle networks. Nothing new, this module combines NEMO Basic Support [19] and Multiple Care-of Address Registration [20]. It must only be present in ITS station IPv6 nodes performing functions to maintain Internet reachability and session continuity (MR and HA)
- **External IPv6 interface module**. This module comprises the mechanisms necessary to transmit IPv6 packets back and forth between the IPv6 layer in the underlying access technologies through the IN-SAP. This module performs the IPv6 address configuration and reports to the ITS station management entity the status of the IPv6 address configuration, through the MN-SAP. There is one instance of such a module for each access technology used to connect the ITS station to other ITS stations or legacy IPv6 nodes.
- **ITS station IPv6 LAN interface module**. This module comprises the mechanisms necessary to transmit IPv6 packets back and forth between the IPv6 layer in the underlying access technologies (e.g., Ethernet) through the IN-SAP. This module performs the IPv6 address configuration and reports to the ITS station management entity the status of the IPv6 address configuration, through the MN-SAP. This module must be present in all IPv6 nodes of a given ITS station when the ITS station functions are distributed into separated IPv6 nodes. ITS station IPv6 nodes are linked through an ITS station internal network, referred to as the IPv6 LAN in ISO 21210.

- **IPv6 security module**. The need for a module in charge of securing IPv6 communications is acknowledged in ISO 21210, but the required features are still under discussion.

9.5 Envisaged IPv6-Enabled Cooperative ITS Scenarios

To justify the convenience of adopting IPv6-related technologies to assist the communication of cooperative ITS systems, this section describes a set of particular scenarios where IPv6 outperforms IPv4 and/or other protocol solutions. In these scenarios, the reference IPv6-enabled ITS station architecture defined in the previous section is considered as the target solution, and the advantages of using particular IPv6 technologies already considered within [11] or under consideration are discussed. It is worth mentioning that these scenarios are proposed by authors in order to better exemplify the needs covered by IPv6 in C-ITS.

9.5.1 Internet Access

A common service in vehicular environments is the Internet access from in-vehicle devices. In this scenario an on-board device installed in the dashboard, a laptop, or any other mobile device carried by a user, requires Internet access. Although this service could be directly provided through IPv4 using 3G, this is not a cost-effective and scalable way of accessing Internet in the next situations (at least):

1. High data volumes can be exchanged, such as file downloading or high-definition multimedia content.
2. Low-delay requirements could be applicable by particular applications, such as the speedy reception of an alert in a particular road stretch.
3. More than one in-vehicle device wanting to access Internet at the same time. This case receives particular attention in public transport means, such as buses or trains.
4. Different access networks should be accessed due to 3G coverage gaps or crowded areas.

In this scenario, the usage of IPv6 technologies can provide further benefits to both drivers and passengers. The NEMO technology can support the seamless mobility of a mobile router providing Internet access to multiple in-vehicle devices. This case is depicted in Fig. 9.5. Providing, for instance, a WiFi access to this mobile router, smartphones, tablets, laptops and the rest of networked vehicular devices can access to the Internet. A mobile router would be in charge selecting the most appropriate communication technology. Not only 3G, but also vehicular WiFi (i.e., IEEE 802.11p or ETSI G5) or WiMax could be also used to better support the requirements of applications and offload the 3G networks. Cost benefits are obvious.

9.5.2 Vehicle-to-Vehicle Infotainment

Some services require communication among vehicles, such as messaging, route information exchange, or audio conference. When IPv4 is used (usually through a 3G connection), these services must deal with:

1. *Network Address Translation* (NAT) issues, since the configured IPv4 address could be a private address within the vehicle domain.
2. Always it is needed a backbone support for allowing the P2P communication, due to the lack of direct IP routes between end devices.
3. Performance and cost issues inherent to the IPv4 protocol operation, which are specially critical when using a cellular connection (e.g., 3G).

IPv6 and the multiple communication technologies that could be transparently provided to in-vehicle devices through NEMO solve the previous problems. First, IPv6 auto-configuration together with prefix delegation protocols can provide a global IPv6 address to each in-vehicle node. NAT will be never necessary. In this way, real P2P communication can be provided for communicating application endpoints in vehicles. On the one hand, an indirect communication between vehicles can be established through Internet, by using the NEMO technology. On the other, a direct communication can also be established by using the Internal Network Prefix Discovery (INPD) protocol (formerly called Mobile Network Prefix Provisioning MNPP). In this last case, INPD enables the exchange of IPv6 routing information among nodes in a vehicular ad-hoc network (VANET) using direct wireless links. Since vehicles are provided with in-vehicle IPv6 prefixes, these could be shared with other vehicles to allow the direct communication. Finally, vehicles may also contact external peers located in the Internet through the infrastructure. These types of communication flows are depicted in Fig. 9.8 by means of red arrows.

9.5.3 Traffic Efficiency Services

Vehicular networks provided with IPv6 can further support traffic efficiency services such as congestion detection, management and notification, route planning, variable speed limit or road tolling, for instance. Most of these services require a vehicle-to-infrastructure communication link that could suffer from next problems when common IPv4 networks are used:

1. Lack of mobile network access within 3G coverage gaps.
2. Connectivity problems in congested road segments due to a limited 3G coverage.
3. Lack of localized notifications according to the road segment. A proper and individualized treatment of vehicles would be needed to provide such information.
4. Lack of interoperability with Internet-based services; for example, when using non-IP communication protocols are used at the roadside (i.e., proprietary communications with road devices or specific ITS protocols for communicating with vehicles).

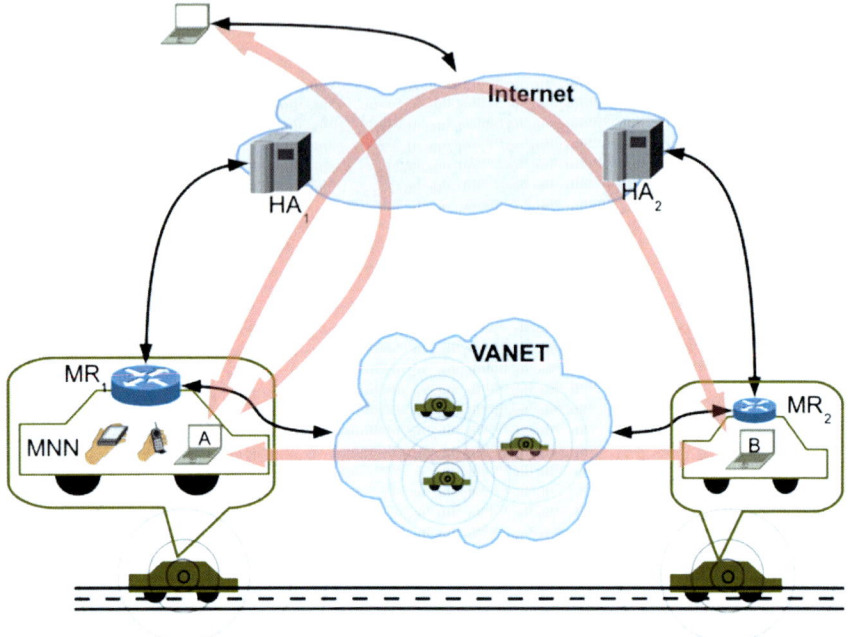

Fig. 9.8 IPv6-powered V2V and V2I communication through internet and direct V2V communication

Apart from the advantages previously described of NEMO, which are also applicable in this kind of scenarios, IPv6 multicast can provide further advantages for distributing information among in-vehicle devices coming from the infrastructure. Novel dissemination strategies can be envisaged also taking into account a mapping between the geographical position of devices and the IPv6 address temporarily assigned. Thanks to the number of available IPv6 addresses, the same group of in-vehicle devices could be provided with a different IPv6 addressing according to the current location. Moreover, given the globally usage of IP technologies on the Internet, remote service points within the Internet could be directly accessed without any protocol translation or adaptation from vehicle nodes or road infrastructure devices such as variable message signs. Only if needed, and during the transition period from IPv4 and IPv6, there are standardized solutions such as NAT64 [26] that can support the access to IPv4-addressed services from an IPv6-based vehicular network.

9.5.4 Safety Services

At first sight, safety applications are better supported by other protocols providing different addressing capabilities, for example, the geographical routing of GeoNetworking. Nevertheless, IPv6 can be also considered as a complement to support many safety services such as crash notification, meteorological alerts, road infrastructure alerts, and vehicle monitoring or emergency calls, among others. However, by using IPv4 or current deployments, the next limitations are found:

1. 3G networks are not able to offer good coverage in remote locations or near mountains.
2. VANET solutions do not warrantee by their own operation of safety services due to, above all, the expected low penetration of equipment (even in the long-term) and the availability of the wireless medium when many vehicles are in the surroundings.
3. IPv4 addressing does not allow the direct access to vehicle devices in general IoT scenarios where a large number of devices require IP addressing capabilities.
4. Solutions exclusively based on wide-area wireless technologies such as 3G do not warrantee that all vehicles within geographical areas are aware of notifications, due to possible availability problems.

The possibility of supporting different communication technologies managed by the IPv6 network layer can solve availability problems, but also the overall performance of the communication could be improved since ITS stations may transparently alternate among the different interfaces. The MCoA technology can support more than one active communication flow between vehicles and infrastructure. In this way, the reception of a critical alert could be further guaranteed by transmitting it through more than one path, and more complex flow management techniques could be applied depending on the data traffic. Moreover, IPv6 addressing and auto-configuration capabilities can support the installation of directly accessible IPv6 devices in the vehicle body. Following this approach, or even providing IPv6 gateways when necessary, advanced infrastructure services could be envisaged to monitor the vehicle status, such as the engine operation, the oil level and quality, the tires pressure, etc.

Applications and facilities specifically designed for GeoNetworking exploit these functionalities, for example to disseminate warning or generic information messages to geographically scoped areas. This approach satisfies the requirements of several ITS services, whose application domain is limited to networks that are disconnected from large existing network infrastructures. However, several ITS applications require the integration of ITS stations with larger networks like private transport networks or the Internet. In order to connect networks based on GeoNetworking to networks running the Internet Protocol (IP), which represent the majority of currently deployed large networks, it is necessary to employ some integration between both technologies. The solution proposed by ETSI [27] allows the transport of IPv6 packets over GeoNetworking, without inflicting modifications on any protocol.

9.6 IPv6 Communications Experimental Assessment

This section presents an experimental evaluation of a reference implementation of an IPv6-based communication architecture that follows the ETSI and ISO recommendations previously presented. This evaluation is included as a success case where IPv6 technologies are successfully applied in a vehicular scenario. Since the communication architecture considered is the reference one standardized by ETSI and ISO as the ITS station reference architecture for C-ITS, the description is summarized and it directly describes the implementation viewpoint with the testbed, followed by the most interesting results gathered.

9.6.1 Testbed Description

As observed in Fig. 9.9, the set-up scenario consists of one vehicle ITS-S, one roadside ITS-S, and the central ITS station for the vehicle. This equipment has been deployed at the University of Murcia, near the Faculty of Computer Engineering and taking advantage of the ring road that surrounds the campus. By means of the three communication interfaces of the MR, the vehicle ITS station can be connected with roadside units supporting 802.11b/g/n, 802.11p, or directly with the central ITS through the 3G/UMTS network. In the last case it is necessary to provide an IPv4 to IPv6 transition solution, since most of the 3G providers (including the used one) still offer IPv4 Internet. In this case, OpenVPN is used between the MR and an AR within the central ITS station. Communications through our 802.11b/g/n and 802.11p roadside units are directly performed using IPv6, since the University of Murcia infrastructure supports this protocol natively.

Home Internal Router and the ARs in Fig. 9.9 execute a common communication stack with IPv6 features. Additionally, the router included in the Central ITS Station also serves as the ending point of the OpenVPN tunnel when 3G is used. The Border Router element uses an IPv4/IPv6 dual stack, since it provides Internet connectivity to the ITS network. Additionally, this border router offers network address translation from IPv6 to IPv4 (NAT64) and a domain name system for getting IPv6 addresses of external IPv4 services (DNS64). In this way, it is possible to provide access to IPv4 resources on the Internet. The Vehicle Home Network is where the vehicle maintains its home addressing. In other words, when any computer outside this network communicates with the vehicle host, it uses the home IPv6 address and packets follow the route towards the home network (within the central ITS station), and the NEMO Home Agent (HA) will redirect these IPv6 packets to the current IPv6 address of the vehicle, which is assigned to the MR by each visited roadside station. The other important part of the central ITS station is the service center, which is in fact distributed in a set of nodes.

The developed prototype implements most of the network entities and modules comprising the station reference architecture designed by ISO and ETSI. More

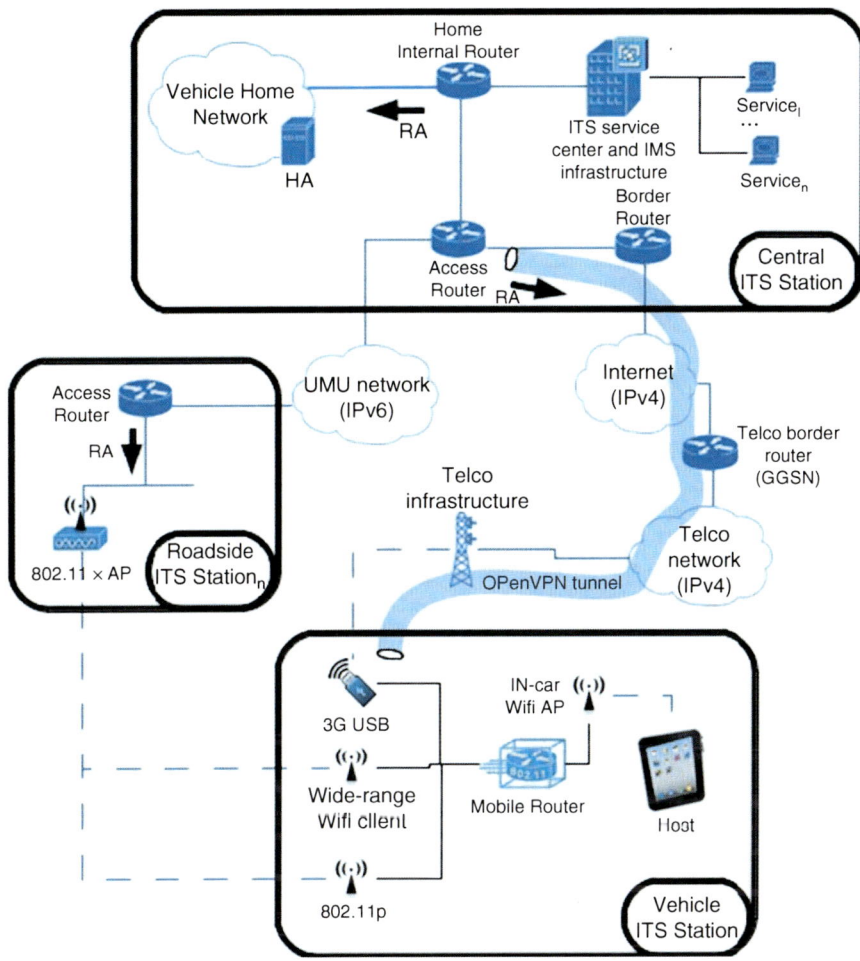

Fig. 9.9 Deployed scenario

details regarding the basic mobility and security capabilities can be found in [28]. Part of the flow management functionality is currently modeled with the BID used by NEMO/MCoA when a new CoA is registered with the HA. Upon the reception of a Router Advertisement message from a new network, the NEMO daemon adds a new routing entry in the Linux kernel of the MR with a priority assigned in accordance with the BID of the communication interface involved in the communication. These BIDs are set in order to assist the handover decision by selecting 802.11p and 3G, in this preference order.

The handover management logic is currently implemented in the form of an automatic CoA registration with the HA upon the detection of a new network

Table 9.1 List of hardware equipment

Networked nodes			
Node	Model	CPU/Mem	Operating system
Vehicle host	PC Viliv X70	Atom 1.3 GHz/1 GB	Windows 7
Mobile router	Commsignia Laguna	Dual Core ARM11/128 MB	ITSSv6 distribution
Access router	Commsignia Laguna	Dual Core ARM11/128 MB	ITSSv6 distribution
Home agent	mini-ITX PC	Via 532 MHz/476 MB	Ubuntu 10.4
Central ITS AR	PC	i5 3.1 GHz/3 GB	Ubuntu 10.4
IMS core x 4	Xen virtual machine	PentiumD 2 GHz/256 MB	Ubuntu 10.4
IMS apps server	Xen virtual machine	PentiumD 2 GHz/1 GB	Ubuntu 10.4
Network interfaces			
Technology	*Hardware*		
3G/UMTS	Ovation MC950D modem		
802.11p	Unex DCMA-86P2 mini-PCI (integrated in AR and MR)		
Relevant software			
Node	Description		
Vehicle host	OSGi Equinox framework 3.6		
Mobile router	NEMO (UMIP 0.4), IKEv2 (OpenIKEv2 0.96) and RA daemon (radvd 1.8)		
Access router	RA daemon (radvd 1.8)		
Home agent	NEMO (UMIP 0.4) and IKEv2 (OpenIKEv2 0.96)		
IMS core	Fraunhoffer Open IMS		
IMS apps server	Kamailio 3.1.2		

when a Router Advertisement message is received from a new AR. Networks are understood to be unavailable when no Router Advertisement messages are received from the corresponding AR after a configurable timeout period.

Several software modules have been also implemented over the network and transport layer for the case of the vehicle host. The most representative ones are the IMS (Internet Multimedia Subsystem) service access layer, the OSGi middleware to host all facilities and applications, and a generic human–machine interface (HMI) for presenting applications in a unified way to the user. These are technologies identified in the facilities layer of the reference ITS-S architecture. More details regarding these functionalities are presented in [29, 30].

The list of hardware and associated software used for each node can be found in Table 9.1. Software coming from the European ITSSv6 (IPv6 ITS Station Stack) project,[3] in which the authors are participated, receives a special mention. Within this project, a reference implementation of the ISO/ETSI specifications of a communication stack has been performed including IPv6 technologies. In this way, a new Linux distribution has been implemented including specific modules for network mobility, management of communication flows, or secure communications with IPsec and IKEv2, among others.

---

[3]http://itssv6.eu/.

9.6.2 Performance Analysis

In the tests included in this section, the vehicle moves within the Espinardo Campus (University of Murcia), using the previous testbed to evaluate the handoff operation and the network performance from the application perspective between 3G (UMTS) and 802.11p. NEMO, IPsec and IKEv2 are enabled. Additionally, the MCoA extensions are disabled in order to better appreciate the performance degradation caused by the handover. A roadside unit connected on the top of the Faculty of Computer Science is used as 802.11p network attachment point. This has been installed to only give 802.11p coverage to a small stretch of a near road. A common vehicle mounting the on-board equipment drives around the building at a urban-like speed between 20 and 40 km/h and, when it is in the communication range of the roadside unit, the MR automatically performs a handoff from 3G to 802.11p.

The next three metrics have been considered in three independent tests:

- Bandwidth, measured in Mbps. It has been evaluated with a TCP flow maintained at the maximum allowable speed from the vehicle host to a CN connected within the central ITS station network.
- Packet Delivery Ratio (PDR), measured in percentage of packets lost. It has been evaluated with a UDP flow in the downlink direction at 500 kbps from the CN to the vehicle host. The UDP packet size has been set to 1,470 bytes.
- Round-trip delay time (RTT), measured in ms. It has been evaluated with ICMPv6 traffic generated from the vehicle host to the CN. ICMPv6 Echo Request messages have been generated at a 1 Hz rate with a size of 64 bytes.

UDP and TCP traffics have been generated with the Iperf utility[4] (version 2.0.4), while the ICMPv6 traffic has been obtained from the common Ping6 Unix command. The period of RA notifications from the AR is set to a random time between 3 and 4 s (to avoid RA collisions with possible nearby ARs). Additionally, the expiration time of the pair CoA–HoA used is set to 30 s in both the MR and the HA.

The bandwidth results obtained in the TCP tests are showed in Fig. 9.10. As can be seen, the slow-start algorithm of TCP tries to adapt to the wireless medium during the whole test, affected by the mobility of the vehicle. The first handoff from 3G to 802.11p occurs at time 310 s (Fig. 9.10b), and the second one, from 802.11p to 3G, at time 445 s (Fig. 9.10c). At these moments the data rate is null for a while, due to the time needed to change the CoA used against the HA. Nevertheless, the connectivity gap is more evident in the second handoff, due to 802.11p technology is preferred when both 3G and 802.11p technologies are present. The handover mechanism waits for a Router Advertisement through 802.11p interface but, if this is not received, the CoA–HoA association timeout indicates that the 802.11p connectivity is over and the handoff to the 3G technology is performed. This effect

---

[4] http://sourceforge.net/projects/iperf/.

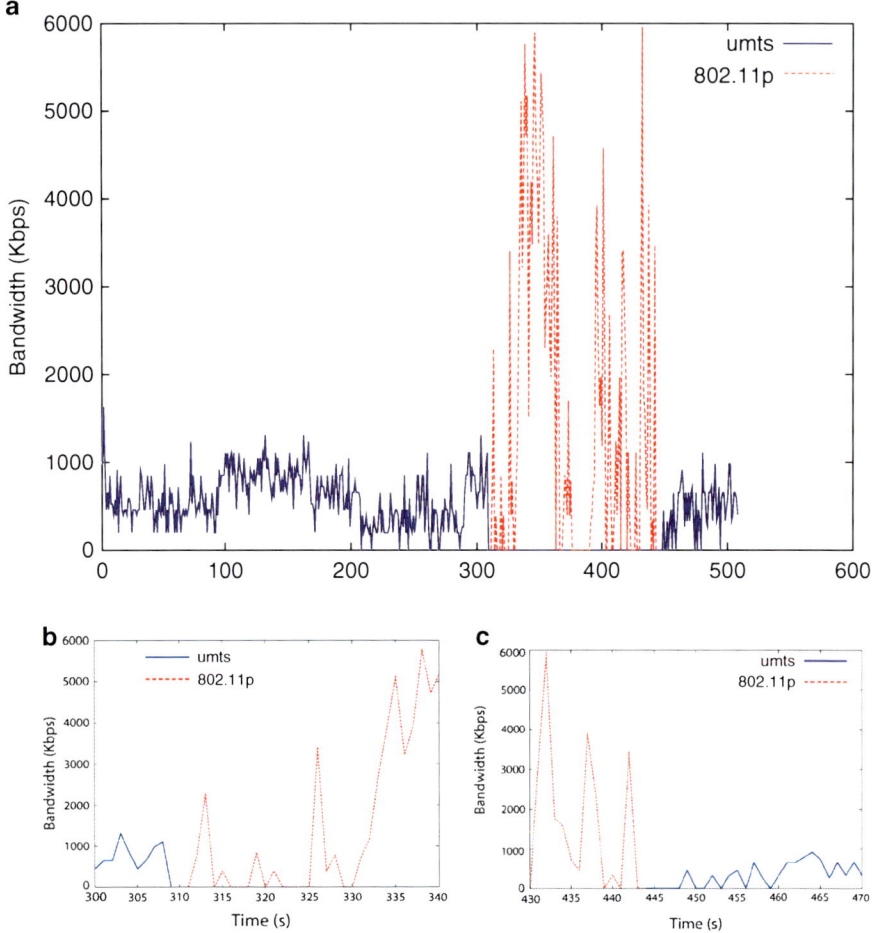

Fig. 9.10 Maximum bandwidth using TCP. (**a**) Bandwidth obtained during the whole test; (**b**) first handoff; (**c**) second handoff

is also noticeable in the rest of the tests. Moreover, it is evident the quite better performance obtained while the 802.11p link is maintained, with peaks near to 6 Mbps. Between times around 100 and 200 s the HSPA (High Speed Packet Access) channel allocation algorithm adapts better, since the vehicle moves near the UMTS base station. It can be noted that the performance of the 3G link is similar at the beginning and the end of each test because the circuit is circular.

Regarding lost packet, the PDR results for the UDP data transfer are plotted in Fig. 9.11. A performance degradation has not been observed in the first handoff, as can be seen in detail in Fig. 9.11b, but the second handoff implies 2 s of null connectivity (Fig. 9.11c). Moreover a great number of lost packets appear while the

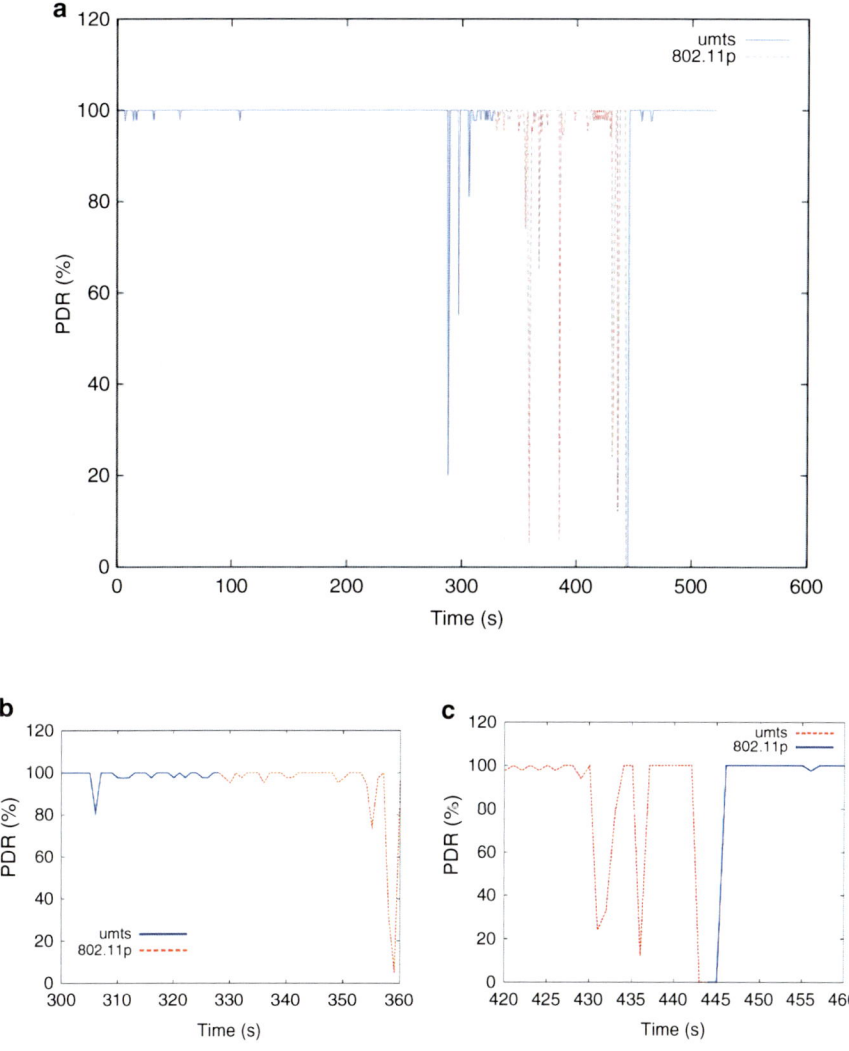

Fig. 9.11 PDR using UDP flow of 500 kbps. (**a**) PDR obtained during the whole test; (**b**) first handoff; (**c**) second handoff

802.11p link is used. This is explained by the fact that the 802.11p channels are located in the 5.9 GHz band, given to suffer more from obstructions to the signal propagation, such as other vehicles, building blocks, or vegetation.

Finally, the network latency has been evaluated and results are given in Fig. 9.12 (note the y-axis has a logarithmic scale). These results have been obtained by generating ICMPv6 traffic from the vehicle host to the CN. As can be seen in the results, the 802.11p area is clearly visible with most of the RTT values under

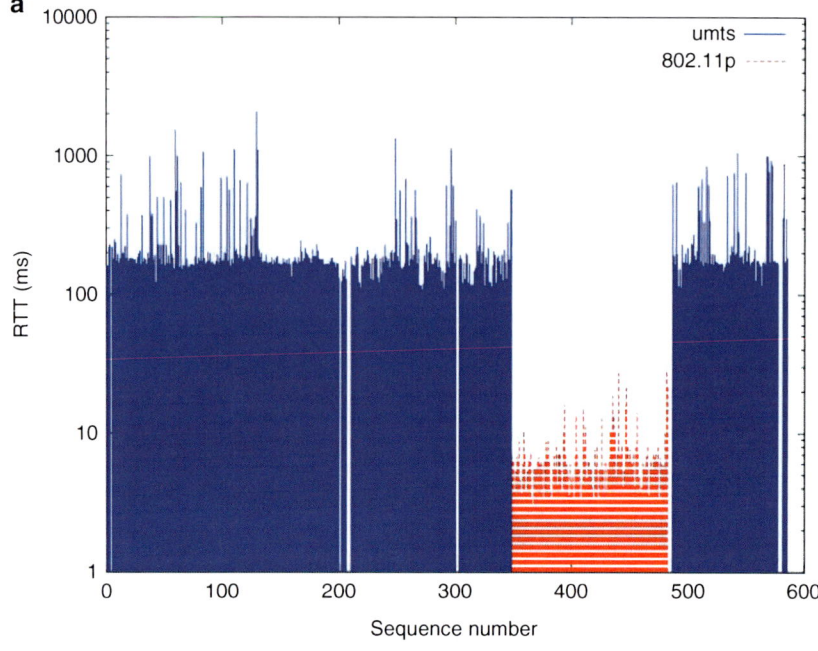

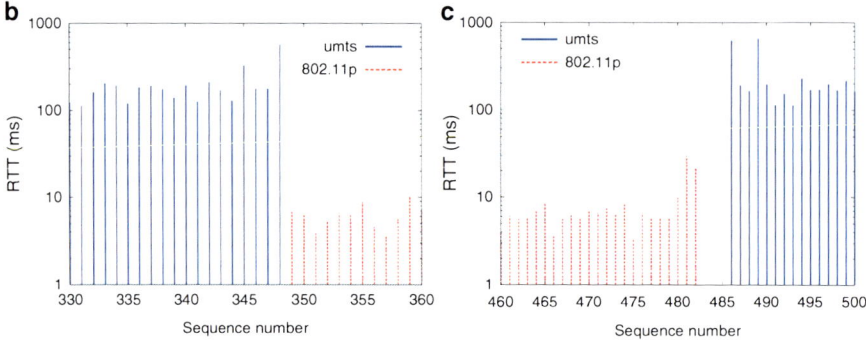

Fig. 9.12 RTT delay using ICMPv6 Echo/Request traffic at 1 Hz. (**a**) RTT delay obtained during the whole test; (**b**) first handoff; (**c**) second handoff

10 ms, while 3G RTT results are more scattered (consider the logarithmic scale in the ordinate). Losses only appear during the second handoff depicted in Fig. 9.12c. Here four messages are lost until the UMTS link is again used.

Considering the results, IPv6-based network mobility functionalities perform efficiently under real inter-technology handoffs, the most difficult to accomplish. The communication stack operates correctly, maintaining the in-vehicle network connectivity in all tests and showing performance results that enable the communication stack to be used in many vehicular services. Unless high-quality multimedia transmissions are required, the bandwidth results indicate that the data rate required by most of the traffic efficiency and comfort cooperative services can be covered and, according to latency tests, even non-critical safety services could be implemented, such as emergency assistance, variable traffic signaling or kamikaze warning. Moreover, it is important to remark that this infrastructure-based vehicular communication approach is a real solution that could be deployed in the short term and it is also capable of allowing indirect V2V communications, thanks to the global IPv6 scope of all in-vehicle hosts.

9.7 Conclusions

By the time ITS services requiring the use of the public IP addresses appear on the market, there will not be enough public IPv4 address available. The use of IPv6 scales to meet the addressing needs of a growing number of vehicles and connected devices, and provides the added functionality necessary in mobile environments. For this reason, the research community has opted for relying on IPv6 to develop the future cooperative ITS architectures.

IPv6 has been found ideally suited for non-time critical applications, like road efficiency and comfort/infotainment ITS applications. There is no question that IPv6 could be a media-agnostic carrier of such non-time critical but safety essential information. In other words, IPv6 can act as the convergence layer ensuring the support of the diversity of access technologies, the diversity of applications, and the diversity of communication scenarios. In fact, this urgent need of using IPv6 has been acknowledged by the European standardization bodies, mainly ISO and ETSI, which are in the process of developing a set of standards describing the application IPv6 technologies to achieve optimized communications for the foreseeable cooperative ITS applications.

Given the relevant role IPv6 is expected to play in the future cooperative ITS, this chapter presents a comprehensive survey of the latest advances achieved in this area. This study takes as reference the standardization efforts jointly developed by ISO and ETSI to develop a harmonized ITS station reference architecture supporting a variety of networking protocols and, in particular, those related to IPv6.

This survey is complemented with the description of a set of valuable cooperative ITS applications requiring the application of IPv6 networking protocols to achieve a correct operation. Finally, the chapter is concluded with a performance assessment of IPv6 communications over a real cooperative ITS scenario. The analysis has demonstrated the ability of IPv6 protocols to achieve an uninterrupted communication between vehicles and the infrastructure under different configurations.

Acknowledgements This work has been sponsored by the European Seventh Framework Program, through FOTsis (contract 270447), GEN6 (contract 297239), and INTER-TRUST (contract 317731) projects.

References

1. Khaled Y, Tsukada M, Santa J, Choi J, Ernst T (2009) A usage oriented analysis of vehicular networks: from technologies to applications. J Commun 4(5):357–368
2. European Telecommunications Standards Institute (2009) Intelligent transport systems (ITS); vehicular communications; basic set of applications; definitions. ETSI TR 102 638, June 2009
3. International Organization for Standardization (2013) Intelligent transport systems - communications access for land mobiles (CALM) - architecture. ISO 21217, Apr 2013
4. European Telecommunications Standards Institute (2010) Intelligent transport systems (ITS); communications architecture. ETSI EN 302 665, Sept 2010
5. Intelligent transport systems (ITS); vehicular communications; GeoNetworking; Part 3: Network architectures, March 2010. ETSI TS 102 636-3 V1.1.1 (2010-03)
6. Deering S, Hinden R (1998) Internet protocol, version 6 (IPv6) specification. RFC 2460 (Draft Standard), Dec 1998. Updated by RFCs 5095, 5722, 5871
7. ISO (2008) Intelligent transport systems – communications access for land mobiles (CALM) – CALM infra red, Apr 2008. ISO/IS 21214:2008
8. ISO (2010) Intelligent transport systems – communications access for land mobiles (CALM) – CALM M5, Apr 2010. ISO/IS 21215:2010
9. ISO (2008) Intelligent transport systems – communications access for land mobiles (CALM) – CALM using 2G cellular systems, Apr 2008. ISO/IS 21212:2008
10. ISO (2008) Intelligent transport systems – communications access for land mobiles (CALM) – CALM using 3G cellular systems, Apr 2008. ISO/IS 21212:2008
11. International Organization for Standardization (2011) Intelligent transport systems - communications access for land mobiles (CALM) - IPv6 networking. ISO 21210, Jan 2011
12. ISO (2011) Intelligent transport systems – communication access for land mobiles (CALM) – Part 2: Fast networking & transport layer protocol (FNTP), June 2011. ISO/CD 29281-2:2011(E)
13. Intelligent transport systems (ITS); vehicular communications; Part 4: Geographical addressing and forwarding for point-to-point and point-to-multipoint communications; Sub-part 1: Media-independent functionality, June 2011. ETSI TS 102 636-4-1 V1.1.1 (2011-06)
14. European Telecommunications Standards Institute (2011) Intelligent transport systems (ITS); vehicular communications; basic set of applications; Part 2: Specification of cooperative awareness basic service. ETSI TS 102 637-2, Mar 2011
15. European Telecommunications Standards Institute (2010) Intelligent transport systems (ITS); vehicular communications; basic set of applications; Part 3: Specifications of decentralized environmental notification basic service. ETSI TS 102 637-3, Sept 2010
16. ETSI (2010) Intelligent transport systems (ITS); security; security services and architecture, Oct 2010
17. Droms R, Bound J, Volz B, Lemon T, Perkins C, Carney M (2003) Dynamic host configuration protocol for IPv6 (DHCPv6). RFC 3315 (Proposed Standard), July 2003. Updated by RFCs 4361, 5494
18. Thomson S, Narten T (1998) IPv6 Stateless address autoconfiguration. RFC 2462 (Standards Track), Dec 1998
19. Devarapalli V, Wakikawa R, Petrescu A, Thubert P (2005) Network mobility (NEMO) basic support protocol. RFC 3963 (Proposed Standard), Jan 2005
20. Wakikawa R, Devarapalli V, Tsirtsis G, Ernst T, Nagami K (2009) Multiple care-of addresses registration. RFC 5648 (Proposed Standard), Oct 2009
21. Kent S, Seo K (2005) Security architecture for the internet protocol. RFC 4301 (Draft Standard), Dec 2005
22. Kauffman C (2005) Internet key exchange (IKEv2) protocol. IETF RFC 4306, Dec 2005
23. Narten T, Nordmark E, Simpson W, Soliman H (2007) Neighbor discovery for IP version 6 (IPv6). RFC 4861 (Draft Standard), Sept 2007

24. Conta A, Deering S, Gupta M (2006) Internet control message protocol (ICMPv6) for the internet protocol version 6 (IPv6) specification. RFC 4443 (Draft Standard), 2006
25. Thomson S, Narten T, Jinmei T (2007) IPv6 stateless address autoconfiguration. RFC 4862 (Draft Standard), Sept 2007
26. Bagnulo M, Matthews P, van Beijnum I (2011) Stateful NAT64: network address and protocol translation from IPv6 clients to IPv4 servers. RFC 6146 (Standards Track), Apr 2011
27. Intelligent Transportation Systems (ITS); vehicular communications; Part 6: Internet integration; Sub-part 1: Transmission of IPv6 packets over GeoNetworking protocols, March 2011. ETSI-TS-102-636-6-1 V1.1.1
28. Fernandez PJ, Skarmeta AFG (2011) Providing security using IKEv2 in a vehicular network based on WiMAX technology. In: 2011 IEEE consumer communications and networking conference (CCNC), Jan 2011, pp 282–286
29. Garcia A, Santa J, Moragon A, Gomez-Skarmeta AF (2011) IMS and presence service integration on intelligent transportation systems for future services. In: Abraham A, Mauri JL, Buford JF, Suzuki J, Thampi SM (eds) Advances in computing and communications. Communications in computer and information science, vol 192. Springer, Berlin/Heidelberg, pp 664–675
30. Santa J, Fernández PJ, Moragón A, García AS, Bernal F, Gómez-Skarmeta AF (2012) Architecture and development of a networking stack for secure and continuous service access in vehicular environments. In: 19th ITS world congress, Oct 2012, pp 1–12

Chapter 10
Security and Privacy for ITS and C-ITS

Scott W. Cadzow

Abstract Intelligent Transport Systems (ITS), and the specialised subset of them represented by Cooperative ITS (C-ITS), are part of the wider machine-to-machine communications, and software driven world. The aims of ITS and C-ITS are multifold but for the purposes of this book are mainly in improving safety of transport users and their transport modes. The chapter that follows covers some of the basics of security, including the contrast with safety and privacy, the role of privacy protection in the ITS context, and a review of the cryptographic countermeasures recommended for C-ITS using the certificate and signature schemes defined in IEEE 1609.2.

Keywords VANET · Security · Privacy · 1609.2 · Pseudonym · Integrity · Trust · Authenticity · Authority · Confidentiality · Pseudonymity · Algorithms · Cryptography · Certification · Verification · Validation · Cryptanalysis · Attack models · Defence strategy · ETSI · Standardisation

Acronyms

CA	Certification Authority
CIA	Confidentiality Integrity Availability
ECC	Elliptical Curve Cryptography
GSM	Global System Mobile (mostly deprecated)
IMEI	International Mobile Equipment Identifier
IMSI	International Mobile Subscriber Identifier
PIN	Personal Identification Number
PKC	Public Key Cryptography
PKI	Public Key Infrastructure
SIM	Subscriber Identity Module

S.W. Cadzow (✉)
Cadzow Communications Consulting Ltd, Sawbridgeworth, UK
e-mail: scott@cadzow.com

TEI	TETRA Equipment Identifier
TETRA	Terrestrial Trunked Radio (largely deprecated in favour of TETRA)
VIN	Vehicle Identification Number

10.1 Introduction

Intelligent Transport Systems (ITS) are a specialised subset of machine-to-machine communications in a software driven and all-connected world. There are a number of dimensions of ITS and different authors may present different lists of them but for the purposes of examining the security issues this list will suffice:

- Advanced Traveller Information Systems (ATIS),
- Advanced Traffic Management Systems (ATMS),
- ITS-Enabled Transportation Pricing Systems,
- Advanced Public Transportation Systems (APTS),
- Vehicle-to-Infrastructure Integration (VII), and,
- Vehicle-to-Vehicle Integration (V2V)

C-ITS sits in this list as a special case of both VII and V2V, with the functionality of C-ITS centred on the exchange of data between co-operating ITS stations.

The root concept behind C-ITS is that of giving machines some degree of spatial awareness, and using that spatial awareness to protect both their own space and the space of all other transport users. If we consider the thing that seems to concern most motorists more than anything else, it is fear of having a collision. This ranges from a little bump in a car park all the way to the catastrophic collision on a motorway. The basic idea of spatial awareness as a defence comes from martial arts—if you know where your opponent is aiming then make sure you're not there when his blow lands. Same in a car—if something is on a path to hit you make sure you're not there when the blow lands or be in a position where the damage is negligible.

Spatial awareness for a driver means being aware of where you and your vehicle are with respect to all other vehicles and the infrastructure in order to ensure that you are not encroaching on the safe zone around any other road user. In a conventional vehicle this means using your senses of sight and hearing to continuously build up a mental three-dimensional map of the other road-users around you and working to ensure you are "safe" with respect to them. This is done through constantly using mirrors, moving around to eliminate blind spots (not moving the vehicle but your body), listening for other vehicles and so forth. However in many ways the design of road vehicles has increasingly limited the ability of a driver to build up this 3-D mental map by being quieter, having more restricted visibility and having a number of distraction technologies to hand (e.g. phones, music, navigation tools). The excuse of "I didn't see you" is the first thing to be reported from far too many accidents and one of the roles of C-ITS is to eliminate the blind spot and allow vehicles to see each other in all environments.

One of the other concerns raised when C-ITS was first conceived is that by transmitting information that allows receivers to build up a spatial model of how your vehicle is interacting with it, the same data could be used to track you. C-ITS has been designed to operate in areas of the world where expectations of privacy are high and therefore the protection of users from being tracked, or of their transmissions being used against them, has been very high on the agenda and C-ITS has been built to maximise privacy protection.

The bulk of ITS types (ATIS, ATMS, APTS, etc.) are data centric, in that they gather and distribute data. What differentiates them is where the data comes from and who its intended recipients are. For VII and V2V, and more specifically C-ITS, data comes from the co-operating vehicles and is intended for those vehicles. In the native environment in which vehicles find themselves the safety of any individual vehicle is determined by a set of factors that include:

- Driver awareness,
- Vehicle road-worthiness,
- Road conditions,
- Weather conditions,
- Traffic signalling,
- Relative velocity,
- Other vehicles

What C-ITS will achieve, at least in intent, is greater driver awareness by giving authoritative information to the driver on their relative velocity and the presence of other vehicles and their vectors. The key term is "authoritative information". The purpose in C-ITS security is to assure the receivers of C-ITS data that it genuinely comes from a vehicle and is an accurate representation of the location and nature of the vehicle. The remainder of this chapter considers how security techniques applied to C-ITS give that assurance. In Sect. 10.2 we first examine the meaning of security expanding this in Sect. 10.3 looking in more detail at trust, and in Sect. 10.4 at privacy. In Sect. 10.5 the topic of identity is examined with Sect. 10.6 looking at symmetric and asymmetric security. Section 10.7 reviews the standards supporting C-ITS security and in Sect. 10.8 some conclusions are drawn.

10.2 Security

First we need to be clear by what we mean by the term security as it is easily confused with safety and with privacy. In order to assist in clarifying this we will quickly look at each of safety and security, privacy (defined as a state in which one is not observed or disturbed by other people) will be dealt with in more detail later.

When looking at the use of transport on the public highway safety is the dominant concern of most users, where safety is defined as the condition of being protected from, or unlikely to cause, danger, risk, or injury. Safety is improved in many ways and the result is essentially less likelihood of danger, risk or injury. In addressing

Fig. 10.1 US DoT mockup of how C-ITS may give awareness

safety improvements we can consider a number of areas that improve fundamental safety and this has, over time, included better road surfaces, better tyres, improved braking and dynamic performance of vehicles, which together allow an aware driver to be able to drive around a risk (e.g. by being able to brake without skidding, or being able to brake and turn at the same time without terminal understeer or oversteer). In addition there is much done to protect the passengers of vehicles when they do crash by designing in crumple zones (allowing the car to deform in a controlled way to absorb impact), improving driver and passenger restraint systems (both seat belts and airbags which prevent the driver or passenger becoming an uncontrolled projectile in the event of an impact), and there have been improvements in road design and lighting to minimise the likelihood of accidents due to the local geography (e.g. identifying and removing adverse camber on corners, improving street lighting at dangerous junctions). Figure 10.1 aims to illustrate how C-ITS can give awareness through all-informed "here I am" transmissions.

It could be argued that greater segregation of drivers will be the next stage in improving safety but giving every vehicle complete segregation from every other is simply not practical. So we are looking at extending the core requirement of driver's being aware of the road and vehicles around them to the vehicles sharing contextual knowledge with each other to give the possibility of virtual segregation (i.e. vehicles declaring a protected geographic zone around them).

In contrast with safety, security (and security technology) is considered as safety enhancing. The paradigm in conventional security is CIA—Confidentiality, Integrity, Availability—and so in the domain of ITS Security we are looking at means of using security technologies to augment safety technologies. Furthermore within the C-ITS Security domain we are seeking means of preventing abuse of the wider technology that may lead to reduction in safety (more risk).

The starting point for the development of ITS security has taken the following as targets:

- Messages must be secure
    - Broadcast messages cannot be spoofed
    - Unicast messages cannot be spoofed or eavesdropped

- Anonymity for end-user vehicles and their occupants
  - Messages (either individually or as sets) do not give away (reveal) identity
- Must be able to remove bad actors

As secondary characteristics for the development of ITS security solutions the following have also been considered:

- Overhead due to security must not be excessive
  - 200 bytes probably okay, 1,000 bytes probably not
- Vehicles may have to receive hundreds of messages per second and must be able to process them

In the security world the primary term to work against is risk, where the risk of something happening is a combination of the impact of it happening and the likelihood of it happening. Reducing the impact of two vehicles colliding is addressed through the design of the vehicles. This is where crumple zones, airbags, seatbelts, Anti-lock Braking Systems (ABS) and all other three-letter acronyms we see in the sales literature for cars play their part. Of course not all parties in collisions are in equally protected vehicles, some of the parties will be cyclists or pedestrians, in some cases the collisions will be with buildings or our road furniture (e.g. lighting systems, lane division bollards), in which case the approach to minimising risk is to minimise the likelihood of collision.

The relationship of security technology to risk is clear: Risk reduction and risk assurance can be managed by the targeted application of security measures (shown in Fig. 10.2) but only if you know about the risk in the first place.

In the development of ETSI's security standards and the services they invoke the means to perform a risk analysis followed the broad outline given in ETSI TS 102 165-1 [1] and whose process is illustrated in Fig. 10.3.

Fig. 10.2 The process of risk management and risk assurance

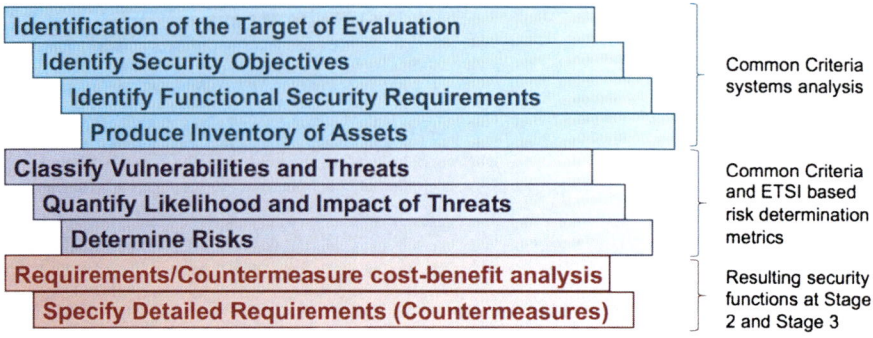

Fig. 10.3 The ETSI standards based Threat Vulnerability Risk Analysis (TVRA) process

In moving towards a safety system that is built on data transmitted from other vehicles we need to ask: Is the data complete? Is the data accurate? It is probably safe to say that the concept of C-ITS is built up by the continuous broadcast from every vehicle of a Cooperative Awareness Message (CAM), the content of which basically says "I am a vehicle of type x and I am at location y at time z". Additional information can be added to this but that is the essence. In addition there is a second form of message transmitted, a Decentralized Event Notification Message (DENM). The second message can take many forms but in essence says "I am reporting event x happened at location y at time z", where the event could be a weather event, a queueing event, an accident event or something else. So it may be that a car has detected ice and reports that. For a detailed discussion about CAMs and DENMs, see Chap. 5.

The way in which the receiver acts on this data determines to some extent the contribution to safety of C-ITS. In addition if a malicious actor can introduce false data to the system how would this impact the system. The problem with data is that it is somewhat intangible and how we react to it depends on two things: Context and Trust.

10.3 Context and Trust

When looking at context, particularly for C-ITS, we need to determine is the information offered of value to me? In other words, is the data relevant to my context?

Trust is rather more complex in that it means do I believe the data to be accurate and does it assist me in building a complete model of the environment? It also means do I trust the sender of the data?

Establishing trust is difficult and has a very human interpretation. Trust is also contextual and this will be examined in more detail later. It may be worth using analogy to examine trust: I trust Bob when playing tennis as my doubles partner so in the context of a tennis court (on my side) I trust Bob's judgement and his instructions; Away from tennis Bob is a car mechanic and offers me financial advice. On what basis should I trust Bob in this new context? For this we mentally create trust relationships which we contextualise and which we modify over time. This leads to some commonly ignored attributes of trust:

- Trust is not a binary operation. There may be various levels of trust that an entity has for another.
- Trust may be relative, not absolute. Alice may trust Bob more than Eve, without trusting either absolutely.
- Trust is rarely symmetric and should never be made artificially so. Alice may trust Bob completely, whereas the amount of trust that Bob has for Alice may be very low. The pupil may trust the master without any requirement for that trust to be reciprocated.
- Trust varies over time and the level of dynamicism may also vary between different relationships. For example, when I first starting partnering Bob in doubles I didn't trust in his every shot, now I trust almost all of his game (with the exception of his backhand so I make sure he plays in the forehand court as much as possible).
- Having a secured communications channel with another entity is never sufficient reason to trust that entity, even if you trust the underlying security primitives on which that communications channel is based.

Quite formally we can define trust as the level of confidence in the reliability and integrity of that entity to fulfil specific responsibilities. In more human terms Alice doesn't need to trust Eve if what Eve does has no direct impact on Alice. However if Alice needs to trust Bob and Bob has to trust Eve then Alice has an indirect trust relationship with Eve through Bob even though Alice and Eve may never be directly aware of this.

Fitting trust to a platform such as ITS is intrinsically complex. However trust is a major part of the transport problem. We choose one make of car over another because we may trust one to be more reliable than another, or we trust we'll get better service from this dealership rather than that one. We also may trust the train to get us to our destination with less stress than the bus, or driving. I may trust Campagnolo gears on my bicycle more that those from Shimano because I have a sneaking worry over a company making fishing tackle fiddling about with bikes (a silly fear really but trust is not meant to be rational). Too many times you can hear the statement "I trust him because he's got a nice face". Not rational. But it does drive our internal decision matrix.

The problem with trust in ITS is that effective trust requires identification—you need to know who you are trusting. So in a system such as C-ITS where most of the actors are unknown to one another we need to provide means for parties to build up trust. In trust networks this is achieved through delegated relational trust, in this case Alice needs to trust Bob but doesn't have a relationship with Bob, however Alice has a trust relationship with Eve and Eve trusts Bob. Bob presents Alice with some proof that Eve trusts him to Alice, and Alice if she accepts this proof essentially delegates the trust decision to Eve. So although Alice doesn't establish a formal direct trust relationship with Bob she has accepted Eve's trust in him as sufficient and makes her own trust based decisions on the basis of Eve's trust.

It should be generally considered that delegated relational trust is weak. If the decision to trust or not to trust is always left to somebody else, there is no ability to build a true contextual trust model. However in cryptographic models used for trust often the model tends towards one of making an assertion, getting a mutually trusted third party to verify it, and then on the basis of what the mutual third party has verified, the relying party acts on the assertion.

For C-ITS in both the CAM and DENM models all the assertions in the message (location, time, identity, event) are self-asserted by an unknown party but the authority to make these assertions is transmitted as a cryptographically strong signed document.

10.4 Privacy

Privacy is quite unlike any other security problem where metrics of the degree of protection can be applied. For example, if a risk analysis identifies that there is a risk that a data object can be manipulated and that such manipulation should be detected it is possible to specify that the data is protected by providing a signed hash of the message which will meet certain conditions.

For C-ITS there are a number of occasions where data that can be considered personal is exposed across interfaces between the system components. One of the main areas of concern is that of locational privacy as C-ITS is based on regular updates of an object's location being broadcast to any party able to receive it and where there is no prior knowledge of who the receivers are. For the present time C-ITS is not fully deployed but it does share similarities with the smartphone and thus the quote from the Article 29 working group opinion on the use of GeoLocation data in the smartphone environment [2]:

> A smart mobile device is very intimately linked to a specific individual. Most people tend to keep their mobile devices very close to themselves, from their pocket or bag to the night table next to their bed.
>
> It seldom happens that a person lends such a device to another person. Most people are aware that their mobile device contains a range of highly intimate information, ranging from e-mail to private pictures, from browsing history to for example a contact list.
>
> This allows the providers of geolocation based services to gain an intimate overview of habits and patterns of the owner of such a device and build extensive profiles. From a pattern of inactivity at night, the sleeping place can be deduced, and from a regular travel pattern in the morning, the location of an employer may be deduced. The pattern may also include data derived from the movement patterns of friends, based on the so-called social graph.
>
> A behavioural pattern may also include special categories of data, if it for example reveals visits to hospitals and religious places, presence at political demonstrations or presence at other specific locations revealing data about for example sex life. These profiles can be used to take decisions that significantly affect the owner.

The technology of smart mobile devices allows for the constant monitoring of location data. Smartphones can permanently collect signals from base stations and wifi access points. Technically, the monitoring can be done secretively, without informing the owner. Monitoring can also be done semi-secretively, when people forget or are not properly informed that location services are switched on, or when the accessibility settings of location data are changed from private to public.

Even when people intentionally make their geolocation data available on the Internet, through whereabout and geotagging services, the unlimited global access creates new risks ranging from data theft to burglary, to even physical aggression and stalking.

As with other new technology, a major risk with the use of location data is function creep, the fact that based on the availability of a new type of data, new purposes are being developed that were not anticipated at the time of the original collection of the data.

Quite simply all this says is that people are sensitive about people knowing where they are. C-ITS requires declaration of where you are and when, so the challenge is to allow this data to be declared without raising the sensitivity. The conventional approaches in technology to give assurances of security are relatively simple:

- Anonymity—ability to use a resource without revealing the user's identity
- Pseudonymity—ability to use a resource without revealing the user's identity but maintaining accountability for the use of the resource
- Unlinkability—a user may make multiple uses of resources or services without others being able to link these uses together
- Unobservability—a user may use a resource or service without others being able to observe that the resource or service is being used.

Of these only Pseudonymity has been fully exploited in C-ITS and takes the form of modifying the identity over the time the vehicle is moving. The intent of providing pseudonymity is to assist in providing privacy protection against a casual observer. There are some difficulties in providing full privacy protection in C-ITS as the vehicle you are driving is traceable back to you as owner or driver (this is a legal requirement), your vehicle is also a visible object on the road and the windscreen is transparent so obviously a keen observer will see you driving the vehicle. Section 10.5 considers identity in much more detail.

10.5 Identity and Identification in C-ITS

Transport, particularly for ITS, has raised lots of flags and worries regarding identity. I find this somewhat difficult to get to grips with and it is at the root of the privacy versus accountability debate. What I will try and outline here are cases where we can make the debate more open and thus understand the steps that have been taken in the C-ITS standardisation world (where I come from) to address both sides of the debate.

It is not possible in the bulk of the industrialised world to simply put together a vehicle in your personal workshop and to drive it on the road. Rather, a vehicle is subject to a long set of tests and approvals before it is allowed on the road. On completion every vehicle is assigned a Vehicle Identification Number (VIN) which is mostly based on ISO-3779 [3] and consists of three parts:

- Manufacturer identifier
- Vehicle descriptor
- Vehicle unique identifier

This is not all that different from the idea and construction of similar identifiers used in mobile phones (the International Mobile Equipment Identifier (IMEI) in 2G/3G, the TETRA Equipment Identifier (TEI) used in TETRA). The VIN can be used to deter fraud and other crime involving the vehicle itself as the VIN is hard coded to the vehicle.

Vehicles are registered for use and this secondary identity is the one we see on the number plate. A large number of, mostly national, standards apply to the construction of vehicle registration numbers and there are standards on the colours used, the fonts and so forth in order to maximise visibility. The laws in most countries state that the vehicle registration number has to be visible at all times the vehicle is on the road. The registration process, in addition to giving the visible identity of the number plate, and making the association of VIN (specific vehicle) to registered vehicle, there is also the association of vehicle to its owner (in the UK it is termed "registered keeper"). In the construction of these identifiers the VIN is structured and managed to be unique, the vehicle registration is structured and managed to be unique. So there is a clear and managed 1-to-1 relationship between VIN and Registration-Id. Any registered keeper of a vehicle may have 1 or many vehicles under their ownership (but of course can only drive one at a time—unless the registered keeper is a corporation and I'll look at that shortly).

Since 1903 in the UK and Germany (only Prussia at the time) driver and vehicle licensing started to become mandatory, and by 1977 when Belgium succumbed and introduced a driving test, almost all of the industrialised world requires mandatory testing of drivers before allowing them to drive a motorised vehicle on the road. One of the results of licensing is the issuing of a unique driver licence identity (note that in some countries, e.g. Spain and Sweden, the driver licence number is the same as the citizen-id). At the root of the thinking is that drivers of vehicles have to be accountable for their actions, it is a crime to withhold the identity of the driver when requested to disclose it by an authority.

Thus we have three (at least) managed identifiers associated with any vehicle on the road:

- VIN
- Registration number
- Driver (licence) identity

The role of these identities in transport management is necessary to consider when we begin to explore the threats to them and in particular their misuse. We also need to consider the privacy angle.

In many cities and roads Automatic Number Plate Recognition (ANPR) is used for access control (more precisely for road pricing but the result is often access control—in order to access certain roads you have to pay a fee). In cities such as London where there is a congestion zone any vehicle crossing into the congestion zone has to pay a fee. The number plate is read by a set of cameras and the recognised number plate is used to determine if the appropriate fee has been paid. The driver has a set time to pay the fee and can do so in a number of ways—in all cases paying the fee against the number. However if the required fee is not paid the system can find the name and address of the registered keeper and make the demand for payment directly.

From a safety perspective each of these identifiers has a specific contribution to make: The VIN allows specific models of vehicle to be identified and specific instances of each model. This becomes important in issuing product recall notices and similar to assure the wider context of vehicle safety. The registration number is visually worn on the vehicle and is used in a number of contexts to control the behaviour of the driver. If a vehicle is seen to do something wrong and needs to be held to account it links to the registered keeper. As vehicles cannot (for now) be directly held to account for their actions it is the driver that is held to account and one of the responsibilities of the registered keeper is to identify who was driving the vehicle at any time.

All of these checks and balances through these identifiers happen in a largely observational system. That is a system where external observations and searches are required to link event to vehicle to registered keeper to driver.

Where ITS, and C-ITS in particular, changes the system is that the vehicle now makes assertions of its own behaviour. This changes a vehicle from not just displaying its registration number on the outside but also allows it to display other attributes on the outside of the vehicle. So prior to C-ITS and the use of regularly updated CAMs an external observer would have to take deliberate action to determine a vehicle's speed and trajectory. Now with ITS any similarly equipped observer can directly calculate the speed and trajectory of all the CAM transmitting vehicles in range.

The impact is now that vehicles, and their drivers, are self-asserting their speed and thus how close it is to the legal limit. Self-declarations of breaking the speed limit are open to prosecution and there have been a number of instances of drivers taking video of their illegal driving, posting them on YouTube, and then having the police knocking on their doors to prosecute. The evidence to prosecute has been self-asserted in the posted video and has the weight of a confession.

There are a number of difficult questions here and the difficult societal one is that of self-assertion of guilt. So there is a scenario in which when driving in an area with a speed limit of (say) 30 mph and you assert you are driving at (say) 35 mph then you are asserting that you are breaking the lawfully set speed limit. There is a line that breaking the law is simply that—breaking the law. If you self-assert you've broken the law, it doesn't change liability. The wider societal question is will you be

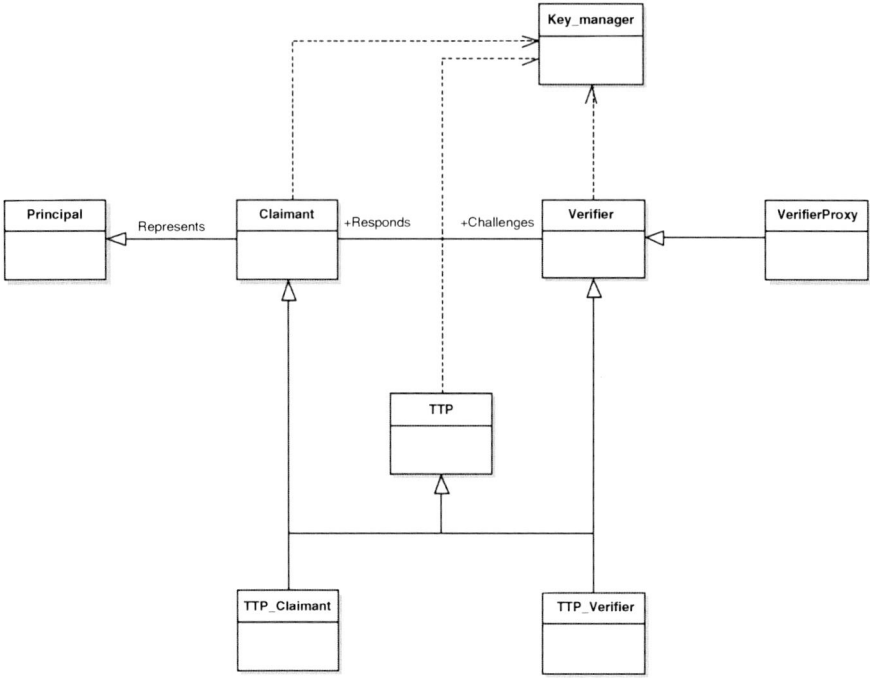

Fig. 10.4 Model for authentication

prosecuted for every self-asserted violation? Happily that is not something this book has to answer. However combining ITS with software control of vehicle dynamics there is no reason why vehicles can't assist drivers in staying legal.

The common security problem associated with identifiers is to masquerade as one. That is Alice claims to be Bob. In order to counter masquerade the security tool is authentication. In formal terms the party being authenticated is referred to as the principal. The entity that requires authentication of the principal is termed the relying party. The model for authentication is given in Fig. 10.4.

Authentication methods rely upon something the principal is, has or knows, where is, has, knows are considered as classes of information. Authentication that uses attributes from only one of these classes is called single factor authentication, and if attributes from two or more of the classes is used in the authentication it is referred to as multi-factor authentication (two-factor if attributes from two classes are used, three-factor if attributes from all three classes are used). The use of two attributes in a class does not make the authentication two-factor. Typically if a human is the principal he is identified by some form of biometric data and the assumption for authentication is that the biometric data is unique and not forgeable.

When the authentication factor uses something the principal has it often refers to a specific hardware module containing some unique and non-forgeable secret. This is the model used in Subscriber Identity Modules (SIM) cards where the key is contained on the SIM held by the user's phone. The third case is where the user has to have knowledge and covers passwords, Personal Identification Numbers (PINs), and pass-phrases.

For all of the existing identifiers authentication is achieved using supporting authoritative documentation. The challenge in C-ITS (and ITS in general) is to first of all identify the principal and the relying parties.

In C-ITS the primary problem is that any pair of vehicles will not have any prior knowledge of each other thus there is no means to establish any credentials for authentication. However in C-ITS the relying party, i.e. the receiver of any CAM or DENM, has to be able to gauge the correctness of the data. So this means data claiming to be from a vehicle has to be verifiable as coming from a vehicle. Similarly data coming from a particular location has to be verifiable as coming from that location.

The role of authentication in C-ITS is inevitably complex. For much of the time we don't really care who you are, but we do care what you are. This means we want to be able to verify attributes and not identity. For example, I need to know you are a car and not a truck but don't need really to know exactly which car or truck. However if I want to determine the likelihood of a collision I do want to be able to link all your transmissions together and this means there has to be some time invariant uniqueness in your transmission (else if there are 20 or 30 or more transmitters all claiming to be cars and I am unable to distinguish one from another then I can't reasonably determine if any are going to collide with me).

10.6 Symmetric and Asymmetric Security

When technologists talk about security they often mean cryptography. Very simply cryptography is the mathematical means of providing security. Where cryptographic methods are used to support security the primary element of achieving security is in the key. The general assumptions for any system relying on cryptology are:

- Knowledge of how algorithms work is in the public domain.
- Knowledge of protocols for authentication and key establishment are in the public domain.

The only means of assuring security remains in place, over and above the known limitations of the algorithm and protocol, is in the secrecy of the key. A secret is by definition not a secret when it is widely known and so a shared secret is not really secret. Symmetric key cryptography works only by control of the number of entities who know the secret and generally, for telecommunications, the intention is to limit this to two parties only. However in public communication where secrecy may be required of communication to a large number of unknown parties the normal definition of secrecy cannot apply. The challenge of this is met by a set of techniques based on non-secret cryptology, or asymmetric keying, whereby a key has two components one of which is private and the other is public. The success is built on the mathematics of the key construction and on the algorithms that make use of the key, but essentially it consists of a pair of one-way functions and the view that is computationally infeasible from knowledge of the public key to find the matching private key. A public key can then be distributed freely to either receive data encrypted by the private key, or to encrypt data to be sent to the holder of the private key.

In symmetric key cryptography there is one mandatory requirement:

- Only two parties have access to the key.

In order to maintain compliance with this requirement there are a number of approaches to key distribution that may be taken. In each case the key should be delivered in a manner that leaves an audit trail. In addition the key should be transmitted in a tamper proof format: tamper proofing may be achieved in either software or hardware. It has often been claimed that symmetric cryptography doesn't scale which can be comprehensively busted with the evidence of the 2G and 3G cellular networks. In this domain symmetric keys are distributed on smart cards (the SIM) and there are some seven billion keys distributed in this way with no known breaches of security. Furthermore the SIM is a tamper proof means of distribution that has been well tested over the years (the same basic form factor and technology is used in all smart cards thus in a very wide range of industries including banking). There are no asymmetric cryptographic key distribution networks of comparable size (the biggest reported PKI is that used internally to the US Department of Defence and has only three million subscribers) (Fig. 10.5).

In asymmetric cryptography a public key can be distributed to either receive data encrypted by the private key, or to encrypt data to be sent to the holder of the private key. However there is a legitimate concern that whilst the mathematical relationship is understood to work there is often only a weak relationship between the two communicating parties hence trust that the data is visible to the correct party has to be assured. The counter to the trust problem is to distribute public keys through a trusted source within public key certificates according to clause 7 of ITU T Recommendation X.509 [13].

A certificate is a signed data object that contains the data elements summarised as follows:

- The identifier of the certification authority.
- The unique identifier of the user of the certificate.
- Some attributes of the user, like address, company, tax code, etc.
- Public key, generated with the private key, to be used to verify digital signature.
- Period of validity of the certificate, defined by a start date and an end date.
- Unique identity code of the certificate.
- Digital signature of the certification authority.
- Environment in which the certificate is valid.

Certificate extensions can be used to provide authorisation as opposed to identification wherein the extensions provide methods for associating additional attributes with users or public keys and for managing the hierarchy.

10 Security and Privacy for ITS and C-ITS

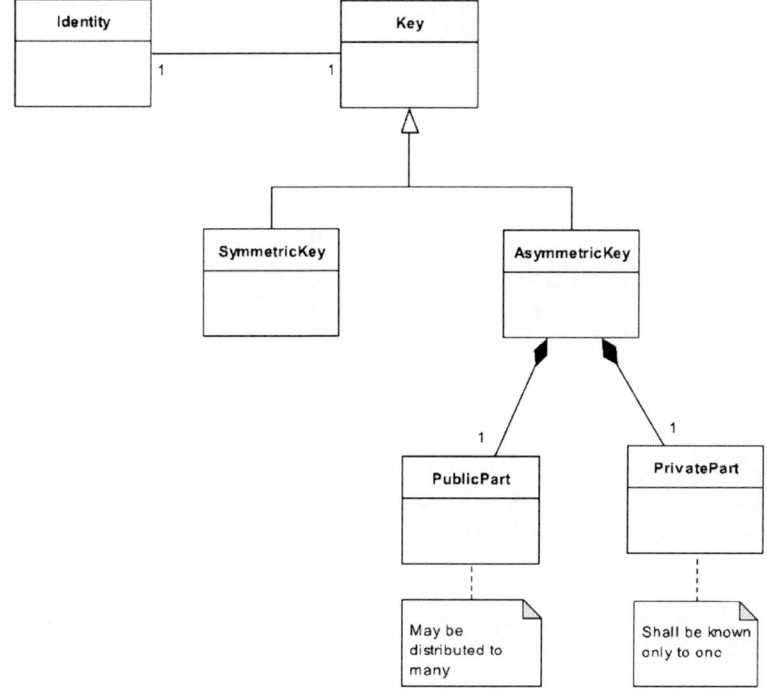

Fig. 10.5 Relationship of identity and keys

A Certification Authority (CA) is a trusted third party that issues certificates. In PKIs the CA verifies identity. CAs (certification authorities) can issue different kinds of certificates:

- Identity.
- Authorisation.
- Transaction.
- Time Stamp.

The standards developed to date (June 2014) provide both identity and authorisation certificates with more than one type of CA in order to give the possibility of pseudonymous operation.

10.7 Security Standards for C-ITS

The global C-ITS market is underpinned by common standards and in this regard the security domain is no different. There are essentially three bodies pushing in this domain:

- International Standards Organization (ISO) (mirrored in European Committee for Standardization (CEN))
- European Telecommunications Standards Institute (ETSI)
- Institute of Electrical and Electronic Engineers (IEEE)

In addition to these bodies there are a number of other parties working to bring C-ITS as a common framework to the market and key amongst these are the Car-to-Car Communications Consortium representing industry and the major Original Equipment Manufacturers (OEMs), the EU and US governments seeking harmonisation across the Atlantic, the International Telecommunications Union Telecommunication and Radio units (ITU-T and ITU-R) who alongside ETSI and the European Conference of Post and Telecommunications Administrations (CEPT) give guarantees of radio access, and there are increasing global standards agreements being developed by ETSI and others to ensure as far as possible that key areas, including C-ITS, have as far as is possible a single set of standards that give assurance of a global market for manufacture, distribution and operation for C-ITS. For a detailed overview of standardisation and harmonisation efforts, see Chap. 2.

The approach to security standards for C-ITS has been fairly conventional:

- Risk Analysis published as ETSI TR 102 893 [4]
- Security Architecture published as TS 102 940 [5]
- Security Countermeasure set published as TS 102 941 [6], TS 102 942 [7] and TS 102 943 [8]
- Data Dictionary for security published as TS 103 097 [9]

In addition these ETSI standards directly reference the IEEE 1609.2 [10] specification for certificate structures and cryptographic measures. Also all of these standards fit into the overall ITS-Station architecture [11, 12] as a set of security services. Graphically this can be seen in Fig. 10.6 which is consistent with the overall approach to risk management that is at the heart of this entire chapter.

The resulting standards that have been developed extend the basic architecture of the ITS Station by describing the application of security services across the stack. This is shown in Fig. 10.7.

In the more visual world this again mimics the core capabilities of C-ITS by mapping specific security services to C-ITS entities and transmission types. This is best shown in Figs. 10.8 and 10.9.

In the standardisation of C-ITS at ETSI and ISO/CEN all groups involved have selected 1609.2 as the core building block for security. Furthermore there is almost complete global acceptance of this selection with the resulting agreements on algorithms and key sizes. The global market is further enhanced by work being done to ensure that the core features have freedom of application and cross border use for C-ITS.

The IEEE 1609.2 standard defines secure message formats and processing for use by Wireless Access in Vehicular Environments (WAVE) devices, including methods to secure WAVE management messages and methods to secure application

10 Security and Privacy for ITS and C-ITS

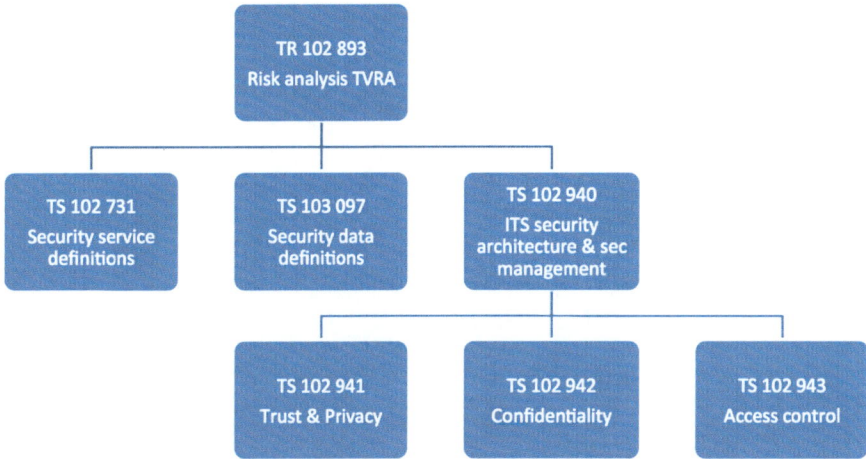

Fig. 10.6 The ETSI standards publications and their relationships

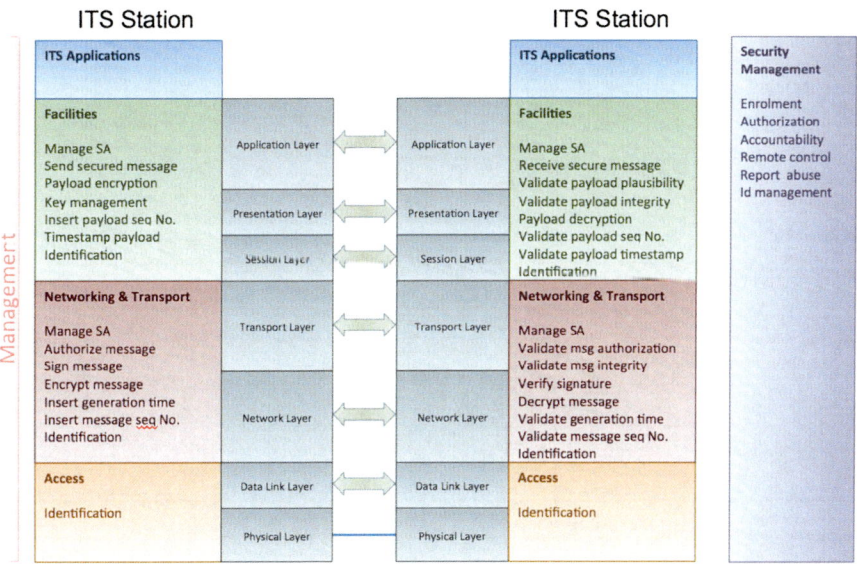

Fig. 10.7 Standardised security services mapped to the OSI protocol stack

messages. Fortunately the IEEE community has worked in close co-operation with ETSI and ISO to ensure that the security functions and features provided are not limited to WAVE and thus those other communities have looked to the 1609.2 capabilities as a toolkit for satisfying the authentication, integrity and confidentiality requirements for ITS in general and C-ITS in particular. Simplifying 1609.2 to its

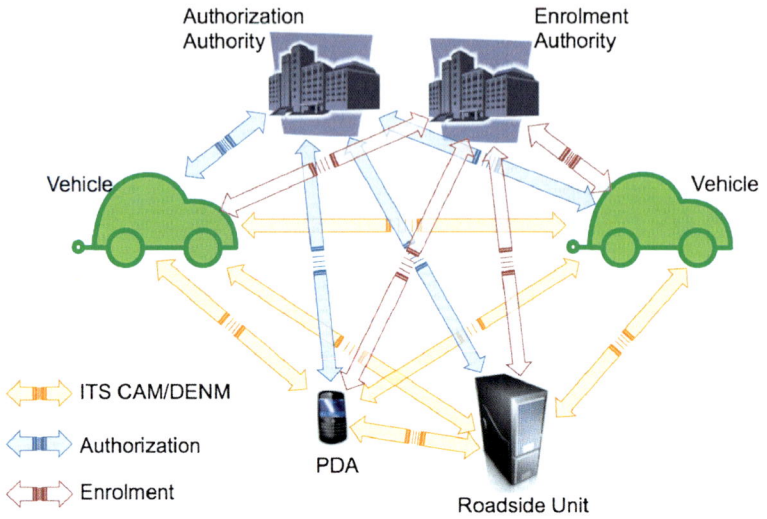

Fig. 10.8 Standardised security services mapped to the C-ITS entities

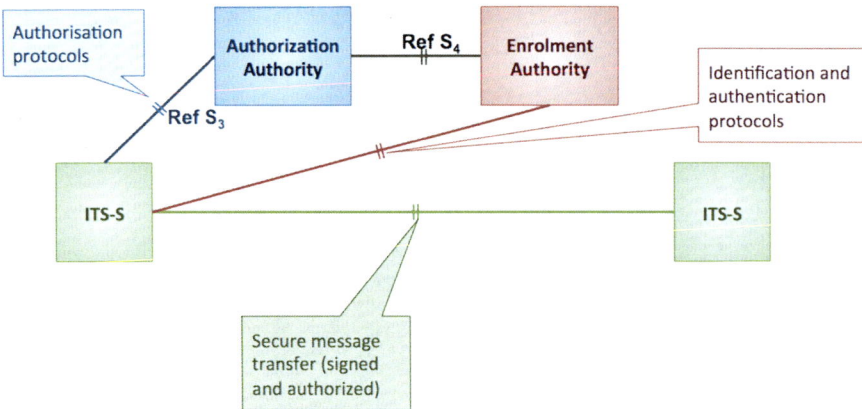

Fig. 10.9 Standardised security services mapped as functional reference points of C-ITS

core services is important as whilst 1609.2 has some novelty in its use of elliptical curve cryptography but that of itself is a natural choice (to give a cryptographic strength of 128-bits using a more conventional RSA[1] approach would require keys of 3,072 bits whereas with elliptical curve the keys are "only" 265 bits long).

---

[1]The approach taken by Rivest–Shamir–Adleman to public key cryptography based on the difficulty in factorising a given set of semi-prime numbers (a set of numbers that have as factors two prime numbers).

10 Security and Privacy for ITS and C-ITS

The security processing services of IEEE 1609.2 consist of the following:

- Secure Data Exchange
  - signing and/or encryption of Protocol Data Units (PDUs) prior to their transmission,
  - decryption and verification (as appropriate) of PDUs on reception.
- Security processing for security management
  - ensuring access to Crypto-material (private keys, public keys and certificates)
  - generating certificate requests and processing responses
  - validating Certificate Revocation Lists (CRLs)
- Storing private keys and their associated certificates.

The end result is that IEEE 1609.2's mechanisms provide support for the core C-ITS capabilities:

- Authorisation verification by certificate verification
- Confidentiality by encryption/decryption
- Integrity proof generation and validation

So in applying 1609.2 to C-ITS, for CAM say, the transmitting station chooses a key pair with a validated pseudonymity and authority public key certificate and signs the CAM prior to transmission with the matching private key. It then transmits the signed CAM (as a PDU or datagram) with the appropriate certificate such that any receiving party can validate that the transmitter did actually send the CAM and that he had authority to do so verified by a shared authority. In this case the application of 1609.2 is much like any other public key scheme with the cryptographic advantages brought by use of elliptical curve cryptography.

We have already looked at the general concepts of cryptographic security and the role of symmetric and asymmetric keying. However we now need to look more closely at what is involved in C-ITS and this requires us to specifically look at IEEE 1609.2 [10] and how it is involved in securing C-ITS.

A pseudonym, or alias, is an alternative identity. In C-ITS the identity given across the radio network is a pseudo-random identifier that is verified by a third party as belonging to an ITS-Station. The authority that does this third party verification doesn't much care which ITS-Station the identifier is attached to.

Every identifier is bound to a key pair and the public key is certified (by the aforementioned third party authority) as belonging to the identifier and the private key. When a package (a C-ITS message) is signed using the participant's private key it can be verified by any receiver holding the public key that it comes from a real ITS-Station.

When using identifier certificates in this way the transmitter is in control of how much they want to reveal of their identity by how often they change their pseudonym. However as each pseudonym has to be independently verified and every verification takes time this is a non-trivial calculation. In C-ITS the scheme used for cryptographic signing is that defined in IEEE 1609.2 [10] and this offers a number of ways to ease the burden of creating multiple certificates from a single authorisation pass.

C-ITS uses the cryptographic certificate scheme defined in IEEE 1609.2 [10]. The cryptographic basis of this is elliptical curve asymmetric cryptography and for the application in C-ITS takes advantage of some of the capabilities of this branch of mathematics to allow autogeneration of new certified identities from a single authorisation.

The rate at which a pseudonym is changed is, as has been mentioned, a black art. It has to change often enough to act as a non-persistent identifier for the ITS-Station (and the vehicle it is associated with) but not too often that any ability of receiving ITS-Stations to determine the vector of movement of the ITS-Station and its associated vehicle. So every transmission is too often, once a month probably not often enough. This problem is not unique to C-ITS and has been faced by many other broadcast radio technologies including GSM and TETRA (Terrestrial Trunked Radio). In the former the unique user identity is exposed once at initial registration and then replaced with a temporary identity at every re-registration, in the latter the identity is similarly replaced with a temporary identity but in this case encrypted such that the value seen by a casual observer is different in every transmission with no means of correlation between values. The C-ITS approach is closer in spirit to the TETRA than the GSM approach but with the fine grain of control left to the transmitting station.

We often mention certificates and signatures, too often as synonyms. A digital signature is included in a certificate and represents a relatively simple process as shown in Fig. 10.10.

There has been a recurring question in the development of the C-ITS security model and that is: Where do you sign? By this it means at which of the protocol layers in the stack between two communicating ITS stations? Digital signatures sign documents and apply only to the document they sign. So we need to identify what is the "document" in C-ITS. The simplest view is that a CAM or DENM is the document—it is the statement prepared by the vehicle for sharing with other co-operating ITS stations. The signature applies to the completed document. So a CAM can be "signed" but only by the generator of the CAM. Given that CAMs are intrinsically mutable (i.e. for a moving vehicle each CAM will be different if the CAM is a statement "I am here now", so *now*, i.e. time, always changes and *here*, i.e. location, may change) and that the sender of the CAM has no prior knowledge of who will receive the CAM the sending station also needs to send the public key required to verify the signature.

The normal process of third party verification that a public key is bound to a corresponding private key is complex and often time consuming. This is a problem in C-ITS as time is in very short supply.

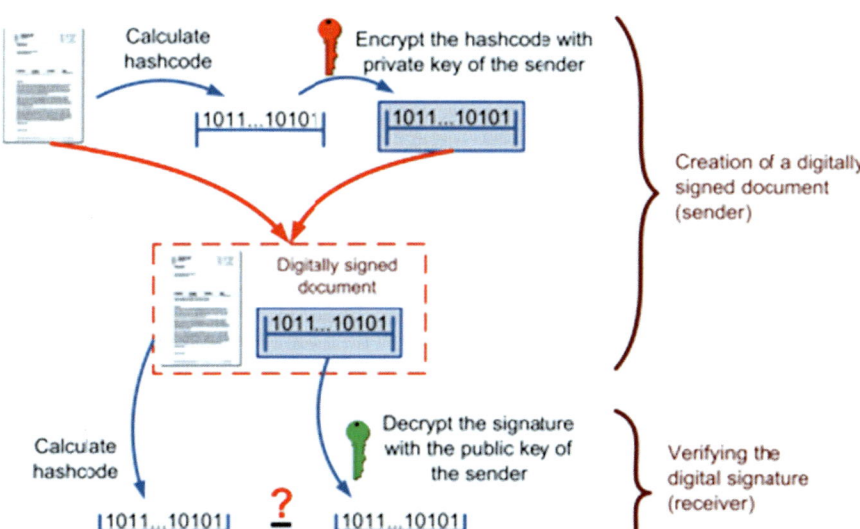

Fig. 10.10 The digital signature process

10.7.1 After the Standards?

The challenge that is still being worked through for global deployment of Secure C-ITS is in building the infrastructure of authorities that distribute the public key certificates. There are some opportunities afforded by the cryptographic mechanisms underpinning 1609.2 that allow for chains of key pairs to be spun off a single validated key pair can be used to advantage. This is important as with several millions of cars active on the roads of Europe at any one time where new key-pairs and certificates are required every few minutes it would simply be infeasible to have a vehicle connected to an infrastructure to go through a complicated and time-consuming certification process for a key that is only going to be used a few times before being discarded.

Beyond this there are other questions to be asked of standards and deployments. Amongst these are how to integrate C-ITS to other systems such as the smart city, to other geo-capabilities and to the inevitable changes in society that require acceptance of all vehicle types as needing to co-operate in ITS. This will need the ITS community to look at how to integrate the standards of today into alternative devices, the most prevalent of which is the smartphone.

10.8 Conclusions

Quite simply providing security in C-ITS is not trivial. The fact that Alice (the transmitting vehicle) and Bob (the receiving vehicle) have not got a previously established relationship, and Eve (the adversary of both Alice and Bob) has equal access to the data sent by Alice, makes any model that assures a secure link between Alice and Bob problematic. Adding to this is the problem that Bob is not necessarily an individual but may be a set of individuals whose only criteria to be Bob is that they are in range of the transmissions from Alice, and furthermore all Bobs are explicitly unknown to Alice.

For now C-ITS security is provided through a validated trust hierarchy. That hierarchy gives a limited degree of assurance that a vehicle is really a vehicle, that its claim to a particular role has been validated by a third party, and ultimately ensures that data received can be traced to its source. The trust model is enforced and reinforced using pseudonymous authorisation certificates following the models of IEEE 1609.2 and its underlying cryptographic model.

We cannot afford to be complacent regarding the security provisions in C-ITS and ITS in the wider domains. Serious efforts have been made to give assurance of global interoperability—all active C-ITS standards are based on a single common model for its cryptographic operations. Great efforts have been made in global standards to assure the industry works to a common set of standards and this is true for security as for other spheres of C-ITS. The long-term success of C-ITS requires that all the aspects discussed in this chapter are maintained.

Finally it is important to remember that security is a process and not a function. That process never stops.

Definitions

Access Control Policy
    A set of privileges representing access control rules that defines which allowed entities for certain operations within specified contexts each entity must comply with, in order to grant access to an object

agency
    Ability and opportunity of the individual to make independent choices

anonymity
    Act of ensuring that a user may use a resource or service without disclosing the user's identity

authentication
    Ensuring that the identity of a subject or resource is the one claimed

confidentiality
    Ensuring that information is accessible only to those authorized to have access

identity
: Set of properties (including identifiers and capabilities) of an entity that distinguishes it from other entities

impact
: Result of an information security incident caused by a threat and which affects assets

integrity
: Safeguarding the accuracy and completeness of information and processing methods

personal data
: Any information relating to an identified or identifiable natural person

privacy
: Right of the individual to have his identity, agency and action protected from any unwanted scrutiny and interference

pseudonymity
: Act of ensuring that a user may use a resource or service without disclosing its user identity, but can still be accountable for that use

risk
: Potential that a given threat will exploit vulnerabilities of an asset or group of assets and thereby cause harm to the attacked system or organization

threat
: Potential cause of an incident that may result in harm to a system or organization

unlinkability
: Act of ensuring that a user may make multiple uses of resources or services without others being able to link these uses together

unobservability
: Act of ensuring that a user may use a resource or service without others, especially third parties, being able to observe that the resource or service is being used

vulnerability
: Weakness of an asset or group of assets that can be exploited by one or more threats

Acknowledgements The work reflected in this chapter has been supported by the i-SCOPE project, and by each of the SUNSHINE and i-locate projects and has incorporated some of the findings from the i-tour project.

i-SCOPE
: The project has received funding from the European Community, and it has been co-funded by the CIP-ICT Policy Support Programme as part of the Competitiveness and Innovation Framework Programme by the European Community, contract number 297284. The author is solely responsible for it and that it does not represent the opinion of the Community and that the Community is not responsible for any use that might be made of information contained therein.

SUNSHINE
: This project is partially funded under the ICT Policy Support Programme (ICT PSP) as part of the Competitiveness and Innovation Framework Programme by the European Community

i-locate
: The project has received funding from the European Community under contract number 621040

References

1. ETSI TS 102 165-1: Telecommunications and Internet converged Services and Protocols for Advanced Networking (TISPAN); Methods and protocols; Part 1: Method and proforma for Threat, Risk, Vulnerability Analysis
2. Article 29 of Directive 95/46/EC Working group Opinion 13/2011 on Geolocation services on smart mobile devices, Adopted on 16 May 2011
3. ISO 3779: Road vehicles - Vehicle identification number (VIN) - Content and structure
4. ETSI TR 102 893: Intelligent Transport Systems (ITS); Security; Threat, Vulnerability and Risk Analysis (TVRA)
5. ETSI TS 102 940: Intelligent Transport Systems (ITS); Security; ITS communications security architecture and security management
6. ETSI TS 102 941: Intelligent Transport Systems (ITS); Security; Trust and Privacy Management
7. ETSI TS 102 942: Intelligent Transport Systems (ITS); Security; Access Control
8. ETSI TS 102 943: Intelligent Transport Systems (ITS); Security; Confidentiality services
9. ETSI TS 103 097: Intelligent Transport Systems (ITS); Security; Security header and certificate formats
10. IEEE STANDARD 1609.2-2013 IEEE Standard for Wireless Access in Vehicular Environments Security Services for Applications and Management Messages
11. ISO 21217:2014: Intelligent transport systems Communications access for land mobiles (CALM) Architecture
12. ETSI EN 302 665: Intelligent Transport Systems (ITS); Communications Architecture
13. ITU-T Recommendation X.509: Information technology - Open Systems Interconnection - The Directory: Public-key and attribute certificate frameworks

Part IV
Evaluation of Vehicular Networks

Chapter 11
Mobility Models for Vehicular Communications

Pietro Manzoni, Marco Fiore, Sandesh Uppoor, Francisco J. Martínez Domínguez, Carlos Tavares Calafate, and Juan Carlos Cano Escriba

Abstract The experimental evaluation of vehicular ad hoc networks (VANETs) implies elevate economic cost and organizational complexity, especially in presence of solutions that target large-scale deployments. As performance evaluation is however mandatory prior to the actual implementation of VANETs, simulation has established as the de-facto standard for the analysis of dedicated network protocols and architectures. The vehicular environment makes network simulation particularly challenging, as it requires the faithful modelling not only of the network stack, but also of all phenomena linked to road traffic dynamics and radio-frequency signal propagation in highly mobile environments. In this chapter, we will focus on the first aspect, and discuss the representation of mobility in VANET simulations. Specifically, we will present the requirements of a dependable simulation, and introduce models of the road infrastructure, of the driver's behaviour, and of the traffic dynamics. We will also outline the evolution of simulation tools implementing such models, and provide a hands-on example of reliable vehicular mobility modelling for VANET simulation.

Keywords Behavioural models • Car following models • Driver behavioral models • Large-scale vehicular networks • Macroscopic mobility • Microscopic mobility • Mobility datasets • Mobility modelling • Mobility scenarios • Mobility traces • Road topologies models • Road traffic flows • Road traffic

---

M. Fiore
CNR - IEIIT, Torino, Italy
e-mail: marco.fiore@ieiit.cnr.it

P. Manzoni (✉) • C.T. Calafate • J.C. Cano Escriba
Universitat Politecnica de Valencia, Valencia, Spain
e-mail: pmanzoni@disca.upv.es; calafate@disca.upv.es; jucano@disca.upv.es

F.J.M. Domínguez
Escuela Universitaria Politécnica de Teruel, Universidad de Zaragoza, Zaragoza, Spain
e-mail: fcomardo@unizar.es

S. Uppoor
INSA Lyon/Inria, Lyon, France

© Springer International Publishing Switzerland 2015
C. Campolo et al. (eds.), *Vehicular ad hoc Networks*,
DOI 10.1007/978-3-319-15497-8_11

simulation • Software tools • Stochastic models • Traffic assignment • Traffic dynamics models • Traffic stream models • Vehicular mobility • Vehicular network simulation

11.1 Introduction

Network solutions designed for vehicular ad hoc networks (VANETs) will play a major role in determining the level of success of vehicular communications. Given the criticality of many applications enabled by VANETs, which include road safety and traffic management services, there is a clear need for validating and testing such network solutions in the real world. However, logistic difficulties and economic issues make experimental evaluations extremely difficult to set up. Even state-of-the-art testbeds that involve car manufacturers and mobile telecommunication operators, such as sim^{TD} in Germany, involve a few hundred vehicles at most. Although that number may appear large, it represents less than 1 % of the total traffic in a medium-sized city.

In fact, many VANET solutions need to be evaluated over large scales (e.g., citywide) and assume nearly 100 % penetration ratios of the vehicle-to-vehicle communication technology. In order to meet these requirements, the only option is resorting to simulation. The computational resources of today's servers allow to simulate VANETs over whole cities, and thus to evaluate even the most demanding protocols and architectures at any sensible scale. The problem shifts instead to the level of realism of the simulation, and the dominating aspect to address is that of mobility. The movement of vehicles in every day's road traffic directly determines the position of nodes in the VANET. Therefore, the use of an unrealistic model for vehicle mobility would lead to simulation results that are biased, unreliable or even completely erroneous.

Mobility models of road traffic have been thoroughly studied in fields such as transportation research, mathematics and physics. From a viewpoint traditionally introduced by physicists, vehicular mobility can be described at three different levels, i.e., macroscopic, mesoscopic or microscopic. Figure 11.1 provides examples of these approaches in a simple scenario where the road traffic is modelled separately over three segments of a same road.

- Macroscopic models, in Fig. 11.1a represent road traffic as a hydrodynamic phenomenon, where flows of cars move along roads similarly to fluids within tubes. They aim at defining fundamental relationship among macroscopic measures such as the speed, density and in-/out-flow of vehicles. Therefore, they do not provide information about individual vehicles, but just an aggregate overview.
- Mesoscopic models, in Fig. 11.1b descend to the individual vehicle level, yet determine the speed of each car using macroscopic measures. As a result, vehicles are not independent of each other in their movement, and all drivers on the same road segment tend to, e.g., travel at the same speed.

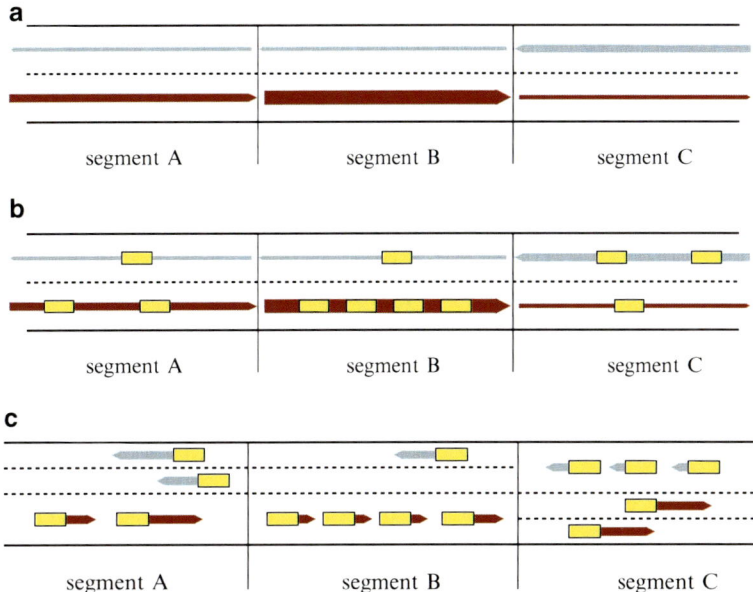

Fig. 11.1 Classification of vehicular mobility models according to their level of detail. (**a**) Macroscopic. (**b**) Mesoscopic. (**c**) Microscopic

- Microscopic models, in Fig. 11.1c treat each vehicle as an autonomous entity, and thus describe the movement of each driver independently. They allow describing complex acceleration and overtaking behaviours, which result in different speeds by cars travelling within the same road segment.

The three classes of models above have advantages and disadvantages. At one end, macroscopic models only provide an aggregate and high-level view of the system, but are mathematically tractable and can be simulated at minimal computational cost. At the other end, microscopic models can be extremely detailed, but they also require significant processing power to be run at large scales. Clearly, mesoscopic models fall in between the other two classes.

In the case of VANET simulation, the choice is mandated by the fact that vehicle-to-vehicle communications have ranges in the order of hundreds of meters. This means that the precision in representing the position of each vehicle needs to be in the order of the meter or less; otherwise, inaccuracies in the mobility representation risk to bias the communication performance. As a result, the only sensible choice is that of microscopic models that consider individual vehicles and can thus output their actual location rather than an approximate one.

Within the context of microscopic modelling of vehicular mobility, several components have to be taken into account in order to obtain a realistic representation of road traffic.

First, one needs a faithful description of the road infrastructure. The term is to be intended in its largest acceptation, i.e., not limited to a graph-like model of the road layout, but including, e.g., speed limits, one-way constraints, traffic lights at intersections and their temporization, stop and yield signs, roundabouts, overpasses, highways ramps, etc. We discuss the representation of the road infrastructure in Sect. 11.2.

Second, the microscopic behaviour of each driver must be modelled in an accurate way. The acceleration and speed of each vehicle must be the result of its interactions with surrounding cars and with the road signalization. Models of the driver's behaviour used in the networking literature are presented in Sect. 11.3.

Third, traffic flows of individual vehicles over the road topology must be properly described. In other words, the trip of each car is to be detailed, in terms of its origin location, its destination, and the time at which the trip starts. This maps to the definition of a so-called travel demand. Moreover, once trips are decided, the precise route taken by drivers needs to be selected, via apt traffic assignment models. Modelling of traffic flows in works on VANETs is introduced in Sect. 11.4.

These three components have been progressively included in tools dedicated to the simulation of microscopic vehicular mobility for networking purposes. We briefly review the evolution of such tools in Sect. 11.5, so as to provide the reader with an overview of the dramatic improvements that have occurred in VANET simulation over the last decade.

Finally, in Sect. 11.6, we provide a thorough description of the generation process of one specific state-of-the-art dataset of vehicular mobility. This hands-on example brings together different models and tools discussed in the previous sections of the chapter, and allows us to stress the challenges encountered during the creation of a large-scale road traffic trace.

11.2 Modeling Road Topologies

The road topology is an important factor accounting for mobility in simulations, since the topology constrains cars' movements, see Cavin et al. [6]. For MANETs, the random waypoint model (RWP) [46] is by far the most popular mobility model. However, in vehicular networks, nodes (vehicles) do not move independently of each other but they move according to a well-established vehicular traffic models, so the results for MANETs cannot be directly applicable. Moreover, vehicles can only move along streets, prompting the need for a road topology model.

Roughly described, an urban topology is a graph where vertices and edges represent, respectively, junction and road elements. Simulated road topologies can be generated ad hoc by users, randomly by applications, or obtained from real roadmap databases. Using complex layouts implies more computational time, but the results obtained are closer to reality.

The two most typical topologies used are: the highway scenarios (the simplest layout, basically a straight line without junctions) and the Manhattan-style street

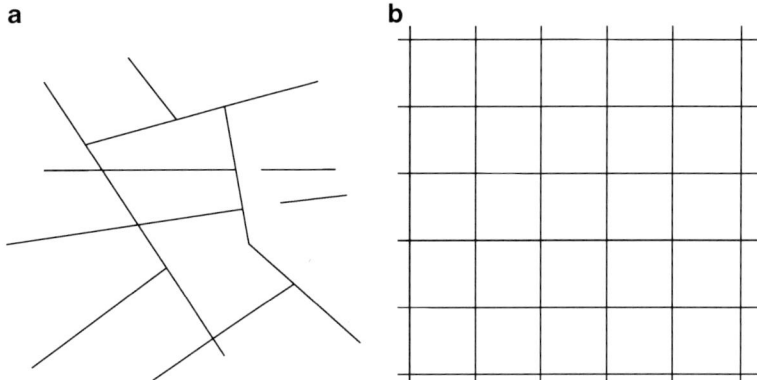

Fig. 11.2 Examples of different roadmap topologies: (**a**) user defined, and (**b**) Manhattan

grids (with streets arranged orthogonally). For example, Huang et al. [18] studied taxi behaviour modelling the city as a Manhattan style grid with a uniform block size across the simulation area. All streets were assumed to be two-way, with one lane in each direction; taxi movements were constrained by these lanes. Figure 11.2 shows two examples of synthetic topologies: (a) used defined, and (b) Manhattan-based, respectively.

These approaches are simple and easy to implement in a simulator. When used, results can give some information about the general performance trends of the different algorithms studied. However, a more realistic layout should be used to ensure that the results are closer to reality.

As proposed by Jardosh et al. [20], a possible solution to randomly generate graphs on a particular simulation area is Voronoi tessellations. With this approach we start by distributing points, representing obstacles (e.g., buildings), over the simulation area. We then draw the Voronoi domains, where the Voronoi edges represent roads and intersections running around obstacles. This way we obtain a planar graph representing a set of urban roads, intersections and obstacles. Figure 11.3 depicts two random Voronoi maps: (a) with uniform density of streets, and (b) clustered density of streets.

Although being an interesting improvement, these graphs lack realism too. Indeed, the distribution of obstacles should be fitted to match specific urban configurations. For instance, dense areas such as city centres have a larger number of obstacles, which in turn increases the number of Voronoi domains. By looking at topological maps, we can see that the density of obstacles is higher in the presence of points of interest. To address these issues, generating clusters of obstacles with different densities is required.

Saha and Johnson [36] modelled vehicular traffic as the random movement of vehicles over real road topologies extracted from the maps of the US Census Bureau TIGER database. In that work, the vehicles selected one point over the graph as their destination and computed the shortest path to get there. The edges sequence is

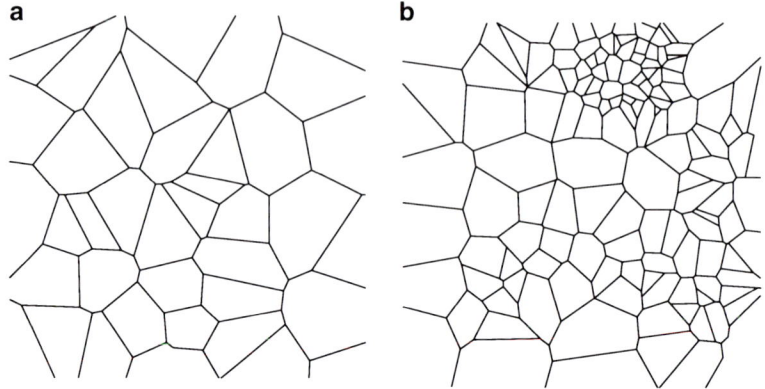

Fig. 11.3 Examples of different roadmap topologies: (**a**) Voronoi uniform, and (**b**) Voronoi clustered

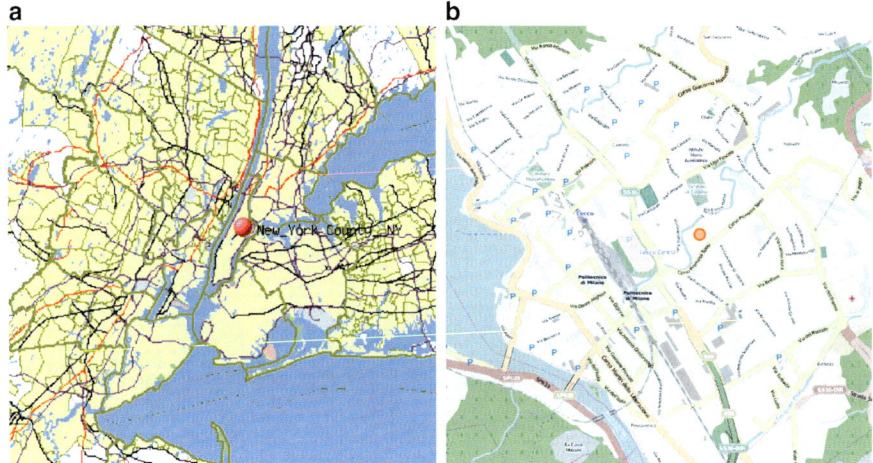

Fig. 11.4 Examples of different roadmap topologies: (**a**) TIGER database, and (**b**) openstreetmap.org

obtained weighting the cost of traveling on each road at its speed limit, including the traffic congestion. Currently, similar approaches are being used by using maps coming from other sources, like OpenStreeMap.org (see Fig. 11.4.)

11.3 Modeling Driver's Behaviour

Major studies have been undertaken in order to develop mathematical models reflecting a realistic physical effect. Fiore et al. [12] wrote a complete survey of models falling into this category. According to Fiore's classification, drivers behaviour can be classified in to five groups: (1) Stochastic models wrapping all models containing purely random motions, (2) Traffic Stream models looking at vehicular mobility as hydrodynamic phenomenon, (3) Car Following models, where the behaviour of each driver is modelled according to vehicles ahead, (4) Queue models, which model roads as FIFO queues and cars as clients, and (5) Behavioural models, where each movement is determined by behavioural rules such as social influences. In the following subsections we will give the most relevant example of mobility model for each sub-category.

11.3.1 Stochastic Models

The Manhattan mobility model [5] is the most widely used example of a Stochastic model which uses a grid road topology (see Fig. 11.2b). The Manhattan model employs a probabilistic approach in the selection of node movements, since, at each intersection, a vehicle chooses to keep moving in the same direction with a 50 % probability, and to turn left or right with a 25 % probability in each case. Vehicles move over the grid with constant speed. The car interaction rules usually employed in the Manhattan model are too simple and do not reproduce a realistic driver behaviour.

11.3.2 Traffic Stream Models

The Fluid Traffic Motion (FTM) [37] is an example of a Traffic Stream model which accounts for the presence of nearby vehicles when calculating the speed of a car. This model describes car mobility on single lanes, but does not consider the case in which multiple vehicular flows have to interact, as in the presence of intersections.

The FTM describes the speed as a monotonically decreasing function of the vehicular density, forcing a lower bound on speed when the traffic congestion reaches a critical state by means of the following equation:

$$s = \max\left[s_{\min}, s_{\max}\left(1 - \frac{k}{k_{\text{jam}}}\right)\right] \quad (11.1)$$

where s is the output speed, $s_{\min}$ and $s_{\max}$ are the minimum and maximum speed, respectively, k_{jam} is the vehicular density for which a traffic jam is detected, and k is the current vehicular density of the road where the node is, whose speed is being computed, is moving on. This last parameter is given by $k = n/l$, where n is the number of cars on the road and l is the length of the road segment itself.

According to this model, cars traveling on very crowded and/or very short streets are forced to slow down, possibly to the minimum speed, if the vehicular density is found to be higher than or equal to the traffic jam density. On the other hand, as less congested and/or longer roads are encountered, the speed of cars is increased towards the maximum speed value. Thus, the Fluid Traffic Model describes traffic congestion scenarios, but still cannot recreate queuing situations, nor can it correctly manage car behaviour in the presence of road intersections. Moreover, no acceleration is considered and it can happen that a very fast vehicle enters a short/congested edge, suddenly changing its speed to a very low value, which is definitely a very unrealistic situation.

11.3.3 Car Following Models

The Krauss Model [26] falls into the Car Following Model subcategory. It takes four input variables (the maximum velocity v_{max}, the maximum acceleration a, the maximum deceleration b, and the noise η that introduces stochastic behavior to the model), discretizes the time with step Δt, and is defined by the following set of equations:

$$v_i^{\text{safe}}(t + \Delta t) = v_{i+1}(t) + \frac{\Delta x_i(t) - v_{i+1}(t)\tau}{(v_i(t) + v_{i+1}(t))/2b + \tau} \quad (11.2)$$

$$v_i^{\text{desired}}(t + \Delta t) = \min\left[v_{max}, v_i(t) + a\Delta t, v_i^{\text{safe}}(t + \Delta t)\right] \quad (11.3)$$

$$v_i(t + \Delta t) = \max\left[0, v_i^{\text{desired}}(t + \Delta t) - \epsilon\, a\, \Delta t\, \eta\right] \quad (11.4)$$

Equation (11.2) computes the speed of vehicle i required to maintain a safety distance from its leading vehicle. The reaction time of the driver is represented by the time τ. Equation (11.3) determines the new desired speed for vehicle i, which is equal to the current speed plus the increment determined by the uniform acceleration, with upper bounds represented by the maximum and safe speeds. Equation (11.4) finally determines the speed of the following vehicle, by adding some randomness, in the measure of a maximum percentage ϵ of the highest achievable speed increment $a\Delta t$ (η is a random variable uniformly distributed in $[0, 1]$).

Another interesting example was the Intelligent Driver model, presented by Treiber et al. [41]. Such model characterizes drivers behaviour depending on their front vehicle. The instantaneous acceleration of a vehicle is computed according to the following equations:

$$\frac{dv}{dt} = a\left[1 - \frac{v^4}{v_0} - \frac{s^{*2}}{s}\right] \quad (11.5)$$

$$s^* = s_0 + vT + \frac{v\Delta v}{2\sqrt{ab}} \quad (11.6)$$

In Eq. (11.5), v is the current speed of the vehicle, v_0 is the desired velocity, s is the distance from preceding vehicle, and s^* is the so-called desired dynamical distance. This last parameter is computed as shown in Eq. (11.6), and is a function of the minimum bumper-to-bumper distance s_0, the minimum safe time headway T, the speed difference with respect to front vehicle velocity Δv, and the maximum acceleration and deceleration values a and b.

When combined, these formulae give the instantaneous acceleration of the car, divided into a "desired" acceleration $[v/v_0^4]$ on a free road, and braking decelerations induced by the preceding vehicle $[s^*/s^2]$. By smoothly varying the instantaneous acceleration, the IDM can realistically mimic car-to-car interactions on a single-lane and straight road. Interesting real-world situations, such as queuing of vehicles behind a slow car, or speed reduction in presence of congested traffic can be reproduced. However, this model alone is not yet sufficient to obtain a realistic vehicular mobility model for urban environments.

Two different extensions were proposed to complete the model: (1) IDM with Intersection Management (IDM-IM), which adds intersection-handling capabilities to the behaviour of vehicles driven by the IDM, and (2) IDM with Lane Changing (IDM-LC), which extends the IDM-IM model with the possibility for vehicles to change lane and overtake each others, taking advantage of the multi-lane capability of the macro-mobility description.

11.3.4 Queue Models

Queue models were introduced in the vehicular traffic field by Gawron [14]. According to the queue paradigm, each road is modelled as a FIFO queue, and each vehicle as a queue client. Each road queue k is characterized by its length l^k and a maximum flow q_{max}^k, determined by the number of lanes. Every time a vehicle enters a road, a travel time is computed, depending on the desired free flow speed of the driver v_{max}, on the number of vehicles on the road n^k and the road length.

The car is then queued in the priority queue of the road, according to the travel time calculated before. At every time step, vehicles whose travel time has expired can be removed from the head of the queue and inserted into the queue representing the next road in their trip. However, when multiple choices are available to exit a road, an intermediate step is necessary, and first-in-first-out queues are added for each outgoing flow. In that case, vehicles at the head of the priority queue are moved to one of the FIFO queues, depending on their destination.

The FIFO queues have a finite capacity, meaning that only a certain number of vehicles per second can access them. Since the movement from one road to another is constrained by the capacity of such next road, a vehicle at the head of an output queue can join the following queue only if there is space on the following road.

The capacity of a road is easily modelled as $c^k = \frac{n^k q_{max}^k}{x_{min}}$, where x_{min} is the distance between the front of two adjacent vehicles in jam conditions. Thus, if the new road has c^k cars already queued, it will not accept further vehicles, and drivers willing to enter the road will have to wait until a spot is freed.

It was shown [14] that even a simple expression of travel time l^k/v_{max}, which neglects the effect of vehicular density on the speed, leads to very good approximations of results obtained with much more complex microscopic mobility models.

Since queue models describe the movement of each vehicle in an independent way, but also with a minimal level of detail, they fall into an intermediate category with respect to macroscopic and microscopic descriptions, which can be referred to as mesoscopic. Queues models have very low computational cost, because they update the status of a vehicle only when a vehicle enters a new priority or FIFO queue. This allows to model very large road topologies, up to hundreds of thousands of vehicles. The drawback is the reduced realism of the outcome, which is less precise than that obtained with other models (e.g., queue models do not reproduce shockwaves caused by periodic perturbations, a common phenomenon in vehicular traffic).

11.3.5 Behavioural Models

Legendre et al. [27] introduced a novel approach to the problem of modelling human mobility, which can be applied to vehicular traffic as well. The approach was called behavioural modelling, and is borrowed from the fields of biological physics and artificial intelligence. The key idea is that every movement is determined by behavioural rules, which are imposed by social influences, rational decisions or actions following a stimulus-reaction process. These rules can be modelled as attractive or repulsive forces. In the case of vehicular mobility, the next intersection towards the trip destination wields an attractive force on the vehicle, whereas other vehicles or obstacles in general exert a repulsive force on it. The result from the composition of these forces determines the acceleration vector driving the car movement. This model is especially expensive under the computational point of view, as every movement requires the elaboration and composition of multiple inter-object forces.

11.4 Modeling Traffic Dynamics

Due to the complexity of modelling vehicular mobility, only few synthetic models are able to come close to a realistic modelling of motion patterns. A different approach could also be followed. Instead of developing complex synthetic models and then calibrating them using mobility traces or surveys, time could be saved by directly extracting generic mobility patterns from movement traces.

Such approach became increasingly popular as mobility traces started to be gathered through the various measurement campaigns launched by activities such as: (1) the project developed in conjunction with the Fleetnet Project—Internet on the Road [13] and the NOW—Network on Wheels Project [33], which is based on traces measured by Daimler AG on a highway section, (2) the UMass DieselNet Project [43], created by the University of Massachusetts, which provides mobility traces of a bus system in the city of Amherst, MA, USA, and (3) the Cabspotting Project [4] which equipped all taxi vehicles in the San Francisco Bay Area and provides a live visualization of the complete taxi system.

The most difficult part of this approach is to extrapolate patterns not observed directly by traces. Complex mathematical models are required to predict mobility patterns, but their limitations are mainly linked to the class of the measurement campaign. For instance, if motion traces have been gathered for bus systems, an extrapolated model cannot be applied to the traffic of personal vehicles. Another limitation for the creation of trace-based vehicular mobility models is the limited availability of vehicular traces.

Surveys are also an important source of macroscopic mobility information. The major large scale surveys are provided by the US Department of Labor (DOL),[1] which gathered extensive statistics of US workers' behaviours, spanning from the commuting time or lunch time, to traveling distance or preferred lunch types. By including such kind of statistics into a mobility model, one is able to develop a generic mobility model able to reproduce the pseudo-random or deterministic behaviour observed in the real urban traffic.

Mobility simulators implementing survey-based models simulate arrival times at work, lunch time, breaks/errands, pedestrian dynamics (e.g., realistic speed-distance relationship and passing dynamics), and workday time-use such as meeting size, frequency, and duration. Vehicle traffic is derived from vehicle traffic data collected by state and local governments and models vehicle dynamics and diurnal street usage.

The UDel Models For Simulation of Urban Mobile Wireless Networks [42] typically falls into the survey-based models category. Its mobility simulator is based on surveys coming from various areas. It includes: (1) time-use studies performed by the US Department of Labor and Statistics, (2) time-use studies by the business research community, (3) pedestrians and vehicle mobility studies by the urban planning and traffic engineering communities. Vehicle traffic is derived from vehicle traffic data collected by state and local governments and models vehicle dynamics and diurnal street usage.

Another Survey-based model is the Agenda-based [47] mobility model, which combines both social activities and geographic movements. The movement of each node is based on an individual agenda, which includes all kinds of activities in a specific day. Data from the US National Household Travel Survey has been used to obtain activity distributions, occupation distributions and dwell time distributions.

---

[1] http://www.dol.gov/dol/topic/.

A complex and computationally demanding vehicular mobility model was proposed by the ETH Laboratory for Software Technology [10], which generates public and private vehicular traffic over real regional roadmaps of Switzerland with a high level of realism within a period of 24 h. The model is calibrated using data from census and other local or national mobility surveys or statistics.

The limitation of the survey-based approaches is that survey or statistical data are only able to provide a coarse grain mobility characterization, modelling global mobility patterns instead of precise movements. Yet, it has the advantage of being able to represent a particular mobility that would be too complex to model by mathematical equations.

11.5 Evolution of Software Tools for Mobility Modelling

The generation of synthetic traces of road traffic is an important requirement in transportation engineering, and a long-studied topic in transportation research. Simulated traces are required in order to understand the weaknesses of transportation systems, and to design and assess potential solutions to the same. However, in many cases, engineering new road infrastructures and devising apt road traffic policies only require a characterization of macroscopic traffic densities and flows. For that objective, simulators focused on traffic assignment, such as MatSim,[2] are sufficient. These tools yield however low detail in the representation of the precise movement of each vehicle, which is instead needed for the evaluation of network solution involving communication-enabled cars. In fact, there exist several fine-grained simulators developed for transportation planning; however, they are typically commercial, including TSIS-CORSIM,[3] Paramics,[4] and VISSIM.[5] Moreover, these tools have a steep learning curve, which represents a major obstacle for many networking researchers who are not willing to spend a significant amount of time on pure road traffic modeling aspects.

The need for high-detail, freely available, and easy-to-use software led at first the networking community to start developing its own vehicular mobility simulation frameworks. Initially, reuse of well-known stochastic models commonly employed in mobile ad hoc network (MANET) scenarios appeared as the easiest choice. However, it was soon clear that the likes of Random Waypoint and Random Direction could not be representative of real-world vehicular mobility.

Thus, early attempts at simulating a more realistic movement of vehicles approximated the road topology with regular grids, and the movement of vehicles with constant-speed random trips constrained over these grids. That is the case, e.g., of the tools introduced by Davies [8].

---

[2]http://www.matsim.org/.

[3]http://mctrans.ce.ufl.edu/featured/tsis/.

[4]http://www.paramics-online.com/.

[5]http://www.ptvamerica.com/vissim.html.

Improvements to the representation of the street layout consisted at first in the possibility of manually defining the road topology. In particular, the CanuMobiSim framework by Tian et al. [39] allowed users to draw graphs where vertices mapped to road intersections and edges to streets joining them. In a second moment, real-world road networks started to be considered, mainly thanks to the public availability of databases such as the US Census TIGER in North America or Ertico GDF in Europe. In particular, Saha and Johnson [36] were the first to employ realistic maps for the study of vehicular networks. However, random trips at constant speed—even if constrained to a realistic road layout—still led to questionable conclusions, e.g., that vehicular mobility could be approximated with unconstrained and fully random movements.

Aware of these problems, the networking research community started to include in their vehicular mobility simulation frameworks more credible speed models. Seminal work was carried out by Bai et al. [1], who introduced the IMPORTANT framework. The latter features an original speed model, named the *freeway* model, based on probabilistic acceleration and bounded speed in order to force each driver to avoid contact with the vehicle ahead. In fact, the freeway model and its extensions were still far from being realistic. Fiore and Härri [11] demonstrated how these models could not pass basic validation tests developed by the transportation research community. The same authors also showed that driver's behaviour models introduced in transportation research (discussed in detail in Sect. 11.3) proved to be much more reliable—and that such a different level of realism had a significant impact on the connectivity properties of the vehicular network.

As a consequence, the networking community begun developing frameworks that integrated, e.g., car-following or cellular automata models borrowed from the rich literature in transportation research. Commonly employed models include those by Treiber et al. [41], Krauss et al. [26], Nagel and Schreckenberg [31]. This resulted in public availability of a number of tools specifically dedicated to the simulation of vehicular mobility for network studies, such as FreeSim by Miller and Horowitz [30], GrooveNet by Mangharam et al. [28], MoVes by Bononi et al.[3], and the City Model by Jaap et al. [19].

Such frameworks did not yet allow to account for overtakings, in- and outflows of vehicles through highway ramps, stop signs or traffic lights at road intersection. All these features are mandatory in complex road traffic simulations of both highway and urban environments. Therefore, in order to include them in the generation process, new tools appeared that also implemented lane changing and intersection management models. The latter were again mostly borrowed from the transportation research literature: common examples are those of the models proposed by Krauss [25], Treiber and Helbing [40], Nagel et al. [32]. Among the most popular vehicular mobility simulation tools that also include such models, we can mention STRAW by Choffnes and Bustamante [7], GMSF by Baumann et al. [2], Udel Models by Kim et al. [21], CityMob by Martinez et al. [29], VanetMobiSim by Härri et al. [16], and SUMO by Krajzewicz et al. [24].

More recently, the most advanced road traffic generators have become part of federated tools that allow run-time interoperability between mobility and network simulators. These tools thus allow to perform vehicular networking simulations that are especially flexible and accurate. On the one hand, using two separate and dedicated tools to reproduce the vehicular mobility as well as the network channel and protocol stack guarantees the accuracy of the representation of each aspect. On the other hand, run-time interaction among the two tools allows (1) the road traffic conditions to trigger network protocol (and overlying service) operation, and (2) the messages received by connected vehicles to influence drivers' behaviour. Overall, federated frameworks represent the current state of the art in the simulation of vehicular networks. Examples of such tools include TraNS by Piorkowski et al. [34], iTetris by Krajzewicz et al. [23], and Veins by Sommer et al. [38].

For a comprehensive discussion of most of the tools mentioned above, we refer the reader to the very complete survey by Haerri et al. [15]. The current state-of-the-art federated tools we mentioned at the end of the discussion will instead be presented in detail in Chap. 13.

11.6 A Hands-on Example: Generating an Urban-Scale Road Traffic Dataset

In this section, we present how the tools introduced previously in this chapter can be brought together so as to generate a comprehensive road traffic dataset, specifically designed for networking studies. The specific dataset we present is a contribution to the TAPASCologne initiative[6] of the Institute of Transportation Systems at the German Aerospace Center (ITS-DLR), which aims at reproducing microscopic car traffic in the greater urban area of the city of Cologne, Germany, with the highest level of realism possible. The detailed description of the process is intended to be useful to researchers who are willing to replicate it and produce their own synthetic traces of vehicular mobility.

As a first step, we detail the data sources employed to generate the dataset, which are listed next.

- **Road infrastructure**. The first source is a detailed description of the road infrastructure. This includes not only the street layout, but also information on the road type and capacity, on per-road speed limits, on intersection signalization, and on the presence of specific structures such as ramps, roundabouts or overpasses. For the dataset under consideration, the road infrastructure data of the Cologne urban area is obtained from the OpenStreetMap (OSM) database.[7] The OSM project provides freely exportable maps of cities worldwide, which are contributed and updated by a vast user community. Maps include most of the needed information, as generated and validated by means of satellite imagery

---

[6] http://sourceforge.net/apps/mediawiki/sumo/index.php?title=TAPASCologne.

[7] http://www.openstreetmap.org.

and GPS traces, and it is commonly regarded as the highest-quality road data publicly available today. The OSM data is filtered with the Osmosis tool[8] so as to extract the road topology information for an area of approximately $400\,km^2$ centred in the urban agglomeration of Cologne, and including around 4,500 km of roads. The Java OSM Editor[9] is used to repair the OSM data and make it compatible with the microscopic mobility simulator, as later detailed in this section. Considering open-source data only, OSM represent the de-facto standard choice to infer the road infrastructure needed for the generation of synthetic microscopic road traffic traces.

- **Microscopic vehicular mobility**. The microscopic mobility of vehicles is simulated with the Simulation of Urban Mobility (SUMO) software.[10] SUMO is an open-source, space-continuous, discrete-time traffic simulator developed by the German Aerospace Center (DLR), capable of accurately modelling the behaviour of individual drivers, accounting for car-to-car and car-to-road signalization interactions. More precisely, SUMO can import road maps and information on traffic lights, roundabouts, stop and yield signs from multiple formats, including OSM. The microscopic mobility models implemented by SUMO are Krauss' car-following model [26] and Krajzewicz's lane-changing model [22], that respectively regulate each driver's acceleration and overtaking decisions, by taking into account a number of factors, such as the distance to the leading vehicle, the traveling speed, and the acceleration and deceleration profiles. These models have been long validated by the transportation research community, a fact that, jointly with the high scalability of the simulator, makes of SUMO the most complete and reliable among today's open-source microscopic vehicular mobility generators. The version we employed for the dataset generation is 12.3, but the simulator has further evolved then since. We refer the reader to Chap. 13 for a detailed introduction to SUMO.

- **Travel demand**. The travel demand information on the macroscopic traffic flows across the Cologne urban area is inferred via the Travel and Activity PAtterns Simulation (TAPAS) methodology [45]. This technique generates an origin-destination matrix of the population mobility by exploiting information on (1) the population itself, i.e., home locations and socio-demographic characteristics, (2) the points of interests in the urban area, i.e., places where work and free-time activities take place, and (3) the time use patterns, i.e., habits of the local residents in organizing their daily schedule [17]. Within the context of the TAPASCologne project, TAPAS is applied to real-world data collected in the Cologne region by the German Federal Statistical Office, including 30,700 daily activity reports from more than 7,000 households [9, 35]. The resulting origin-destination matrix faithfully mimics the daily movements of inhabitants of the area over 24 h, for a total of 1.2 million individual trips. Among the data sources needed for to

---

[8] http://wiki.openstreetmap.org/wiki/Osmosis.

[9] http://josm.openstreetmap.de.

[10] http://sumo.sourceforge.net.

complete the generation process, the travel demand is without doubt that most difficult to retrieve. Within that context, the TAPASCologne origin-destination matrix is, up to now, the only realistic traffic demand dataset of a large urban region that has been disclosed.

- **Traffic assignment**. The actual assignment of the vehicular traffic flows described by the TAPASCologne origin-destination matrix over the road topology is performed by means of Gawron's algorithm [14]. This traffic assignment technique computes the fastest route for each vehicle, and then assigns to each road segment a cost reflecting the intensity of traffic over it. By iteratively moving part of the traffic to alternate, less congested paths, and recomputing the road costs, the scheme finally achieves a so-called user equilibrium. Additionally, since the intensity of the traffic demand varies over a day, the traffic assignment model must also be able to adapt to the time-varying traffic conditions. Indeed, Gawron's algorithm satisfies such a requirement, thus attaining a so-called dynamic user equilibrium. Gawron's is one of the most popular traffic assignment techniques developed within the transportation research community, and allows to reach a road capacity utilization close to reality and significantly higher than that obtained with, e.g., a standard weighted Dijkstra algorithm. Moreover, an implementation is embedded in SUMO, which eases its adoption by the research community.

The individual components presented above are combined as depicted in Fig. 11.5 in order to generate the vehicular mobility dataset. First, the information contained in the TAPASCologne origin-destination matrix are used to identify the boundaries of the exact simulation region, extract the associated map from OSM and filter it so as to remove unneeded content that does not concern the road layout. Then the OSM map is converted to a format readable by SUMO, and fed to the microscopic mobility simulator. The TAPASCologne origin-destination matrix is also used as an input to Gawron's algorithm, which, in turn, determines an initial traffic assignment and provides it to SUMO. At this point, a first vehicular mobility simulation can be started with SUMO. Once the first run finished, a feedback on the resulting traffic density over the road topology is sent back to Gawron's algorithm. Based on this new information, a new traffic assignment is computed, and a second SUMO simulation is run. The process is iterated until we obtain a traffic assignment that allows to sustain the whole volume of the traffic demand, and further iterations do not bring advantage in terms of the aggregate travel time of all vehicles.

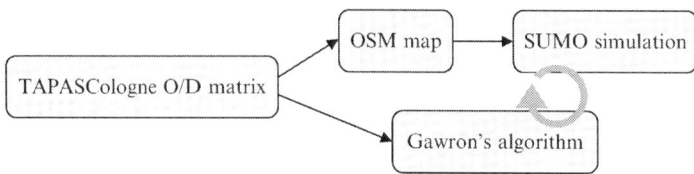

Fig. 11.5 Simulation workflow (figure appeared in Uppoor et al. [44])

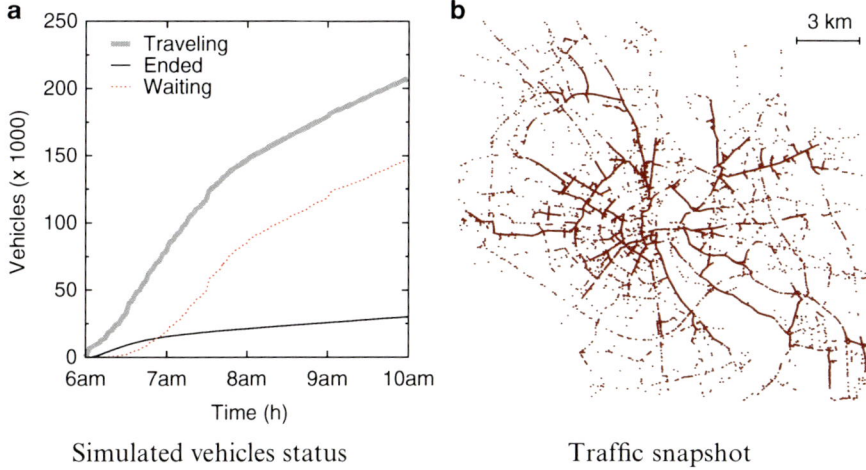

Fig. 11.6 Original TAPASCologne dataset. (**a**) Traffic features over time. (**b**) Snapshot of the traffic status at 7:00 a.m., in a 400 km^2 region centred on the city of Cologne (figure appeared in Uppoor et al. [44])

Unfortunately, making all of the previous components work together is not a straight procedure. Figure 11.6a shows the result obtained by simply running the SUMO simulation with the data sources made available by OSM and TAPAS-Cologne. The plot details the temporal evolution of the number of vehicles that (1) are traveling on the road topology, (2) have successfully ended their trip by reaching their destination, (3) are waiting to enter the road topology due to excessive congestion at entry points. The latter is an undesirable simulation artefact, identifying situations where the road topology cannot accommodate all the travel demand. The number of traveling vehicles present in the simulation rapidly grows up to exceed a hundred thousands units, a figure completely unrealistic for a city the size of Cologne. Additionally, such a number does not tend to decrease as one could expect once the morning traffic peak ends; instead, it keeps growing indefinitely. We also observe that the number of vehicles that end their trip grows very slowly over time, as only a very small fraction of the cars that are present on the road topology can reach their destination. Finally, the number of vehicles that are waiting to enter the road topology, which we would like to stay as close as possible to zero, grows to hundreds of thousands of units.

These results are clear symptoms of how the road topology cannot sustain the volume of cars injected according to the traffic demand model. Indeed, when looking at a snapshot of the car traffic in the region, it is evident how the simulation quickly reduces to a huge traffic jam. As an example, Fig. 11.6b depicts the map of the road traffic at 7:00 a.m., with each dot representing one vehicle. The road topology presents a very high number of dots gathered together along major arteries, representing cars stuck in heavily congested traffic. Next, we discuss the reasons for such a result, and present solutions to them.

Fig. 11.7 Original data sources. (**a**) Volume of traffic injected in the road network between 6:00 a.m. and 12:00 p.m., according to the originalTAPASCologne origin-destination matrix. (**b**) Example of wrong restriction in OSM. (**c**) Example of continuous restriction in OSM (figure appeared in Uppoor et al. [44])

11.6.1 Repairing the Dataset

The poor outcome of the road traffic simulation is the result of the combination of a number of undesirable effects emerging when the different tools discussed before are coupled into a single generation process, as follows.

- **Over-comprehensive and bursty traffic demand**. The original TAPASCologne origin-destination matrix yields the traffic demand volume depicted in Fig. 11.7a, which shows the number of vehicles injected in the whole road network every second. There, we identified and fixed three major problems. First, the demand is not limited to vehicular traffic, but also includes information on the daily trips of all Cologne inhabitants. We thus pruned it, considering that around 50 % of the overall trips are performed by drivers in the region [17, Fig. 4]. Second, the demand presents sudden peaks, unrealistic given that the traffic is aggregated over a very large area. We smoothed down the original matrix, by introducing a random jitter in departure times that allows to remove the injection bursts, yet retaining the demand properties over larger time scales. Third, the demand only includes trips starting or ending within the 400 km^2 simulated region. We employed historic data from the Nordrhein–Westfalen Ministry of Transport[11] to introduce the missing highway traffic in the demand.
- **Inconsistent road information**. Although very complete from a topological viewpoint, the OSM map embeds information at times inconsistent with respect to reality. The impact of such inconsistencies, albeit negligible on most of the usages of OSM, can be dramatic for the simulation of vehicular mobility. A first type of inconsistency is represented by wrong traffic movement restrictions enforced on some road segment, e.g., in Fig. 11.7b: there, a restriction is present

---

[11]http://www.autobahn.nrw.de.

that prevents cars from turning left. At times, wrong restrictions of this kind are present in the OSM data, forcing vehicles to perform long detours or to wait indefinitely for a possibility to turn and continue their journey. Visual inspection against Google Street View allowed us to fix approximately one thousand erroneous restrictions in the area under study. A second type of inconsistency is that of correct movement restrictions being enforced on the whole road extent, whereas they only apply to road portions. Figure 11.7c portrays an example where two one-way roads cross each other. In the real world, the roads pass one over the other, and vehicles traveling on one road can join the other by means of the slanting relief route. In the OSM road representation, the horizontal road is represented by a sequence of segments all featuring an only-straight traffic restriction. This prevents vehicles from considering the overpass as an intersection, but also from taking the relief route. Such situations occur in most of the interchange nodes among high-speed roads. We solved the problem by separating the segments of approximately 800 roads, to which we assigned correct restrictions.

- **Flawed road topology conversion**. The OSM road information is natively imported by SUMO through an automated conversion process that proves not to be error-free. First, attributes with multiple values are considered as incorrect by the converter, as in Fig. 11.8a, and the associated roads are removed from the SUMO topology, as in Fig. 11.8d. Second OSM rendering of complex intersections is unfit for direct conversion to SUMO street layout. As an example, the crossroad in Fig. 11.8b is modeled in OSM as a sequence of multiple junction, each regulated by a different traffic light. Conversion of the latter within SUMO results in an exceedingly intricate intersection, where vehicles get stuck and rapidly form a permanent traffic jam, Fig. 11.8e. Third, the OSM data contains at times traffic lights that are not present in the real world; moreover, the SUMO converter employs by default a technique to deploy additional traffic lights over the street layout. We corrected the OSM data by making all attributes compatible with the SUMO converter, by joining road segment links that refer to the same physical intersection, and by removing unrealistic traffic lights from the OSM data.

- **Simplistic default traffic assignment**. Running the microscopic mobility simulation, with the corrected travel demand and road topology still results in widespread road traffic congestion. The reason lies in the traffic assignment, i.e., the way drivers choose the route to reach their intended destination. By employing Gawron's traffic assignment algorithm, a dynamic user equilibrium is reached after 35 iterations. Figure 11.8c shows the evolution of the number of vehicles traveling at the same time over the road topology, which tends to explode during the first iterations, but is reduced already at the 10th iteration. Figure 11.8f confirms that iterations increase the number of vehicles that can successfully reach their destination.

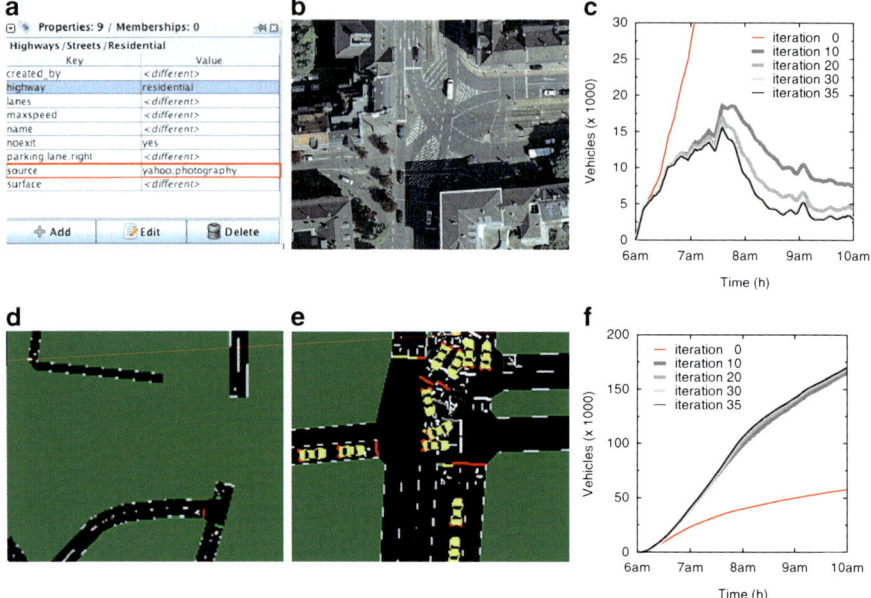

Fig. 11.8 Original data sources. (**a, d**) Example of unrecognized information ignored during map conversion. (**b, e**) Example of topological information unfitness during map conversion. (**c, f**) Traffic evolution over multiple iterations of the assignment algorithm (figure appeared in [44])

11.6.2 Final Dataset

The resulting dataset comprises more than 700,000 car trips in the Cologne larger metropolitan area, over a period of 24 h. The simulated traffic mimics the normal daily road activity in the region, as the fixed road topology can accommodate the updated traffic demand and assignment. Evidences of the correct behaviour of the simulated mobility are given in Fig. 11.9a. By comparing it to the equivalent plot before repair, in Fig. 11.6a, it is clear that the number of traveling cars now follows the traffic demand, with peaks during the morning (from 7:00 a.m. to 9:00 a.m.) and afternoon (from 4:30 p.m. to 6:00 p.m.) rush hours. An approximate maximum of 15,000 vehicles travel at the same time over the road topology, at around 8:00 a.m. Real-world behaviours, such as very low traffic at night and a lower traffic peak at around noon, can also be observed. Also, the number of ended trips now grows over time, as more and more drivers reach their destinations, and the number of vehicles waiting to enter the simulation is reduced to values close to zero. As a result, the road traffic at 7:00 a.m., in Fig. 11.9b, looks significantly better than the original one, in Fig. 11.6b. Indeed, large portions of the urban road layout present a sparse density of points, indicating fluid traffic conditions. The traffic appears only congested in the

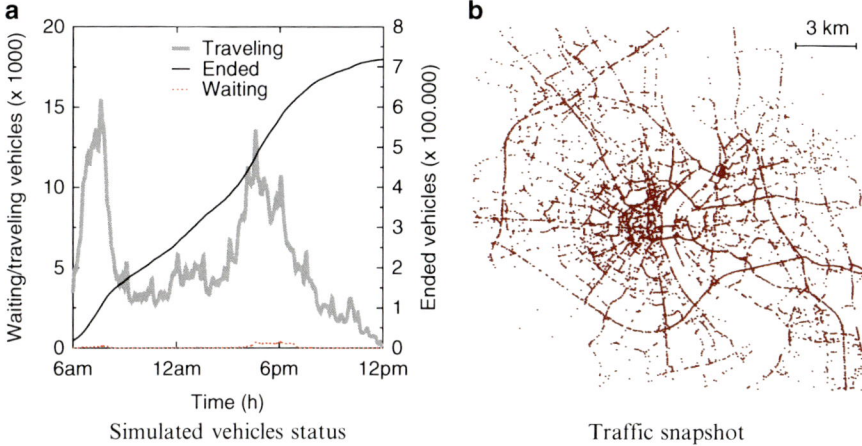

Fig. 11.9 Original TAPASCologne dataset. (**a**) Traffic features over time. (**b**) Snapshot of the traffic status at 7:00 a.m., in a 400 km^2 region centred on the city of Cologne (figure appeared in Uppoor et al. [44])

city centre, where higher concentrations of dots are visible. However, this represents a normal condition in Cologne at that time, and, moreover, the congestion level is much lower than that recorded over vast regions in Fig. 11.6b.

11.7 Future Perspective

The modelling of mobility is still an active area and various aspects must still be considered. The final goal is finding the best approximation of real mobility pattern to achieve that modelling or simulation results of VANETs scenario are as close as possible to reality. With this aim various factors are to be introduced and represented in the future solution.

First of all, the structure of the roads layout and the streets configuration should provide the possibility to represent different categories (rural, highway, etc.), multiple lanes, and different maximum velocities. Also, the road crossing average or density should be considered. For certain types of applications, reaching a higher level of detail, thus including possible obstacles on the road, like bumpers, could result to be very useful.

Moreover, the driving manner of users should be considered, like the way they decelerate or accelerate or brake, since it impacts on the interaction with the environments. Also, regarding the chosen route, different drivers may have different needs, which affect their route selection. Therefore, the mobility model should control the interactions between vehicles, especially in situations like traffic jam or overtaking. The modelling of the behaviour at the intersections should also be improved. The driver behaves differently whether he finds a stop signs, a yield sign, or a traffic light.

The type and characteristics of the vehicle should be modelled. It is not the same to consider a truck on a rural road than on a highway. Moreover, acceleration, deceleration and speed capabilities of a car or a truck are different. Accounting for these characteristics alters the traffic generation engine when modelling realistic vehicular motions.

The so-called Attraction points should be included in a path like time patterns. The final destinations of many road trips are typically shared among various users, likewise the initial locations, called repulsion points (e.g. main entrance avenues). Traffic density strongly depends on the time of the day, the day of the week and the period in the year. Traffic is not the same on a summer Sunday morning than a Monday morning in February. Driver can even change their trip preferences depending of the time and date.

Finally, random and external events should be considered to include the influence of accidents, temporary road works, or real-time knowledge of the traffic status on the motion constraints and the traffic generator blocks.

A vehicular mobility model will be more precise as the number of factors it includes will increase. Parameters defining the different major building blocks such as topological maps, car generation engine, or driver behaviour engine cannot be randomly chosen but must reflect realistic configurations. Therefore, due to the large complexity to obtain such kind of information, the research community took more simplistic assumptions and neglected several factors. Currently most models available include a topological map, or at least a graph, as motion constraints. However, they do not include speed constraints or more generally attraction or repulsion points. The car generation engine block is also widely absent from all models, and the driver behaviour engine is limited to smooth accelerations or decelerations.

11.8 Summary

The increasing popularity and attention in VANETs has prompted researchers to develop accurate and realistic simulation tools. In this chapter, we introduced some of the different available mobility models for VANETs which reproduce the complex vehicular motion patterns, presented a classification of them, and discussed some important concepts related with mobility such as the road topology and the mobility models' validation. As shown, different solutions were proposed, from mathematical to behavioural models. The choice between the different approaches highly depends on the application requirements. For example, if the application is a vehicular safety protocol, the mobility model must represent the real motion at a high level of precision, and thus must be generated by a synthetic model. In contrast, when testing a data dissemination protocol, the gross motion patterns are sufficient and a trace or survey-based model may therefore be envisioned.

We also made a survey of several publicly available mobility generators, network simulators, and VANET simulators. While each of the studied simulators provides a good simulation environment for VANETs, refinements and further contributions are needed before they can be widely used by the research community.

References

1. Bai F, Sadagopan N, Helmy A (2003) The IMPORTANT framework for analyzing the impact of mobility on performance of routing protocols for adhoc networks. Elsevier Ad Hoc Netw1:383–403
2. Baumann R, Legendre F, Sommer P (2008) Generic mobility simulation framework (GMSF). In: ACM mobility models
3. Bononi L, Di Felice M, D'Angelo G, Bracuto M, Donatiello L (2008) MoVES: A framework for parallel and distributed simulation of wireless vehicular ad hoc networks. Comput Netw 52(1):155–179
4. Cabspotting Project (2006) San Francisco exploratorium's invisible dynamics initiative. http://cabspotting.org/index.html
5. Camp T, Boleng J, Davies V (2002) A survey of mobility models for ad hoc network research. Wirel Commun Mobile Comput 2(5):483–502. Special issue on Mobile Ad Hoc Networking: Research, Trends and Applications
6. Cavin D, Sasson Y, Schiper A (2002) On the accuracy of MANET simulators. In: Proceedings of the second ACM international workshop on principles of mobile computing. ACM, New York, pp 38–43
7. Choffnes D, Bustamante F (2005) An integrated mobility and traffic model for vehicular wireless networks. In: ACM VANET
8. Davies V (2000) Evaluating mobility models within an ad hoc network. Master's thesis, Colorado School of Mines, Boulder, Etats-Unis
9. Ehling M, Bihler W (1996) Zeit im Blickfeld. Ergebnisse einer repräsentativen Zeitbudgeterhebung. In: Blanke K, Ehling M, Schwarz N (eds) Schriftenreihe des Bundesministeriums für Familie, Senioren, Frauen und Jugend, vol 121. W. Kohlhammer, Stuttgart, pp 237–274
10. ETH Laboratory for Software Technology (2009) K. Nagel. http://www.lst.inf.ethz.ch/research/ad-hoc/car-traces/
11. Fiore M, Härri J (2008) The networking shape of vehicular mobility. In: ACM MobiHoc, Hong Kong, China
12. Fiore M, Haerri J, Filali F, Bonnet C (2007) Vehicular mobility simulation for VANETS. In: Proceedings of the 40th annual simulation symposium (ANSS 2007), Norfolk, VA
13. Fleetnet Project - Internet on the Road (2000) NEC Laboratories Europe. http://www.neclab.eu/Projects/fleetnet.htm
14. Gawron C (1998) An iterative algorithm to determine the dynamic user equilibrium in a traffic simulation model. Int J Mod Phys C 9(3):393–407
15. Haerri J, Filali F, Bonnet C (2009) Mobility models for vehicular ad hoc networks: a survey and taxonomy. IEEE Commun Surv Tutorials 11(4):19–41. doi:10.1109/SURV.2009.090403. http://dx.doi.org/10.1109/SURV.2009.090403
16. Härri J, Fiore M, Filali F, Bonnet C (2011) Vehicular mobility simulation with VanetMobiSim. Simulation 87(4):275–300. doi:10.1177/0037549709345997. http://dx.doi.org/10.1177/0037549709345997
17. Hertkorn G, Wagner P (2004) The application of microscopic activity based travel demand modelling in large scale simulations. In: World conference on transport research
18. Huang E, Hu W, Crowcroft J, Wassell I (2005) Towards commercial mobile ad hoc network applications: a radio dispatch system. In: Sixth ACM international symposium on mobile ad hoc networking and computing (MobiHoc 2005), Urbana-Champaign, IL

19. Jaap S, Bechler M, Wolf L (2005) Evaluation of routing protocols for vehicular ad hoc networks in city traffic scenarios. In: ITST
20. Jardosh A, Belding-Royer E, Almeroth K, Suri S (2003) Towards realistic mobility models for mobile ad hoc networks. In: ACM/IEEE international conference on mobile computing and networking (MobiCom 2003), San Diego, CA
21. Kim J, Sridhara V, Bohacek S (2009) Realistic mobility simulation of urban mesh networks. Ad Hoc Netw 7(2):411–430
22. Krajzewicz D (2009) Kombination von taktischen und strategischen Einflüssen in einer mikroskopischen Verkehrsflusssimulation. In: Jürgensohn T, Kolrep H (eds) Fahrermodellierung in Wissenschaft und Wirtschaft. VDI-Verlag, Düsseldorf, pp 104–115
23. Krajzewicz D, Blokpoel RJ, Cartolano F, Cataldi P, Gonzalez A, Lazaro O, Leguay J, Lin L, Maneros J, Rondinone M (2010) iTETRIS - a system for the evaluation of cooperative traffic management solutions. In: Advanced microsystems for automotive applications 2010, VDI-Buch. Springer, Berlin, pp 399–410
24. Krajzewicz D, Erdmann J, Behrisch M, Bieker L (2012) Recent development and applications of SUMO—simulation of urban mobility. Int J Adv Syst Measur 5(3/4):128–138
25. Krauss S (1998) Microscopic modeling of traffic flow: investigation of collision free vehicle dynamics. Ph.D. thesis, Universität zu Köln
26. Krauss S, Wagner P, Gawron C (1997) Metastable states in a microscopic model of traffic flow. Phys Rev E 55(304):55–97
27. Legendre F, Borrel V, Dias de Amorim M, Fdida S (2006) Reconsidering microscopic mobility modeling for self-organizing networks. Network IEEE 20(6):4–12. doi:10.1109/MNET.2006.273114
28. Mangharam R, Weller D, Rajkumar R, Mudalige P (2006) GrooveNet: a hybrid simulator for vehicle-to-vehicle networks. In: IEEE Mobiquitous
29. Martinez FJ, Cano JC, Calafate CT, Manzoni P (2008) Citymob: a mobility model pattern generator for VANETs. In: IEEE vehicular networks and applications workshop (Vehi-Mobi, held with ICC), Beijing
30. Miller J, Horowitz E (2007) FreeSim: a free real-time freeway traffic simulator. In: IEEE ITSC
31. Nagel K, Schreckenberg M (1992) A cellular automaton model for freeway traffic. J Phys I 2(12):2221–2229
32. Nagel K, Wolf D, Wagner P, Simon P (1998) Two-lane traffic rules for cellular automata: a systematic approach. Phys Rev E 58:1425–1437
33. NOW - Network on Wheels Project (2008) Hartenstein H, Härri J, Torrent-Moreno M. https://dsn.tm.kit.edu/english/projects_now-project.php
34. Piorkowski M, Raya M, Lugo A, Papadimitratos P, Grossglauser M, Hubaux JP (2008) TraNS: realistic joint traffic and network simulator for VANETs. ACM Mobile Comput Commun Rev 12(1):31–33
35. Rindsfüser G, Ansorge J, Mühlhans H (2002) Aktivitätenvorhaben. In: Beckmann K (ed) SimVV Mobilität verstehen und lenken—zu einer integrierten quantitativen Gesamtsicht und Mikrosimulation von Verkehr, Ministry of School, Science and Research of Nordrhein-Westfalen
36. Saha A, Johnson D (2004) Modeling mobility for vehicular ad hoc networks. In: ACM VANET
37. Seskar I, Maric S, Holtzman J, Wasserman J (1992) Rate of location area updates in cellular systems. In: IEEE 42nd vehicular technology conference, 1992, vol 2, pp 694–697. doi:10.1109/VETEC.1992.245478
38. Sommer C, German R, Dressler F (2011) Bidirectionally coupled network and road traffic simulation for improved ivc analysis. IEEE Trans Mobile Comput 10(1):3–15
39. Tian J, Haehner J, Becker C, Stepanov I, Rothermel K (2002) Graph-based mobility model for mobile ad hoc network simulation. In: SCS ANSS, San Diego
40. Treiber M, Helbing D (2002) Realistische mikrosimulation von strassenverkehr mit einem einfachen modell. In: ASIM, Rostock, Allemagne
41. Treiber M, Hennecke A, Helbing D (2000) Congested traffic states in empirical observations and microscopic simulations. Phys Rev E 62(2):1805–1824

42. UDel Models for Simulation of Urban Mobile Wireless Networks (2009) Stephan Bohacek. http://www.udelmodels.eecis.udel.edu
43. UMass DieselNet Project (2009) UMass diverse outdoor mobile environment (DOME). https://dome.cs.umass.edu/umassdieselnet
44. Uppoor S, Trullols-Cruces O, Fiore M, Barcelo-Ordinas JM (2015) Generation and analysis of a large-scale urban vehicular mobility dataset. IEEE Trans Mobile Comput 1:1. PrePrints. doi:10.1109/TMC.2013.27
45. Varschen C, Wagner P (2006) Mikroskopische Modellierung der Personenverkehrsnachfrage auf Basis von Zeitverwendungstagebuchern. Stadt Region Land 81:63–69
46. Yoon J, Liu M, Noble B (2003) Random waypoint considered harmful. In: Proceedings of IEEE INFOCOMM 2003, San Francisco, CA
47. Zheng Q, Hong X, Liu J (2006) An agenda-based mobility model. In: 39th IEEE annual simulation symposium (ANSS-39-2006), Huntsville, AL

Chapter 12
Channel Models for Vehicular Communications

Mate Boban and Wantanee Viriyasitavat

Abstract Recent empirical studies have shown that correctly modeling the vehicular channel is imperative for realistic evaluation of VANET applications (Gozalvez et al., Telecommun Syst:1–19, 2010; Dhoutaut et al., Impact of radio propagation models in vehicular ad hoc networks simulations. VANET 06: Proceedings of the 3rd international workshop on Vehicular ad hoc networks, 2006). This is particularly the case for safety applications, where the correct reception of a single message can help avoiding an accident. With this in mind, this section focuses on vehicular channel and propagation models. We start by describing the basic propagation mechanisms that enable wireless communication. Next, we elaborate on specific considerations for vehicular channel modeling, including diverse environments where the communication takes place and the objects that impact channel modeling. We then classify the models based on the propagation mechanism scale, modeling approach, and suitability for a particular environment, among others. Using this classification, we overview the current state of the art in vehicular channel and propagation modeling and make a qualitative comparison between the models. Finally, to address the aspects of vehicular channel modeling that are not sufficiently explored, we provide some directions for future work.

Keywords VANET • Propagation • Free-space • Reflection • Diffraction • Fading • Scattering • Line of sight (LOS) • Non-line of sight (NLOS) • LOS obstruction • Ray-tracing • V2V • Multipath • Doppler spread

---

M. Boban (✉)
NEC Laboratories Europe, NEC Europe Ltd., Kurfürsten-Anlage 36, 69115 Heidelberg, Germany
e-mail: mate.boban@neclab.eu

W. Viriyasitavat
Faculty of Information and Communication Technology, Mahidol University, Nakhon Pathom, Thailand

Department of Telematics, Norwegian University of Science and Technology, Trondheim, Norway
e-mail: wantanee.vir@mahidol.ac.th

12.1 Wireless Channel Basics

12.1.1 Wireless Propagation Primitives

As the wireless signal, in the form of an electromagnetic wave, travels or propagates through a medium, several mechanisms take place that affect the intensity and characteristic of the electric field of the transmitted wave. The mechanisms behind such propagation greatly affect the electric field of the electromagnetic wave observed at the receiving antenna. In general, such mechanisms can be attributed to free space propagation, reflection, diffraction, scattering, and penetration through material [2].

12.1.1.1 Free Space Propagation

Free space propagation describes the propagation mechanism of an electromagnetic wave in the scenario where a transmitter and a receiver are separated but have an *unobstructed, line of sight (LOS)* path. Free space loss depicts the decay of the signal as it propagates to the receiver and is usually expressed in terms of separation distance, signal frequency or wavelength, parameters associated with antennas, *but not* factors that are related to propagation environment. For instance, in a free space propagation model, the power received at a receiver antenna is given by the Friis free space equation:

$$P_r = \frac{P_t G_t G_r \lambda^2}{(4\pi)^2 d^2 L}, \qquad (12.1)$$

where P_t is the transmitted power, G_t and G_r are the transmitter antenna gain and receiver antenna gain, respectively, λ is signal wavelength in meters, d is the separation distance in meters, and L is the system loss factor.

12.1.1.2 Reflection

In addition to attenuation caused by the propagation distance, the transmitted wave can also be affected by surrounding objects. Reflection describes a phenomenon that takes place when the radio wave impinges upon a medium that has different electrical properties and has large dimensions compared to the wavelength of the propagating wave. Therefore, the electromagnetic wave may be reflected from the surface of the ground, walls, etc. While part of the wave energy reflects, some of the energy penetrates into the second medium; and the amount of reflected and transmitted energy depend on reflection coefficients, R, which can be computed given material properties of the two mediums (i.e., relative permittivity, ϵ_r and permeability, μ_i), angle of incidence (θ_i), and signal frequency (f) or wavelength.

12 Channel Models for Vehicular Communications

In the propagation scenario where the first medium is free space and both mediums have the same permeability (i.e., $\mu_1 = \mu_2$), the reflection coefficients can be simplified as follows for vertical ($R_\parallel$) and horizontal ($R_\perp$) polarization, respectively [2]:

$$R_\parallel = \frac{-\epsilon_r \sin \theta_i + \sqrt{\epsilon_r - \cos^2 \theta_i}}{\epsilon_r \sin \theta_i + \sqrt{\epsilon_r - \cos^2 \theta_i}} \quad (12.2)$$

and

$$R_\perp = \frac{\sin \theta_i - \sqrt{\epsilon_r - \cos^2 \theta_i}}{\sin \theta_i + \sqrt{\epsilon_r - \cos^2 \theta_i}}. \quad (12.3)$$

Note that in the systems with low antennas and hence, small incidence angle (θ_i), both reflection coefficients, $R_\parallel$ and $R_\perp$ approach 1 regardless of ϵ_r for perfectly smooth surfaces. In other words, the earth (i.e., the ground) can be abstracted as a perfect reflector: as the propagating signal grazes the earth, the reflected wave will be equal in magnitude and between 0° and 180° out of phase with the incident wave, thus resulting in constructive or destructive interference. This phenomenon gives rise to the *two-ray ground reflection* model [2]. In practice, V2V measurements have shown that, while V2V communication in LOS conditions exhibits a behavior that can be modeled by the two-ray ground reflection model, the magnitude of the ground-reflected ray is considerably lower than that predicted by the theoretical model [3, 4].

12.1.1.3 Diffraction

While reflection describes how a wave behaves when it impinges upon an object such as the ground, diffraction describes the phenomena in which the signal propagation path between a transmitter and a receiver is obstructed by objects. In this situation, the wave diffracts and propagates around the surfaces of the objects (i.e., propagates behind the obstacles). Diffraction mechanism can be explained by the *Huygens–Fresnel principle* which states that the propagation of a wave can be visualized by considering every point on a wavefront as a point source for a secondary spherical wave [2]. Electric field magnitude of the diffracted wave is thus the vector sum of the electric field components of these secondary waves and in some cases, it is sufficiently strong to produce a useful signal.

In general, diffraction loss can be calculated based on the difference between the direct path and the diffracted path (i.e., secondary waves). These differences are described by the concept known as *Fresnel zone*. The nth Fresnel zone is defined as the region where path length of secondary waves are $n\lambda/2$ greater than length of the direct LOS path. As a rule of thumb, only diffraction rays caused by obstacles in the first Fresnel zone attribute to the electric field of the wave received at the receiving antenna. Furthermore, if 55 % of the first Fresnel zone is kept clear, further Fresnel zone clearance does not significantly alter the diffraction loss [2].

The simplest model used to estimate signal loss due to diffraction is the *knife-edge* diffraction model [5]. Instead of modeling the diffraction loss over complex terrains, this model assumes that the obstruction can be estimated by treating them as a diffracting knife-edge. As a result, the signal attenuation caused by the diffraction over a knife-edge can be computed as follows:

$$\frac{E_d}{E_0} = F(v) = \frac{(1+j)}{2} \int_v^\infty \exp((-j\pi t^2)/2) dt, \quad (12.4)$$

where E_d is the electric field strength of a diffracted wave, E_0 is the free space field strength, and v is *Fresnel–Kirchoff* diffraction parameter which is given by

$$v = h \sqrt{\frac{2d}{\lambda(d - d_{\text{obs}})d_{\text{obs}}}}, \quad (12.5)$$

where h is the obstructing height of the objects, d_{obs} is the distance between the transmitter and the obstacle, and d is the separation distance between the transmitter and the receiver. While the single knife-edge diffraction model is applicable only to a scenario with a single obstructing object, the extended multiple knife-edge model can model signal attenuation due to multiple obstructions and is usually used in practice [5].

12.1.1.4 Scattering

Scattering is used to describe a phenomena in which the transmitted wave encounters an object that has rough surface or object with dimensions that are small compared to the wavelength of the propagating wave. Surfaces of objects such as foliage, trees, street signs, lampposts, and vegetation can cause the reflected energy to scatter in all directions and provide additional signal energy at the receiver. Roughness of the surface is usually measured relative to a critical height, h_c, defined by the Rayleigh criterion [6]. A surface is considered smooth if its minimum to maximum disturbance is less than h_c and is considered rough otherwise. The critical height is expressed in terms of the signal wavelength, λ and incidence angle, θ_i:

$$h_c = \frac{\lambda}{8 \sin \theta_i}. \quad (12.6)$$

For a rough surface, electric field intensity of the scattered waves can be computed in a similar manner as that of the reflected wave but with a modified reflection coefficient [7, 8]:

$$R_{\text{rough}} = \rho_S R, \quad (12.7)$$

where

$$\rho_S = \exp\left[-8\left(\frac{\pi\sigma_h \sin\theta_i}{\lambda}\right)^2\right] I_0\left[8\left(\frac{\pi\sigma_h \sin\theta_i}{\lambda}\right)^2\right], \quad (12.8)$$

where σ_h is the standard deviation of the surface height and I_0 is the Bessel function of the first kind and zeroth order.

12.1.1.5 Penetration Through Material

In addition to LOS, reflected, diffracted, and scattered rays, electric field intensity at the receiving antenna is also attributed to the waves that penetrate through materials (e.g., walls, buildings, foliage, etc.), noting that free space propagation and penetration through material are mutually exclusive. Models used to describe penetration loss are derived using empirical results, which can vary greatly depending on the type of environment (e.g., indoor or outdoor), wavelength, geometry, and the properties of the penetrated material. For instance, measurements have shown that windows can cause 6 dB attenuation loss on average and the presence of tinted metal in the windows could cause up to 30 dB additional penetration loss [9]. Transmission through trees also cause attenuation depending on the signal frequency and the penetration distance [10].

12.1.2 Wireless Channel Modeling

While it is impossible to precisely estimate signal attenuation caused by all of the aforementioned primitives (free space propagation, reflection, diffraction, scattering, and signal penetration loss), a number of models have been introduced that can reasonably predict the received signal strength. These models can be classified into three main types depending on the cause of signal attenuation: (1) a model to estimate *average* signal loss due to propagation distance, (2) a model to estimate *large-scale* variation caused by propagation environment, and (3) a model to estimate *small-scale* rapid fluctuation over a short period of time or distance. These models are often in the form of theoretical approximations that are adjusted and extensively validated by empirical measurements.

12.1.2.1 Path Loss

Path loss (PL) is a measure of the average RF attenuation as the electromagnetic signal travels from the transmitter to the receiver and is usually expressed in dB scale:

$$\mathrm{PL(dB)} = 10\log\frac{P_t}{P_r}, \quad (12.9)$$

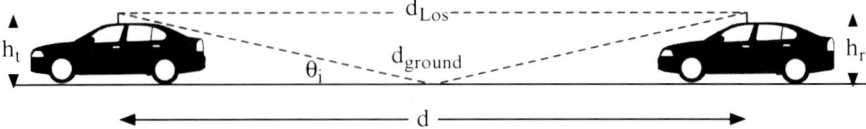

Fig. 12.1 Two-ray ground reflection model

where P_t and P_r are the transmitted and received signal power, respectively. While the Friis free space given in Eq. (12.1) provides a path loss estimate when the signal propagates in free space, measurements and theoretical studies have shown that in mobile radio channels, the average received power does not always follow the Friis space formula. Instead, the received power decreases *logarithmically* with separation distance. In other words, in the *log-distance* model, the average path loss for a given separation distance d, $\overline{PL}(d)$ in dB, can be expressed as:

$$\overline{PL}(d) = \overline{PL}(d_0) + 10\gamma \log(d/d_0), \tag{12.10}$$

where $\overline{PL}(d_0)$ is the average path loss (in dB) at a reference distance d_0, and γ is the path loss exponent, which denotes the power-law relationship between the separation distance and the received power. The value of γ is most often obtained from measured data and is usually in the range of 2–6, depending on the propagation environment.

Another widely used model in predicting path loss in mobile radio channels is the *two-ray ground reflection* model [2]. In contrast to a single direct path assumed in the log-distance model, in the two-ray ground reflection model, the signal received at the receiving antenna consists of *two rays*: the direct LOS ray and the ground-reflected ray (see Fig. 12.1). E-field (in volts per meter) for the two-ray ground reflection model is given by the following equation [2, Chap. 3]:

$$E_{\text{TwoRay}} = \frac{E_0 d_0}{d_{\text{LOS}}} \cos\left(\omega_c \left(t - \frac{d_{\text{LOS}}}{c}\right)\right) + R_{\text{ground}} \frac{E_0 d_0}{d_{\text{ground}}} \cos\left(\omega_c \left(t - \frac{d_{\text{ground}}}{c}\right)\right), \tag{12.11}$$

where $\frac{E_0 d_0}{d_{\text{LOS}}}$ is the envelope E-field at a reference distance d_0, ω_c is the angular frequency ($\omega_c = 2\pi f$, where f is frequency), t is the time at which the E-field is evaluated, d_x represents distance traversed by ray x, and R_{ground} is the reflection coefficient of the ground-reflected ray. When the originating medium is free space, R_{ground} is calculated for vertical and horizontal polarization using Eqs. (12.2) and (12.3), respectively. The resulting received power Pr is equal to

$$\text{Pr(dB)} = 20 \log(E_{\text{TwoRay}}) + \text{Gr}_{\text{dB}} + 20 \log\left(\frac{c}{\sqrt{480\pi f}}\right), \tag{12.12}$$

where Gr_{dB} is antenna gain at the receiver and c is the speed of light. Note that the two-ray ground reflection model takes into account signal loss due to both signal propagation and reflection. Equation (12.12) assumes that the signal is reflected from the ground which is in the first Fresnel zone (as defined in Sect. 12.1.1.2).

12.1.2.2 Shadowing

The path loss model is one of the main ingredients of a propagation model. However, path loss model alone is not sufficient for predicting the received signal strength, since it does not take into account the rapid change in propagation conditions inherent in mobile systems and the resulting shadowing, diffraction, and scattering created by the environment.

Measurements have indicated that the average received power at the receiver antennas can be significantly different when measured at different locations despite having the same separation distance. This phenomenon is referred to as the *shadowing* effect.

The model that is commonly used to predict signal attenuation caused by the shadowing effect stochastically is the *log-normal* shadowing model. This model is based on empirical measurements which indicate that the path loss at a given location is random and distributed log-normally [11, 12]. The total path loss of Eq. (12.10) can be re-written as:

$$\overline{\text{PL}}(d) = \overline{\text{PL}}(d_0) + 10\gamma \log(d/d_0) + X_\sigma, \qquad (12.13)$$

where X_σ is a zero-mean Gaussian distributed random variable (in dB) with standard deviation, σ (in dB). Similar to path loss exponent γ, the value of σ is usually obtained from measured data.

In vehicular networks, where both transmitter and receiver can be mobile, shadowing is more severe and dynamic compared to virtually any other network. For this reason, efforts have been made to calculate shadowing in a deterministic manner, using the information about the objects in the vicinity of the transmitter and receiver (e.g., [13, 39]).

12.1.2.3 Small-Scale Fading

In mobile radio channels, the signal received at the receiver antenna usually consists of multiple waves which are copies of the same transmitted wave but arrive at the receiver at different times and may have different amplitudes and phases. These multipath waves create the small-scale *fading* effects which cause the rapid fluctuation of the received signal over a short period of time or distance.

Fading is most pronounced in vehicular networks when there is no LOS path between the transmitter and the receiver. However, even in a scenario where a direct LOS path exists, the multipath phenomenon could also occur due to reflections from the ground and/or buildings. As a result, the severity of fading varies depending on the existence of a LOS path, structure of the surrounding environment, speed of mobile stations and surrounding objects, etc.

In general, the small-scale fading effect in mobile radio channels is described by either the Rayleigh or the Ricean distributions. The Rayleigh distribution is commonly used to describe the channel when there is no dominant LOS signal components and the random multipath waves may arrive at any angle. Probability distribution function (pdf) of the Rayleigh distribution is given by

$$p(r) = \begin{cases} \frac{r}{\sigma^2} \exp\left(-\frac{r^2}{2\sigma^2}\right), & \text{if } 0 \leq r \leq \infty, \\ 0, & \text{otherwise}, \end{cases} \quad (12.14)$$

where σ is the root-mean-square value of the received voltage and σ^2 is the time-average power of the received signal.

On the other hand, when a strong LOS path is present, the Ricean distributed signal envelope is used instead and the Ricean distribution:

$$p(r) = \begin{cases} \frac{r}{\sigma^2} e^{-\frac{r^2+A^2}{2\sigma^2}} I_0\left(\frac{Ar}{\sigma^2}\right), & \text{if } 0 \leq r \leq \infty, \\ 0, & \text{otherwise}, \end{cases} \quad (12.15)$$

where A is the peak amplitude of the dominant signal and I_0 is the modified Bessel function of the first kind and zeroth order.

Note that the measurements have shown that, on average, the received amplitude distribution gradually transitions from near-Ricean to Rayleigh as the separation distance increases and dominant path begins to fade away [14].

12.1.3 *Propagation and Channel Modeling for Mobile Cellular Systems*

A number of outdoor propagation models that take into account all the above factors have been introduced. Here, outdoor propagation models that are widely used in practice are presented. For example, the Longley–Rice model is a commonly used propagation model in the frequency range from 40 MHz to 100 GHz [15, 16]. In the Longley–Rice model, the two-ray ground reflection model is used to predict signal attenuation within the radio horizon and the knife-edge models are used to further account for diffraction loss caused by obstacles.

Okumura and Hata models are popular models for estimating signal attenuation in cellular mobile systems in city areas [17, 18]. Both of these models are based primarily on the classical free space path loss. In addition, correction factors are added to account for different terrains, antenna height, etc. Although Okumura and

Hata models provide very good signal loss predictions, both models are well suited only for the transmission between a stationary and elevated base station and a mobile station in cellular mobile systems.

Modeling channels in case of existence of LOS path (i.e., when the optical and electromagnetic path between the transmitter and receiver is unobstructed) is arguably a less difficult task than modeling non-LOS channels. A large number of studies tackled outdoor propagation modeling for mobile communication (for an extensive survey, see Sarkar et al. [19]). In terms of deterministic propagation modeling of non-LOS channels, research efforts often rely on Uniform Geometrical Theory of Diffraction [20]. One example is work by Anderson [21], where the author analyzed path loss induced by around-the-corner communication.

Erceg et al. [22] proposed a deterministic model for non-LOS communication in urban areas and validated it against measurements, whereas Durgin et al. [23] performed measurements and developed path loss models for non-LOS communication caused by residential buildings.

However, these models might not be best suitable for V2V communication, where both transmitter and receiver can be mobile and of similar, low height. To that end, the early work on mobile-to-mobile channels by Akki and Haber [24] is more relevant for modeling V2V communication.

12.2 Specific Considerations for Vehicular Channel Modeling

While a number of existing mobile channel models have been extensively used for cellular systems, they are not well suited for the vehicular systems, due to unique features of vehicular channels. For instance, difference in the relative height of transmitter and receiver antennas could lead to significant difference in the signal propagation behavior. Operating frequency and communication distance assumed in vehicular communications also differ from those of the cellular systems; i.e., vehicular communications systems operate mostly at 5.9 GHz and over short distance (100–500 m) whereas cellular systems operate at 700–2,100 MHz over a long distance (up to tens of kilometers) [25].

Because of the aforementioned differences, in this section we elaborate on specific issues that need to be considered for vehicular channel modeling. As depicted in Fig. 12.2, vehicular communication has distinct characteristics, such as varying surroundings that can include obstructing objects, thus creating a rich propagation environment, low height of both transmitting and receiving antenna, and potential mobility of the transmitter, receiver, and the surrounding objects. These characteristics result in highly variable quality of communication links. Figure 12.3 sheds light on the complexity of vehicular environment; even for single bounce (e.g., first-order) reflections and diffractions, the number of resulting rays is large. Calculating multiple-order rays results in nearly exponential increase of computational complexity. This example shows that capturing the complexities of vehicular channels is far from trivial.

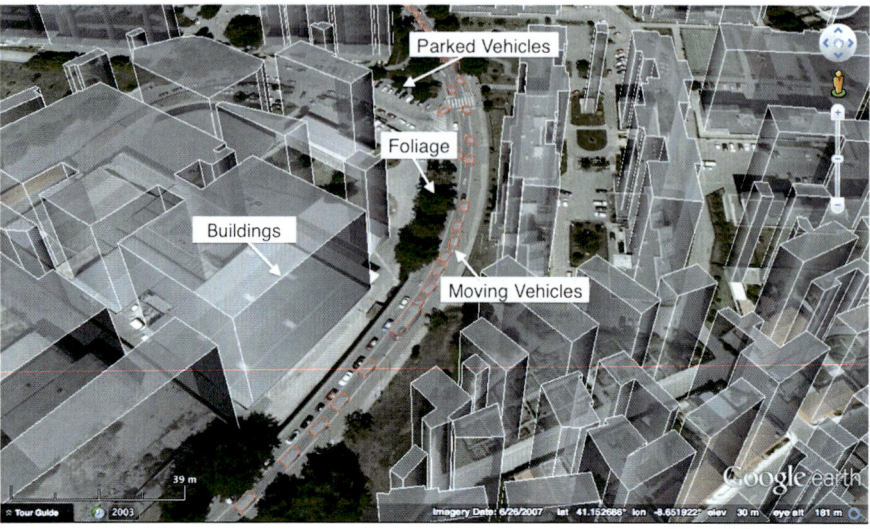

Fig. 12.2 Typical vehicle-to-vehicle communication environment. Building and vehicle overlays are generated using GEMV^2 [26]

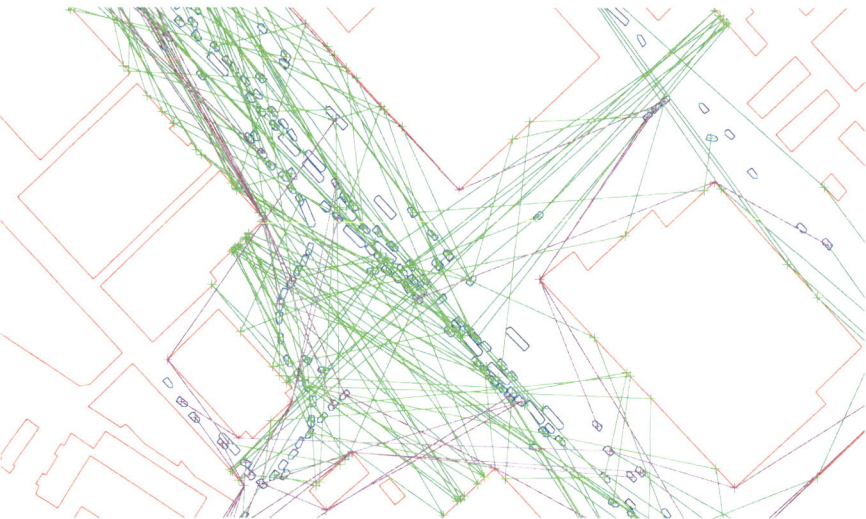

Fig. 12.3 Single bounce reflections (*green*) and diffractions (*magenta*) off buildings (*red*) and vehicles (*blue*) for a set of randomly selected transmit-receive vehicles. Results are generated using GEMV^2 [26]

12.2.1 Environments

Theoretical and measurement-based studies have indicated that environment has a tremendous impact on the characteristics of the mobile radio channels. For vehicular channel models, the propagation environments generally considered in the literature are rural areas, urban canyons, and highways. These environments are characterized by varying presence, locations, and density of roadside objects such as buildings, trees, parked cars, etc., as well as the velocity and density of traffic on the road. Considerably different propagation characteristics of these environments require that the channel models are designed for each of them separately.

For instance, V2V measurement campaigns have shown that the path loss exponents [i.e., γ in Eq. (12.10)] differ across various environment; measured path loss exponents were $2.3 \leq \gamma \leq 2.75$ in a suburban environment [27], $2.44 \leq \gamma \leq 3.39$ in an urban environment [28], and 3 in a parking garage [29].

12.2.2 Objects

As depicted in Fig. 12.2, the vehicular propagation environment may consist of a number of objects of different types and characteristics. We categorize these objects into two groups: (1) static objects such as buildings, trees, road signs, parked vehicles, etc.; and (2) mobile objects such as vehicles on the street. While both types of objects generally cause signal attenuation, the level of impact varies depending on the environment. For instance, mobile objects (i.e., vehicles on the roads) are more important objects to consider for modeling vehicular channels in highway environment, because communication between transmitting and receiving vehicles on highways usually happens over the road surface. On the other hand, in urban environment with two-dimensional topology, the communicating vehicles are likely to be on different streets. In this case, along with mobile objects, accounting for static objects is critical for modeling vehicular channels, since both types of objects are sources of shadowing, reflections, and diffractions [30].

12.2.3 Link Types

In addition to the nature of propagation environment, it is also important to distinguish between vehicle-to-vehicle (V2V) and vehicle-to-infrastructure (V2I) channels as they exhibit vastly different propagation properties. In V2V channels, the transmitter and receiver antennas are usually mounted on the vehicle rooftop, whereas in the V2I channels, the base station (or access point) may be elevated.

The difference in relative height of the transmitter and receiver antennas poses significant difference in reflection, diffraction, and scattering patterns of the transmitted waves [31].

For example, in the V2I channels with elevated base stations, a transmitter and a receiver usually have a dominant LOS path which might not exist in the V2V channel, especially in a crowded urban area. In addition, since the elevated base station is usually not surrounded by scatterers, scattering effects leading to small-scale fading is reduced. For these reasons, the propagation channel in the elevated V2I system may be estimated using the existing cellular propagation models. Nevertheless, in some cases where the base stations are installed at the street level, the V2I channel experiences unique behavior [32].

A number of measurement campaigns have also indicated that the LOS condition is a key factor in modeling the V2V propagation channels. Measurements performed by Tan et al. [33] show that, regardless of propagation environment (e.g., highway or urban scenarios), non-LOS channel has noticeably larger delay spread than that of the LOS channels. This is due to stronger signal attenuation and multipath effects caused by an increasing number of reflections and scatterers. A detailed investigation on different link types and how they affect the vehicular channel modeling is given in the next sections.

12.3 Classification of Models

In an attempt to classify the models according to their most important characteristics, in this section we introduce and describe different "dimensions" we use to classify the models. While models may not necessarily fit into these categories perfectly, the categorization helps in understanding which model can be used for a particular purpose. Figure 12.4 shows different dimensions we use to classify the models.

12.3.1 Propagation Mechanism Scale

As described in Sect. 12.1.2, propagation models are typically divided by their scale into:

- *Path loss*, which is defined as distance-dependent signal attenuation;
- *Large-scale fading*, which includes signal variations due to shadowing by objects significantly larger than the carrier wavelength;
- *Small-scale fading*, which includes variations due to multipath and/or Doppler spread.

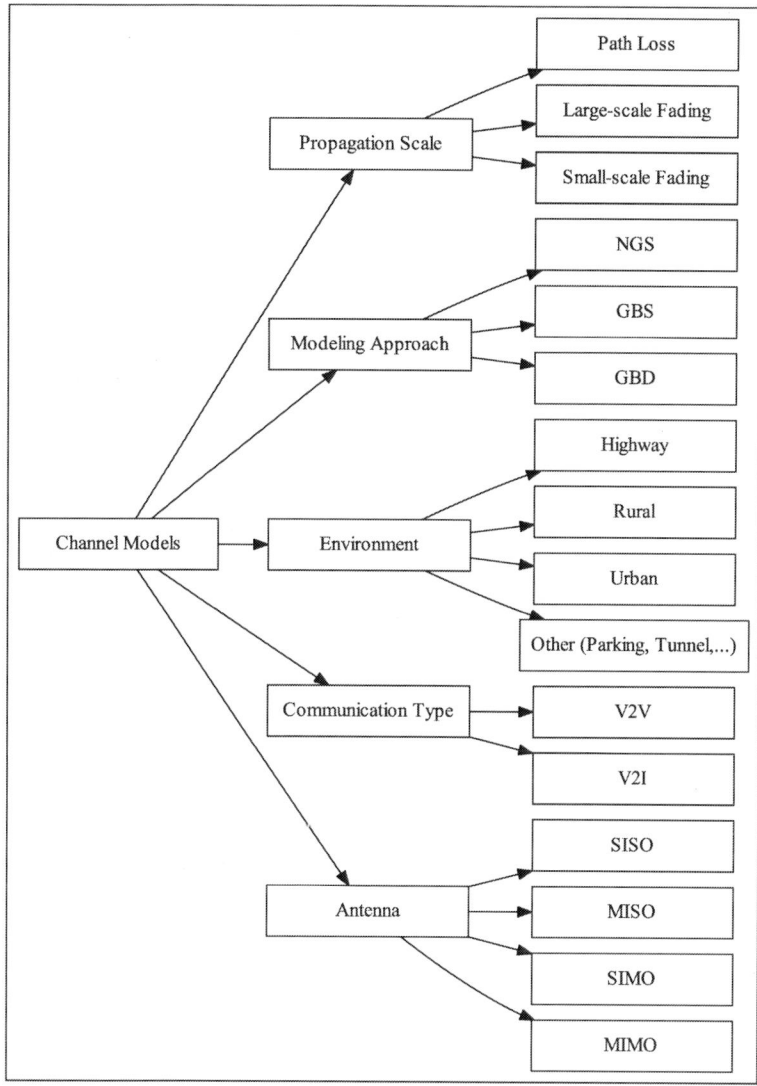

Fig. 12.4 Channel model classification

12.3.2 Modeling Approach

Depending on the use of geographical descriptors of the simulated area and on the approach to modeling the signal variations, the models can be divided into [14]:

- Geometry-based deterministic (GBD) models incorporate relevant objects in the simulated area and calculate the channel statistics in a completely deterministic manner. Ray-tracing method is an example of GBD models that requires a detailed description of propagation environment in order to reproduce the actual physical propagation process for a given environment [14]. By considering propagation factors caused by all possible paths, ray-tracing models can estimate the actual receive power within 3 dB standard deviation [39]. However, ray-tracing models are computationally expensive and not suitable for large-scale simulations. GBD models that are more scalable than ray-tracing have been proposed recently (for details, see Sect. 12.4).

- Geometry-based stochastic (GBS) models take into account the geometrical properties of the surroundings, but calculate the channel statistics according to the statistics extracted either from measurements or obtained through simulations. The simple GBS model used in V2V channel model is the *two-ring* model which assumes that scatterers are randomly placed in a *two-ring* structure, with one ring around the transmitter and one around the receiver [34]. A simplified ray-tracing method is then applied to this topology. In order to account for more realistic locations of scatterers in vehicular environment, another GBS model assumes that most of the scatterers such as buildings, trees, and houses are positioned along the sides of the road or the transmitter and receiver path. Furthermore, the model also takes into account different properties of these scattering objects [35].

- Non-geometrical stochastic (NGS) models generate channel statistics in a completely stochastic fashion, where both the geometrical properties and the channel statistics are generated stochastically. Examples of NGS models used for V2V channel modeling include a tapped-delay-line model. Each tap in this model represents signal received from several propagation paths; each with different delay and different type of Doppler spectrum [36].

12.3.3 Antenna and Small-Scale Fading Characteristics

Small-scale fading occurs due to multipath time delay spread and Doppler spread. If a channel model has the ability to model the statistics related to multipath time delay spread, including both flat and frequency-selective fading, we consider it as being able to model multipath delay spread. Similarly, if it is able to model the effects of Doppler spread, including both slow and fast fading, we consider it as being able to model Doppler spread.

Related to channel model's ability to incorporate small-scale fading is the ability to support different types of antenna configurations that exploit the positive and counter the negative effects of small-scale fading. Therefore, we include the information about the model's ability to support different antenna configurations (e.g., SISO, SIMO, MISO, MIMO).

12 Channel Models for Vehicular Communications

12.3.4 Communication Type

Communication links considered in vehicular channels are usually classified into two groups: links between vehicles and links between a vehicle and a stationary (and potentially elevated) base station (i.e., infrastructure). Both of these models further distinguish communication links according to three conditions: line of sight (LOS), Non-LOS due to static objects (NLOSb), and Non-LOS due to vehicles (NLOSv).

The three LOS conditions lead to very different physical propagation behaviors. For a LOS link, power at the receiver is usually dominated by the direct LOS ray and the ground-reflected wave. In the absence of LOS, however, the most important rays in the NLOSb links are rays which are diffracted or reflected from *stationary* objects such as building, roadsigns, and streetlights. These diffraction and reflection phenomena are always present since the buildings are significantly taller than vehicles and can reflect the signal for any communicating pairs. While the waves can be reflected from vehicles, the reflected rays from stationary objects are often the dominating mechanism [37]. On the other hand, for a NLOSv link where the communication is blocked by vehicles only, reflection rays caused by tall vehicles should also be considered.

In addition to the large-scale variation, small-scale fading effects for different LOS conditions also vary depending on the richness of reflection environment (e.g., reflections in case of LOS, the number of obstructing vehicles in case of NLOSv, and deep or slight building obstruction in case of NLOSb [26]).

12.3.5 Environment

Vehicular channel models are mainly classified into the following categories based on the roadside environments and traffic characteristics:

- *Open space and highway* environment is characterized by one-directional, high-speed motion of vehicles. Roadside environment contains mostly vegetation with a few houses and street signs which are usually located far from the road.
- *Suburban* environment is a mixture of low-rise buildings and open spaces such as park areas and parking lots. These roadside objects are usually set further back from the curb as compared to the urban environment. Low to medium vehicle density and few pedestrians and bicyclists are assumed in this scenario.
- *Urban canyon* describes a scenario with high traffic densities and a higher density of pedestrians and cyclists. In contrast to open space and suburban street environments, urban vehicular channel models assume two-dimensional streets and consider the possibility that the communicating vehicles may be on different streets. In urban environments, objects such as buildings, houses, and street signs are densely scattered along the side of the road and tend to be located very close to the streets.

With regard to the applicability of a model to different environments, we distinguish between channel models that were calibrated by extracting the pertinent parameters from measurements at a specific set of locations and those that have the ability to model effects beyond those captured at particular locations. Since the former category depends on the measurements, it can be used to model the channel in locations similar to those where measurements were performed. However, these models can give no accuracy guarantees for locations with considerably different characteristics. On the other hand, models that take into account geometry-specific information of the simulated area can give some insights for environments beyond those characterized by the measurements. For this reason, we add another classification category named "Fitted to measurements," to describe the ability of the model to generalize to environments beyond those (similar to) the measured environments used to fit the model.

12.4 State-of-the-Art Vehicular Channel Models

Key distinguishing aspects of vehicular channels are varying path loss across space (e.g., different environments) and time (e.g., different time of day), potentially high Doppler shift, non-stationarity, and shadowing by both mobile objects (surrounding vehicles) and static objects (e.g., buildings, foliage) [26, 38–40]. Because modeling all of these aspects is a complex task, the most common approach thus far has been piecemeal modeling, wherein the problem is split into manageable parts and modeling is done on one or some of those parts. Therefore, certain models suit certain applications better than the other. For example, if the goal of a study is to evaluate system-wide performance of an ITS application involving hundreds or more vehicles, it is infeasible to use a detailed modeling approach such as the one described by Mittag et al. [41], where fine-grained statistics are calculated for each transmitted message (e.g., OFDM modulation and modulation, interleaving, convolutional decoding, etc.). On the other hand, such approach is well suited for precise modeling of intra-packet statistics in a small network over a short period of time.

Based on the classification described in Sect. 12.3, in this section we describe relevant channel models for vehicular communication. Specifically, in line with vehicular channel model surveys by Molisch et al. [25], Wang et al. [14], and Mecklenbräuker et al. [42], we group the models based on their modeling approach into NGS models, GBS models, and GBD models. Furthermore, we include both propagation models (which are concerned with physical characterization of radio wave propagation) and channel models (which also include the transmitting and receiving antenna characteristics, including any diversity methods). Table 12.1 summarizes the state-of-the-art vehicular propagation and channel models that are further described in the subsequent sections.

Table 12.1 Classification of channel models

Model	Propagation mechanism scale	Modeling approach	Ability to classify link types	Differentiation btw LOS/NLOS models	Antenna	Environment	Fitted to measurement	Per-link comput. complexity
Cheng et al. [27]	Large- and small-scale	NGS	No	No	SISO	Suburban	Yes	$O(1)$
Acosta and Ingram [43]	Small-scale	NGS	No	No	SISO	Urban, rural, highway	Yes	$O(1)$
Bernadó et al. [44]	Small-scale	NGS	No	No	MIMO	Urban, highway, suburban, tunnel, bridge	Yes	$O(1)$
Sen and Matolak [45]	Small-scale	NGS	No	No	SISO	Urban, highway	Yes	$O(1)$
Otto et al. [46]	Path loss and large-scale	NGS	No	Yes	N/A	Urban, suburban, open road	Yes	$O(1)$
Karedal et al. [30]	Path loss	NGS	No	No	N/A	Rural, highway, urban	Yes	$O(1)$
Matolak et al. [47]	Small-scale	NGS	No	No	SISO	Urban, highway	Yes	$O(1)$
Karedal et al. [35]	Path loss and small-scale	GBS	Yes (Stoch.)	Yes	MIMO	Rural, highway	Yes	$O(R+V)$
Cheng et al. [48]	Large- and small-scale	GBS	Yes (Stoch.)	Yes	MIMO	Rural, highway	N/A	$O((R+V)^2)$
Abbas et al. [49]	Path loss, large- and small-scale	GBS	Yes (Stoch.)	Yes	N/A	Highway	No	$O(1)$
Wang et al. [50]	Small-scale	GBS	Yes (Stoch.)	Yes	N/A	All	No	$O(1)$
Maurer et al. [39]	Path loss, large- and small-scale	GBD	Yes (Det.)	Yes	All	All	No	At least $O((R+V)^2)$

(continued)

Table 12.1 (continued)

Model	Propagation mechanism scale	Modeling approach	Ability to classify link types	Differentiation btw LOS/NLOS models	Antenna	Environment	Fitted to measurement	Per-link comput. complexity
Biddlestone et al. [51]	Path loss, large- and small-scale	GBD	Yes (Det.)	Yes	All	Urban	No	At least $O((R+V)^2)$
Mangel et al. [52]	Path loss and large-scale	GBD	No	Yes	N/A	Urban intersections	Yes	$O(1)$
Boban et al. [13]	Path loss and large-scale	GBD	Yes (Det.)	Yes	N/A	Highway	No	$O(V)$
Giordano et al. [53]	Path loss and large-scale	GBD	Yes (Det.)	Yes	N/A	Urban grid	No	$O(R)$
Cozzetti et al. [54]	Path loss	GBD	Yes (Det.)	Yes	N/A	Urban grid	No	$O(1)$
Pilosu et al. [55]	Path loss	GBD	No	No	N/A	Urban	N/A	$O(1)$
Boban et al. [26]	Path loss, large- and small-scale	GBD	Yes (Det.)	Yes	N/A	All	No	$O(R+V)$

R and V denote the number of roadside objects and vehicles, respectively. "Ability to classify link types" indicates whether or not the model includes a mechanism that can detect the LOS property of a link while "Differentiation between LOS/NLOS models" indicates whether or not the channel models use different modeling mechanisms for LOS and NLOS links

12.4.1 Non-geometrical Stochastic Models

Most NGS models conform to the following recipe: measuring the channel characteristics in a specific environment and adjusting the parameters of well-known path loss, shadowing, and small-scale fading models (e.g., log-distance path loss [56], two-ray ground reflection, Rayleigh/Rice/Nakagami fading [2], etc.). The studies below are not an exception: their computational complexity is usually low (e.g., $O(1)$ per link.)

Cheng et al. [27] performed narrowband measurements of the V2V channel in the 5.9 GHz frequency band in a suburban environment. In the study, the measurement data was fitted to a dual slope piecewise log-distance path loss model; different fading statistics were also extracted from measurements. Acosta-Marum and Ingram [43] developed small-scale channel models that capture delay and Doppler characteristics of V2V communication. The models are based on extensive measurements for urban, suburban, and highway environments in the 5.9 GHz frequency band. The authors also develop packet error rate models for each of the channels. Bernadó et al. [44] performed extensive measurements in urban, highway, suburban environments, as well as measurements in a tunnel and on a bridge. Delay and Doppler statistics were extracted from measurements for each of the environments.

An extensive measurement campaign was performed by Sen and Matolak [45] in urban, suburban, and highway environments with two levels of traffic density (high and low). Based on the measurements, the authors proposed several V2V channel models that apply to a specific environment and vehicle traffic density. The study also points out the effect that the antenna location has on the channel characteristics. This leads to several antenna diversity techniques proposed to improve the packet reception in vehicular environment [57, 58]. Karedal et al. in [30] estimated path loss by performing measurements in rural, highway, suburban, and urban environments. Two-ray ground reflection model [2] was found to be the best fit for path loss in rural environment with low traffic density; in other environments, higher traffic density often created non-LOS conditions, thus the results did not conform to the two-ray model. Similar study was performed for V2V communication in the 2.4 GHz frequency band by Otto et al. in [46], where the authors perform measurements in urban, suburban, and open road environments at different times of day. Based on the measurements, the authors extract the path loss exponent and shadowing deviation for each environment. Due to the varying density of vehicles in different times of day, both path loss exponent and shadowing deviation were higher in case of measurements during daytime, when there were more surrounding vehicles. Different from other NGS models, the work in [46] makes a distinction between a LOS and non-LOS links. Based on the link condition, different channel parameters (i.e., median path loss exponent and shadowing standard deviation) are then estimated for each environment. It is important to note that the model is unable to detect the actual LOS condition of the link; rather, it assumes that the link type is known a priori. Lack of the geometry consideration and thus the inability to classify/detect the LOS condition of a link is one distinguishing factor between the NGS and geometry-based models, as discussed in the next subsections and shown in Table 12.1.

12.4.2 Geometry-Based Stochastic Models

Karedal et al. [35] designed a model for the V2V channel based on extensive measurements performed in highway and suburban environment in the 5.2 GHz frequency band. The model distributes the mobile scatterers (vehicles) and static scatterers at random locations and analyzes four distinct signal components: LOS, discrete components from mobile objects, discrete components from static objects, and diffuse scattering. While path loss, multipath, and Doppler spread are modeled, the existence of LOS component is assumed; therefore, the model does not distinguish between LOS and non-LOS conditions.

With regard to characterizing small-scale fading using actual scatterer locations, Wang et al. [50] employ aerial photography to determine the location and the density of scatterers in the environment. The scatterer density serves as the indicator of small-scale signal variation. The authors point out that aerial photography can be used to model static scatterers, whereas mobile scatterers need to be incorporated using a complementary technique.

Cheng et al. [48] proposed a MIMO V2V channel model that takes into account the LOS, single-bounced rays, and double-bounced rays by abstracting the scatterer positions using a combined two-ring and ellipse model. The model can be used in different V2V environments, provided that the appropriate parameters for a given environment are available.

Abbas et al. [49] designed a model that incorporates shadow fading for V2V communication. The distinction between LOS and non-LOS conditions is modeled using a probabilistic model based on Markov chains; transition probability between conditions is extracted from the probability distributions of the LOS and non-LOS conditions measured in different environments. The model demonstrates the importance of differentiating a LOS link from a non-LOS link as well as energy contributed from LOS and non-LOS rays.

12.4.3 Geometry-Based Deterministic Models

One of the first efforts to describe V2V channels in a fully deterministic manner was by Maurer et al. [39]. The authors propose an optical ray tracing model that uses a geographic database of all relevant objects in the simulated area. It calculates the channel statistics by analyzing the 50 strongest propagation paths between the transmitter and receiver. The model showed close agreement with the measurements performed in the same location. However, the model is computationally complex and requires a precise and detailed geographical database. A similar study was performed by Biddlestone et al. [51], where authors propose a propagation model that employs ray-tracing. Outlines of buildings are used to determine LOS conditions and perform reflections and diffractions to estimate the received power in LOS and non-LOS conditions. The model matches well the small-scale measurements performed by the authors. However, the computational complexity of the model remains an issue.

Several models were proposed with the goal of reducing the computational complexity and the need for complex geographical databases, at the same time keeping the beneficial characteristics of deterministic modeling (e.g., modeling effects of shadowing by actual objects). Below we overview these models.

The largest variation in path loss arises due to changing LOS conditions [28, 49, 59]. In particular, the rapid transitions between LOS and non-LOS conditions create considerably different channel statistics in terms of path loss as well as small-scale fading (e.g., delay spread). This has a significant impact on the performance of V2V applications [4, 32]. To that end, the following propagation models attempt at modeling the LOS obstruction in an efficient manner.

Urban intersections are a particularly interesting scenario when it comes to channel modeling. From the application point of view, the vehicles that are in shadowed, non-LOS region are arguably the most interested in receiving a message, particularly in case of safety messages, where receiving the message can prevent an accident. To that end, Mangel et al. [52] developed VirtualSource11p, a model that incorporates the relevant information about street intersections (e.g., street width, existence of buildings on intersection corners, etc.). The model is fitted to the measurements that the authors performed at representative intersections. Cozzetti et al. in [54] extend this model to account for propagation across multiple intersections in an urban grid environment. The model is able to emulate the hidden terminal phenomenon that leads to an unexpected drop in packet receptions at the center of intersections.

In highways, static objects rarely obstruct LOS for V2V communication. However, LOS is often blocked by surrounding vehicles, in particular large commercial vans and trucks. Measurement results showed that the LOS obstruction due to a van could cause up to 20 dB attenuation, whereas a large truck can cause in excess of 30 dB attenuation [59]. Boban et al. [13] developed a model that deterministically calculates the additional attenuation due to obstructing vehicles by abstracting vehicles as diffracting objects using the multiple knife-edge diffraction model [5]. The model was validated against measurements in open space and on highways.

In terms of channel modeling on a city-wide scale, Giordano et al. propose CORNER [53], an efficient propagation model for a grid-like urban environment. CORNER separates the links into three categories, based on the LOS obstruction level caused by buildings near the road intersections. The authors compared CORNER to measurements in terms of packet success ratio and found good agreement. Pilosu et al. [55] propose RADII, a propagation model that incorporates a preprocessing technique for ray-tracing simulations. RADII can simulate propagation for an arbitrary urban geometry at different levels of granularity. RADII was implemented in the NS-2 network simulator [60]. In an attempt to model signal propagation in the complete set of environments where V2V communication can occur (e.g., highway, rural, urban, complex intersections, etc.), Boban et al. developed GEMV^2 [26], a computationally efficient propagation model that uses outlines of vehicles, buildings, and foliage to distinguish the following three types of links: LOS, non-LOS due to vehicles, and non-LOS due to static objects. For each link, GEMV^2 calculates the large-scale signal variations deterministically, whereas the small-scale signal variations are calculated stochastically based on the number and size of surrounding objects. For links whose LOS is obstructed by other vehicles, GEMV^2 implements vehicles-as-obstacles model [13]. GEMV^2 can simulate city-wide vehicular networks with thousands of communicating vehicles. It was validated against extensive measurements performed in urban, suburban, highway, and open space environment.

Note that the propagation models above can serve as a basis for a more finegrained channel modeling, where small-scale effects are incorporated. This can be achieved by assigning small-scale channel statistics (e.g., delay and Doppler spreads) to each link based on the detected link properties (e.g., LOS obstruction, environment, etc.). The per-link-type statistics can be obtained from measurements (e.g., [61–63]).

12.4.4 Which Model to Use and When?

Models listed in Table 12.1 differ in many aspects: from stochastic models that do not include any information about the specific propagation environment under investigation, to environment-specific models with parameters extracted from measurements, to geometric models that can give good estimates of channel statistic even in locations that have not yet been surveyed with measurements. The decision on which model to use should ultimately depend on the type of application and/or protocol that needs to be evaluated [1]. However, practical issues of the availability of the required data (be it geographical or measurements data) and required processing power also dictate which models can be used in practice. Below we discuss several use-cases and give recommendation on which type of model to use.

- Application requirements
  - If system-wide networking performance statistics are of interest (e.g., overall packet delivery rate, average end-to-end delay, etc.) AND simulation speed is important, then measurement-derived NGS models may be used.
  - If network topology statistics are of interest (e.g., determining the number and size of clusters of directly communicating vehicles, determining average neighborhood size, etc.), then either GBS or GBD models may be used. Conversely, NGS models are not able to generate such statistics correctly, because of their location-agnostic channel estimation, which leads to "averaging" of the resulting network statistics (e.g., number of vehicles per cluster would be roughly equal in a built-up urban area and in an open space containing twodimensional roads of similar structure to the urban area).

12 Channel Models for Vehicular Communications

- If analyzing performance of a routing protocol in a large area with rapid channel fluctuations (e.g., city with high-rises), simplified GBD models are best suited for the purpose, due to their ability to quickly and correctly distinguish where and when messages can be relayed between two vehicles as a result of the surroundings.
- If simulating safety-critical application that disseminates information about a specific safety event (e.g., emergency breaking, blind intersection warning, etc.), then GBD models should be used.

- Geographic data and processing limitations

  - If detailed geographic information is available (e.g., locations, dimensions, and material properties of vehicles, buildings, foliage, and other roadside objects) AND processing speed is not an issue (either due to small simulation area or availability of computing power), very detailed channel model based on ray tracing method can be used (e.g., Maurer et al. [39]).
  - If limited geographic information is available (e.g., outlines of vehicles and objects surrounding the road), with processing speed being important (though not critical), then simplified GBD models can be used (e.g., Boban et al. [26]).
  - If geographic information is not available, but the qualitative type of simulated environment are known (e.g., performing highway, urban, or rural simulations), GBS models [35, 48, 49] may be used—apart from environment type, these models require only the information about the density of the scatterers and roadside objects.

However, as the evaluation of vehicular communication moves from the academic sphere (where certain level of lower-layer abstraction might be allowed to increase simulation performance) into the real world (where application and protocols are simulated to assess their suitability for deployment), the necessity to use realistic channel models increases. Therefore, using non-geometric models is only suitable for applications constrained to a single real-world propagation environment whose statistics do not change considerably over space and time (e.g., tunnel with low density traffic, unobstructed open space communication, etc.). In all other situations, geometry-based models should be used that are, at minimum, able to account for dynamic link transitions from LOS to NLOS conditions (large-scale signal variations) and that can model dynamic small-scale variations based on the transmitter and receiver surroundings.

12.5 Future Directions

This section has provided an overview of mechanisms that govern wireless vehicular communication and its modeling. We introduced the basic wireless communication primitives and we discussed the characteristics that distinguish the vehicular communication from other types of wireless communication. We then classified and summarized the state-of-the-art vehicular channel and propagation models. While there has been a lot of work in the area of vehicular channel modeling, below are some areas where further research efforts are welcome.

12.5.1 Measurements and Models for Diverse Vehicular Environments

While there exist numerous channel measurement and modeling studies dealing with the most common environments (e.g., urban, suburban, highway), there is a need for more systematic studies in other environments. For example, V2V signal propagation measurement in a parking garage has been performed in one study to date [29]; similar applies to V2V communication in tunnels, where one of the rare studies was done by Maier et al. [64] and bridge environment (a few studies, including one by Bernadó et al. [44]). Overpasses, multi-level highways (see Fig. 12.5 for an illustrative example), as well as more complex parts of known environments (e.g., roundabouts in urban environment, shadowed on-ramp highway access, etc.) are not well explored.

12.5.2 Measurements and Models for Different Vehicle Types

Research community has been performing channel measurements and modeling primarily focused around personal cars. Studies dealing with other types of vehicles (e.g., commercial vans, trucks, scooters, and public transportation vehicles) are rare, despite them having considerably different dimensions and road dynamics. For example, the mobility of scooters and motorcycles is notably different than that of personal cars [65]. Combined with their smaller dimensions and lack of roof for antenna placement, the mobility of scooters indicates that the propagation characteristics for scooters can be significantly different than that of personal cars. Similarly, recent studies have shown that, in the same environment, commercial vans and trucks experience different channel propagation characteristics than the personal cars. This resulted in different reliable communication range and packet error rates [66, 67]. Therefore, further studies are needed that investigate channel characteristics for vehicles other than personal cars.

12.5.3 V2I

Apart from some recent efforts (e.g., by Chelli et al. [68]), V2I channel modeling is not nearly as well-researched as V2V—for example, all models described in Table 12.1 focus on V2V communication. Part of the reason is that V2I resembles

Fig. 12.5 The High Five Interchange in Dallas represents a complex environment where vehicular communication can improve safety and efficiency. Modeling this and similar environments is a complicated, but important task. Photo by austrini, available under a Creative Commons Attribution license

existing cellular systems, where one of communicating entities (base-station) is stationary while the other (user equipment) is mobile. However, typical positioning of static (infrastructure) nodes in V2I communication is unique for vehicular communication: in highways, road side units (RSUs) will be placed close to the road and at heights considerably lower than that of cellular base stations (see, e.g., current efforts within the Amsterdam Group: https://www.amsterdamgroup.mett.nl). In urban areas, the most beneficial locations are near large intersections. Furthermore, a study performed by Gozalvez et al. [32] showed that V2I communication in urban areas is highly variable, with both static and mobile objects creating a considerably changing channel over both space and time. Therefore, there exists a need for further studies investigating the V2I channels.

12.5.4 Tools for Realistic Large-Scale Simulation

As the deployment phase in main markets is getting closer [69], realistic channel modeling for large-scale simulations is necessary for effective evaluation of applications before they are deployed in the real world. However, channel and propagation models currently used to simulate V2V and V2I communication in VANET simulators (e.g., NS-3 [70]) are based on simple statistical models (e.g., free space, log-distance path loss [2], etc.) that are used indiscriminately for all environments where communication occurs. These models cannot capture the complexities of the vehicular channel, namely rapid transitions between LOS and non-LOS conditions, changes in delay and Doppler spread, etc. Consequently, simple models were shown to exhibit poor performance in terms of link-level modeling, particularly in complex environments [71]. A way forward in this respect would be to combine geometry-based scalable propagation models (e.g., [26, 49]), which are able to distinguish between different LOS conditions and environments, with small-scale channel models, which are able to provide appropriate delay and Doppler statistics for each representative environment (e.g. [27, 44]). Finally, attempts should be made to implement such realistic models in large-scale network simulators in order to enable realistic evaluation of protocols and applications.

References

1. Gozalvez J, Sepulcre M, Bauza R (2012) Impact of the radio channel modelling on the performance of VANET communication protocols. Telecommun Syst 50(3):149–167
2. Rappaport TS (1996) Wireless communications: principles and practice. Prentice Hall, Upper Saddle River
3. Kunisch J, Pamp J (2008) Wideband car-to-car radio channel measurements and model at 5.9 GHz. In: IEEE vehicular technology conference, 2008. VTC 2008-Fall, Sept 2008, pp 1–5
4. Boban M, Viriyasitavat W, Tonguz OK (2013) Modeling vehicle-to-vehicle line of sight channels and its impact on application-layer performance. In: Proceeding of the 10th ACM international workshop on vehicular inter-networking, systems, and applications (VANET '13). ACM, New York, pp 91–94
5. ITU-R (2007) Propagation by diffraction. In: International Telecommunication Union Radiocommunication Sector, Geneva, Recommendation P. 526, Feb 2007
6. Rayleigh L (1879) XXXI. Investigations in optics, with special reference to the spectroscope. Lond Edinb Dublin Philos Mag J Sci 8(49):261–274
7. Boithias L (1987) Radio wave propagation. McGraw-Hill, New York
8. Ament WS (1953) Toward a theory of reflection by a rough surface. Proc IRE 41(1):142–146
9. Horikoshi J, Tanaka K, Morinaga T (1986) 1.2 GHz band wave propagation measurements in concrete building for indoor radio communications. IEEE Trans Veh Technol 35(4):146–152
10. Benzair B, Smith H, Norbury J (1991) Tree attenuation measurements at 1–4 GHz for mobile radio systems. In: Sixth international conference on mobile radio and personal communications, Dec 1991, pp 16–20
11. Cox D, Arnold H, Porter P (1987) Universal digital portable communications: a system perspective. IEEE J Sel Areas Commun 5(5):764–773

12. Bernhardt R (1987) Macroscopic diversity in frequency reuse radio systems. IEEE J Sel Areas Commun 5(5):862–870
13. Boban M, Vinhoza TTV, Ferreira M, Barros J, Tonguz OK (2011) Impact of vehicles as obstacles in vehicular ad hoc networks. IEEE J Sel Areas Commun 29(1):15–28
14. Wang C-X, Cheng X, Laurenson DI (2009) Vehicle-to-vehicle channel modeling and measurements: recent advances and future challenges. IEEE Commun Mag 47(11):96–103
15. Rice PL, Longley AG, Norton KA, Barsis AP (1967) Transmission loss prediction for tropospheric communication circuits. NBS Tech Note 101; two volumes, Jan 1967
16. Longley AG, Rice PL (1968) Prediction of tropospheric radio transmission loss over irregular terrain; a computer method. ESSA Technical Report, ERL 79-ITS 67
17. Okumura T, Ohmori E, Fukuda K (1968) Field strength and its variability in VHF and UHF land mobile service. Rev Electr Commun Lab 16(9–10):825–873
18. Masaharu H (1980) Empirical formula for propagation loss in land mobile radio services. IEEE Trans Veh Technol 29(3):317–325
19. Sarkar T, Ji Z, Kim K, Medouri A, Salazar-Palma M (2003) A survey of various propagation models for mobile communication. IEEE Antennas Propag Mag 45(3):51–82
20. Kouyoumjian RG, Pathak PH (1974) A uniform geometrical theory of diffraction for an edge in a perfectly conducting surface. Proc IEEE 62(11):1448–1461
21. Anderson H (1998) Building corner diffraction measurements and predictions using UTD. IEEE Trans Antennas Propag 46(2):292–293
22. Erceg V, Ghassemzadeh S, Taylor M, Li D, Schilling DL (1992) Urban/suburban out-of-sight propagation modeling. IEEE Commun Mag 30(6):56–61
23. Durgin G, Rappaport T, Xu H (1998) Measurements and models for radio path loss and penetration loss in and around homes and trees at 5.85 GHz. IEEE Trans Commun 46(11):1484–1496
24. Akki A, Haber F (1986) A statistical model of mobile-to-mobile land communication channel. IEEE Trans Veh Technol 35(1):2–7
25. Molisch A, Tufvesson F, Karedal J, Mecklenbrauker C (2009) A survey on vehicle-to-vehicle propagation channels. IEEE Wireless Commun 16(6):12–22
26. Boban M, Barros J, Tonguz O (2014) Geometry-based vehicle-to-vehicle channel modeling for large-scale simulation. IEEE Trans Veh Technol 63(9):4146–4164
27. Cheng L, Henty BE, Stancil DD, Bai F, Mudalige P (2007) Mobile vehicle-to-vehicle narrow-band channel measurement and characterization of the 5.9 GHz dedicated short range communication (DSRC) frequency band. IEEE J Sel Areas Commun 25(8):1501–1516
28. Paschalidis P, Mahler K, Kortke A, Peter M, Keusgen W (2011) Pathloss and multipath power decay of the wideband car-to-car channel at 5.7 GHz. In: IEEE Vehicular technology conference (VTC Spring), May 2011, pp 1–5
29. Sun R, Matolak DW, Liu P (2013) Parking garage channel characteristics at 5 GHz for v2v applications. In: IEEE 78th vehicular technology conference (VTC Fall), 2013, pp 1–5
30. Karedal J, Czink N, Paier A, Tufvesson F, Molisch A (2011) Path loss modeling for vehicle-to-vehicle communications. IEEE Trans Veh Technol 60(1):323–328
31. Paier A, Faetani D, Mecklenbräuker C (2010) Performance evaluation of IEEE 802.11p physical layer infrastructure-to-vehicle real-world measurements. In: Proceedings of ISABEL 2010, Nov 2010, Rome
32. Gozalvez J, Sepulcre M, Bauza R (2012) IEEE 802.11p vehicle to infrastructure communications in urban environments. IEEE Commun Mag 50(5):176–183
33. Tan I, Tang W, Laberteaux K, Bahai A (2008) Measurement and analysis of wireless channel impairments in DSRC vehicular communications. In: IEEE international conference on communications, ICC '08, May 2008, pp 4882–4888
34. Wang L-C, Cheng Y-H (2005) A statistical mobile-to-mobile rician fading channel model. In: IEEE 61st vehicular technology conference, May 2005, vol 1, pp 63–67
35. Karedal J, Tufvesson F, Czink N, Paier A, Dumard C, Zemen T, Mecklenbrauker C, Molisch A (2009) A geometry-based stochastic MIMO model for vehicle-to-vehicle communications. IEEE Trans Wireless Commun 8(7):3646–3657

36. Molisch A, Tufvesson F, Karedal J, Mecklenbrauker C (2009) Propagation aspects of vehicle-to-vehicle communications - an overview. In: IEEE radio and wireless symposium, 2009 (RWS '09), Jan 2009, pp 179–182
37. Abbas T, Karedal J, Tufvesson F, Paier A, Bernado L, Molisch A (2011) Directional analysis of vehicle-to-vehicle propagation channels. In: 73rd IEEE vehicular technology conference (VTC Spring), May 2011, pp 1–5
38. Acosta-Marum G, Ingram M (2006) Doubly selective vehicle-to-vehicle channel measurements and modeling at 5.9 GHz. In: Proceedings of international symposium on wireless personal multimedia communication, 2006
39. Maurer J, Fugen T, Schafer T, Wiesbeck W (2004) A new inter-vehicle communications (IVC) channel model. In: IEEE vehicular technology conference, 2004, VTC2004-Fall, Sept 2004, pp 9–13
40. Boban M, Vinhoza TTV (2011) Modeling and simulation of vehicular networks: towards realistic and efficient models. In: Mobile ad-hoc networks: applications. InTech, Rijeka
41. Mittag J, Papanastasiou S, Hartenstein H, Strom EG (2011) Enabling accurate cross-layer phy/mac/net simulation studies of vehicular communication networks. Proc IEEE 99(7):1311–1326
42. Mecklenbrauker CF, Molisch AF, Karedal J, Tufvesson F, Paier A, Bernado L, Zemen T, Klemp O, Czink N (2011) Vehicular channel characterization and its implications for wireless system design and performance. Proc IEEE 99(7):1189–1212
43. Acosta-Marum G, Ingram M (2007) Six time- and frequency-selective empirical channel models for vehicular wireless LANs. IEEE Veh Technol Mag 2(4):4–11
44. Bernadó L, Zemen T, Tufvesson F, Molisch AF, Mecklenbrauker CF (2014) Delay and Doppler spreads of nonstationary vehicular channels for safety-relevant scenarios. IEEE Trans Vehicular Technol 63(1):82–93
45. Sen I, Matolak D (2008) Vehicle-vehicle channel models for the 5-GHz band. IEEE Trans Intell Transp Syst 9(2):235–245
46. Otto J, Bustamante F, Berry R. Down the block and around the corner the impact of radio propagation on inter-vehicle wireless communication. In: 29th IEEE international conference on distributed computing systems, ICDCS '09, June 2009, pp 605–614
47. Matolak D, Sen I, Xiong W, Yaskoff N (2005) 5 GHz wireless channel characterization for vehicle to vehicle communications. In: IEEE military communications conference (MILCOM 2005), Oct 2005, pp 3016–3022
48. Cheng X, Wang C-X, Laurenson DI, Salous S, Vasilakos AV (2009) An adaptive geometry-based stochastic model for non-isotropic mimo mobile-to-mobile channels. IEEE Trans Wireless Commun 8(9):4824–4835
49. Abbas T, Tufvesson F, Karedal J (2012) Measurement based shadow fading model for vehicle-to-vehicle network simulations. arXiv preprint. arXiv:1203.3370v2
50. Wang X, Anderson E, Steenkiste P, Bai F (2012) Improving the accuracy of environment-specific vehicular channel modeling. In: Proceedings of the 7th ACM international workshop on wireless network testbeds, experimental evaluation and characterization, WiNTECH '12. ACM, New York, pp 43–50
51. Biddlestone S, Redmill K, Miucic R, Ozguner U (2012) An integrated 802.11p wave dsrc and vehicle traffic simulator with experimentally validated urban (los and nlos) propagation models. IEEE Trans Intell Transp Syst 13(4):1792–1802
52. Mangel T, Klemp O, Hartenstein H (2011) A validated 5.9 GHz Non-Line-of-Sight path-loss and fading model for inter-vehicle communication. In: 11th international conference on ITS telecommunications (ITST), Aug 2011, pp 75–80
53. Giordano E, Frank R, Pau G, Gerla M (2010) CORNER: a realistic urban propagation model for VANET. In: Proceedings of the 7th international conference on Wireless on-demand network systems and services (WONS), 2010, pp 57–60
54. Cozzetti H, Campolo C, Scopigno R, Molinaro A (2012) Urban VANETs and hidden terminals: evaluation through a realistic urban grid propagation model. In: IEEE international conference on vehicular electronics and safety (ICVES), July 2012, pp 93–98

55. Pilosu L, Fileppo F, Scopigno R (2011) Radii: a computationally affordable method to summarize urban ray-tracing data for vanets. In: 7th international conference on wireless communications, networking and mobile computing (WiCOM), 2011. IEEE, New York, pp 1–6
56. Parsons JD (2000) The mobile radio propagation channel. Wiley, New York
57. Kaul S, Ramachandran K, Shankar P, Oh S, Gruteser M, Seskar I, Nadeem T (2007) Effect of antenna placement and diversity on vehicular network communications. In: 4th annual IEEE communications society conference on sensor, mesh and ad hoc communications and networks, SECON '07, June 2007, pp 112–121
58. Oh S, Kaul S, Gruteser M (2009) Exploiting vertical diversity in vehicular channel environments. In: IEEE 20th international symposium on personal, indoor and mobile radio communications, Sept 2009, pp 958–962
59. Meireles R, Boban M, Steenkiste P, Tonguz OK, Barros J (2010) Experimental study on the impact of vehicular obstructions in VANETs. In: IEEE vehicular networking conference (VNC 2010), Jersey City, Dec 2010, pp 338–345
60. Scopigno R, Cozzetti HA, Pilosu L, Fileppo F, Gupta S, Vázquez-Castro M-A, Imadali S, Karanasiou A, Petrescu A, Sifniadis I et al (2012) Advances in the analysis of urban VANETS: scalable integration of radii in a network simulator. In: WiMob, 2012, pp 563–570
61. Maurer J, Fugen T, Wiesbeck W (2002) Narrow-band measurement and analysis of the inter-vehicle transmission channel at 5.2 GHz. In: IEEE 55th vehicular technology conference (VTC Spring), 2002, pp 1274–1278
62. Renaudin O, Kolmonen V, Vainikainen P, Oestges C (2008) Wideband mimo car-to-car radio channel measurements at 5.3 GHz. In: IEEE 68th vehicular technology conference (VTC Fall), 2008, pp 1–5
63. Alexander P, Haley D, Grant A (2011) Cooperative intelligent transport systems: 5.9 GHz field trials. Proc IEEE 99(7):1213–1235
64. Maier G, Paier A, Mecklenbrauker C (2012) Channel tracking for a multi-antenna its system based on vehicle-to-vehicle tunnel measurements. In: 19th IEEE symposium on communications and vehicular technology in the Benelux (SCVT), 2012, pp 1–6
65. Shih O, Tsai H, Lin H, Pang A (2011) A rule-based mixed mobility model for cars and scooters (poster). In: IEEE vehicular networking conference (VNC), 2011, pp 198–205
66. Boban M, Meireles R, Barros J, Tonguz OK, Steenkiste P (2011) Exploiting the height of vehicles in vehicular communication. In: IEEE vehicular networking conference (VNC 2011), Amsterdam, Nov 2011, pp 284–291
67. Boban M, Meireles R, Barros J, Steenkiste PA, Tonguz OK (2014) TVR - tall vehicle relaying in vehicular networks. IEEE Trans Mob Comput 13(5):1118–1131
68. Chelli A, Hamdi R, Alouini M (2013) A vehicle-to-infrastructure channel model for blind corner scattering environments. In: 78th IEEE vehicular technology conference (VTC Fall 2013), Sept 2013, pp 1–6
69. U.S. Department of Transportation announces decision to move forward with vehicle-to-vehicle communication technology for light vehicles, Feb 2014. http://www.nhtsa.gov/About+NHTSA/Press+Releases/2014/USDOT+to+Move+Forward+with+Vehicle-to-Vehicle+Communication+Technology+for+Light+Vehicles
70. Network Simulator 3. http://www.nsnam.org
71. Dhoutaut D, Regis A, Spies F (2006) Impact of radio propagation models in vehicular ad hoc networks simulations. In: VANET 06: Proceedings of the 3rd international workshop on vehicular ad hoc networks, 2006, pp 69–78

Chapter 13
Simulation Tools and Techniques for Vehicular Communications and Applications

Christoph Sommer, Jérôme Härri, Fatma Hrizi, Björn Schünemann, and Falko Dressler

Abstract In the domain of Inter-Vehicle Communication (IVC), even though first field operational tests are already going on, performance evaluation is still dominated by simulation experiments. Yet, they require a very specific methodology as well as adapted tools and models not straightforwardly found in other domains. In this chapter, we first describe the required methodology in terms of scalability and applicability to select the right models and their interactions. In particular, we classify each class of models as in increasing level of granularity, and discuss in detail the trade-off between scalability and applicability typical to IVC simulations. We then introduce some of the most widely used and openly available simulation frameworks applicable to the domain of IVC, and emphasize their capabilities related to the required methodology. In particular, we present the IVC simulation toolkits Veins, iTETRIS, and VSimRTI, three prominent simulation platforms openly available for IVC simulations. To provide guidelines for efficient and scalable simulations of IVC applications, we discuss the appropriate selection of models and their level of granularity as function of the IVC application requirements, and provide an overview of their corresponding support in each of toolkit.

Keywords Channel access model • Channel model • Heterogeneous vehicular network • Intelligent transportation system • Inter-vehicle communication • iTETRIS • ITS • IVC • JiST/SWANS • Mobility model • Network simulation • ns-3 • OMNeT++ • Road traffic simulation • Simulation model • Simulation tool • SUMO • V2V • V2X • VANET • Vehicle-to-vehicle • Vehicle-to-X • Veins • VISSIM • VSimRTI

C. Sommer • F. Dressler
CCS Group, Univ. Paderborn, Paderborn, Germany
e-mail: sommer@ccs-labs.org; dressler@ccs-labs.org

J. Härri (✉) • F. Hrizi
Mobile Communications EURECOM, Sophia Antipolis, France
e-mail: haerri@eurecom.fr; hrizi@eurecom.fr

B. Schünemann
OKS/Daimler Center for Automotive IT Innovations, Technische Universität Berlin, Berlin, Germany
e-mail: bjoern.schuenemann@dcaiti.com

© Springer International Publishing Switzerland 2015
C. Campolo et al. (eds.), *Vehicular ad hoc Networks*,
DOI 10.1007/978-3-319-15497-8_13

13.1 Introduction

Vehicular networks, Intelligent Transportation Systems (ITS) supported by wireless communication, have become one of the key research areas in the wireless networking community. There have been major achievements in this field including new concepts for efficient IVC [19, 25, 52]. Aside from the obvious use of cellular networks, first standards for short range radio communication have been defined in this field based on DSRC, most prominently the IEEE 802.11p protocol [20]. Furthermore, standardization bodies are working on higher layer protocols such as those based on the IEEE 1609.4 Wireless Access in Vehicular Environments (WAVE) [21] or ETSI ITS-G5 [14] standards for IVC. First concepts such as single-hop broadcasts using CAMs/Basic Safety Messages (BSMs) or ETSI ITS G5 geocast ideas provide the floor for more mature applications and protocols [51]. The relevant industry players are now ready for deploying short range broadcasting systems based on DSRC/WAVE in addition to cellular networks. Yet, many open questions and challenges remain that have to be worked on. Examples include protocol design on all layers and especially scalability of proposed concepts.

In general, such performance assessment can be based on analytical models, simulation, or experimentation. In the very complex vehicular environment, analytical models are frequently based on too many unrealistic assumptions. Experimentation is extremely important and field operational tests are currently being conducted in Germany, France, Austria, and new ones are just starting in the USA. Yet, those tests are limited in space and time, i.e., only a certain number of applications and protocol variants can be tested. Smaller scale experiments primarily help validating available models. Thus, simulation is and will be the primary tool for performance assessment [12, 43].

In the last few years, substantial progress has been achieved in the field of simulation tools and techniques helping to assess the performance of IVC protocols. Thus (and in contrast to early stages of IVC research) there are now a good number of tools and models ready for use and openly available for the research community. Still, choosing exactly the right set of techniques, tools, and models is very complex. *Consolidation efforts* and recent research on *integrated IVC simulation frameworks* led to the development of three promising candidates: Veins [46], iTETRIS [38], and VSimRTI [42].

In this survey chapter, we study what specific aspects of IVC have an influence on simulation experiments. Essentially, we discuss all the models that are specific to performance evaluation of ITS, i.e., wireless networking aspects and mobility modeling. We introduce three of the best known and openly available simulation frameworks, Veins, iTETRIS, and VSimRTI, to provide an overview of the various aspects that need to be considered for producing realistic results, both for researchers already active in the field and those interested in moving into vehicular networking research.

Our main motivation for this chapter is to make researchers working on IVC and ITS applications as well as those moving into the field aware that the mentioned consolidation efforts successfully converged and the most widely used simulation frameworks Veins, iTETRIS, and VSimRTI meanwhile offer all the components and models needed for performance evaluation of vehicular networking protocols and applications—and that matching realistic setups very closely.

Our contributions can be summarized as follows:

- We give a broad overview of necessary simulation models that are specific to the IVC domain and discuss the level of granularity and applicability required by IVC applications. In particular, we provide guidelines for the selection of the right models at the right level of granularity to optimize the trade-off between simulation scalability, granularity, and applicability.
- We discuss the current state of simulation technology that is available for applying these models for investigating IVC approaches and present three fully featured toolkits that are openly available: Veins, iTETRIS, and VSimRTI.

The remainder of this chapter is organized as follows. We first briefly study typical IVC applications and their requirements in terms of simulation granularity and scale in Sect. 13.2. We then discuss models that are specific to IVC simulations, in particular those used for the wireless communication and networking part (Sect. 13.3) and those to simulate vehicles' mobility (Sect. 13.4). In the main part of the chapter, we introduce network and mobility simulators (Sect. 13.5) that build the basis for all the leading integrated simulation frameworks. We outline the capabilities of the most widely used frameworks in Sect. 13.6. We finally discuss the outcome of this chapter in Sect. 13.7.

13.2 IVC Applications

The most distinctive aspect of the design of IVC systems is the role of the application, leading to a strong bi-polarity in the communication design: IVC applications dictate the requirements, whereas the access technology and the shared mobile wireless medium formulate the capabilities. The challenge is therefore to find a trade-off between such a Yin and Yang. Accordingly, the design of a communication solution *fitting all* relates to an engineering myth, and dedicated communication solutions must be developed for mostly each IVC application. This impacts not only the protocol design at different layers, but also the simulation toolkits employed for their evaluations.

In order to bring order to such a chaotic situation, IVC applications have been classified [19, 25] in three classes as function of their scale, infrastructure support, and communication sensitivity [19, 52]:

- *Safe Mobility* contains safety-of-life communications to notify drivers of dangerous situations [24]. Such applications are localized, and are very sensitive to transmission impairments, viz. delay and loss. They must also work without the help of any form of infrastructure.
- *Smart Mobility* gathers and disseminates information related to traffic state or navigation guidelines in a decentralized way [26]. As the precision of traffic

information decays with distance and time, the scale is small to medium, and communications are delay bounded.
- *Connected Mobility* encompasses all applications providing Internet access, content exchange, or commercial advertisements [2]. Applications in this class are fundamentally large scale, and focus mostly on throughput maximization.

Such classification also impacts the models and toolkits required to evaluate the performance (or conformance) of IVC applications. Given the class of the IVC application, the selection of the models is a constant trade-off between performance and precision. Thus, given the constrained simulation capability, the modeling of only those behaviors that are effectively impacting the IVC application should be sought.

13.2.1 Required Criteria

In order to help select the right model with the right level of abstraction required for a particular IVC application, we provide in Table 13.1 a synthetic view of the required granularity in key criteria for the three main types of IVC applications. We provide in this section a list of criteria guiding the IVC's requirements.

- scale—represents how large the scenario should be both spatially and in number of actors (vehicles). It can range from small scale when an area spanning a couple of vehicles is required to large scale when a city-wide network is required.
- applicability—represents the accuracy of the models used by simulations. It can range from a high level of accuracy when models need to be as close as possible to real conditions, to low when abstractions can be tolerated.
- infrastructure—represents the support of communication infrastructures, which can either be inexistent, homogeneous (metro-scale or micro-scale) or heterogeneous (metro-scale or micro-scale)
- COM pattern—represents the communication pattern for IVC. It can range from pure direct patterns (WiFi-direct, IEEE DSRC, ETSI ITS-G5, LTE-Direct, etc.) to relay patterns (4G/5G cellular networks, WiFi, etc.).
- NET protocol type—represents the support of network protocols, such as IPv6 or non-IP (e.g., ETSI Geonet, WAVE WSMP, or application defined).

Safe Mobility application simulation is critically reliant on accurate modeling of transmission impairments. Thus, these must be precisely reproduced. Accordingly, the communication models should come as close as possible to the real condition

Table 13.1 Overview of different granularity/applicability requirements for IVC applications

IVC application	Scale	Applicability	Infrastructure	COM pattern	NET protocol
Safe mobility	Small	High	No	Direct	None
Smart mobility	Medium	Medium	Yes	Direct and relay	IPv6 and non-IP
Connected mobility	Large	Low	Yes	Direct and relay	IPv6

found by the application, and the network models must be as realistic as the ones found in chipsets and computers. The scale relates not only to the geographic area but also to the number of actors. Constrained by finite computational resources, the level of realism used by the models required by traffic safety applications limits their evaluation to a few actors in a small area.

Smart Mobility requires a large amount of actors to gather trustworthy traffic states. Here, the precision of the traffic states is more important than the communication or networking models employed to provide it. Therefore, the level of precision of the models must frequently be reduced to cope with the scale of the network.

Connected Mobility are typically characterized by their large scale. This forces a further abstraction in the level of precision of the models employed to evaluate them.

Accordingly, the performance evaluation of IVC applications requires models to be carefully selected for an evaluation scenario. Models may be categorized into different layers (PHY and Channel, MAC, Network, Mobility, etc.). Literature and the vehicular community developed different levels of abstractions in each layers, each of them adapted to different requirements from the IVC applications. A trustworthy evaluation scenario is therefore an appropriate selection of models and their level of abstraction, all of which need to be carefully considered—and described [23].

In the next sections, we will describe various levels of accuracy found in the different models employed in a typical simulation-based evaluation scenario. We will provide guidelines for their efficient selections as a function of the required level of applicability to IVC applications. In particular, for each modeling class, we will provide a classification table containing *safe*, *smart*, *cnt* entries (i.e., corresponding, respectively, to safe, smart, and connected mobility classes) and indicating which models are required and at which level of granularity and applicability they should be used.

13.3 Wireless Communications and Networking

Simulation tools used to assess the performance of IVC protocols and applications must support a variety of simulation models that is very specific to this application scenario [12]. In particular, signal propagation models need to incorporate the environmental conditions of communicating vehicles, models for channel access need to be able to treat the CSMA behavior, congestion control, or even multi-channel operations. Also, information dissemination schemes for IVC, such as beaconing [13] or geocasting [31], or even facilities-layer protocols such as BSM or CAM might also be required for IVC applications. In the following, we discuss both domains and give examples for models that are frequently used in this area.

13.3.1 Signal Propagation and Fading

In application-layer centered simulations, the most commonly considered physical layer effects are throughput, delay, and Bit Error Rate (BER). As the first two are straightforward to model in any discrete event simulation, in the following, we will focus on considerations of BER calculations only.

We can distinguish three types of fading models that need to be considered depending on the granularity of the simulated applications:

- *Distance* based models simply help understanding free space radio transmissions to a certain extent. But even in freeway scenarios, shadowing by other vehicles and fast fading need to be investigated.
- *Shadowing* can be modeled in a very abstract way using stochastic models, or using very accurate shapes of the obstacles, either using geometry-based approaches or fine-grained ray tracing.
- *Fast fading* is typically caused by multi-path propagation and is frequently either ignored or modeled using stochastic models.

In their simplest case, *Distance based* models the success or failure of any transmission by p_{err} in a fully deterministic fashion, by comparing the distance d of sender and receiver with a fixed threshold distance d_{max}, as follows.

$$p_{err} = \begin{cases} 1 & \text{if } d > d_{max}, \\ 0 & \text{else.} \end{cases} \quad (13.1)$$

This approach is widely known under the name of *unit disk* graph connectivity model, however, is only appropriate for simulations that model large-scale data flows rather than individual transmissions [48]. Fading models may include a higher granularity by considering the Signal to Interference plus Noise Ratio (SINR) of signals to arrive at a BER of transmissions, making the decision whether a packet can be received probabilistic. The SINR is calculated at each potential receiver by weighting the received power level of any interfering transmission (and the noise floor) against the received power level of the signal to be decoded. For this calculation of the receiving power P_r, any such model will need to rely on a set of radio propagation models to predict losses between transmitting and receiving station, where P_t is the transmit power, G_t and G_r are the transmit (and receive) antenna gains, and L_x are terms capturing loss effects during transmission [1, 34]. The terms L_d, L_s, L_f represent the loss effect from distance, shadowing and fast fading, respectively.

$$P_r[\text{dBm}] = P_t[\text{dBm}] + G_t[\text{dB}] + G_r[\text{dB}] - L_d[\text{dB}] - L_s[\text{dB}] - L_f[\text{dB}] \quad (13.2)$$

The simplest of such propagation models are empirical adaptations of a *free space* model, taking into account only the path loss L_d, i.e., loss in transmission power over distance, that is caused by the radiation pattern of ideal omni-directional antennae, depending on the wavelength λ and an empirically determined path loss coefficient α [37]. One step up are *two ray interference* models [40], which deterministically derive path loss caused by constructive/destructive self-interference effects of the signal component that is reflected on the ground, taking into account the height over ground of transmitter h_t and receiver antenna h_r together with an empirically determined relative permittivity of the ground ϵ_r [44].

Although providing an increasing level of realism toward large-scale fading impacts, they do not include the critical fading effects found in a vehicular environment, i.e. shadowing or fast fading. To consider them, a shadowing model L_s, and a fast fading model L_f must also be considered in the overall received power P_r.

Shadowing models help determining additional factors that contribute to path loss such as obstacles, which shield a receiver from all or part of the radiated power. If all transmission attempts can be assumed to be uncorrelated in time (on the order of seconds) and space (on the order of tens of meters), such obstacles can be modeled using purely stochastic models, e.g., a log-normal shadowing model [10]. However, if these assumptions do not hold, i.e., if multiple transmission attempts are made at roughly the same point in time or space, correlated (space or time) models are required, as well as deterministic classification on how positions of buildings [45] and location of vehicles [6] impact line-of-sight and Fresnel zones, and as such the path loss. Also, shadowing models can be made arbitrarily complex, such as building precise fading database from measurements campaigns or by relying on *ray tracing* techniques [47], yet at the cost of extreme computational overhead or reduced spatial resolution. Simulations requiring this class of models also require a very specific simulation framework that is seldom found in available tools.

Additional small-scale or fast fading effects can be added as further loss terms, to model Rayleigh or Rician fading. This typically would depend on the environment and mobility of vehicles. For example, a simple Nakagami-m model can stochastically capture narrow-band multi-path fading effects, considering the received power level and an empirically determined value of m, which can be used to adapt the model for differently pronounced line-of-sight components [40]. For further details related to vehicular channel and propagation models for IVC, we refer the reader to Chap. 12.

As depicted in Fig. 13.1 narrow-band fading, shadowing, and path loss models can then be cumulatively used to derive the SINR and, as described, the packet reception probability. The choice of the class of fading model and their level of granularity significantly depend on the IVC application requirements, as summarized in Table 13.2.

13.3.2 Channel Access

Channel Access protocols have to answer the basic question of how to efficiently share the radio channel resources between multiple transmitters. A strong

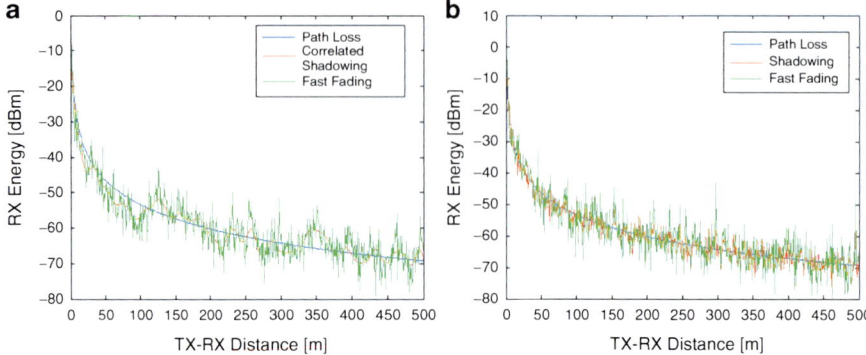

Fig. 13.1 Illustration of the cumulative impact of the three types of fading. (**a**) Correlated shadowing. (**b**) Uncorrelated shadowing

Table 13.2 Overview of different granularity/applicability propagation models and possible applications

Mobility	Granularity	Applicability	Safe	Smart	Cnt	Application
Distance	Unit disk	Low			✓	Analytical/asymptotic
	Free space	Medium			✓	Long range radio tx
	Self-interference	High			✓	Short range radio tx
Shadow	Ignored	Low			✓	No buildings or vehicles
	Stochastic	Medium		✓	✓	Signals uncorrelated in time and space
	Stochastic correlations	High	✓			Signal correlated in time and/or space
	Geometry-based	High	✓			Location-dependent fading
	Ray tracing	Very high	✓			Specific real-world modeling
Fast	Ignored	Medium		✓	✓	Isolated protocol aspects
	Stochastic	High		✓		System behavior

particularity of IVC comes from the strict *distributed* requirement for channel access, opposed to personal communications where a centralized approach is mostly found (WLAN, 4G).

Conceptually speaking, distributed channel access mechanisms for IVC have to provide the following three functions:

- *Multiple Access Control*—Provides fair access to the vehicular channel. It determines when a channel is busy or idle, models when a transmission has failed, and properly reacts.
- *Congestion Control*—Restricts the spatial or temporal usage of the vehicular channel to avoid degrading channel conditions.
- *Multi-Channel Control*—Distributes communications across multiple available IVC channels for efficient spectrum usage.

Table 13.3 Overview of different granularity/applicability channel access models and possible applications

Mobility	Granularity	Applicability	Safe	Smart	Cnt	Application
MAC	Unit disk	Low			✓	Asymptotic performance
	Stochastic	Medium		✓	✓	Analytical/asymptotical
	CSMA-CA	Medium		✓	✓	Packet level tests
	ITS-G5/ DSRC	High	✓	✓	✓	Validation and conformance
Congestion control	Ignored	Low			✓	Large-scale test
	Generic model	Medium		✓	✓	Channel load impact
	ITS DCC	High	✓	✓	✓	Validation and conformance
Multi-Channel	Ignored	Medium	✓	✓	✓	Isolated examination
	Off-loading	High	✓			Multi-channel congestion
	Services	High		✓	✓	System behavior

Their applicability depends on the use case, as outlined in Table 13.3.

If packet reception probability is modeled independent of channel conditions (idle/busy) or if the probability of failed transmissions may be abstracted by a stochastic process, the simplest access control solution is to abstract packet reception by a packet reception probability modeled by either a deterministic function (e.g., *unit disk* model) or a stochastic model. The latter can take into account various communication parameters, such as transmit range, number of neighbors in range, transmit rate, or packet size, in order to model a decreasing reception probability with increasing load on the channel. It may also be noted that such approach is not strictly linked to a particular technology, as the stochastic reception model may be tuned to the communication specificities of LTE or an ITS G5 technology. Simulations using such an *idealistic* model can certainly yield highly instructive results and can be performed extremely fast [5, 27]. It is in particular preferred for smart mobility applications.

The performance of safe mobility applications is critically reliant on the reception of a particular packet or even a signal, which requires fine grained multiple access models [49]. In such condition, estimating if and when a station might detect the channel to be busy is already non-trivial [8]. Accordingly, multiple access modeling may again have different levels of abstraction, ranging from a simple *CSMA/CA* listen-before-talk protocol to full-fledged IEEE 802.11-2012 or even higher layer control protocols such as those in ITS G5. This granularity relates to a conformance level to a standard describing IVC protocols. A low conformance level will not reflect realistically how an IVC protocol would truly behave once deployed, whereas an IVC protocol with high level of conformance to a standard will provide a high level of confidence in measured effects. The selection of the appropriate conformance level for a multiple access model is a trade-off between confidence and performance, as modeling a high level of conformance requires also more computational resources.

Providing fair multi-access control does not specifically mean that the need of all transmitters may be supported. A higher congestion level impacts the delay,

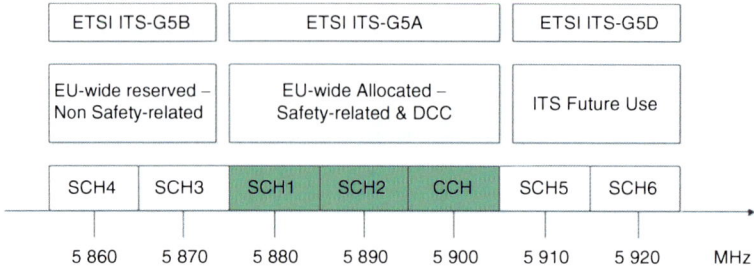

Fig. 13.2 ETSI channel allocation in EU

throughput, and reliability of IVC protocols. In a vehicular context, distributed multi-channel mechanisms cannot react to the transmissions they cannot sense (e.g., hidden nodes). Congestion control mechanisms are therefore critical to maintain a particular Quality-of-Service (QoS), which can take the shape of a reduced transmit power, rate, or packet size in order to control the load on the channel. As depicted in Fig. 13.2, ETSI channel access regulations restrict the usage of the ITS-G5A channels for safety-related communications under a congestion control mechanism. If the objective of the IVC protocol or application is conformance, then such congestion control mechanism should be included and as close as possible to the standards, such as ETSI DCC. Congestion control mechanisms are also not specific to a particular access technology, as long as the transmit parameters may be controlled and altered on a per-packet basis. For further details related to DCC mechanisms for IVC, we refer the reader to Chap. 4.

Directives from EU ETSI and the US FCC allocate multiple channels to IVC applications. So far, most IVC focused on one channel, but IVC traffic should benefit from multiple channels as function of their types, either by traffic off-loading on alternative channels, or for IVC service management. Various standards for multi-channel systems exist, such as those based on the IEEE 1609.4 WAVE [21] or ETSI ITS-G5 [14] standards for IVC. With more than one channel to choose from (and potentially less transceivers than channels), the MAC layer has to pick which channel to listen to and which to pick for transmitting [9, 28]. Accordingly, packets cannot be received if a transmitter is away on a different channel, which generates delay and unreliability.

13.3.3 Networking

Network protocols have the responsibility of disseminating IVC information through vehicular networks or to a back-end or Internet service. Another objective of a networking layer is to provide a network-wide view of particular IVC parameters. The objective of this section is not to exhaustively describe vehicular networking protocols, but illustrate the link between the class of IVC applications and the required IVC networking stack.

Fig. 13.3 Conceptual overview of the different NET layer stack for IVC applications

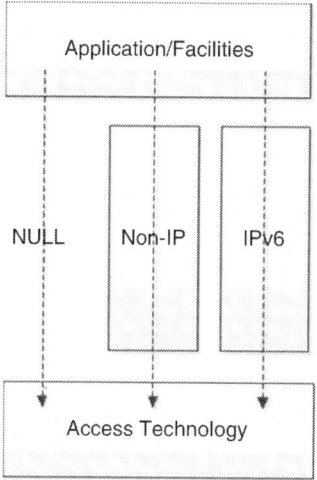

Following Fig. 13.3, the vehicular networking stack is situated between IVC applications/facilities and IVC access technologies. Irrespective of the various IVC standards, network protocols may be classified into three types:

- *Null Net*—A network layer is or limited to transport port numbers only. This is typically the case in the ISO CALM standard, where IVC packets contain all required information to be transmitted to IVC access technologies.
- *non-IP Net*—A non-IP network layer is available. This is found in the WSMP stack and the ETSI Geo-networking stack, although contrary to ETSI Geo-networking, WSMP are mainly headers for IVC service management, and do not provide multi-hop networking support.
- *IPv6 Net*—An IPv6 stack is available, including the required IPv6 mobility management functions.

Safe mobility applications are typically restricted to single-hop, potentially including multi-hop piggybacking. Accordingly, a networking stack is most likely not required. For instance, both CAM and BSM are application/facilities layer messages and contain sufficient addressing information to support safety-related IVC. ETSI provides network support for multi-hop dissemination of emergency Decentralized Emergency Notification Message (DENM). Although a networking layer might not be required for most of the safe mobility applications, a Facilities layer stack would be required for high conformity with ETSI standards.

Smart mobility applications require data dissemination either horizontally (directly between vehicles) or vertically (to back-end or Internet services). In that case, the selection of the networking stack depends on the IVC protocols. As the Internet Engineering Task Force (IETF) does not currently provide IPv6 multi-hop support for ITS applications over a vehicular network, a non-IP stack would be required if smart mobility data needs to be disseminated between vehicles. If

Table 13.4 Overview of different granularity/applicability network models and possible applications

Mobility	Granularity	Applicability	Safe	Smart	Cnt	Application
NULL		Low	✓	✓	✓	Cooperative awareness
Non-IP	1-hop broadcast	Low	✓	✓	✓	Neighborhood monitor
	Geo-routing	Medium	✓	✓	✓	Isolated IVC protocols
	DTN	High		✓	✓	Traffic offloading
IPv6	Isolated	Low			✓	Static, or IP protocol test
	Mobility mgnt	High		✓	✓	Impact of mobility

smart mobility traffic is mostly vertical, an IPv6 stack is required as vehicles need to interconnect to Internet.

As connected mobility applications are mostly based on keeping vehicles connected to the back-end or Internet, an IPv6 stack is critical to evaluate the performance of this class of IVC applications. Considering the potential high scale of vehicular traffic on the IPv6 back-end, efficient IPv6 mobility and traffic management might also be required.

It may also be noted that the selection of the networking stack also depends on the selected IVC as described in Table 13.4. For instance, if the selected IVC access technology is 4G/5G, then an IPv6 networking support is required. With the advent of the Internet-of-Things (IoT) paradigm, it is expected that future new networking layers will become available, including IoT-specific addressing and routing.

13.4 Mobility Modeling

When simulating IVC applications, the crucial role of vehicular mobility models has been illustrated by multiple studies [12, 15].

13.4.1 Overview and Constraints

As illustrated in Fig. 13.4, vehicular mobility models consist of three major blocks: *Motion Constraints*, *Mobility Demand*, and *Traffic Demands*, mutually influencing each other. Mobility and traffic demands provide a description of the desired mobility, whereas motion constraints limit them. The level of accuracy of a particular model depends on the granularity in each of these two blocks.

Considering motion constraints, we can distinguish three major levels—An increasing the level of accuracy in the motion constraint block shows a significant impact on mobility patterns, but the impact of traffic demand should also be carefully considered:

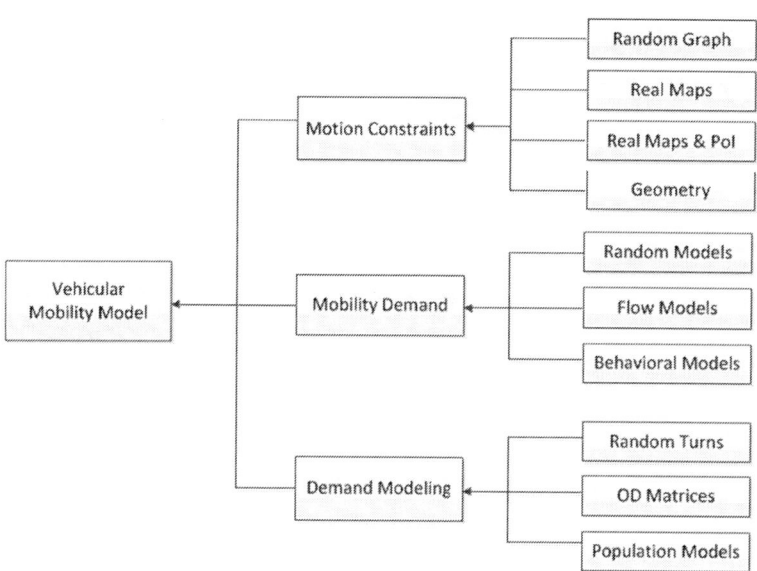

Fig. 13.4 Concept map for vehicular mobility models, where the two main blocks are enriched with added blocks and features, each of them providing an increased accuracy, but at an increased cost and complexity

- *Random Topology*—Random graph representations, such as star, honeycombs, or Voronoi tessellations. They can be rapidly generated, but only provide an abstract topology representation.
- *Real Topological Maps*—Real street and cadastral data provide a higher accuracy to the graph representation, and aim at reflecting motion constraints found in urban areas.
- *Maps with added features*—Speed, access restrictions, turn limitations, and Points-of-Interest (POI) are added to the graph representation for an increased accuracy in motion constraints. Traffic lights, variable message signs, and similar technologies act as dynamic influence on road topology.
- *Maps with precise geometry*—Streets are no longer represented by a graph, but as a precise geometrical shape, including lane, sidewalk, and any street-level geometry. This enables the precise positioning of vehicles and pedestrians in the network, in particular their interactions.

Mobility demands may also be classified as a function of their scope, granularity level, and characteristics. For many years, *random models* have been the preferred and easiest solutions to model mobility for communication networks. Their simplicity and stochastic properties are counter-balancing their limited precision to model vehicular mobility.

Depending on the scale of the evaluation, *flow models* such as microscopic models are needed, exactly representing the interaction of a vehicle with other cars, traffic regulations, street topologies, etc. For large-scale evaluation, mesoscopic models are usually a good compromise between accuracy and simulation performance. Finally, when pure physical laws cannot represent vehicular motions, *behavioral models* allow to describe patterns based on stimuli and responses, with the objective to represent the human behavior rather than flow or traffic equations [11].

Traffic Demands represent large-scale mobility patterns of vehicular mobility. A *Random turns* approach operates a random selection of a new direction at each intersection. For an increased granularity, *Origin-Destination (O-D) matrices* are used to represent initial and destination areas, as well as optimal paths between them. For an even finer granularity, area-wide or city-wide *Population Models* may be obtained in order to adjust the O-D Matrices, as well as the preferences in paths and mobility options (bus, car, etc.).

A concept map for vehicular mobility models, where the three main blocks are enriched with added blocks and features, each of them providing an increased accuracy, but at an increased cost and complexity is shown in Fig. 13.4. The figure outlines the many aspects influencing motion constraints, mobility, and traffic demand, all defining the vehicular mobility model. The relationship between the various levels of granularity in the mobility models and the IVC applications are illustrated in Table 13.5.

Table 13.5 Overview of different granularity/applicability of mobility models and possible applications

Mobility	Granularity	Applicability	Safe	Smart	Cnt	Application
Motion Constraints	Random	Low			✓	Analytical evaluations
	Maps	Low			✓	Impact of mobility on IVC
	Maps and PoI	Medium		✓		Dynamic navigation
	Street geometry	High	✓			Safety and autonomous drive
Mobility Demands	Random	Low			✓	Analytical evaluations
	Flow	Medium		✓	✓	Traffic jam modeling
	Behavioral	High	✓			Accident modeling
Traffic Demands	Random	Low			✓	Analytical evaluations
	OD Matrix	Medium		✓	✓	Social impact on smart mob.
	Population	High		✓		Multi-modal IVC app.

13.4.2 Towards More Realistic Mobility Modeling

One of the challenges [15] is to develop a traffic demand providing an accurate vehicular mobility description at both macroscopic and microscopic levels. For instance, it could be shown that a flow model was an absolute minimum to reproduce the realistic microscopic Vehicle-to-Vehicle (V2V) interactions, typically observed when evaluating traffic safety applications. Traffic models are also required when large-scale motion patterns are necessary, for instance when evaluating traffic efficiency applications.

Flow and traffic models are based on physical equations developed to provide accident-free and optimal traffic spreading. Given a specific situation, cars will always react the same way. In real life, drivers are not perfect and react to stimuli with different reactions as function of the context. Such patterns are represented by behavioral models, where reactions are not following any strict rule, but adapted to given contexts. These models are, for instance, the preferred choice to model accidents.

Finally, trace/survey models are experiencing an increasing interest from the community in building large-scale calibrated urban mobility models. Although providing a level of accuracy very close to the calibration data, the limitations are their complexity and the spatial and temporal constraints from the calibration data.

The choice of a mobility model and its level of accuracy depend on the application and protocol requirements, and should carefully be selected for accurate and scalable mobility modeling. Although all of the common network simulators have, by now, integrated support for node mobility, their mobility models' level of sophistication varies widely. It has long been established that the quality of results obtained from mobile ad hoc network simulations is heavily influenced by the quality of the employed mobility model [7]. Furthermore, the impact of mobility models on IVC simulation results, as well as the inadequacy of simple random mobility models, is well documented in the literature [39, 43].

For a detailed comprehensive discussion and current state-of-the-art of vehicular mobility models, we further refer the reader to Chap. 11.

13.5 Network and Mobility Simulation Tools

In the following, we introduce and briefly describe simulators frequently used for performance evaluation of IVC applications and protocols [23]. For a comprehensive discussion of most of the mobility models mentioned below, we refer the reader to the very complete survey by Harri et al. [17].

13.5.1 The ns-3 Network Simulator

The network simulator 3 (ns-3) [16] is an open discrete-event simulation environment that was designed to be the successor of the popular simulator ns-2. Aiming to be more scalable and more open for extension, it significantly differs from ns-2 with its novel structural and modular implementation.

The core architecture is object-oriented, it has been developed in C++, contrary to ns-2 (which is written in OTcl and C++), ns-3 optionally uses python scripts for simulations. Some C++ based models have been ported from ns-2 to ns-3. Being open also to commercial use, its base architecture has been designed to support network virtualization and real testbed integration.

The object aggregation model is the main feature of ns-3. Multiple objects could be linked together at run-time such as nodes, applications, and protocol stacks. This mechanism has been proposed as solution to the "weak base class" problem in C++ where the base class should be modified each time the programmer needs to reuse it with different configuration. Moreover, the aggregation model handles the access between the different aggregated objects and eases the automatic memory control.

Although less wide than ns-2, ns-3 supports an increasing amount of models, from IEEE 802.11, 3GPP LTE, to IPv6 and selected MANET routing protocols. ns-3 is a packet-level simulator, but if the precise simulation of wireless transmissions at a signal level is required, ns-3 has been extended by Mittag et al. [33] with a physical layer implementation, integrating OFDM symbol processing, space-time channel modeling as well as a precise reception model. This however requires significant computational resources, which restricts this module to be used on specific small-scale wireless communication scenarios.

ns-3 has also been extended by the iTETRIS [38] and the subsequent COLOMBO[1] projects with the support of an ETSI ITS compliant IVC protocol stack as depicted in Fig. 13.5. Major communication-related facilities, such as CAM, DENM, and SAM have been integrated, as well as a non-IP ETSI Geo-networking stack supporting various geographic routing protocols (geo-unicast,

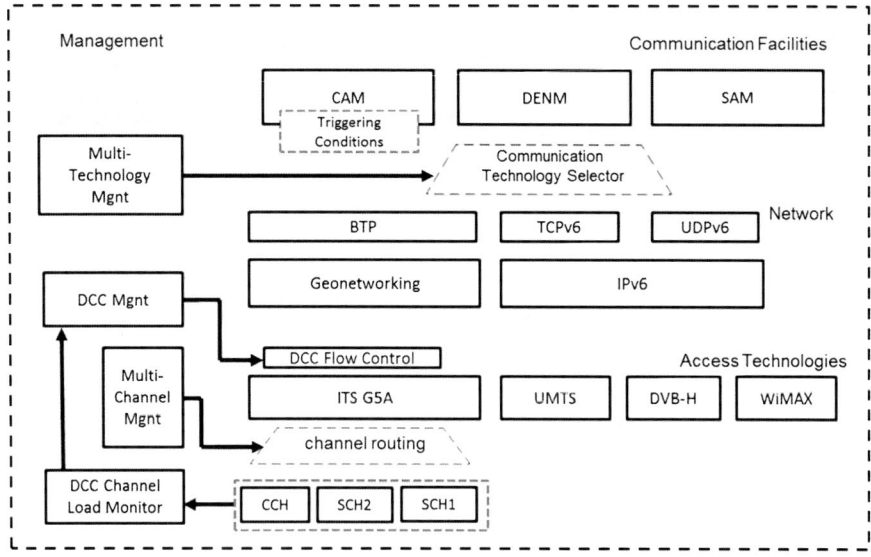

Fig. 13.5 ETSI ITS extensions on ns3

---

[1]http://www.colombo-fp7.eu/.

geo-broadcast, geo-anycast) and Delay Tolerant Networks (DTN) functions. On the lower layers, heterogeneous access technologies are available, including ETSI ITS-G5. The management side of iTETRIS supports *multi-channel operations*, as well as a *multi-technology selections*. Whereas the former one is capable of offloading traffic between ITS G5 channels, the latter is capable of off-loading to multiple the access technology and the network stack as function of the IVC application requirements. An ETSI compliant DCC is also available, first with a *DCC channel load monitor* function at the management layer, as well as a *DCC flow control* function in the data plane.

13.5.2 The OMNeT++ Simulation Environment

OMNeT++, now at version 4, is an Open Source simulation environment that is distributed free for non-commercial use [50]. A separate version of the same simulation environment which is licensed for commercial use is sold by *Simulcraft, Inc.* under the *OMNEST* brand. OMNeT++ comprises an IDE, an execution environment, and a simulation kernel. The IDE is based on Eclipse, enhanced with facilities to graphically assemble and configure simulations, as an alternative to editing the plain text files. The execution environment exists in two flavors. The command line based environment targets unattended batch runs on dedicated machines. The graphical environment better supports interactive interactions with components of a running simulation, allowing to directly monitor or alter internal states.

OMNeT++ enforces a strict separation of behavioral and descriptive code. All behavioral code (i.e., code specifying how simple modules handle and send messages, as well as how channels handle messages) is written as C++ code linking to the OMNeT++ kernel. All descriptive code (i.e., code declaring the structure of modules/channels and messages) is stored in plain-text *Message Definition* (msg) and *Network Description* (ned) files, respectively. All run-time configuration of modules is achieved by an *Initialization File* (ini). With all behavioral code being contained in a C++ program, OMNeT++ components can easily interface with third-party libraries and can be debugged using off-the-shelf utilities; thus, it lends itself equally well to rapid prototyping and developing production quality applications.

Building on the discrete event simulation kernel are several module libraries modeling various protocol stacks. One popular example is the *MiXiM* [29] module library, which is focusing on accurate channel modeling and signal processing. Signals at a certain location are modeled as three-dimensional entities whose power level varies over both time and frequency. Calculating how such signals propagate in a simulation, as well as how they interfere with each other, is handled by MiXiM itself with no further effort from the model developer required. Thus, MiXiM lends itself very well to IVC simulation, where accurate models of common upper-layer Internet protocols matter less than precise simulation of wireless transmissions. If this is desired, another module library, the *INET Framework*, focuses on accurate representations of IPv4 and IPv6 as well as Internet transport layer protocols and applications, while overlay networks and peer-to-peer networking is the focus of the *Oversim* module library extension. Cellular networks are handled by yet other module libraries, like *SimuLTE* for simulating LTE and LTE advanced. Among the IVC specific models that are configured for use in OMNET++ are path loss models such as Two Ray Interference, shadowing by buildings and vehicles, IEEE 802.11p DSRC, and IEEE 1609.4 WAVE as well as auxiliary models such as driver behavior and emission computation. Models for the ETSI ITS G5 protocol stack, most importantly the DCC have been implemented. OMNeT++ acts as a framework for executing simulations that are assembled from these module libraries.

13.5.3 JiST/SWANS

The scalable wireless network simulator SWANS is built atop the JiST platform [3]. The JiST design is used to achieve high simulation throughput, save memory, and run standard Java network applications over simulated networks. JiST/SWANS programming code is open source and released under the Cornell Research Foundation license.

Particularly for vehicular ad hoc networks, a number of extensions and improvements for JiST/SWANS were developed at Ulm University [41]. For example, the DUCKS tool uses configuration files to define a complete setup of a simulation study. Moreover, the DUCKS tool supports an extensible model for storing result data. A configuration file is used to set up how results are saved. Thus, simulation results can be stored, for instance, in a MySQL database.

13.5.4 The SUMO Simulation Environment

Today, a huge number of simulation environments exist which implement traffic microsimulation models. In the interest of comparability of research results, however, it is evidently more beneficial to use readily available Free and Open Source Software simulation environments. In this section, we briefly introduce the most popular of those [23], the SUMO [30] microscopic road traffic simulation environment. This simulator is in widespread use in the research community, which makes it easy to compare results from different simulations.

SUMO is Free and Open Source Software licensed under the *GNU General Public License* (version 2 or later), is highly portable, and allows high-performance simulations of multi-modal traffic in city-scale networks. Simulations in SUMO can be run both with and without the OpenGL-based GUI, which allows for direct interaction with a running simulation. In order to afford accurate simulations of a large number of vehicles, SUMO was designed to incorporate an adaptation of the aforementioned microscopic vehicle mobility model described by Kraußand, more recently, more complex mobility models such as the IDM model. The parameterization of vehicles can be freely chosen with each vehicle following a statically assigned route, a dynamically generated route, or driving according to a configured timetable. Traffic flows can be assigned manually, computed based on demand data, or generated completely at random. Each road in SUMO can consist of multiple lanes, each of which can be restricted to be usable only by certain vehicle classes. Individual lanes can have any shape and can be interconnected with junctions, with inter-junction traffic being regulated by simple right-of-way rules, by fixed-program traffic lights, or by demand-actuated traffic lights. For individual traffic, vehicle trips are generated either from Origin/Destination (OD) matrices or following random turns at intersections and a wide variety of mobility related metrics are collected.

The *TraCI* interface allows the interconnection of external control functions to SUMO. TraCI provides flexible open APIs to retrieve metrics (e.g., for combining them with network metrics into ITS-specific metrics) and to control most of the SUMO parameters, from the mobility model, road network, to traffic light control. Accordingly, the TraCI interface allows to add new functionalities to SUMO or connect it to other tools. For instance, it is possible to alter the microscopic model or even implement a completely new one, or interconnect an external traffic light control module to SUMO.

13.5.5 VISSIM

The commercial simulation software VISSIM [32] is a microscopic traffic simulator based on a psycho-physical driver behavior model for vehicles and a social force model for pedestrians. Simulated traffic scenarios can consist of cars, trucks, buses, two-wheelers (bicycles, motorcycles), and public transport (bus, tram, underground). Furthermore, VISSIM includes a dedicated pedestrian model. For individual traffic, vehicle trips are generated by OD matrices and for public transport, schedules are defined. The routing through the traffic network is done by predefined routes.

VISSIM provides interfaces to add new functionality or connect other tools. A DLL-interface allows to implement new driver-behavior models. A COM-interface can be used to have read and write access to simulation data during runtime of a simulation. As a result, a coupling with other simulation tools is possible allowing an interaction during the runtime of the simulation.

13.6 Integrated IVC Simulation Toolkits

A key difference between the evaluation of wireless communication application and IVC application is that in the former, mobility is considered simply as a perturbation, whereas in the latter mobility is the application and must be able to be dynamically

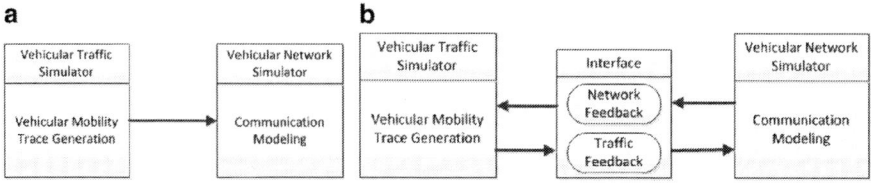

Fig. 13.6 Interactions between network and traffic simulators. (**a**) Isolated case. (**b**) Federated case

altered. Moreover, the various specific expertise in the required models for IVC applications, in particular when considering compliance to standards, makes it difficult to integrate all required models in a single simulator. Considering that each community (mobility, application, communication, etc.) traditionally relies on their specific simulators, an efficient strategy is to federate the required simulators in an IVC simulation framework.

In particular, when communication and mobility must interact, the traditional approach is to extract synthetic traces from traffic simulators in the shape of mobility files that may later be integrated into the network simulator using a dedicated parser (see Fig. 13.6a). This approach remains a favorite choice when mobility does not need to be influenced by vehicular communication or networking. Considering IVC applications, mobility shall be influenced, either to avoid accident or to reduce congestion. A bi-directional interaction is created between two (or more) simulators to be able to exchange mobility data or to influence the control of one or the other (see Fig. 13.6b). This approach is currently the favorite choice for evaluation of IVC applications.

There has been substantial effort leading to the development of quite a number of integrated IVC simulation frameworks. In the last years, consolidation efforts helped these efforts to converge. In the following, we introduce three promising candidates: Veins [46], iTETRIS [38], and VSimRTI [42].

13.6.1 Veins

Veins (Vehicles in Network Simulation) [46] is a simulation framework that is built on the aforementioned OMNeT++ simulation environment. It employs the OMNeT++ simulation kernel for discrete event simulation, that is, all simulation control and data collection is performed by OMNeT++. Veins instantiates SUMO to model vehicle movement and it provides a modular framework for the simulation of custom applications. In order to abstract away from discrete event simulation of wireless channels, e.g., managing event routing between nodes and modeling signal processing, the aforementioned MiXiM model suite is used, whereas dedicated model libraries are used for simulating, e.g., Internet protocols or cellular network communication.

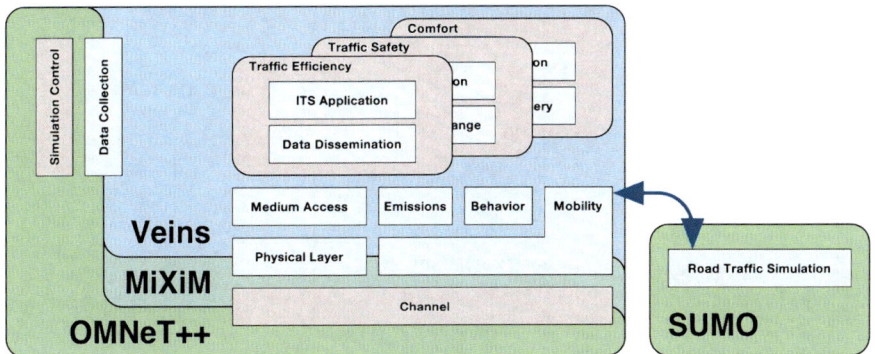

Fig. 13.7 The Veins simulation framework

As illustrated in Fig. 13.7, Veins builds on this basis to provide a suite of models that can then, in turn, serve as a modular framework for simulating applications. Based on the suite of IVC models available in OMNET++, custom and application-specific data generation and dissemination protocols can be implemented, e.g., for traffic safety, traffic efficiency, and infotainment applications.

Such applications' modules, and all used modules of Veins, are compiled and linked into an executable that can be run as a GUI application or as a command line batch simulation. Vehicle movement is simulated by a separate instance of the aforementioned SUMO road traffic simulator, which is instantiated, then controlled by the running simulation. In order to improve efficiency, Veins makes use of *object subscriptions* integrated [46] in SUMO, which allow it to request push notifications and updates from a running simulation, e.g., when vehicles are created or their state changes. The combination of precise channel and access models, behavior, and mobility feedback enables capturing a wide range of aspects necessary for, e.g., investigating intersection collision avoidance approaches [24].

More information, full source code, a beginner's tutorial, and related publications are available from the Veins website.[2]

13.6.2 iTETRIS

The architecture of iTETRIS [38], illustrated in Fig. 13.8, is centered around the iTETRIS Control System (iCS) interface. Although being simulator agnostic, it federates the traffic simulator SUMO, the network simulator ns-3, and one or multiple instances of an ITS application simulator.

---

[2]http://veins.car2x.org/.

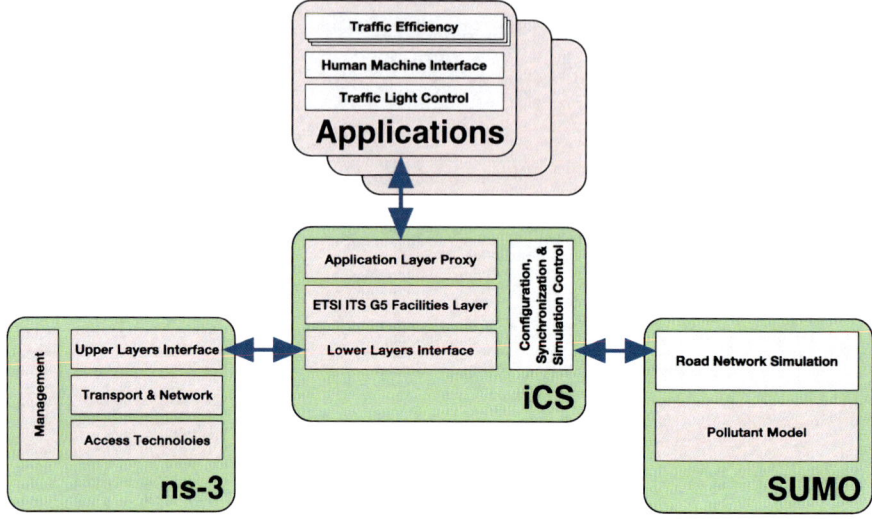

Fig. 13.8 The iTETRIS simulation platform

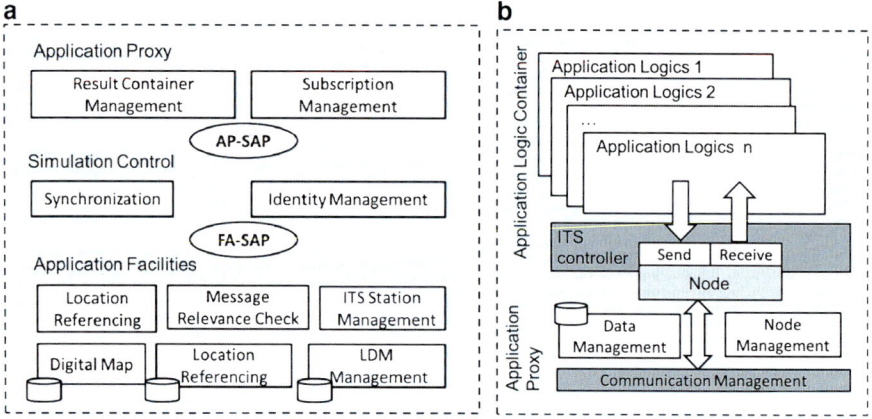

Fig. 13.9 Illustration of the architecture of the iTETRIS (**a**) iCS and (**b**) IVC applications [4]

The role of the iCS is yet more important, as it also integrates the IVC application aspects of the ETSI ITS Facilities (see Fig. 13.9a), such as the *ITS station management* or *local dynamic maps*. It also contains a *result container* used to store and exchange generic data between multiple instances of the IVC application module. The most important module of the iCS is yet the *subscription* module, which implements a set of open APIs controlling the exchange of information between ns-3, SUMO, and the IVC applications. This particularity allows the simulation of the dynamic interactions between mobility and networking on IVC applications.

13 Simulation Tools for VANET

A very unique aspect of iTETRIS is the IVC application simulator (see Fig. 13.9b). It allows the evaluation of ITS applications with minimal efforts by embedding the application logics outside of the main other simulators. The architecture of the IVC simulator is separated in two layers. The first layer handles all connection primitives between the IVC application logics, the iCS, SUMO, and ns3. A *Payload Storage* block is also available to keep IVC data local to the IVC application simulator. This is typically used to increase scalability, as the packet-level granularity of ns-3 makes that it does not need real data in the payload of the simulated packets. The second layer is a container for the IVC application logics. This layer has been extended by Bellavista et al. [4] with an innovative flexible higher-layer *Node* architecture that can support basic *send* and *receive* primitives very similarly to ns-3, as well as an *ITS controller* playing the role of an interface and a coordinator of IVC applications logics. This architecture allows an easy integration and interaction between the IVC application logics, the application simulator, and the other iTETRIS modules, both from modeling and simulation runtime perspective.

Finally, as IVC application logics running on multiple instances of the IVC application simulator need to exchange information, a specific open API allows for their interactions over the iCS. This is typically critical for parallel development of complex IVC applications, where, for instance, one IVC application is in charge of monitoring traffic, while another one provides personalized navigation services.

More details related to the iCS and the IVC application simulator may be found in[18], and we refer interested readers to the iTETRIS community website[3] for more details on the iTETRIS platform.

13.6.3 VSimRTI

VSimRTI (Vehicle-to-X (V2X) Simulation Runtime Infrastructure) [42] goes one step further in decoupling individual components. It is a generalized framework coupling different simulators, each for a particular domain, following an ambassador concept inspired by some fundamental concepts of the IEEE Standard for Modeling and Simulation (M&S) High Level Architecture (HLA) [22]. All management tasks, such as synchronization, interaction, and life-cycle management are handled completely by VSimRTI (see lower part of Fig. 13.10). Several optimization techniques, such as optimistic synchronization, target high performance simulations [35].

By implementing the generic VSimRTI interfaces (see upper part of Fig. 13.10), an easy integration and exchange of simulators is possible. Consequently, the deployment of simulators is enabled for each particular domain—allowing a realistic presentation of vehicular traffic, emissions, wireless communication (cellular and ad-hoc), user behavior, and the modeling of mobility applications [42].

---

[3]http://www.ict-itetris.eu/.

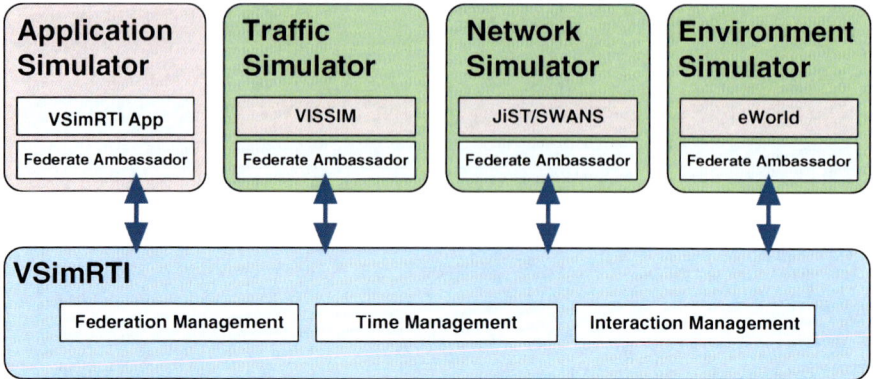

Fig. 13.10 VSimRTI sample simulator coupling

For immediate use, a set of simulators is already coupled with VSimRTI: the traffic simulators VISSIM and SUMO; the communication simulators ns-3, OMNeT++, JiST/SWANS, and a cellular communication simulator; a Java-based application simulator; and several visualization and analysis tools. Furthermore, VSimRTI is currently being extended to allow the simulation of electric mobility scenarios.

The application simulator of VSimRTI is optimized for the simulation of V2X applications. Applications run in a sandbox which offers vehicle-like interfaces, e.g., for requesting sensor data or interacting with communication modules. Data provided by traffic, communication network, and further connected simulators is transformed into a format used by components of real vehicles. To run an application, its logic is implemented in Java. Additionally, the VSimRTI application simulator supports various settings to specify the characteristics of an application and to configure its behavior, e.g., the CAM sending rate or the conditions for broadcasting a DENM can be defined [36].

Further information can be found on the VSimRTI website.[4]

13.7 Discussion

We provided in this chapter a description of the specific simulation modeling requirements from the three different classes of IVC applications: *safe*, *smart*, and *connected* mobility. Considering propagation, channel access, networking, and mobility, we classified the various available levels of granularity available and classified them as function of their applicability to each class of IVC applications.

---

[4]http://www.dcaiti.tu-berlin.de/research/simulation/.

One challenge is to identify the right trade-off between scalability, granularity, and applicability. To this objective, we provided insight of the models critically required or not necessary to conduct simulations of IVC according to the right trade-off.

Another challenge is to use simulators that contain the required models and support the necessary level of granularity for each specific class of IVC applications. We summarized the current state of the art in simulation models and techniques and presented available integrated IVC simulation toolkits. The primary lesson learned is that the quality and comparability of simulation studies in the ITS world clearly improves over time, given that most of the described simulators include all major models at various levels of granularity. This and the tight integration of simulation components to create a unified bi-directionally coupled simulation of aspects has become a cornerstone of modern simulative performance studies. The consolidation efforts led to the emergence of three integrated and well-accepted toolkits: Veins, iTETRIS, and VSimRTI. From our investigations, there is no clear winner among the three, all cover the aforementioned required aspects; the choice mainly depends on the protocols or applications under study.

In general, the credibility of IVC simulation studies, and most importantly, the reproducibility can substantially be increased by using just one of these toolkits. The main reason is that validated and openly available models that are unique in IVC can be used. This includes specific propagation, channel access, networking and realistic mobility models, which may be tightly and dynamically coupled and integrated with IVC applications.

Acknowledgements EURECOM acknowledges the support of its industrial members: SFR, Orange, ST Microelectronics, BMW Group, SAP, Monaco Telecom, Symantec, IABG.

References

1. Ahmed S, Karmakar GC, Kamruzzaman J (2010) An environment-aware mobility model for wireless ad hoc network. Elsevier Comput Netw 54(9):1470–1489
2. Baiocchi A, Cuomo F (2013) Infotainment services based on push-mode dissemination in an integrated VANET and 3G architecture. J Commun Netw 15(2):179–190. doi:10.1109/JCN.2013.000031
3. Barr R, Haas ZJ, van Renesse R (2005) JiST: an efficient approach to simulation using virtual machines. Softw Pract Experience 35(6):539–576. doi:10.1002/spe.647
4. Bellavista P, Caselli F, Foschini L (2014) Implementing and evaluating v2x protocols over itetris: traffic estimation in the colombo project. In: Proceedings of the 4th ACM international symposium on development and analysis of intelligent vehicular networks and applications, DIVANet '14, pp 25–32
5. Bieker L, Krajzewicz D, Röckl M, Cappelle H (2010) Derivation of a fast, approximating 802.11p simulation model. In: 10th International conference on intelligent transport systems telecommunications (ITST 2010), Kyoto
6. Boban M, Vinhosa T, Barros J, Ferreira M, Tonguz OK (2011) Impact of vehicles as obstacles in vehicular Networks. IEEE J Sel Areas Commun 29(1):15–28
7. Camp T, Boleng J, Davies V (2002) A survey of mobility models for ad goc network research. Wirel Commun Mob Comput 2(5):483–502. Special Issue on Mobile Ad Hoc Networking: Research, Trends and Applications

8. Chen Q, Jiang D, Taliwal V, Delgrossi L (2006) IEEE 802.11 based vehicular communication simulation design for NS-2. In: 3rd ACM international workshop on vehicular ad hoc networks (VANET 2006). ACM, Los Angeles, CA, pp 50–56. doi:10.1145/1161064.1161073
9. Cheng W, Cheng X, Znati T, Lu X, Lu Z (2009) The complexity of channel scheduling in multi-radio multi-channel wireless Networks. In: 28th IEEE conference on computer communications (INFOCOM 2009), Rio de Janeiro, pp 1512–1520. doi:10.1109/INFCOM.2009.5062068
10. Chrysanthou C, Bertoni HL (1990) Variability of sector averaged signals for UHF propagation in cities. IEEE Trans Veh Technol 39(4):352–358. doi:10.1109/25.61356
11. Dressler F, Sommer C (2010) On the impact of human driver behavior on intelligent transportation systems. In: 71st IEEE vehicular technology conference (VTC2010-Spring). IEEE, Taipei, pp 1–5. doi:10.1109/VETECS.2010.5493964
12. Dressler F, Sommer C, Eckhoff D, Tonguz OK (2011) Towards realistic simulation of inter-vehicle communication: models, techniques and pitfalls. IEEE Veh Technol Mag 6(3):43–51. doi:10.1109/MVT.2011.941898
13. van Eenennaam M, Wolterink W, Karagiannis G, Heijenk G (2009) Exploring the solution space of beaconing in VANETs. In: 1st IEEE vehicular networking conference (VNC 2009). IEEE, Tokyo. doi:10.1109/VNC.2009.5416370
14. European Telecommunications Standards Institute (2009) Intelligent transport systems (ITS); European profile standard for the physical and medium access control layer of Intelligent Transport Systems operating in the 5 GHz frequency band. ES 202 663 V1.1.0, ETSI
15. Fiore M, Härri J (2008) The networking shape of vehicular mobility. In: 9th ACM international symposium on mobile ad hoc networking and computing (Mobihoc 2008). ACM, Hong Kong, China, pp 261–272
16. Font J, Iñigo P, Domínguez M, Sevillano JL, Amaya C (2010) Architecture, design and source code comparison of ns-2 and ns-3 network simulators. In: 2010 spring simulation multiconference (SpringSim 2010). SCS, Orlando, FL
17. Harri J, Filali F, Bonnet C (2009) Mobility models for vehicular ad hoc networks: a survey and taxonomy. IEEE Commun Surv Tutorials 11(4):19–41
18. Härri J, Cataldi P, Krajzewicz D, Blokpoel RJ, Lopez Y, Leguay J (2011) Modeling and simulating its applications with itetris. In: MSWiM'11, 14th ACM international conference on modeling, analysis and simulation of wireless and mobile systems
19. Hartenstein H, Laberteaux KP (2008) A tutorial survey on vehicular ad hoc networks. IEEE Commun Mag 46(6):164–171
20. Institute of Electrical and Electronics Engineers (2010) Wireless access in vehicular environments. Std 802.11p-2010, IEEE
21. Institute of Electrical and Electronics Engineers (2011) IEEE trial-use standard for wireless access in vehicular environments (WAVE) - multi-channel operation. Std 1609.4, IEEE
22. Institute of Electrical and Electronics Engineers and IEEE-SA Standards Board (2000) IEEE standard for modeling and simulation (M&S) high level architecture (HLA): framework and rules. IEEE Standard 1516. IEEE standard, Institute of Electrical and Electronics Engineers
23. Joerer S, Dressler F, Sommer C (2012) Comparing apples and oranges? Trends in IVC simulations. In: 9th ACM international workshop on vehicular internetworking (VANET 2012). ACM, Low Wood Bay, pp 27–32. doi:10.1145/2307888.2307895
24. Joerer S, Segata M, Bloessl B, Lo Cigno R, Sommer C, Dressler F (2014) A vehicular networking perspective on estimating vehicle collision probability at intersections. IEEE Trans Veh Technol 63(4):1802–1812. doi:10.1109/TVT.2013.2287343
25. Karagiannis G, Altintas O, Ekici E, Heijenk G, Jarupan B, Lin K, Weil T (2011) Vehicular networking: a survey and tutorial on requirements, architectures, challenges, standards and solutions. IEEE Commun Surv Tutorials 13(4):584–616. doi:10.1109/SURV.2011.061411.00019
26. Katsaros K, Kernchen R, Dianati M, Rieck D, Zinoviou C (2011) Application of vehicular communications for improving the efficiency of traffic in urban areas. Wirel Commun Mob Comput 11(12):1657–1667. doi:10.1002/wcm.1233

27. Killat M, Hartenstein H (2009) An empirical model for probability of packet reception in vehicular ad hoc networks. EURASIP J Wirel Commun Netw 2009(1):721301
28. Klingler F, Dressler F, Cao J, Sommer C (2013) Use both lanes: multi-channel beaconing for message dissemination in vehicular networks. In: 10th IEEE/IFIP conference on wireless on demand network systems and services (WONS 2013). IEEE, Banff, pp 162–169
29. Köpke A, Swigulski M, Wessel K, Willkomm D, Haneveld PTK, Parker TEV, Visser OW, Lichte HS, Valentin S (2008) Simulating wireless and mobile networks in OMNeT++ – the MiXiM vision. In: 1st ACM/ICST international conference on simulation tools and techniques for communications, networks and systems (SIMUTools 2008): 1st ACM/ICST international workshop on OMNeT++ (OMNeT++ 2008). ACM, Marseille
30. Krajzewicz D, Hertkorn G, Rössel C, Wagner P (2002) SUMO (Simulation of Urban MObility); An open-source traffic simulation. In: 4th Middle east symposium on simulation and modelling (MESM 2002), Sharjah, pp 183–187
31. Lee KC, Lee U, Gerla M (2009) TO-GO: topology-assist geo-opportunistic routing in urban Vehicular Grids. In: 6th IEEE/IFIP conference on wireless on demand network systems and services (WONS 2009). IEEE, Snowbird, UT, pp 11–18
32. Lownes NE, Machemehl RB (2006) VISSIM: a multi-parameter sensitivity analysis. In: 38th winter simulation conference (WSC '06). IEEE, Monterey, CA, pp 1406–1413. doi:10.1109/WSC.2006.323241
33. Mittag J, Papanastasiou S, Hartenstein H, Strom E (2011) Enabling accurate cross-layer phy/mac/net simulation studies of vehicular communication networks. Proc IEEE 99(7):1311–1326
34. Nagel R, Eichler S (2008) Efficient and realistic mobility and channel modeling for VANET scenarios using OMNeT++ and INET-Framework. In: 1st ACM/ICST international conference on simulation tools and techniques for communications, networks and systems (SIMUTools 2008). ICST, Marseille, pp 1–8
35. Naumann N, Schünemann B, Radusch I, Meinel C (2009) Improving v2x simulation performance with optimistic synchronization. In: Services computing conference, 2009. APSCC 2009. IEEE Asia-Pacific, pp 52–57. doi:10.1109/APSCC.2009.5394142
36. Protzmann R, Schünemann B, Radusch I (2014) A sensitive metric for the assessment of vehicular communication applications. In: IEEE 28th international conference on advanced information networking and applications (AINA), 2014, pp 697–703. doi:10.1109/AINA.2014.86
37. Rappaport TS (2009) Wireless communications: principles and practice, 2nd edn. Prentice Hall, Upper Saddle River
38. Rondinone M, Maneros J, Krajzewicz D, Bauza R, Cataldi P, Hrizi F, Gozalvez J, Kumar V, Röckl M, Lin L, Lazaro O, Leguay J, Härri J, Vaz S, Lopez Y, Sepulcre M, Wetterwald M, Blokpoel R, Cartolano F (2013) iTETRIS: a modular simulation platform for the large scale evaluation of cooperative ITS applications. Simul Modell Pract Theory 34:99–125. doi:10.1016/j.simpat.2013.01.007
39. Saha AK, Johnson DB (2004) Modeling mobility for vehicular ad-hoc networks. In: 1st ACM workshop on vehicular ad hoc networks (VANET 2004), Philadelphia, PA, pp 91–92
40. Saunders SR, Aragón-Zavala A (2007) Antennas and propagation for wireless communication systems, 2nd edn. Wiley, Ney York
41. Schoch E, Feiri M, Kargl F, Weber M (2008) Simulation of ad hoc networks: ns-2 compared to JiST/SWANS. In: 1st ACM/ICST international conference on simulation tools and techniques for communications, networks and systems (SIMUTools 2008). ICST, Marseille
42. Schünemann B (2011) V2X simulation runtime infrastructure VSimRTI: an assessment tool to design smart traffic management systems. Elsevier Comput Netw 55(14):3189–3198. doi:10.1016/j.comnet.2011.05.005
43. Sommer C, Dressler F (2008) Progressing toward realistic mobility models in VANET simulations. IEEE Commun Mag 46(11):132–137. doi:10.1109/MCOM.2008.4689256
44. Sommer C, Dressler F (2011) Using the right two-ray model? A measurement based evaluation of PHY models in VANETs. In: 17th ACM international conference on mobile computing and networking (MobiCom 2011), Poster Session. ACM, Las Vegas, NV

45. Sommer C, Eckhoff D, German R, Dressler F (2011) A computationally inexpensive empirical model of IEEE 802.11p radio shadowing in Urban Environments. In: 8th IEEE/IFIP conference on wireless on demand network systems and services (WONS 2011). IEEE, Bardonecchia, pp 84–90. doi:10.1109/WONS.2011.5720204
46. Sommer C, German R, Dressler F (2011) Bidirectionally coupled network and road traffic simulation for improved IVC analysis. IEEE Trans Mobile Comput 10(1):3–15. doi:10.1109/TMC.2010.133
47. Stepanov I, Rothermel K (2008) On the impact of a more realistic physical layer on MANET simulations results. Elsevier Ad Hoc Netw 6(1):61–78
48. Torrent-Moreno M, Schmidt-Eisenlohr F, Füßler H, Hartenstein H (2006) Effects of a realistic channel model on packet forwarding in vehicular ad hoc networks. In: IEEE wireless communications and networking conference (WCNC 2006). IEEE, Las Vegas, NV, pp 385–391. doi:10.1109/WCNC.2006.1683495
49. Torrent-Moreno M, Mittag J, Santi P, Hartenstein H (2009) Vehicle-to-vehicle communication: fair transmit power control for safety-critical information. IEEE Trans Veh Technol 58(7):3684–3703. doi:10.1109/TVT.2009.2024873
50. Varga A, Hornig R (2008) An overview of the OMNeT++ simulation environment. In: 1st ACM/ICST international conference on simulation tools and techniques for communications, networks and systems (SIMUTools 2008). ACM, Marseille
51. Werner M, Lupoaie R, Subramanian S, Jose J (2012) MAC layer performance of ITS G5 - optimized DCC and advanced transmitter coordination. In: 4th ETSI TC ITS Workshop, Doha, Qatar
52. Willke TL, Tientrakool P, Maxemchuk NF (2009) A survey of inter-vehicle communication protocols and their applications. IEEE Commun Surv Tutorials 11(2):3–20

Chapter 14
Field Operational Tests and Deployment Plans

Yvonne Barnard, François Fischer, and Maxime Flament

Abstract In this chapter an explanation is given of Field Operational Tests (FOT), studies to evaluate the impact of Intelligent Transport Systems (ITS) in the real world. The methodology for designing and conducting these tests is described. Different types of FOT can be distinguished, testing advanced driver assistance systems and nomadic devices, as well as cooperative systems allowing communication between vehicles and between vehicles and infrastructure. This last type of test is discussed in more detail, addressing the question of how they can be used for testing vehicular ad hoc network (VANET) technologies, and examples are given of both European and US projects. Specifically the DRIVE C2X project is described, as it explicitly addresses these technologies. Conducting FOTs is an important step on the way to the deployment of ITS. Also, data deriving from these studies can be deployed, in new projects, to answer new research questions. The chapter concludes with a description of the networking and community building activities in the Field Operational Test domain.

Keywords VANET • FOT • FESTA • Deployment • Cooperative systems • Data sharing • Data re-use • FOT-Net • FOT-Net Data • European projects • DRIVE C2X • ITS Spot • Connected Vehicle Safety Pilot • Stakeholders • Standardisation • Evaluation

14.1 Introduction

In order to better understand the impact of the introduction of Intelligent Transport Systems (ITS), Field Operational Tests (FOTs), where ITS services are tried out in the real world, are an important tool [1]. Although other research tools such as

---

Y. Barnard (✉)
Institute for Intelligent Transport Studies, University of Leeds, Leeds, UK

ERTICO - ITS Europe, Bruxelles, Belgium
e-mail: y.barnard@leeds.ac.uk

F. Fischer • M. Flament
ERTICO - ITS Europe, Bruxelles, Belgium
e-mail: f.fischer@mail.ertico.com; m.flament@mail.ertico.com

simulators and test-tracks may provide insight in the functioning of ITS, trying them out in everyday traffic is an essential step in the evaluation of their performance, their acceptance by users, their influence on the behaviour of other traffic participants, and the impact they may have on the general network and on society. In FOTs, a wide variety of services can be tested, varying from advanced driver assistance systems (ADAS) such as intelligent speed adaptation, to nomadic devices such as navigation apps on a smart phone, to services such as traffic information.

Over the last decades many new technologies for ITS have been researched and developed, in Europe often supported by the European Commission (EC) Research Framework Programme as well as by national programmes. There is a growing need to understand the short- and long-term impact of these systems to answer questions which are crucial for market introduction and penetration. By testing the systems on a large scale, in real driving conditions during a significant period of time, FOTs may provide answers to these questions. FOTs are expected to test close-to-the-market systems in sufficiently large vehicle fleets, in order to provide enough real operational data to carry out robust impact assessments, and thus obtain sound statistical conclusions at the European level [2].

The results of FOTs enable policy makers to establish the right policy framework for the deployment of these systems, and industry to make informed decisions about their market introduction.

In FOTs, ITS are implemented in the vehicles of ordinary drivers and/or in the infrastructure. Drivers use their vehicles under normal conditions, such as they are used to, while all kinds of data are being collected. Data may concern the behaviour of the vehicle, the system, and the drivers themselves, as well as the interactions between them. Data may be collected by recording equipment in the vehicle such as video cameras or special data acquisition systems recording Controller Area Network (CAN) bus data, Global Positioning System (GPS) position, status of the ITS, etc. By collecting and analysing data about driving under normal conditions with and without the system tested, it becomes possible to evaluate the impact of the system.

The more formal definition of a FOT is: "A study undertaken to evaluate a function, or functions, under normal operating conditions in road traffic environments typically encountered by the participants using study design so as to identify real world effect and benefits". This definition is from the so-called FESTA handbook [3], which describes the FOT methodology that was developed by a large number of experts and stakeholders involved in ITS. We will describe this methodology in more detail below.

There are a few elements in this definition that should be noted. Firstly, the definition talks about a function or functions, and not about systems. Functions are focussed on what an ITS will do for a driver, for example, the navigation function or a speed alert function. These functions can be implemented using different systems, for example, there are different systems, from different suppliers, that provide navigation functions. A system is just a combination of hardware and software that enables functions. FOTs are usually interested in the effects of making a function available to users and less in the specific implementation of a system or its user interface. Questions like "how does the use of advanced cruise control influence the driving behaviour?" are more the focus of a FOT than differences in the manner of implementation of advanced cruise control in systems A and B.

The element "under normal operating conditions in road traffic environments typically encountered by the participants" expresses the difference with studies using driving simulators, test-tracks or professional test-drivers. If the functions to be tested are aimed at the general public, non-professional drivers will be asked to drive in their own vehicles (or their lease cars) that are instrumented with the functions and data acquisition equipment, and to drive as they would normally do, going to work, visit friends and family, do their courses, etc. Drivers do not get specific instructions and their data is recorded automatically. Participants usually drive over longer periods of time, at least several weeks, but often months or even a year. If the functions are aimed at specific types of drivers or vehicles, such as taxis or trucks, the professional drivers will drive as they normally do on their job. So FOTs are usually not using employees from the system manufacturer asked to drive specific roads under specific weather conditions, etc.

The third element "using study design so as to identify real world effect and benefits" is especially noteworthy. In this definition FOTs are seen as a scientific evaluation. It is not just about letting participants drive around and making a subjective statement about the effects. A FOT should have a carefully planned study design, allowing for comparison between situations with and without using the function, and use scientifically and statistically sound methods for analysing the data, and for determining the impact of the function.

FOTs have become an important tool in Europe, in the form of large-scale testing programmes aiming at a comprehensive assessment of the efficiency, quality, robustness and acceptance of ITS used for smarter, safer and cleaner, and more comfortable transport solutions. A number of such FOTs have been conducted at national and European levels, to evaluate a range of systems, particularly a variety of driver support systems. These FOTs have been identified as an important means of verifying the real-world impacts of new functions and as a means to verify that previously conducted research and development have the potential to deliver identifiable benefits.

FOTs are often focused on cars and trucks, but other vehicles are used as well, such as powered two-wheelers and even bicycles.

In this chapter we will address the question of how FOTs can be used to study the impact of ITS systems, and especially systems dealing with communication between vehicles, and between vehicles and the infrastructure. First the methodology for designing and conducting these tests will be described. Different types of FOT are distinguished in the next section, testing advanced driver assistance systems and nomadic devices, as well as cooperative systems. For each type some example projects are given. FOTs on cooperative systems are discussed in more detail, addressing the question of how they can be used for testing vehicular ad hoc network (VANET) technologies. Examples are given of both European and US projects. Specifically the DRIVE C2X project is described, as it explicitly addresses these technologies. Next the question of how the results of FOTs can be used for deployment purposes is addressed. Not only results but also the data gathered in FOTs may be deployed, in new projects, to answer new research questions. The chapter concludes with a description of the networking and community building activities in the FOT domain.

14.2 The FESTA Methodology for Designing and Conducting FOTs

As FOTs aim at providing results that are evidence-based, it is necessary to follow a structured methodology. This is also important if we want to compare results from different FOTs. A common and well-agreed methodology allows delivering insight into how ITS in general contribute to policy objectives such as safety, sustainability and mobility.

The EC-funded Field opErational teSt supporT Action (FESTA) project in 2008 developed a methodology for planning and executing FOTs. In this project a large number of European organisations, Original Equipment Manufacturers (OEMs), research organisations and other stakeholders collaborated to develop a structured approach, and to reach consensus about the methods to be recommended. This methodology has been updated several times based on the experiences from FOTs that used the methodology; in 2014 version 5 was completed [3]. The methodology is described in a handbook which is available on-line. The main purpose of this handbook is to provide guidelines for conducting FOTs. It takes the reader through the whole process of planning, preparing, executing, analysing and reporting a FOT, and it gives information about aspects that are especially relevant for this type of large study, such as administrative, logistic, legal and ethical issues. Another aspect of the handbook is to pave the way for standardisation of some aspects of FOTs, which would be helpful for cross-FOT comparisons. It has to be kept in mind, though, that many traffic parameters in different European countries differ substantially. The steps of the methodology are depicted as a V diagram (see Fig. 14.1).

The left top of the V covers setting up a goal for the study, and the last steps on the right include an overall analysis of the systems and functions tested, and the socio-economic impact assessment dealing with the more general aspects of a FOT, and with the aggregation of the results. The further down the FOT Chain V-Shape the steps are located, the more they focus on aspects with a high level of detail, such as which performance indicators to choose or how to store the data in a database. Ethical and legal issues have the strongest impact on those aspects where the actual contact with the participants, data handling and potential data sharing takes place.

In order to make the picture more complete a horizontal bar has been added on top of the diagram, summarising in principle the context in which the FOT is supposed to take place. For instance, the choice of a function to be tested implies that there is either a problem to be addressed that the chosen function is defined to

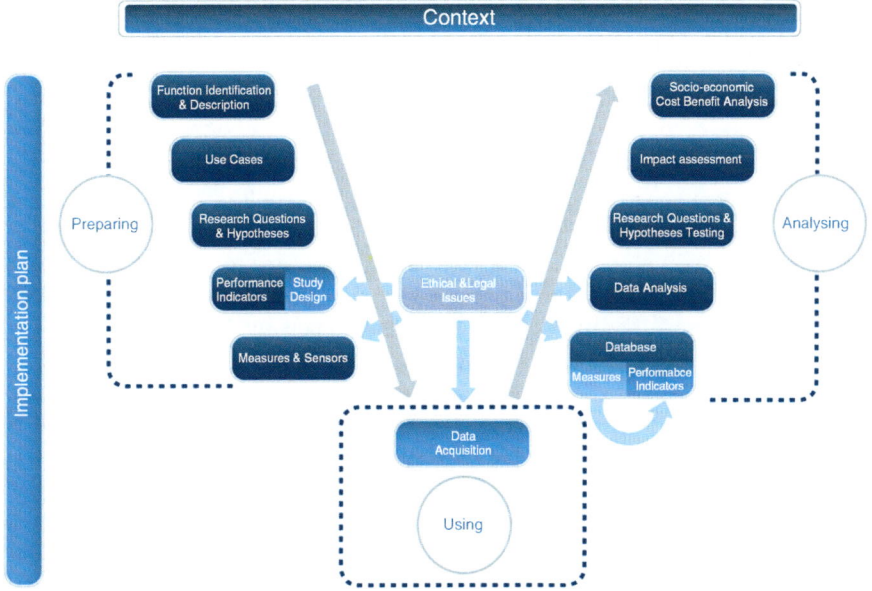

Fig. 14.1 The FESTA methodology

solve, or that a policy objective is stated and that the function tested can be used to reach the objective. A FOT can always be related to a wider perspective than is defined by just a description of the function to be tested.

The representation of the FESTA methodology in the form of a V does not mean that designing and performing a FOT is always a linear process. Decisions made at a certain stage of the FESTA-V influence the next steps, and it is likely that there will sometimes be a need to go back and review some steps. The FESTA-V provides a static picture of the complex design and execution process of a FOT, but in reality a more iterative process will be needed. The FOT Implementation Plan (FOTIP) integrates all steps into one big table which can be used as a reference when actually carrying out a FOT.

14.3 Different Types of Field Operational Tests

Originally the focus of FOTs was on the testing of in-vehicle systems, and especially driver support systems. Examples of these systems are speed alert, advanced cruise control and lane departure warning. These systems are usually developed by OEMs. Car manufactures are involved in these types of FOT, collecting valuable information about how their systems influence driver behaviour, how and when drivers use the systems, and what the effects would be of large-scale implementation. ADAS FOTs usually test mature systems that are close-to-market. The basic methodology is to compare driving with the system active, with a baseline driving without the system being available. The euroFOT project (see below) was a large-scale project evaluating ADAS systems [4].

A second type of system was tested by another large European project, TeleFOT (see below) that evaluated nomadic devices [5]. Nomadic devices are brought into the vehicle after its purchase, and may not even be connected to the vehicle. Examples are navigation systems and apps on smart phones. Manufacturers are often not OEMs, and functions may be hard to separate. For example, a navigation system may not only provide route advice but also actual traffic information. Since the FESTA methodology was developed and the first large European FOTs started, developments in functions of nomadic devices have been rapidly accelerating.

A different type of study also emerged: naturalistic driving studies (NDS). This started with the 100 car study in the USA [6], and is now also taking place in other countries in and outside Europe. Examples are the very large SHRP2 NDS [7–9], which followed the 100-car study, the European UDRIVE project [10, 11], and the Australian large-scale NDS [12]. In NDS the objective is not testing systems but observing driver behaviour. Drivers drive in the way they normally would, and their driving behaviour is being recorded. NDS may serve as a source of information for the baseline for other FOTs, providing answers to the question of how drivers behave without a system. However, as ITS become more and more commonly used, NDS often involve drivers using systems as part of their everyday driving. NDS share many characteristics with FOTs, being large-scale studies involving ordinary drivers. In the update of the FESTA methodology these studies were given their own place.

The most recent type of FOT concerns the evaluation of Cooperative Systems, including systems for vehicle-to-vehicle communication and infrastructure-to-vehicle (and vice-versa) communication. Two large-scale European projects concern this type of systems, DRIVE C2X [13] and FOTsis [14]. Cooperative systems may comprise a large variety of services, examples being information about traffic, accidents and road-works from traffic management, tolling services, and slippery-road warnings from other vehicles.

FOTs on Cooperative Systems bring their own specific problems. It is not always possible to have participants drive completely naturalistically; for example, the chances that two instrumented cars will be on the same spot so that car-to-car communication can take place may be few. Also new tools, new types of data, and new ways of performing impact assessment are needed. The latest revisions of FESTA especially addressed this type of FOT. Stakeholders in Cooperative Systems are usually manifold, including road operators, information and service providers, telecommunication companies, and emergency services.

14.4 Examples of Field Operational Test Projects

In Europe FOTs are part of the EC Research Framework Programme. Two large pan-European FOTs were conducted (co-funded by the EC) from 2008 to 2012, using the methodology: euroFOT and TeleFOT.

In euroFOT [4, 15, 16] for over 12 months, 1,000 cars and trucks equipped with advanced driver assistance systems travelled European roads. Their movements were tracked and parameters such as acceleration and lane changes recorded. The FOT focused on eight distinct vehicle functions that assisted drivers in detecting hazards and avoiding accidents: Adaptive Cruise Control (ACC), Forward Collision Warning (FCW), Speed Regulation System (SRS), Blind Spot Information System (BLIS), Lane Departure Warning (LDW), Curve Speed Warning (CSW), safe human/machine interface, and Fuel Efficiency Advisor (FEA). More than 100 TB of data were collected and analysed.

The TeleFOT project [5, 17] evaluated in-vehicle aftermarket and nomadic devices. Almost 3,000 drivers covering a combined distance of more than 10 million kilometres in eight European countries participated. Different functions were tested, covering two main areas: safe driving, and economic and fuel-efficient driving. The services tested included Static and Dynamic Navigation Support, Green Driving Support, Speed Limit Information, and Traffic Information. TeleFOT not only collected a huge amount of data through in-vehicle data loggers but also subjective data such as participant questionnaires. The data was further enriched by map matching and metadata which provided specific contextual details.

Next to these two large pan-European FOTs many FOT projects have been conducted at national or regional levels in Europe, testing a variety of systems. Also outside Europe, such as in the USA (the Safety Pilot is described in the next section) and Japan, but also in many other countries FOTs are used as an important tool to evaluate the effects of ITS. For example in South Korea, the Smart Highway Project addresses road-based, automobile-related and traffic management technologies, using wireless communications. The European support action FOT-Net [18] has set up a wiki in which information about currently 147 FOT and NDS projects is to be found from all over the world [19].

In recent years the so-called European pilot projects have gained momentum. These are projects that also test ICT systems under realistic traffic conditions, but they specifically concern systems that are close to deployment, paving the way to full implementation and deployment of the systems under evaluation. Ensuring interoperability of systems and services is also an important goal. Examples are the pilots on ICT-services for electromobility such as the projects smartCEM, MOLECULES, MOBI.Europe, and ICT4EVEU [20–23].

In the US Connected Vehicle Test Beds provide the opportunity for testing and certifying activities [24]. Test-beds are available in multiple locations such as Michigan, Virginia, California, New York, Minnesota and Maricopa County. In these test-beds, research, testing, and demonstration of connected transportation system concepts are performed with the aim to lead the way to deployment of systems and services using wireless communication between vehicles. Applications that are being tested concern (1) safety, providing warnings and advice, for example, warnings for slippery patches of roadway ahead, (2) mobility, providing a connected, data-rich travel environment based on information transmitted anonymously from thousands of cars and (3) environment, such as real-time information about traffic congestion and other travel conditions, helping to make trips more fuel-efficient.

14.5 How Can FOTs Be Used for Testing VANET Technologies?

Over the last few years a shift has been taking place from performing FOTs on single systems such as Intelligent Speed Adaptation systems, to more complex integrated systems, especially Cooperative Systems. Cooperative Systems are vehicle systems based on vehicle-to-vehicle (V2V) or car-to-car (C2C), vehicle-to-infrastructure (V2I) or car-to-infrastructure (C2I) and infrastructure-to-vehicle (I2V) or infrastructure-to-car (I2C) communication technology. The development of safety-critical V2V systems in Europe has mainly been promoted by the Car-to-Car Communication Consortium (C2C-CC) [25]. More recently, the EC has placed emphasis on the application of Cooperative Systems to achieve environmental and efficiency impacts.

Cooperative Systems differ from other FOTs, and this directly influences their planning and operation. Cooperative systems are not yet common and drivers are not familiar with their functionality. Another difficulty lies in the essence of cooperation: all V2V functions rely on there being more than one vehicle in communication range, and V2I relies on roadside-stations. Before starting the FOT, it is hard to estimate the number of vehicles in one area, and how often vehicles will pass a Road-Side Unit (RSU). This penetration rate is a crucial factor in designing and evaluating a FOT. While some functions work with only two vehicles (e.g. a slow vehicle warning), other functions require several more (e.g. traffic-jam-ahead warning). For certain functions the combined frequency of events might make a naturalistic FOT not feasible, and require additional more controlled or even experimental methods, such as requiring the presence of drivers at a certain time and place, and the use of simulation. A specific problem with Cooperative Systems is due to the distributed nature of the components involved. It is not sufficient just to log the state of the vehicle and the functions, but it is also necessary to handle all incoming and outgoing communications as well as the status of the local dynamic map. These various sources provide a huge amount of data. The revised version of the FESTA handbook provides many recommendations on planning and conducting cooperative system FOTs.

14.6 Examples from Cooperative System FOTs

In 2011 two major European cooperative system FOTs started: DRIVE C2X and FOTsis. These projects were also related to smaller national FOTs, e.g. the German simTD [26] and the French SCORE@F [27] projects to DRIVE C2X.

14.6.1 DRIVE C2X

The DRIVE C2X project [13] focuses on communication among vehicles (C2C), and between vehicles, a roadside and backend infrastructure system (C2I). The story of DRIVE C2X started with an EU funded preparation project named PRE DRIVE-C2X (2008–2010), which aimed at specifying a common European C2X communication system as well as the tools to carry out FOTs for cooperative systems. DRIVE C2X started in 2011 building on the results of this project. DRIVE C2X is to be considered as a Cooperative driving pre-deployment and assessment project. The project mainly targets the evaluation of cooperative driving functions at different levels: technical evaluation, impact assessment and user acceptance.

The cooperative systems used in the DRIVE C2X project provided Safety and Traffic efficiency functions to the drivers. The functions triggered warnings to the driver and displayed information about the corresponding safety or traffic relevant events (e.g. distance to the event or actual speed limit). The triggering of the events resulted from corresponding information received from other vehicles or from the road infrastructure, for instance: slippery road, motorcycle or emergency vehicle approaching, speed limit, traffic light phases, road works, etc. The organisation of the FOT for cooperative systems required specific attention, because the relevant event information was not issued by the driving vehicle but rather by the environment (other C-ITS equipped vehicle or the infrastructure). The FOT operation was using the two types of test: naturalistic and controlled. However, for the naturalistic tests it was necessary to ensure users drove through "C-ITS corridors". For instance, portions of frequently used roads the proposed subject drivers were assumed to use (for example, on the way to their workplace) were equipped with C-ITS road side equipment.

For the controlled test it was necessary to ensure that all vehicles involved in the triggering of an event—as, for instance, slow vehicle, approaching emergency vehicle or traffic jam detected at the vehicle level—were driving in a coordinated way to create a corresponding event in the test vehicle. DRIVE C2X has used a tool named Web Scenario Editor to guide all the vehicles involved in the test sessions.

14.6.2 FOTsis

FOTsis [14] is a FOT on Safe, Intelligent and Sustainable Highway Operation. It is centred on road operations and the interaction between drivers, administrations, communication operators, and emergency services. Seven close-to-market cooperative services are tested, dealing with emergency and safety incident management, intelligent congestion control, dynamic route planning, special vehicle tracking, advanced enforcement, and infrastructure safety assessment. Services are tested at nine test-sites in four European countries. This project also strongly focuses on the interoperability of systems and services.

Both projects put an important emphasis on performing cost–benefit analyses and on determining the socio-economic impact of Cooperative Systems. Cooperative systems should bring benefits in terms of improvements in safety and efficiency, reducing costs of accidents, travel time, and emissions.

14.6.3 Connected Vehicle Safety Pilot

In the USA the Connected Vehicle Safety Pilot Program also started in 2011 [28]. It is part of a major scientific research program run jointly by the US Department of Transportation (DoT) and the industry. The Connected Vehicle Safety Research Program supports the development of safety applications based on V2V and V2I communication systems, using dedicated short-range communication (DSRC) technology. The Safety Pilot is designed to determine the effectiveness of these safety applications in reducing crashes, and to find out how real-world drivers will respond to these safety applications in their vehicles. Systems tested include preinstalled and aftermarket technology for crash avoidance, blind spots, anticipation of upcoming traffic congestion, lane changes and distances from other cars. The test included 73 miles of instrumented roadway, 29 roadside units, around 3,000 vehicles (cars and trucks) and 1.5 years of data collection.

14.7 Results and Benefits of Conducting Field Operational Tests

FOTs are complex studies; not only do they look at the direct outcomes of the tests, for example, at whether the average speed was influenced by the use of a speed warning system, but they also aim to answer impact questions such as what would happen if a majority of drivers would use such a function, and what that would mean for the safety and efficiency of the network. An example of an outcome from euroFOT is: "if widely deployed across the EU, the systems studied by euroFOT could potentially reduce accidents and use of resources. The socio-economic assessment reveals a cost benefit ratio of 1.3 to 1.8 for ACC in trucks. Using the ACC and FCW systems for cars and trucks, euroFOT determined that the costs of equipping the passenger cars and heavy trucks with the combined system leads to annual savings of approximately 1.2 billion EUR (passenger cars) and approximately 180 million EUR for heavy goods trucks" [29].

As might be expected from the complexity of FOTs, it is not always easy to find such straightforward outcomes. However, testing systems in real life conditions provides a huge advantage over all other methods: only if drivers use these systems for a longer period of time is it possible to understand their influence on driving behaviour, and the acceptance of such systems. For example, TeleFOT found that drivers became somewhat disappointed in the systems they tested, having started with high expectations. However, the longer they used the services, the more satisfied they became, having been able to experience the benefits [17].

Evaluation of Cooperative Systems is even more complex. The evaluation of Cooperative Systems suffers from the difficulty of collecting a large enough number of events. Also, as systems are not always mature, and concern potentially safety critical situations, it is not always possible to have a completely naturalistic study. However, FOTs do provide valuable results. In a workshop dedicated to this topic [30] in particular driver behaviour, user acceptance and the identification of potentially negative impacts due to inappropriate warnings were considered as the most important points to be explored in the FOTs.

In a stakeholder needs analysis study by FOT-Net 41 stakeholders who had been involved in European FOTs answered questions about their expectations and experiences with FOTs [30]. These stakeholders were from industry, public authorities and research organisations. A majority (59%) indicated that in fact their expectations had been met, 24% indicated that they were only partly met, and only 5% stated that this was not the case at all. Only 29% indicated that they saw problems regarding the use of FOT-results and 63% explicitly stated not expecting any problems.

Not only the direct outcomes of a FOT provide benefits for stakeholders, increased collaboration with other organisations is also seen as a positive outcome.

14.8 Deployment of Results of Field Operational Tests

The aim of FOTs is to reach results that can be used for the deployment of ITS, either in the form of bringing systems and services to the market, or to make decisions about promoting and regulating them.

14.8.1 Deployment by Stakeholders

The FOT-Net project [18] explored the needs and expectations of stakeholders (both public and private) concerning FOTs, and the ways in which results can be deployed [30]. As case studies two stakeholders, a road operator and a car manufacturer, described in [30] how FOTs contributed to the deployment of cooperative ITS. The participation of the road operator in various projects made possible the testing of concepts, technical solutions, and driver acceptance of services which are now deployed. They identified several factors of success for the deployment of Cooperative Systems: a coherent integration of Cooperative Systems with existing dynamic traffic and incident management is needed. New actors involved in safety and traffic management (automotive industry, telecom industry, equipment and content providers, insurance companies, etc.) should collaborate with road operators, and their responsibilities should be clearly defined. Handover and homogeneity of safety and information services on the Trans-European network should be considered as well as the use of common standards. The business model should recognise economic value anywhere along the chain of information production. The interface in the vehicle should be ergonomically sound, and standardisation and interoperability should allow receiving messages independent of the device that is being used.

The car manufacturer is involved in FOTs because this allows them to make technical assessments (tests of compliance with the latest standards, interoperability tests, performance tests) in open environments, so that they can refine the evaluation of the investments to be made [30]. FOTs also are a means to ensure that the system does indeed produce safety benefits, and does not introduce any further risk—whether real or perceived by drivers. Safety is of course a major concern for car manufacturers, but also their potential customers' perception, and the potential value these systems may have for them.

Public authorities and policy decision makers are interested in the outcomes of FOTs because they allow them to make decisions about public investments, regulations, and the role ITS can play in their mobility plans, both at a national and a local level [30].

14.8.2 Deployment Issues in Projects

The DRIVE C2X project has tackled several deployment issues:

- Design and implementation of a cooperative system based on EU standards
- Implementation, test and validation of the whole system at a system test site
- Deployment of the cooperative functions at 7 EU test sites
- Execution of FOT to collect logs and user feedbacks for the evaluation
- Evaluation and promotion of cooperative driving

The project started at the time when the European Telecommunications Standards Institute (ETSI) was providing the first set of communication standards, answering the requirements of the EU mandate M/453. For deploying cooperative systems in the vehicles and on the seven European test sites, the DRIVE C2X partners decided to develop a "reference system" based on the EU mandated standards, and in particular on the communication standards published by the ETSI technical committee ITS. The DRIVE C2X reference system is a kind of software toolkit, providing the communication, application and logging functionalities, compatible with several existing hardware platforms.

DRIVE C2X has thoroughly addressed the interoperability challenges by testing the reference system at different levels. At component level during several integration test workshops, at communication level while applying the conformance test suites from the ETSI TC ITS, and by supporting the organisation of three ERTICO/ETSI Cooperative ITS Plugtests events, at the system level during several driving test workshops at the TNO System Test Site in Helmond (NL). In this way DRIVE C2X contributed to the development of a common C-ITS test bed. The interoperability of the infrastructure and vehicle systems was tested during four public test events. Vehicles were also exchanged among different test sites.

DRIVE C2X has proven the possibility to deploy efficiently an EU wide interoperable cooperative system. The first results of the impact assessment, as well as the user acceptance, have showed concrete benefits for road users. The suitability of the DRIVE C2X cooperative functions was evaluated for private drivers, but also in the context of public vehicles, like emergency vehicles. The powered two-wheelers safety application "Motorcycle Approaching Information" also showed the usefulness of the DRIVE C2X cooperative systems for vulnerable road users.

An example of a European pilot project focusing on the deployment of cooperative systems is Compass4D (Cooperative Mobility Pilot on Safety and Sustainability Services for Deployment) [31]. Seven European cities/regions are working together in piloting cooperative services addressing road safety issues, traffic efficiency problems and environmental impacts. Dedicated short range communication (ETSI G5) and cellular networks (3G/LTE) are used. The project explores and analyses the business viability of the Compass4D piloted services. It defines the state of the market, and studies the benefits to the stakeholders involved in the deployment and management of the infrastructure and services. There exists a close collaboration with the USA and Japan to harmonise the specification of services.

In Japan, there is a direct relation between FOTs and deployment. The Ministry of Land, Infrastructure, Transport and Tourism (MLIT) has been conducting research and development in the cooperative system project Smartway, in collaboration with academia and industry [32]. ITS Spot, the high-speed, large-capacity road-to-vehicle communication system, was implemented in 2011, as a deployment of the results [33]. ITS Spot provides various ITS services to a single on-board unit. There are currently three basic services: dynamic route guidance, safe driving assistance and electronic toll collection. ITS Spot also enables vehicles to transmit their probe information to the road infrastructure.

A dedicated European support action is focusing on the deployment of cooperative ITS. The COMeSafety2 project [34] aims at the coordination of activities to support the implementation of cooperative systems. COMeSafety2 is developing several actions such as supporting progress of the standardisation mandate at ETSI and Comité Européen de Standardization (CEN); assisting the EU-US cooperative systems task force; developing a European multimodal cooperative ITS architecture, and updating the European cooperative systems communication architecture; and contributing towards the European research agenda planning process. A particular goal of COMeSafety2 is to ensure synergies among intercontinental FOTs on cooperative systems with the help of an "Intercontinental FOT Exchange Platform". This platform addresses the need to gather co-operative system FOT information, focusing on test methodologies, data collection and evaluation as well as consolidating the main FOT results.

14.9 Deployment of Data from Field Operational Tests

In FOTs huge amounts of data are gathered. In the first place these data are used to answer the questions asked in the project. However, often more data were collected than could be used for analysis in a project. Sometimes this is due to lack of time and resources, sometimes not all the data collected were useful for the analysis at hand. In projects data is used for a specific purpose. Data, however, may also be of value for other purposes. In most FOTs data is gathered about the speed of vehicles. These speed data can be used for a variety of purposes. If a project is focussed on driver behaviour, this data may be used by the developer of a system to see how an ITS influences the speed at which drivers drive. This data may, however, also be valuable for a completely different purpose and a different stakeholder, for example, a road operator wanting to know the speed patterns in parts of their network.

Re-using data may sound simple, but it isn't. In order to re-use data in a new project, the original project must be willing to share the dataset, and make it available in a way that ensures it can be re-used. Sharing data poses many challenges.

Technical challenges concern issues such as the quality of metadata, descriptions on implementations, how field tests were run, how the data was collected, the tools used to collect and store the data, the standards and formats used. Legal and organisational issues concern ownership, data protection and privacy issues. For example, permission from the FOT participants is needed to allow third parties access to the data. Re-engineering tested services or used sensor systems is also seen as problematic. Practical and financial issues concern questions about who is paying for the access, the training of new data analysts to understand the data, its limitations, tools, and physical access to the data. And finally, there are ill-defined issues concerning attitudes like trust and willingness to share, and resources to be allocated to support data sharing [35].

14.10 Networking and Community Building: FOT-Net

From 2008 to 2014, the European support action FOT-Net has worked on building a community of people involved in FOTs, both at national and European levels, with connections to further international ones [36]. These people are from research organisations, car manufacturers, system and service providers, user organisations, public authorities, and road operators. FOT-Net promotes and maintains the FESTA-methodology and the corresponding handbook. Also, workshops, a website, and a wiki containing a catalogue of all known FOTs have been developed to ensure wide sharing of knowledge and experiences.

In the current FOT-Net project, FOT-Net Data [18, 35] that started in January 2014, the technical, practical and organisational issues of data sharing are addressed. International cooperation, with the USA and other countries, is sought to gain consensus on a global level for a data sharing strategy and platform.

14.11 Concluding Remarks

FOTs have been conducted in a variety of regions and countries [19]. Although they are tests that are costly and time-consuming, they have provided interesting insights in the use of ITS, and further analysis of their data will bring more knowledge about the behaviour of vehicles, systems and drivers. In the future we might not always see the large-scale FOTs as described in this chapter, but smaller and dedicated ones as part of projects in which systems are to be tested in real-life environments. Where in the past there was a strong emphasis on studying passenger cars and trucks, now FOTs with powered two-wheelers and bicycles are also conducted. The last few years more sources of data are becoming available, such as probe data, originating from vehicles, but also from nomadic devices such as smart phones. International collaboration on FOTs, on probe data, and on frameworks to share data is taking place and will lead to a better understanding of the impacts of a large-scale implementation of ITS in vehicles and on the roads world-wide.

References

1. Barnard Y, Krems J, Risser R (eds) (2011) Safety of intelligent driver support systems: design, evaluation, and social perspectives. Ashgate, Farnham
2. Barnard Y, Carsten O (2010) Field operational tests: challenges and methods. In: Krems J, Petzoldt T, Henning M (eds) Proceedings of European conference on human centred design for intelligent transport systems. HUMANIST, Lyon, pp 323–332
3. FESTA (2014) FESTA Handbook, Version 5 (Field opErational teSt supporT Action). Available at: http://wiki.fot-net.eu/index.php?title=FESTA_handbook
4. http://www.eurofot-ip.eu/
5. http://www.telefot.eu/

6. Dingus TA, Klauer SG, Neale VL, Petersen A, Lee SE, Sudweeks J, Perez MA, Hankey J, Ramsey D, Gupta S, Bucher C, Doerzaph ZR, Jermeland J, Knipling RR (2006) The 100-car naturalistic driving study, phase II: results of the 100-car field experiment. NHTSA report DOT HS 809 593. NHTSA, Washington
7. http://www.trb.org/StrategicHighwayResearchProgram2SHRP2/Public/Blank2.aspx
8. Antin JF (2011) Design of the in-vehicle driving behavior and crash risk study: in support of the SHRP 2 naturalistic driving study. Transportation Research Board, Washington
9. Boyle LN, Lee JD, Neyens DM, McGehee DV, Hallmark S, Ward NJ (2009) SHRP2 S02 integration of analysis methods and development of analysis plan. Phase 1 Report. University of Iowa, Iowa City
10. www.udrive.eu
11. Eenink R, Barnard Y, Baumann M, Augros X, Utesch F (2014) UDRIVE: the European naturalistic driving study. In: Proceedings of the TRA 2014 - transport research arena conference in Paris
12. Regan MA, Williamson A, Grzebieta R, Charlton J, Lenneb M, Watson B, Haworth N, Rakotonirainy A, Woolley J, Anderson R, Senserrick T, Young K (2013) The Australian 400-car naturalistic driving study: innovation in road safety research and policy. In: Proceedings of the 2013 Australasian road safety research, policing & education conference, Brisbane, Queensland
13. http://www.drive-c2x.eu/project
14. http://www.fotsis.com
15. Benmimoun M, Pütz A, Ljung Aust M, Faber F, Sánchez D, Metz B, Saint Pierre G, Geißler T, Guidotti L, Malta L (2012) Deliverable D6.1 Final evaluation results. euroFOT Consortium
16. Kessler C, Etemad A, Alessandretti G, Heinig K, Selpi, Brouwer R, Cserpinszky A, Hagleitner W, Benmimoun M (2012) Deliverable D11.3 Final Report, euroFOT Consortium
17. Mononen P, Franzen S, Pagle K, Morris A, Innamaa S, Karlsson M, Touliou K, Montanari R, Fruttaldo S (2012) D1.15 TeleFOT Final Report
18. http://www.fot-net.eu
19. http://wiki.fot-net.eu
20. http://www.smartcem-project.eu/
21. http://www.molecules-project.eu/
22. http://www.mobieurope.eu/
23. http://www.ict4eveu.eu/
24. http://www.its.dot.gov/testbed.htm
25. http://www.car-to-car.org/
26. http://www.simtd.de/index.dhtml/enEN/index.html
27. https://project.inria.fr/scoref/en
28. http://www.its.dot.gov/research/safety_pilot_overview.htm
29. euroFOT press release 26-06-2012, http://www.eurofot-ip.eu/download/events/eurofot_press_release_26062012_english.pdf
30. Schmidt E, Barnard Y, Alkim T, Zennaro G, Toulminet G, Barbier C, Friis G (2013) FOT-Net 2 Deliverable D62. Stakeholder Needs Analysis
31. http://www.compass4d.eu/
32. Kanazawa F, Kanoshima H, Sakai K, Suzuki K (2010) Field operational tests of smartway in Japan. IATSS Res 34:31–34
33. http://www.mlit.go.jp/kokusai/itf/kokusai_itf_000006.html
34. http://www.comesafety.org/
35. Barnard Y, Koskinen S, Gellerman H (2014) A platform for sharing data from field operational tests. In: Proceedings of the ITS world congress 2014 in Detroit
36. Barnard Y (ed) (2014) FOT-Net 2 Deliverable D2.2 Report on FOT Network activities

Part V
The Evolution of Vehicular Networks

Chapter 15
Insights into Possible VANET 2.0 Directions

Xinzhou Wu, Junyi Li, Riccardo M. Scopigno, and Hector Agustin Cozzetti

Abstract Over the last decade, there have been vigorous joint efforts from the Industry, Academia, and Governments to validate the Dedicated Short Range Communications (DSRC) technology and also to identify and address key technical and business challenges. These efforts have confirmed the applicability of DSRC to improve vehicular safety. They also point to several areas for further improvements. In this chapter, we will discuss potential improvements that can be beneficial to future generations of DSRC.

Keywords VANET • Hidden terminal • Adjacent channel interference • Vehicle-to-pedestrian • Li-Fi • TPEG • White space • Synchronous protocols • MS-Aloha

15.1 Introduction

In February 2014, USDOT announced their intention to mandate the Dedicated Short Range Communications (DSRC) technology for all new light vehicles. Prior to this, there have been decade-long joint efforts from the Industry, Academia, and Governments to validate the DSRC technology: several pilots and field trials have been carried out worldwide, as already discussed in Chap. 14 (just to mention two initiatives: the 1-year safety pilot, with around 3,000 equipped vehicles, held in Ann Arbor Michigan between August 2012 and August 2013; the manifolds European

---

X. Wu (✉) • J. Li
Qualcomm, New York, NY, USA
e-mail: xinzhouw@qti.qualcomm.com; junyil@qti.qualcomm.com

R.M. Scopigno
Istituto Superiore Mario Boella, Via P.C. Boggio 61, Torino, Italy
e-mail: scopigno@ismb.it

H.A. Cozzetti
McAfee, 7 Rue Soutrane, Valbonne, Sophia Antipolis, France
e-mail: agustin_cozzetti@mcafee.com

© Springer International Publishing Switzerland 2015
C. Campolo et al. (eds.), *Vehicular ad hoc Networks*,
DOI 10.1007/978-3-319-15497-8_15

initiatives networked by the FOTNET initiative). Such efforts have confirmed the applicability of DSRC to improve vehicle safety; however, they also permitted to identify and address key technical and business challenges.

Actually, DSRC is based on IEEE 802.11p at the PHY and MAC layer, which dates back to more than 10 years ago. Meanwhile, our understanding of wireless communications has been advanced a lot: there are definitely several areas for further improving [108] DSRC by leveraging the recent wireless technology advances. In this chapter, we will discuss potential PHY and MAC layer improvements that can benefit future generations of DSRC.

The chapter is organized in the following sections: in Sect. 15.2 the *challenges*, that is the improvable features and open issues are presented, split among physical layer (Sect. 15.2.1), MAC layer (Sect. 15.2.2), and the multi-channel environment (Sect. 15.2.3); then, in Sect. 15.3 some reflections are proposed, about the improvement of the physical layer, based on the ideas collected in the recent scientific literature; similarly, in Sect. 15.4 the proposals emerged for the improvement of MAC layer (both asynchronous and asynchronous ones) are shortly summarized. The chapter is concluded by two *side* topics somehow complementary: the use of VANETs also for vulnerable road users (Sect. 15.5)—hence the possible overloading of VANETs—and the use of complementary wireless media to unload the channels (Sect. 15.6).

15.2 Key Challenges of the Current VANETs

Before reasoning on possible evolutions of VANETs, it is useful to pinpoint once more the technical areas which deserve more attention. The following analysis will highlight those aspects which at physical (Sect. 15.2.1) and MAC layer (Sect. 15.2.2) have shown some sub-optimal performance.

15.2.1 PHY Layer

Vehicular communications are set in a challenging environment: vehicles have strong mutual speeds and their communications are degraded by a plethora of obstacles—buildings, intersections, tunnels, surrounding big vehicles, bridges. As widely discussed in Chap. 13, the channel models for vehicular communications has been extensively studied [21, 90, 98] and show, in general, a deeply different behavior compared to the stationary environments typical of IEEE 802.11 use cases.

In particular, there are two key differentiating characteristics of vehicular communication environments: multi-path delay spread and mobility (see Chap. 3). Delay spread is related to the frequency selectivity of the channel, while mobility causes time-selective fading channels.

15 Insights into Possible VANET 2.0 Directions

The characterization of the delay spread in vehicular communications have been addressed by several studies. In particular, the effect of reflections in Urban LOS environments, is described in [98]: the results indicate that, in addition to a strong initial tap observed at 250 ns, a number of other reflections are present. The Root Mean Squared (RMS) delay spread (which is inversely proportional to the coherence bandwidth) is about 500 ns, demonstrating the significant effect of the reflections from objects. On the other hand, on highways, the power delay profile has fewer taps—due to reduced reflections—and an RMS delay spread of 190 ns.

Field trials have also reported the empirical distributions for the delay and the Doppler spread [3]. In urban NLOS conditions, the RMS delay spread is more than 200 ns with only 1 % probability; besides, the maximum delay spread typically does not exceed 1.4–1.6 μs. NLOS situations on the highways are likely to cause an increased Doppler spread; conversely, in urban situation, the Doppler spread is typically less than 1,000 Hz. The RMS Doppler spreads in LOS highway scenario gets smaller, since the high frequency Doppler taps have reduced power. This is confirmed since the maximum Doppler spread is high for both LOS and NLOS highway scenarios.

All in all, the vehicular communication channel can be highly frequency-selective—due to multipath delay spread—and highly time-selective—due to vehicle mobility. It can be estimated that, in some scenarios, the 50 % coherence bandwidth is in the order of 1 MHz and 50 % coherence time can be as short as 0.2 ms. This implies two major challenges for the adaptation of the legacy IEEE 802.11 to vehicular communications:

- channel estimation error in vehicular environment, due to time-selective and frequency-selective fading,
- lack of time-interleaving in fast time-selective fading channels.

15.2.1.1 The Issue of Channel Estimation in Vehicular Environments

To illustrate the channel estimation problem, let us look at the pilot structure in IEEE 802.11a PHY shown in Fig. 15.1. An IEEE 802.11p frame starts with a few wideband pilot symbols and then the data symbol and SIGNAL symbol, which conveys the PLCP header. Only 4 out of the 52 sub-carriers are used for pilots.[1] In other words, for the 10 MHz operation in IEEE 802.11p, two adjacent sub-carriers

---

[1] As discussed in Chap. 3, pilot sub-carriers are used to prevent frequency and phase shift errors in the OFDM receiver (only the *nominal*—not the *actual*—frequencies of OFDM sub-carriers are known) and for supporting equalization.

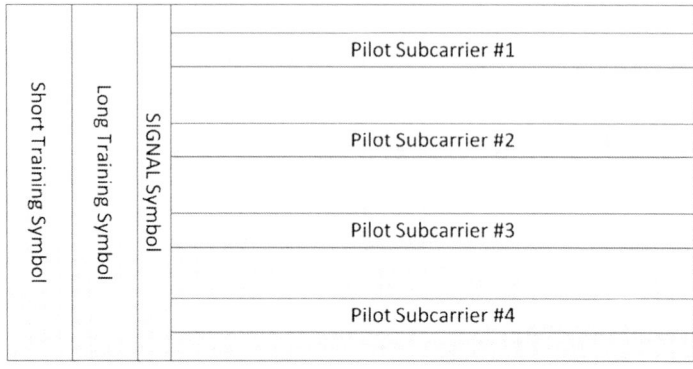

Fig. 15.1 IEEE 802.11p pilot structure over time and frequency

are 2.4 MHz away. Further, a typical safety broadcast message can take up to 0.5 ms to transmit. Two observations arise:

(a) The spacing between two pilot sub-carriers is significantly larger than the coherence bandwidth.
(b) The frame transmission time (e.g., 300 bytes in approximately 0.5 ms) can be longer than the coherence time, which, depending on channel conditions, can drop to 0.2 ms

Typical IEEE 802.11p devices first obtain a wide-band channel estimate from pilot symbols; then, they monitor the residual channel variation using the pilot sub-carriers. The latter is usually referred to as *pilot tracking*. In benign channel models, such algorithms are sufficient. However, after the coherence time, the channel estimates obtained from the pilot symbol become obsolete and, unfortunately, the sparse pilot sub-carriers are not sufficient to track the channel. This may happen with VANETs: as a result, the frame reception can fail even when the received power of the frame is well above the noise. Figure 15.2 shows the simulated results of the performance of a typical IEEE 802.11p receiver with a conventional channel estimation implementation at an outdoor mobile environment; the frame drop rates are mostly caused by channel estimation errors which rise sharply with a combination of delay spread and mobility.

There are a few tools from modern coding theory to deal with channel estimation errors. One such method is to introduce a Turbo receiver [39], which iterates between channel estimation and decoding, instead of carrying out the two steps in a sequential order, as conventional receivers do. The nature of the IEEE 802.11p PHY coding and modulation enables a natural way of implementing such a scheme. In particular, IEEE 802.11p uses convolutional codes with a short constraint length (7) and no time interleaving. This means that the receiver can reliably decode a particular information bit coded and transmitted earlier in the frame without waiting for the reception of the whole frame. Further, one can use the already decoded bits

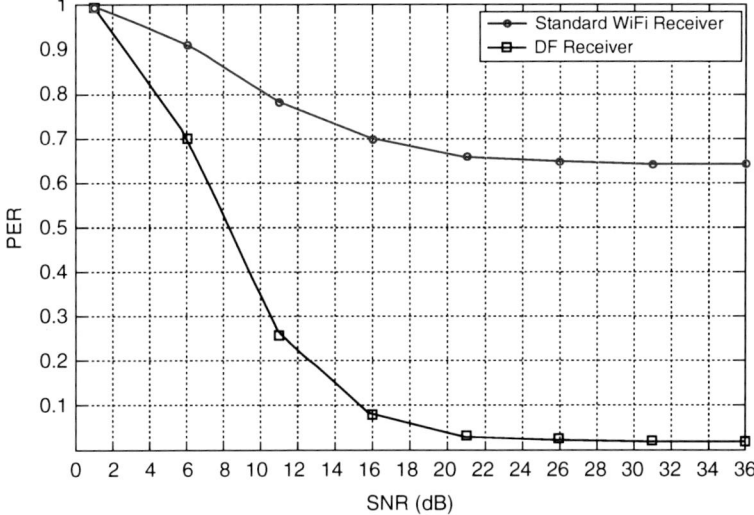

Fig. 15.2 Performance comparison of conventional IEEE 802.11a implementations and decision feedback (DF) receiver against mobility and delay spread

as pilots for the remaining frames to improve channel estimation. Such a scheme is referred to as *decision feedback* and can track the channel variations significantly better than standard non-iterative schemes. As shown in Fig. 15.2, it almost removes entirely the channel estimation error caused by time and frequency selective fading.[2] Notably, both Turbo receiver and decision feedback receiver are receiver algorithm enhancements which are fully compatible with the existing IEEE 802.11p standards.

15.2.1.2 Time-Selective Fading Without Time Interleaving

Now let's look at the other side of the coin: no time interleaving leads to significant performance degradation in a time-selective channel, even without channel estimation errors. This is an outcome of using convolutional code with short memory without time interleaving. An information bit is only reflected by a small number of coded bits determined by the constraint length (7). Without time interleaving, most likely the encoded bits will be modulated and transmitted on the same or

---

[2]These advantages could get enforced even more if a variable pilot distribution were used, though with the same channel efficiency. For instance, in [76] the author proposes to adopt a non-standard encoding where the position of pilots varies symbol by symbol: they propose to switch between the position at sub-carrier ±4 and ±11 (instead of keeping it fixed at ±7), and between ±18 and ±25, instead of ±21: this solution is claimed to enforce channel estimation also under conditions of poor coherence time and coherence bandwidth but, conversely, it would not be compatible with the current standard.

adjacent time symbols. In time-selective channels, if a symbol (or a few adjacent symbols) is in a deep fade, there is a very low probability that the receiver can recover the information bits encoded in the symbol. As a result, the frame reception rate is determined by the minimum signal-to-noise (SNR) across all symbols in the frame rather than the average SNR. On the other hand, in IEEE 802.11p, the frame length can be longer than the coherence time, which indicates that deep fades can happen within the transmission time of a frame with significant probability. This problem is well understood in the wireless communication community and could be resolved by a better coding scheme (e.g., Turbo or LDPC code). However, this indeed requests a standard change and is better to be addressed in future versions of DSRC. We will elaborate more on this in Sect. 15.3.

15.2.2 MAC Layer

The primary issue at MAC layer is the frame loss due to collisions when a large number of devices are in each other's radio transmission ranges and/or due to hostile and obstructed propagation. They are, respectively, analyzed in Sect. 15.2.2.1 and in Sect. 15.2.2.2.

15.2.2.1 CSMA Behavior at High Node Density: Scalability and Fairness

Vehicle safety communication applications rely heavily on periodic broadcast of basic safety messages (BSM) which contain the positions, velocities, and other information about the vehicles. These messages, including the PHY layer overheads, typically measure around 300 bytes and are expected to be transmitted up to once every 100 ms, so to meet latency and accuracy requirements of vehicle safety applications.

IEEE 802.11p uses CSMA/CA and a random back-off procedure to reduce collisions of over-the-air frames. At high densities of vehicles, the wireless medium can quickly become highly congested and collisions among frame transmissions may occur with high probability. The MAC layer congestion issues have been observed and reported in numerous studies in the literature [44, 62, 100]. The main source for the collisions at high densities is the synchronous countdown of the back-off timers by any two transmitters within the carrier sense range of each other. At high densities, many devices observe the channel to become idle at the same moment, start counting down their timers to zero synchronously, and then transmit at the same time, as shown in Fig. 15.3.

Figure 15.4 shows the frame reception rate performance of the standard IEEE 802.11p MAC protocol for different vehicle densities. Frame reception rate, also commonly referred to as the discovery probability, measures the percentage of frames that can be successfully received by a receiver. The plots demonstrate that, as the density increases, the discovery probability degrades even at close-by

Fig. 15.3 Illustration of a collision scenario. Nodes A and B count their timers down synchronously and transmit simultaneously

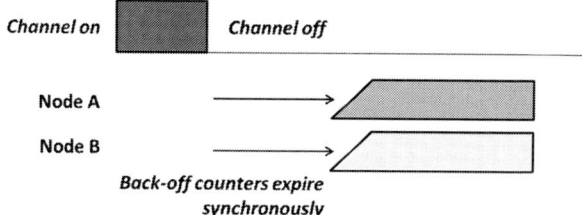

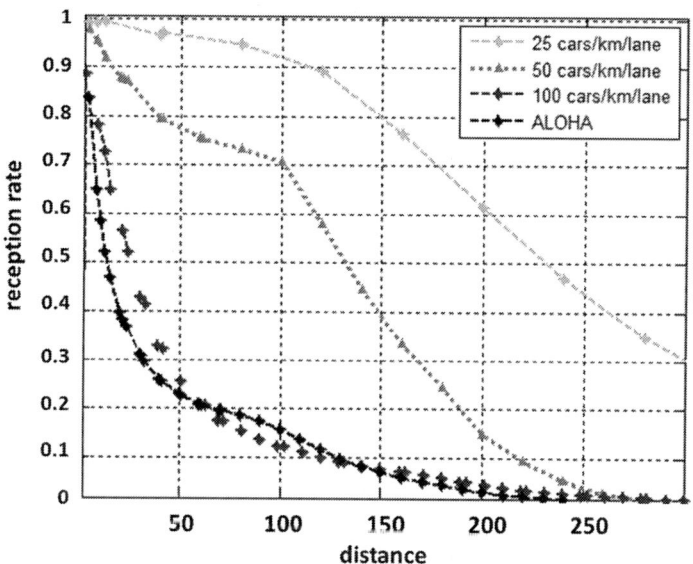

Fig. 15.4 Reception rate vs. distance. The plots demonstrate that at higher densities, the performance tends towards an ALOHA-like performance. Adapted from [108]

distances. The performance trends towards an ALOHA-like behavior, due to a lack of protection against simultaneous transmissions by nearby nodes. As the vehicle density increases, the discovery range will also reduce for the same frame length, periodicity and code-rate. The discovery range is defined as the longest distance between a transmitter and a receiver, so that the receiver is able to receive frames with a given reception rate. Ideally, the discovery range should shrink in proportion to the density increase, such that the number of discovered devices remains constant at all densities. However, the ALOHA type of collisions can lead to significant performance loss, as demonstrated by the normalized discovery in Fig. 15.5: the x-axis is the discovery distance in inter-car spacing (to normalize for a linear density increase with distance), and the y-axis is the probability of discovery. This degrading reception with increasing vehicle density can have a significant impact on the performance of vehicle safety applications in high vehicle density areas such as crowded intersections and highway segments.

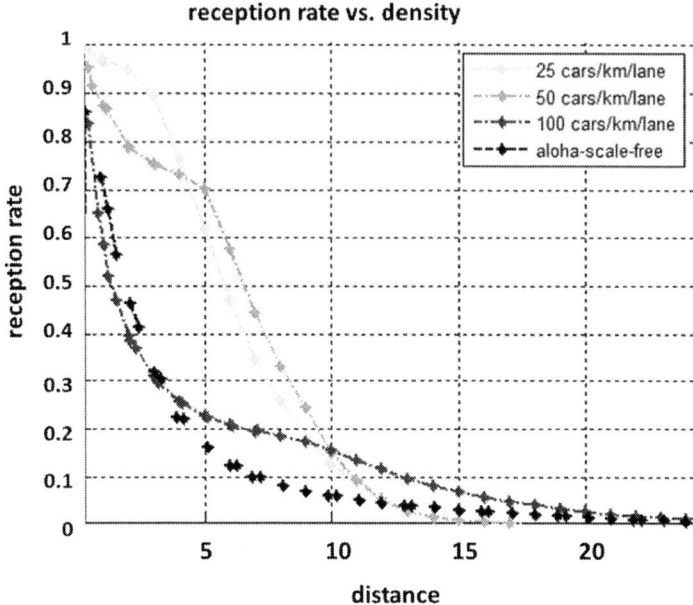

Fig. 15.5 Reception rate vs. normalized distance for a four-lane road of length 2 km under a typical two-slope radio path loss model and a transmit power of 20 dBm and a BSM frequency of 10 Hz. The frame size is assumed to be 300 bytes. Adapted from [108]

Last but not least, as highlighted in [13, 14] and due to the periodic nature of CAM, most of the VANET traffic will be generated periodically as well: each vehicle will start its broadcast transmissions when switched on and, since then, it will keep the generation of messages with the same initial phase ϕ. This means also that if the transmissions by two nodes collide once, they are likely to collide again and again. In other words, also a problem of fairness arises among the nodes in a VANET. Obviously, also this phenomenon gets worsened by a higher density of nodes: the more the nodes, the more likely the event that the phases ϕ_1 and ϕ_2 of two nearby nodes is almost the same.

All in all, congestion leads to a harsh worsening of reception, or, in other words, the current MAC of VANETs is prone to joint scalability and reliability problems. Being well known, this issue has been widely addressed by the scientific community: a plethora of proposed solutions can be found in literature and they can be mainly grouped into three classes.

- A natural approach to reduce congestion is to reduce the number of transmitters within the carrier sense range of each device. A typical scheme to balance collisions and channel utilization is to use a distributed congestion control mechanism, as described in [62]. These solutions have been discussed in Chap. 5 under the umbrella of DCC (Decentralized Congestion Control): their rationale is that each vehicle dynamically adjusts its transmission parameters (power,

periodicity, etc.) in response to the observed channel load, so to prevent congestion and reacting as soon as the perceived channel load exceeds a threshold. DCC algorithms are still being investigated and future VANET solutions are likely to benefit from enriched versions of DCC.
- Another promising method is to use a time-slotted synchronous protocols (as, for instance, [26, 65, 80]). Synchronicity is already mandated as part of IEEE 1609.4 for channel switching and it is feasible to use a Global Positioning System (GPS) clock to synchronize MAC-layer frame transmissions as well. Synchronous MACs will be discussed in Sect. 15.4.2.
- An intuitive alternative is, eventually, the use of other communication media. It has already been discussed in Chaps. 1 and 2: the ISO CALM architecture of the OBU foresees the integration of multiple communication channels, not only the VANET one. One possibility is hence to unload VANET by sending some information on other channels, such as LTE (Chap. 15), or some new emerging ones—as further discussed in Sect. 15.6.

15.2.2.2 Impact of Hidden Terminals

Hidden terminals (HTs) constitute a subtle and often neglected problem of VANETs which, as such, deserves some deeper insight.

It is well known, from the theory, when hidden terminals occur: they are met when you have at least three stations, say A, B, and C, which communicate in wireless and are positioned in such a way that B can sense both A and C, but A and C cannot sense each other. For this reason, following the CSMA/CA approach, A and C will not be able to coordinate their transmission and will frequently collide— not (only) due to simultaneous count-down (as discussed in Sect. 15.2.2.1) but, even more frequently, because one cannot know when the other is transmitting. Last but not least, considering the mostly broadcast nature of VANETs traffic, the Request-to-Send (RTS)/Clear-to-Send (CTS) mechanism cannot be used either.

The most obvious case in which HTs occur is due to distance: A and C are so far apart that, due to attenuation, they cannot sense each other. This case happens under line-of-sight conditions and is depicted in Fig. 15.6a. This first case is not particularly dangerous: if A is much closer to B than C (as in the figure), even in case of overlapping transmissions by A and C, B is more likely to receive from the closest node and this is important for safety (the most dangerous hazards are those by the closest nodes). Instead, if A and C are both far from B, then they might hamper the reception by B from each other, but, being both far, it would not be particularly dangerous.

Some studies have focused on this case, with simulations including also the effect of fading, either with a theoretical approach [38], or by simulations [94], concluding that HTs should not be an issue. However this analysis seems partial and optimistic, because it does not consider the case (Fig. 15.6b) when the nodes A and C are very close but, due to obstructions, they cannot sense each other. This situation is particularly detrimental: simply due to the extra attenuation caused

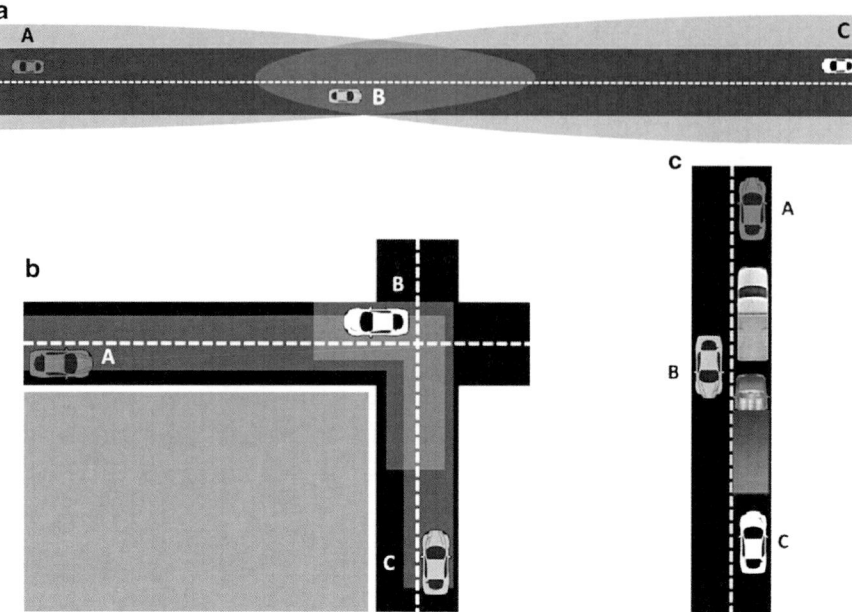

Fig. 15.6 Hidden terminals in multiple scenarios: (**a**) due to distance; (**b**) caused by fixed obstructions; (**c**) due to vehicular obstructions. In (**a**) and (**b**), the *shaded areas* represent the radio coverage by the two hidden nodes (A and C)

by the obstruction in the corner (more details about the propagation phenomena and models can be found in Chap. 13), A and C, despite close, will not be able to coordinate their transmissions. It means also that, in case of uncoordinated overlapping transmissions by the HTs A and C, the Signal-to-Noise-Interference Ratio (SINR) at the receiver B will drop (due to the high interference) and B will not be able to receive by either of them.

Some significant results on HT have been achieved thanks to the emerging realistic propagation models in [27] and, partially, in [92]. The results are only preliminary, however, they already highlight the following facts:

- the heavier the traffic load, the most dramatic effects on the reception in the center of the crossroads. Intuitively, if the channel is unloaded, two stations will probably not overlap their transmissions even if they cannot sense each other; conversely, when the medium gets mostly engaged, transmissions need to be carefully coordinated and, if this is not possible, the interference may get disruptive;
- the results by simulations show that in all the scenarios, in spite of a traffic load which is less than one third of the maximum load, in the centers of the crossroads nodes will not be able to receive more than 10–50 % of the traffic sent by nodes as close as 20 m (depending on the specific traffic hypotheses). Altogether, the effect might be dramatic.

One may wonder why such phenomenon tends to be neglected: the reason is probably that field operational tests (see Chap. 14) have not considered these scenarios as the first ones (several nodes would be needed to load the channel) and, on the other hand, realistic simulation models are very recent.

Importantly, concerning the latter aspect, also the obstructions by large vehicles (e.g., SUVs and trucks) are being modeled and simulated (see Chap. 12) and, probably, they will unveil many more cases of HTs (Fig. 15.6c). Obstructions by buildings and by vehicle, together, are likely to cause many more unintentional overlapped transmissions than currently perceived.

At the moment HTs are not being addressed yet by the standardization: most likely, at least at the beginning of VANET era, the networks will be so unloaded not to be significantly affected by HTs; additionally at least initially, it will be possible to rely on DCC to prevent HTs effect also, by keeping the channel congestion sufficiently under control. These however are assumptions that will be certainly further studied in the future.

15.2.3 Multi-Channel Operations

The 5.9 GHz DSRC spectrum is divided into seven 10 MHz bands—though with some regional differences, as discussed in Chap. 3. Such channelization was chosen because it permitted to utilize the existing WiFi chipsets (operating over 20 MHz channels) but running "half-clocked," so to achieve a 10 MHz bandwidth and be more suitable for the highly mobile and frequency-selective vehicular channels.

With the availability of multiple channels, specific mechanisms are needed to manage them. The IEEE 1609.4 MAC extension layer (multi-channel operations—see Chap. 7) was created to enable the use of multiple channels by the upper layers of a DSRC device, so to support both safety and other applications—for example, it provides mechanisms for the synchronous channel switching. Actually, the investigation into multi-channel operation has been so far limited, basically because the attention of the industry has been mainly focused on the validation and improvement of the safety channel performance. In this section, we discuss the challenges of the IEEE 1609.4 multi-channel operation approach and introduce some few potential improvements.

Initially, when DSRC systems were envisioned, each vehicle was expected to be equipped with a single 10 MHz DSRC radio. At that time, it was proposed to use one of the 10 MHz channels—Channel 178, the common control channel (CCH)—for safety messages and the remaining ones as service channels (SCH) for point-to-point communications. This implied that a common time period was used for safety message transmissions, so to allow all vehicles to hear each other's safety messages: for this purpose, a sync-interval of 100 ms was to be divided into CCH and SCH intervals, via time domain orthogonalization, and all devices should tune to the CCH channel in the CCH interval. In the SCH interval, the devices could use any of the service channels. The main issue with the above scheme was the reduced capacity to support the broadcasting of safety messages, since all devices could use a 10 MHz channel for only a fraction of the time (CCH/sync-interval). Many simulations have shown that to support vehicle safety broadcasts in typical vehicle densities, most or all of the sync-interval would be required: furthermore, some studies [19, 103] indicate that even a fully dedicated 10 MHz channel, just for safety and control, would not fully prevent the channel congestion; besides, the above method is highly spectrum inefficient (six of the seven channels are underutilized during the CCH interval).

Due to these issues, the industry is moving towards a "1609.4 optional" route [42], with all devices using the Channel 172 (different from the CCH channel) for safety broadcasts. Vehicles with dual radios can use one radio on Channel 172 for safety message broadcast and the other radio for control transactions and service transactions in a completely different channel. This approach recognizes the highest priority to safety applications and makes more capacity available to them.

However, the simultaneous operation of multiple radios on the same vehicle imposes further challenges. One issue to address is the influence of separate radio channels on the reception in the safety channel. Consider, for instance, the case of a device using Channel 172 for safety and Channel 174 for other applications: due to the energy spillage from Channel 174 to Channel 172, the reception of safety data in Channel 172 can be affected while transmissions over Channel 174 occur.

As shown in Fig. 15.7, the transmit power mask specification in the IEEE 802.11p standard only requires that the adjacent channels are −50 dB compared

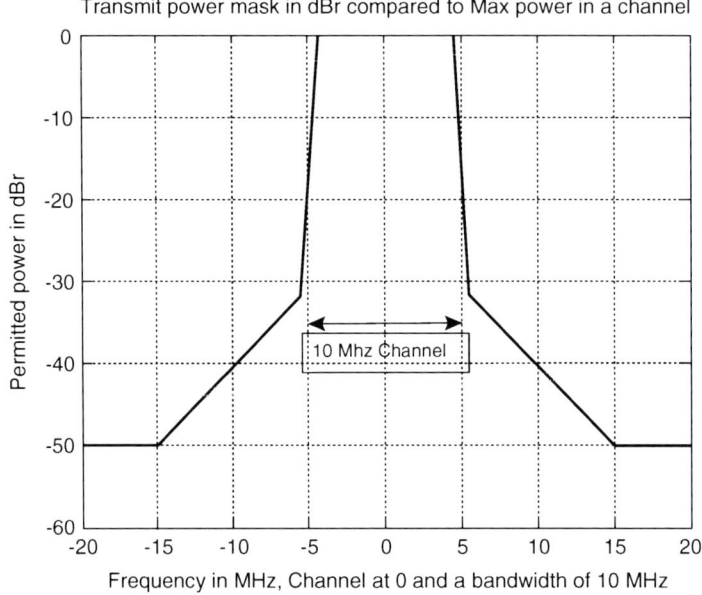

Fig. 15.7 DSRC multi-channel power mask for a Class C transmitter

15 Insights into Possible VANET 2.0 Directions

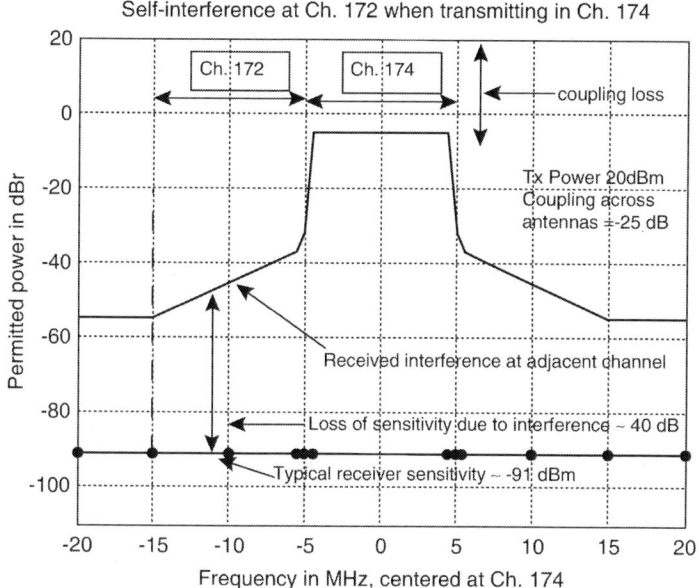

Fig. 15.8 The transmission in an adjacent channel (174) can spill into safety channel (172) increasing the interference level by more than 40 dB from the thermal noise floor, significantly reducing the discovery range

to the original transmitted power. Actually, the dual radios in a vehicle may be located in close proximity: with coupling losses between the radios in the order of 20 dB, the interference magnitude could be as high as −70 dB. For a transmit power of 20 dBm, this would account for a noise level of −50 dBm at the collocated radio in an adjacent channel. With typical Wi-Fi receiver sensitivities at −91 dBm, the interference level can impact the receiver sensitivity, reducing the discovering range. Roughly speaking, raising the sensitivity by 10 dB leads to losing 50 % of the discovery range; with a difference of 40 dB the discovery range gets reduced by a factor as high as 16: the spillage of power into adjacent bands when transmitting is depicted in Fig. 15.8.

The problem of self-interference cancellation is attracting a lot of attention in the academic society in the context of a full duplex modem [22, 57, 89], where concurrent transmitting and receiving are enabled in the same band. The adjacent band leakage problem here in DSRC is a much simpler problem: certain techniques, which are well studied in the full duplex context, including analog cancellation and digital cancellation, could be even more beneficial to DSRC. However, these algorithms are still in prototype stage and are not included in the off-the-shelf chip solutions. Future study of cost-effective implementations in DSRC chips still requires attention from the industry.

> As the industries and governments worldwide are considering the deployment of DSRC-based vehicle safety communication capabilities, a natural question is—What is next, i.e. what will be the evolution path of DSRC? In Sects. 15.3–15.6, we introduce a few key preferable components of the evolution and discuss the possible paths to include these components into future version of DSRC.

15.3 Key VANET 2.0 Features at PHY Layer

In Sect. 15.2, we discussed a few key technical challenges related to the application of IEEE 802.11 solutions to vehicular communications and the standard compliant enhancements to PHY/MAC which can address these challenges. Although the DSRC solution with these enhancements seems to be adequate to address safety applications, there are still opportunities for further improvement.

Here are a few such enhancements we believe are important for future versions of DSRC.

15.3.1 Better Channel Interleaving and Channel Coding

In vehicular communication environments, the current channel coding and interleaving mechanisms defined in IEEE 802.11p lead to performance loss, since the frame reception error is determined by the worst channel condition during the frame transmission (rather than the average channel condition). A straightforward improvement is to enforce time-domain interleaving. The downside of such an approach is longer frame decoding delays, since the receiver might have to wait for the reception of the entire frame before starting decoding. The delay can be a concern for unicast frames because, according to the standard, the ACK signal has to send back by a receiver rather quickly ($10 \mu s$), in an atomic way after receiving each frame. But such a concern does not exist for the majority of the DSRC safety applications that rely on broadcast rather than unicast. Thus, a simple approach would be to enforce time interleaving for broadcast frame only. Meanwhile, LDPC code has been developed and implemented in the more modern versions of IEEE 802.11 family standards, e.g. 802.11n, which has significant performance gain against convolutional code used in IEEE 802.11 providing an effective solution for improving DSRC PHY performance.

15.3.2 Migration to More Modern PHY Technologies Including MIMO Support (IEEE 802.11n) and Multiple Stream Support (IEEE 802.11ac)

Multiple antenna support can bring a lot of value to DSRC systems. For example, both transmitter and receiver beamforming can increase the link budget and the transmission range. This is especially valuable for non-line-of-sight communication between vehicles, since the modems can track and beam form to the non-LoS path to enhance the reception reliability. Further, the support of multiple streams, similar to IEEE 802.11ac, can enable the vehicle to receive multiple frames simultaneously, further enhancing frame reception rate.

15.3.3 More Flexibility in Channelization

The 10 MHz channelization is chosen at the very early stage of DSRC development mainly to find the most economical way to increase cyclic prefix using the technology available at that time. However, such an approach underutilizes the capability of current technology. The latest generation of IEEE 802.11 chips mostly supports wideband operations to provide higher data throughput. For example, IEEE 802.11n supports 40 MHz channelization and 802.11ac supports 80 and 160 MHz channelization. Meanwhile, the 10 MHz channelization used by the current DSRC also limits the efficiency of the DSRC frequency band (75 MHz in US): one needs multiple DSRC radios to be able to use multiple channels. The most common DSRC devices on the market today or on the roadmap have one or two DSRC radios, which means it can support vehicle safety communications in a dedicated safety channel and potentially switch between the other channels to transmit or receive. This, however, limits the data throughput for DSRC communications and introduces unnecessary additional complexities for managing channel switching between different channels.

15.3.4 Decoupling of DSRC Upper Layer with 802.11p PHY/MAC to Support Multiple Potential Technologies

Cellular data services can be used to support an increasing range of vehicular communication applications [74]. Feasibility of certain safety applications, such as road hazard warning, has also been demonstrated in today's 3G and 4G networks [24]. An analysis of long-term evolution (LTE) network capacity for vehicle communications showed the feasibility of using LTE to support event-triggered vehicle-to-vehicle hazard warning messages but also revealed significant overload due to periodic messages that were sent every 100 ms [84]; other possible technologies for vehicular data communication are WiFi technologies; many other ones could be mentioned and are shortly recalled in Sect. 15.6. Thus, for DSRC evolution, there is value to decouple the upper layers (1609.x protocol stacks) and applications from the lower layers so that the actual data transmission can use any radio technology that best fit the application needs rather than being limited to IEEE 802.11p.

Another important challenge to evolve DSRC is backward compatibility: future DSRC devices have to be able to communicate with the legacy DSRC devices. How do we ensure that legacy devices and future devices can communicate with each other? This backward compatibility issue is significantly different from the compatibility issues that people faced in the conventional enterprise networks, telecommunications networks, or the Internet. In the conventional networks, the mobile devices communicate through infrastructure networks and servers, which are controlled by the operators. An upgrade of the infrastructure network can be sufficient to allow legacy devices and new devices to communicate with each other. Further, the newer devices usually support multiple radio technologies.

Vehicle safety applications rely heavily on V2V message broadcast directly among vehicles. All broadcast packets are meant to be decodable by all neighboring devices around you. No intervening infrastructure networks or devices will be there to assist communication between legacy and new devices. Therefore, one must carefully plan the hardware and software to be used for the initial DSRC deployment. The WiFi industry has moved much further ahead of IEEE 802.11a or 802.11p after 802.11p was selected as THE technology for DSRC a decade ago. The current IEEE 802.11a compliant chipsets on the market are much more powerful and at very affordable prices. Many chips support other newer generation of IEEE 802.11 standards in addition to 802.11a. Thus, when the DSRC market is ready for commercial deployment, the 802.11 chips will likely have much more hardware in them beyond what is required for the current IEEE 802.11p. Therefore, rather than using IEEE 802.11a chipsets for the initial DSRC deployment as most existing DSRC implementations do, a better approach will be to use, for example, IEEE 802.11n or 802.11ac chipsets so that advanced capabilities, such as LDPC code and decode and MIMO capabilities, can be available in the initial DSRC deployment. Although the initial DSRC communication is carried out using the current 802.11p standards, the chips will be ready for many advanced features that are likely to be needed in the future.

15.4 Key VANET 2.0 Features at MAC layer

It is difficult to comprehensively overview all the MAC variants and the non-compatible alternatives which have been proposed to improve the performance and scalability of the current IEEE 802.11 for VANETs. Some papers [66, 75] proposed to classify MACs into two categories: CSMA-based and Time Division Multiple Access (TDMA)-based solutions.

15 Insights into Possible VANET 2.0 Directions

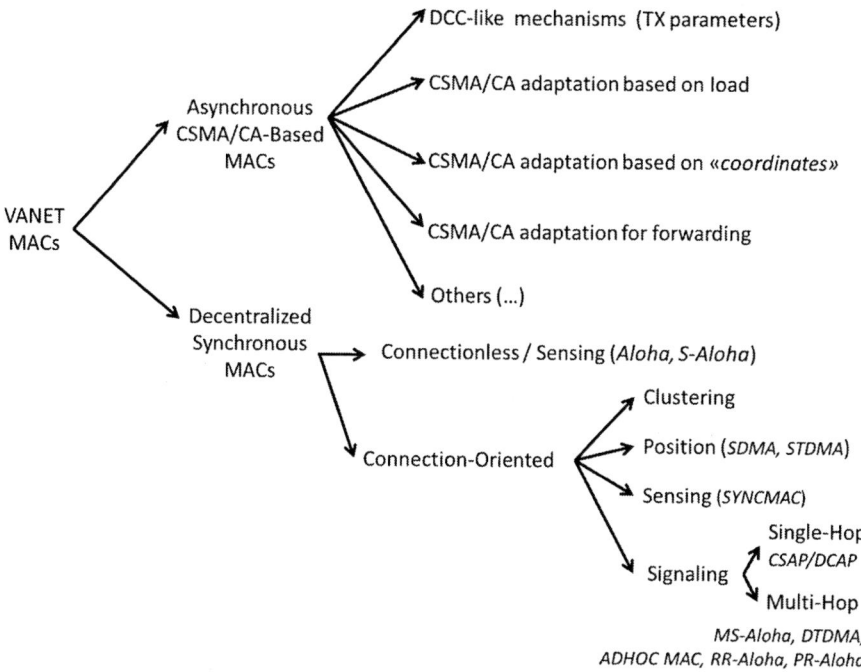

Fig. 15.9 A possible taxonomy of the MAC found in literature for medium access in VANETs

Obviously things are never strictly dichotomic: you can find also mixed types, for instance mixed synchronous and asynchronous MAC for multichannel environment in [7, 81], or MAC optimizations aimed at specific purposes (as in the case of the IEEE 802.11p geo-forwarding based on contention timer, described in Chap. 8).

However, a classification helps the general understanding, so it will be followed also here: more room will be devoted to synchronous MACs (Sect. 15.4.2) because they are the ones which may bring the most significant novelty and lead to very important benefits. Far from the intention of proposing a unique and unequivocal taxonomy, in Fig. 15.9 a scheme is proposed to visualize and summarize the proposed analysis.

15.4.1 (Asynchronous) CSMA-Based MACs

Almost the totality of asynchronous MAC for VANETs found in literature are based on CSMA/CA: this confirms that CSMA/CA is a good option and that, in spite of its well-known limitations, it offers significant benefits. CSMA/CA is simple, does not imply either cumbersome protocol overheads or a central coordinator and effectively copes with mobility and topology changes; even more, its weaknesses are supposed to become significant only when the VANETs will start to get crowded. Altogether, it made (and still makes sense) to identify possible tricks to improve CSMA/CA.

One possible way to classify the *plethora* of CSMA/CA-based protocols is to take into account what is the purpose of their optimization and what are the mechanisms which they act on. The most relevant categories are listed in the following (the list is not exhaustive): they will be shortly discussed in within this section.

- DCC-like mechanisms aimed at congestion prevention (hence reacting to channel load) and not acting on CSMA parameters (rather acting on parameters such as the reception sensitivity, the transmitted power, the generation rate, and the transfer rate);
- mechanisms still based on channel load analysis and acting on CSMA/CA parameters (such as Contention Window, Enhanced Distributed Channel Access, EDCA, priorities, etc.) to improve the reception rate and further prevent the occurrence of collisions;
- protocol adaptations pursuing more effective transmissions (less collisions and better spatial multiplexing) leveraging the information about nodes' coordinates (e.g., their position, time, or speed);
- solutions aimed at the improvement of expedite forwarding—as in the case of media-dependent geo-forwarding.

The discussion of DCC-like algorithms would not add much to what already presented in Chap. 5 and analyzed in [8, 9]: we just want to mention (outside the current DCC standards) the approach proposed in [63], that is to transmit setting a randomly distributed transmit power (Random Transmit Power Control—RTPC), with the purpose of minimizing the inter-frame reception (thanks to the background noise decrease).

In the same way, the category of MAC modifications aimed at making the forwarding more effective (and efficient) corresponds to the media-dependent techniques already discussed in Chap. 8: several papers have proposed similar approaches, exploiting the information on the distance from the transmitter to set the contention timer of the potential forwarders, so that the furthest node gets more likely to forward—thus reducing the number of hops. In this area it is worth mentioning the case of Mapcast [91], which proposes to include also city-maps in the forwarding decisions, so to avoid the forwarding of messages in non relevant areas.

We will devote more space something about the remaining two classes.

In the category of proposals operating CSMA/CA parameters, several acts on the contention window (CW): in [104] CW is adapted according to the number of nearby vehicles in a centralized way and, the higher the number of competing nodes, the wider the CW. In order to overcome the need for a roadside collecting and relaying information about network load, in [58, 105] it is proposed to estimate channel load using a mechanism similar to that used in the current DCC (see Chap. 6).

Also probabilistic approaches are met in literature: for instance, in [2] channel access parameters follow a *p*-persistent CSMA/CA-based mechanism, with the back-off interval based on a geometric distribution which is configured according to the number of neighboring nodes.

Finally some methods act on the differentiation of traffic, following the EDCA priority classes, so to upper-bound the percentage of high-priority traffic (for example, the first and superseded version of DCC [33] and some scientific papers, such as [107]).

Maybe some improvements in this class will be encompassed in future VANET standards: for sure they should be evaluated also in terms of their compatibility with other VANET mechanisms leveraging the same parameters, such as the media-dependent forwarding (Chap. 8).

The other class of CSMA/CA-based protocol aims to improve the protocol performance by leveraging information on position, speed, direction, and time: the topic has been recently revitalized by the European FP7 project GLOVE [25]. A first interesting idea [64] is to adopt a CW depending on the *position*, avoiding that in nearby locations similar CWs are used: the proposed technique was demonstrated to improve the reception rate, but mainly due to the wider CW being used.

Concerning *speed*, in [60] the authors address the communications between nodes and fixed road-side units (RSUs): in order to give all the nodes the same probability to exchange data with the RSU, they adapt the minimum CW size of each node according to its speed (the higher the speed, the shorter the window). Instead, in [2] the V2V communications are addressed and a scheme is described to assign EDCA parameters (channel access priority and CWmin/CWmax values) based on the relative speed of vehicles.

Concerning the possible role of *time*, two proposals will be mentioned: the first one [14] aims at avoiding iterated collisions between nodes whose CAMs generation is casually synchronous, by introducing a random and variable delay (Periodic Phase Changing Application—PPCA). The latter (named TD-uCSMA [102]) could be (but has not been yet) applied to VANETs: it proposes to periodically switch the EDCA parameters, so that only one node has high priority at a given time; in so doing, collisions are further prevented, but, actually, the technique may be applied only to limited number of nodes (e.g., RSUs and emergency vehicles).

15.4.2 Decentralized Synchronous MACs

At the dawn of the debate about VANET MACs, several studies were published about potentially suitable (wireless) decentralized synchronous MACs (W-DSM)—the most prominent ones are shortly recalled in Sect. 15.4.2.1. Manifolds are the reasons why, in the end, the WiFi approach was chosen: the maturity and widespread diffusion of IEEE 802.11, the intrinsic simplicity and decentralized approach of CSMA/CA and, last but not least, its asynchronous nature, which permits to skip the potential issue of synchronization.

Even if, from a theoretical point of view, not all the synchronous protocols need to be *slotted*, as a matter of fact almost all of them are so and rely on the following entities and concepts:

- a periodic frame structure divided into a fix number of slots;
- the slots are interleaved by guard-times (T_g), so to counteract propagation delays or a non-accurate synchronization;
- all the nodes share a common *absolute* synchronization;
- each node sends a packet within a slot (which needs to be as long as the longest packet).

W-DSMs paid a cultural penalty at the time of WAVE definition: up to then, few decentralized synchronous protocols had been studied for wireless and most of them required a central coordinator—this would not be viable for VANETs; additionally, synchronization had not been extensively studied and was perceived as a threat; finally, it is still commonly believed that a synchronous and connection-oriented approach will be hampered by mobility and will necessarily lead to a blocking performance: as soon as the number of nodes exceeds the number of available slots, some nodes will not be allowed to transmit.

Actually, over the years, the category of synchronous wireless MAC protocols have been demonstrated to be flexible and to have a strong potential—obviously case by case, depending on each protocol design criteria. To mention the most significant features, some W-DSMs can reuse (i) the same physical layer as CSMA/CA and adopt (ii) decentralized mechanisms; they can (iii) work at vehicular speed and (iv) guarantee a non-blocking performance; W-DSMs have also the (v) potential to solve the main issues of CSMA/CA (scalability and proneness to hidden terminals) thanks to a clever spatial multiplexing—especially if coordination is performed over more than one hop, as conversely CSMA/CA does. Furthermore, it has been shown that (vi) backward compatibility and co-existence between W-DSM and WiFi can be achieved, even using the same physical layer. Finally, (vii) synchronization is already foreseen in OBUs (in particular for channel switching—see Chap. 7) and can be easily achieved by Global Navigation Satellite System (GNSS) receivers (GPS, Galileo, GLONASS, Beidou, etc.) with satisfactory precision also under hold-on (Sect. 15.4.2.3).

All in all, at least it makes sense to reason once more on a possible synchronous evolution of current VANET MAC, to solve its current limitations: if this will not happen, at least the analysis will accomplish the goal of dropping the prejudices on W-DSMs, so to unveil their potential and give them the green light for other suitable applications (also other than VANETs).

In next sections a survey on synchronous MACs from literature is provided in Sect. 15.4.2.1, while some more details are provided on a significant example (MSAloha) in Sect. 15.4.2.2 and, eventually the technical side-issue of on-board synchronization is shortly recapped in Sect. 15.4.2.3.

15.4.2.1 Survey of Synchronous MACs

When talking about W-DSMs for VANETs, in most cases you inherently refer to a connection-oriented (or *soft* connection-oriented) protocol. The reason is that CSMA/CA already represents an improvement over the (connectionless) slotted Aloha (which, in turn, is better than pure Aloha): thus, synchronization is worth only if it permits to spread transmissions over space better than CSMA/CA, with an ordered approach—by *assigning* slots; this, intuitively, makes sense only if one node can make its *decentralized* decisions based on a quite stable perception of the channel—that is, based on some kind of perceived reservation. So, a distinctive point of W-DSMs is how they manage reservations; namely, reservations may be:

- *soft*—if they vanish after some time—or *hard*—when they are hung up at the end of the transmissions;
- *blocking* or *non-blocking*, depending on the possibility to re-use some slots, thus having more transmitters than slots; the latter is strictly needed for VANETs, so to cope with the *scalability* and *fairness*;
- based on *sensing* and/or *signaling*; in the former case, one could select a slot just checking that no other node is using it at a one-hop distance; in the latter case the information about the status (e.g., empty, busy) of each slot can be shared: this introduces a further distinction since the signaling may span over one or *multiple-hops*. Intuitively, an explicit signaling over multiple hops would serve as a countermeasure against HTs; however, the signaling should not represent an excessive burden and should work effectively also under mobility;
- compatible at different levels with coexisting CSMA/CA nodes: frames can be mutually received by the two MACs; even, W-DSMs reservation can be robust against CSMA/CA interference and/or W-DSM can be disruptive or not for CSMA/CA.

In a historical perspective, the first slotted protocol was based on the well-known Aloha [1], just introducing a slotted structure to prevent partially overlapping transmissions. This led to the definition of the MAC named slotted Aloha (S-Aloha) [87], in which frames are sent immediately and, in the event of a missing acknowledgment, a retransmission is made after a random backoff. The efficiency of S-Aloha is higher than pure Aloha, but it still lacks the concept of reservation. It was first introduced with the Reservation S-Aloha (R-Aloha) protocol [28], in which nodes continuously listen to the channel to identify free slots and, when one needs to transmit, it will just pick a free slot and keep it as long as it needs it. Then the reservation of R-Aloha is soft-blocking and just based on sensing/receiving (not on signaling).

The next step consisted in the introduction of some explicit signaling, but still keeping the decentralized approach; in literature, three main families are found:

- clustering protocols;
- position-based protocols (the signaling carries position-information);
- slot-state protocols (the signaling propagates slot-related events and state).

The first class does not count many popular solutions [4, 37, 78] and has been somehow dismissed. The common idea which they subtend is the dynamic and decentralized creation of clusters for the optimal and scalable management of resources (in particular for slot allotment and re-use): but the idea of creating clusters before slot reservation is considered today too challenging and critical in a very dynamic environment and with rapidly variable topologies.

The position-based category includes space division multiple access (SDMA) and self-organizing TDMA (STDMA). In SDMA [11, 15, 61], the strategy for channel access is based on the location of the vehicles, which needs to be broadcasted: the rationale is to divide the roads into segments used for a given access policy—for instance in [15] a segment corresponds to a time/slot of a TDMA frame structure. The SDMA keeps the protocol overhead very limited but has the drawback of depending on the absolute position whose precision in urban areas, also with GNSS (GPS), is quite likely to exceed the width of a segment (5 m in [15]). As a result, both due to limited precision in their positioning or to a high density of cars, more nodes might use the same slot, leading to a poor reception.

STDMA [56] aims to solve these issues by exploiting the information about position in a different way: the nodes in STDMA sense the channel during one frame and then select a free slot for transmission; if no slots are available, the one used by the node situated furthest away is reused (not the one causing the lowest interference, as conversely proposed in [96]). STDMA has been simulated in some line-of-sight scenarios with encouraging results, so that it was proposed within ETSI as an alternative to CSMA/CA for VANETs [31, 32]. However, as an SDMA method, STDMA does not solve the issue of hidden terminals (it was originally meant for the naval line-of-sight environments [56]) and does not include any mechanism for collision detection. Additionally, recently, STDMA has been demonstrated to be sub-optimal at a distance exceeding the radio range [36]: if nodes B and C cannot receive by each other, they will not mutually coordinate and might transmit on the same slot and a node in the middle (say A) could not receive. Specifically, in [36], the authors define a *Packet Level Incoordination* and show that STDMA is not competitive against CSMA when the frames cannot be received: in fact, CSMA just requires physical sensing, while STDMA needs to receive the signaling frames (and cannot count on multi-hop propagation). The results presented in [31] may deserve some additional analyses in the light of [36]. Altogether, also considering the lack of mechanisms for coexistence with CSMA/CA, STDMA still needs to fill some gaps.

The category of multi-hop protocols dates back to 1988, with the concurrent slot assignment protocol (CSAP) [73], in which each node has the possibility to announce collisions (when no frame can be received in spite of a high sensed power), so that the colliding node can change the currently used slot: the prevention of collision assumes a probabilistic nature and does not require a third node in the middle. Decentral channel access protocol (DCAP) [111] proposes a similar approach and, in addition, it also considers periodic reports on the SINR perceived in the neighborhood. SYNCMAC [96] behaves in a similar way: energy sensing is the technique chosen for slot selection, while collisions are not announced but rather self-discovered by each node, by checking its slot state after muting for a while. In CSAP, DCAP, and SYNCMAC the signaling does not exceed on hop, so they do not address hidden terminals and strong fading. Additionally, CSAP and DCAP are blocking—unlike STDMA, there is no way to force slot re-use.

A more "controlled" slot allotment over multiple hops was achieved by Reliable RAloha (RR-Aloha) [16] and ADHOCMAC [17]: both protocols propose that nodes append to each transmission an allocation chart describing how a slot is perceived by this particular node, so to perform multi-hop signaling and preventing hidden terminals; Priority R-Aloha (PR-Aloha) [5] and Decentralized TDMA (D-TDMA) [67] subtend, more or less, the same rationale and represent variations over RR-Aloha. Also VeMAC [81] elaborates on ADHOCMAC, proposing to make slot reservations robust against rapid mobility, by assigning slots depending on the direction of speed: for the first time, VeMAC also copes with the multichannel environment of VANETs. Unfortunately, none of them deals with the problem of slot re-use (just propose pre-emption, that is priority among connections).

Mobile Slotted Aloha (MS-Aloha) [26, 27] moved its steps also from [17]—hence inherently prevents collisions by hidden terminals: after fixing some signaling issues for to the VANET environment, it was enriched by a solution for slot re-use in a scalable and decentralized way. The most recent, theoretical results on it have not been published yet but are available in [93].

In the following section MS-Aloha is shortly recalled, as a significant example being, at the moment, the only W-DSM solution addressing all the VANET MAC issues which have been mentioned before: as further discussed it also led to very encouraging comparative results versus CSMA/CA.

15.4.2.2 An Example of Synchronous Protocol: MS-Aloha

MS-Aloha is one of the two protocols which were studied in STF395 and published in the resulting ETSI reports [31, 32] (actually, the titles of the reports are misleading and refer only to STDMA). In this section a brief overview of the protocol and its main properties is provided—for a detailed description of its mechanisms, the bibliographic references will fill the gaps. The building principles of MS-Aloha are:

(a) the protocol should use the same physical layer and frame formats as DSRC;
(b) reservations are soft and announced (together with collisions) over multiple (3) hops, so to achieve a better spatial multiplexing and prevent HTs (the principle is shown in Fig. 15.10); such signaling is appended to each transmission (or on alternate frames) describing the full perception of the channel (how each slot is perceived and, if engaged, by what node);
(c) there is not a dedicated set-up phase: a node just transmits in a free slot and checks if it works by controlling the other nodes' feedback; the reservation fades after one period;

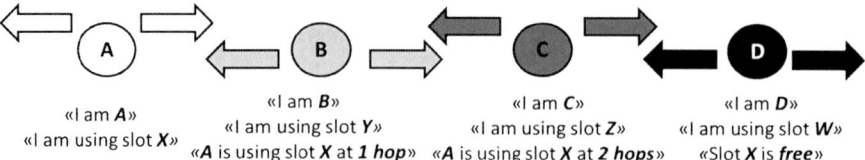

Fig. 15.10 The mechanism of signaling in MS-Aloha: the information is not propagated beyond the third hop; the length of each hop may vary in a decentralized way, depending on the perceived traffic load

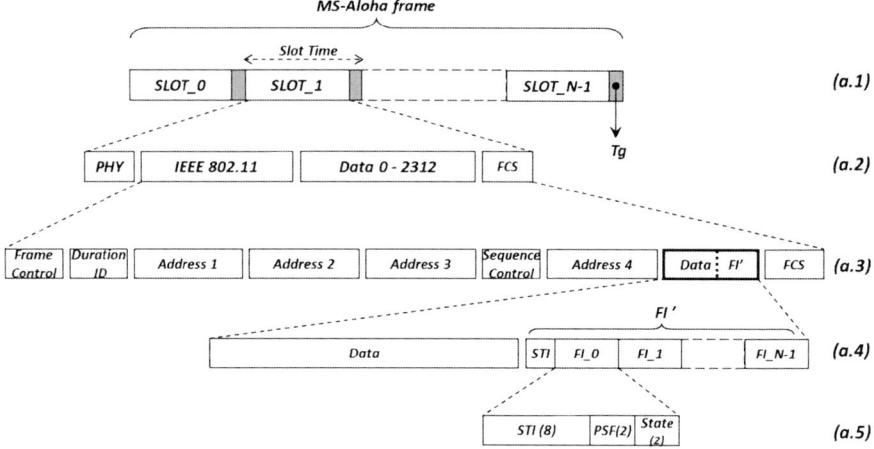

Fig. 15.11 MS-Aloha: the proposed frame format met in [32]

(d) the signaling overhead is kept limited, thanks to a trick delivering unambiguous *Short Temporary Identifiers* (STI) instead of MAC addresses, as shown in Fig. 15.11(a.4)–(a.5): STI are continuously swapped over time;

(e) slot re-use is granted, thanks to a power-based (decentralized) mechanism which permits to use a slot closer, as the number of available ones starts to starve (the length of each signaling hop gets shrunk).

Altogether, considering the classification of Sect. 15.4.2.1, MS-Aloha is based on a hard reservation, leveraging a 2-hop signaling with collision detection and adaptive slot re-use. It has been demonstrated to have the capability to coexist with CSMA/CA. From a different perspective, in designing MS-Aloha the authors have pursued a protocol in which a slot is used at a number of hops sufficient to prevent mutual interferences (*multi-hop signaling*) but, in case of starving resources, each slot can be re-used closer. Simulations[3] tend to confirm that the stated design

---

[3] In all the simulations here presented, the same propagation model used in [31] has been adopted (dual slope decay with Nakagami fading depending on distance) and the nodes (from 400 to 600) move in a ring (radius = 2 km) with a distribution between 40 and 120 km/h, half per direction;

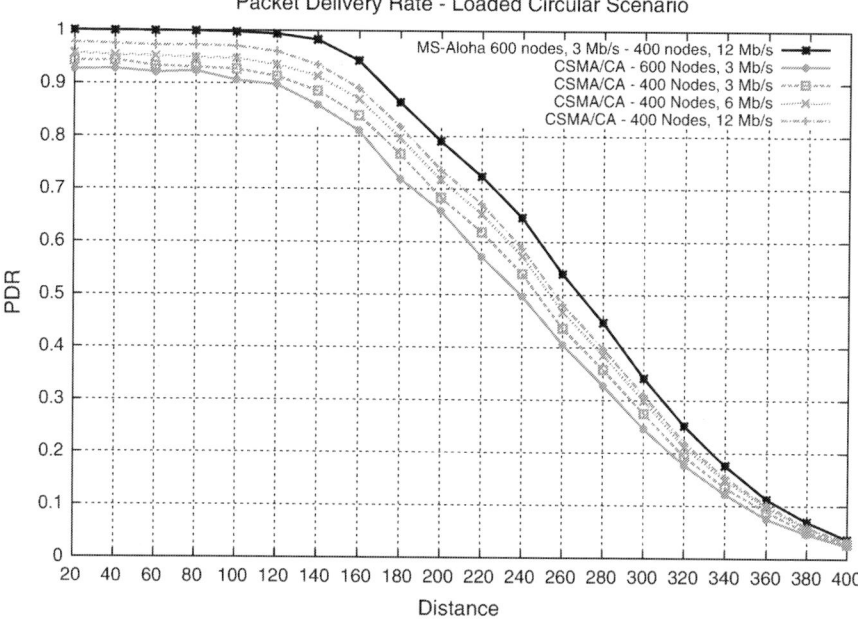

Fig. 15.12 The PDR in a loaded but not congested circular scenarios. A higher percentage of load is achieved by decreasing the transfer rate (from 12 Mb/s down to 3 Mb/s) and by increasing the number of nodes. MS-Aloha almost unaffected (the graphs are overlapped)

objectives have been fulfilled. Due to the limited room only some graphs will be shown: at the time of the current writing some comprehensive papers have been updated and will be announced in the website [77]; the interested reader, for more details, can refer to the official documents [31, 32, 93].

1. The reception keeps ideal also under *heavy traffic*, because a slot is reused only after multiple hops; this is shown in Fig. 15.12 where the packet delivery ratio (PDR) of MS-Aloha is almost ideal and always better than CSMA/CA (without DCC); the graphs refer to multiple simulations with growing load;
2. under heavier congestion (results available in [26]), MS-Aloha still manages to keep the ideal reception in the first meters (100% in close proximity of the transmitter, when CSMA/CA cannot reach 85%): the price paid by MS-Aloha is a drop of PDR at higher distance (over 70 m in [26]).
3. due to its protocol overhead, MS-Aloha may meet more simultaneous transmissions than CSMA/CA but, thanks to the ordered re-use, they will be segregated at a longer distance: the distribution of simultaneous transmissions over distance

---

CSMA/CA does not use any DCC features and the transmitted power is set to 20 dBm for both MS-Aloha and CSMA/CA. VANET frames are periodically generated at 10 Hz and are long 300 bytes.

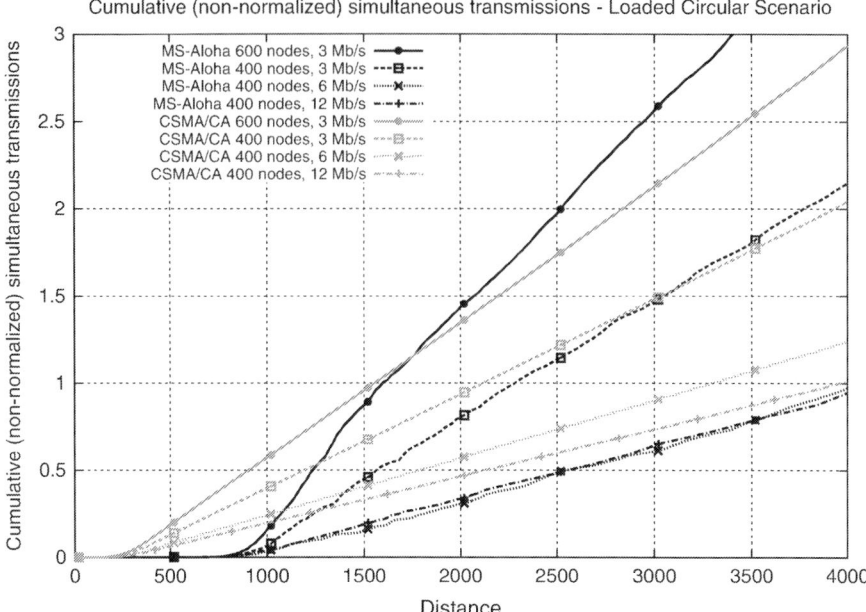

Fig. 15.13 The distribution of simultaneous (even *partially* overlapping) transmissions in CSMA/CA and MS-Aloha, for the same scenarios of Fig. 15.12. The network is not fully crowded so that MS-Aloha does not force slot re-use at a distance shorter than 750 m

in Fig. 15.13 (cumulative function for the same simulations of item 1) witnesses this phenomenon. In fact, the graph of MS-Aloha, compared with CSMA/CA, reaches higher values at longer distances but involves a higher cut-off for the minimum distance of simultaneous transmissions: with MS-Aloha it corresponds to the three propagation hops (about 750 m), while in CSMA/CA it corresponds to the sensing distance (about 250 m);

4. the multi-hop signaling will keep the reception ideal also in presence of obstructions facilitating the occurrence of hidden terminals. In [27] the reception rate was analyzed for nodes in the centers of crossroads obstructed by buildings: MS-Aloha could still guarantee an almost ideal reception, while CSMA/CA—even if still far from congestion—dropped dramatically (between 60 and 20 % in the simulations); CSMA/CA mis-reception was ascribed to the effect of uncoordinated transmissions by the hidden nodes in the legs of the crossroads.

Furthermore, in [93] it has demonstrated that MS-Aloha can achieve an effective coexistence with CSMA/CA, thanks to its pre-emption capabilities.[4] In fact, pre-emption can be used also to prevent CSMA/CA from breaking the reservations

---

[4] With pre-emption, a high-priority connection can override a low-priority one.

15 Insights into Possible VANET 2.0 Directions   437

by MS-Aloha.[5] Importantly, coexistence can be also leveraged for the fall-back to CSMA/CA in case of missing synchronization. Finally, MS-Aloha inherently gives ACKs to any transmissions, also in broadcast—in fact every nodes will communicate in its signal what nodes could receive from in each slot. This may constitute a criterion for retransmissions of packets which have been scarcely received.

Two final remarks may conclude this overview:

- as shown in Fig. 15.11, MS-Aloha implies a significant protocol overhead: if the number of slots is N, then you will have an overhead equal to $12 * N^2$. So one might conclude that it cannot be worth. But the overhead itself is not a significant parameter, especially if it does not limit the number of transmissions which are possible (as in MS-Aloha). Even more, the overhead becomes a value if thanks to the signaling conveyed by it, one can transmit more effectively or, the other way around, the safety information can be received by more nearby nodes. This is what happens with MS-Aloha (see Fig. 15.12)
- the number of slots N, given the quadratic relation between the overhead and N (previous bullet) is quadratic as well (it has a square-root dependence [32]). Typical settings used in the simulations adopt slots able to house the PLCP and 300-byte-long packets, thus getting: 200 slots at 12 Mb/s, 130 at 6 Mb/s or 80 at 3 Mb/s (in all the cases $T_g > 50\,\mu s$).

All in all, MS-Aloha is an interesting candidate to improve reception rate in VANETs and can also guarantee a strong coexistence with CSMA/CA; synchronization will not be a critical aspect either (see Sect. 15.4.2.3).

On the other hand, CSMA/CA is counteracting congestion quite well with the new version of DCC, but it is not addressing hidden terminals yet: perhaps, depending on how much the effect of HTs will really be detrimental for VANETs, the Community will be more or less motivated to consider MS-Aloha as a sensible alternative. In any case, VANETs have given the spark to define a new flexible MAC with deterministic behavior, which might be useful for other purposes, such as wireless automation.

15.4.2.3   The Issue of Synchronization

This section is about synchronization in vehicular environments (accuracy, hold-on, countermeasures for mobility and missing signal) and is quite technical: for those

---

[5]The rationale for coexistence is to let MS-Aloha reserve two slots every time he needs one: the second is really used for transmissions, the first one is used just to let CSMA/CA sense the channel busy; this *fake* reservation will have the lowest priority (so that other synchronous nodes can pre-empt it) and will involve a low-power transmission of fake data (so to interfere as little as possible any other synchronous transmissions).

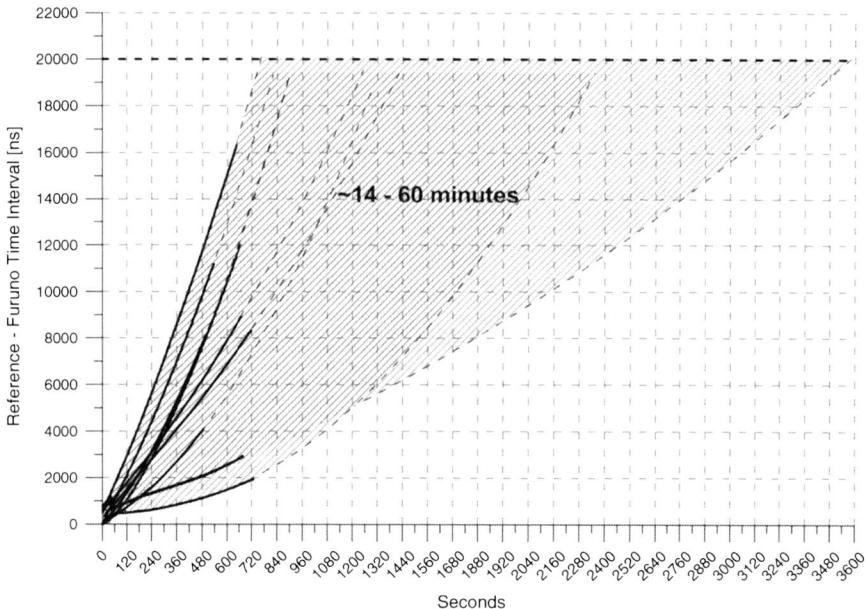

Fig. 15.14 The accuracy of synchronization during hold on, as from GLOVE project [79]: after achieving GNSS synchronization and moving into a under-ground parking (no GNSS signal available), the synchronization could be kept for about a quarter of an hour with an error not exceeding 20 μs. If one accepts such hold-on, a $T_g = 50\,\mu s$ would be sufficient for a slotted protocol: $40\,\mu s = 20 \times 2\,\mu s$ would counteract the possible two opposite clock drifts of 20 μs, while 10 μs would be more than sufficient to balance the propagation delay

who may be not interested in such a vertical insight, the caption of Fig. 15.14 shortly recaps the main concepts and results.

One of the reasons which discouraged the use of slotted protocols for VANETs was certainly a deep worry about what precision in the absolute[6] synchronization of nodes could be achieved under mobility and what would happen in case it got lost—for example due to urban canyons or other causes of GNSS signal loss. Actually, synchronization issues may be counteracted with several expedients, namely:

1. the first trick is to adopt a sufficient guard-time (T_g in Fig. 15.11). A guard-time is, in any case, required to cope with propagation delays: even under the hypothesis of an *ideal* synchronization of nodes, a frame sent by node A will take some time to fly to B and, as a result, it will be perceived by B as non-perfectly

---

[6]In this chapter we will call *absolute synchronization* the synchronization (i) both in frequency and phase and (ii) achieved by means of an external source (such as GPS, Galileo or other GNSS systems), not by a clock tree distributing it among the communicating nodes.

synchronized.[7] Consequently, in setting the duration of T_g one should consider the impact of propagation. To make an example, in most of the simulations of MS-Aloha $T_g = 50\,\mu s$; in the hypothesis of significant interference up to a distance of 1.5 km, it would take 5 μs to reach that end and one would keep 45 μs for balancing two opposite clock drifts of up to 22.5 μs; with a clock accuracy of ±0.1 ppm, it would correspond to 4 min hold-on;

2. another option would be to build a solution to fall-back to an asynchronous MAC when the synchronization starts to deteriorate (for instance, based on how long the GNSS signal could not be received); for instance, MS-Aloha has already been provided with such mechanisms and studied by simulations, as discussed in Sect. 15.4.2.2;
3. synchronization should be made as robust as possible, while gaining a deeper awareness on the performance achievable with off-the-shelf components (one could not count on expensive atomic clocks), also under mobility and paying particular attention to the hold-on performance, in case of missing signal;
4. theoretically, one could also investigate how to extend the hold-on by cooperative positioning [82], that is, correlating position and synchronization by the surrounding nodes, or using other communication channels (such as a radio signal in the white spaces or a Li-Fi signal), as further discussed in Sect. 15.6. However this hypothesis is not mature enough today and is left for future investigations.

The clause 3 was preliminary studied in ETSI's Specialist Task Force STF395, leading to the technical report [32], but has been more extensively studied in the GLOVE project, also carrying out practical experiments.

In general, when talking about synchronization, the most influential parameters are the precision and the effectiveness of the oscillator in holding-on. Concerning the former aspect, typically, still GNSS receivers (i.e., with a good and stable reception) can achieve errors lower than 25 ns; even under strong mobility (as high as 1,440 km/h) there are results claiming errors not exceeding 250 ns [41, 43].

Further, it is known that, thanks to GPS Disciplined Oscillators (GPSDOs) [29, 40, 72, 83], the GNSS signal can be exploited not only to adjust the phase of the oscillator but also to continuously adjust the frequency. This, intuitively, can stabilize the frequency, because the oscillator is steered with a controller retaining the knowledge of the past: as a result, also in the event of a missing GNSS signal, the oscillator will be able to keep the synchronization error low for a long time. In [71] they show that, after a long lasting steering (weeks) and afterwards detaching the antenna, a GPSDO could keep the offset lower than 100 μs for over a week. This is very promising in the perspective of adopting a guard-time of 50 μs.

One step still misses: What is the hold on when the receiver is moving and when the GPSDO cannot be locked for such a long time as in the experiment [41]? For this purpose, some focused experiments were carried out within GLOVE

---

[7] In TDMA methods, for these reasons, frame reception cannot be strictly based on synchronization, but rather needs to exploit the data-link preamble and start of frame delimiter.

project, checking the precision of the synchronization under mobility and in hold-on, leading to very interesting results (see Fig. 15.14 and [79]). The experiments considered very different environments—including open-sky conditions, urban canyons, woods, underpasses, and underground parking lots. In urban canyons and in motion, the 1 PPS (pulse per second, that is synchronization output from the GNSS receiver) varied up to 120 ns against reference time generated by the portable rubidium atomic clock; in the highway, the synchronization pulses from the receiver varied up to several tens of ns. Tests also allowed to estimate the threshold time of 15 min, as the minimum time after which the precision of synchronization signal resulting from GNSS signal loss exceeds the condition of accuracy of 20 μs which we set before.

All in all, precise synchronization for VANETs may require more extensive investigations, but should not be perceived any more as an insurmountable threat.

15.5 Extend V2V/V2I to V2P

Pedestrian and, more generally, vulnerable road users (VRU) casualties on the road has been a major factor in the overall road casualties around the world, especially in developing countries. In particular, the world road traffic crashes kill nearly 1.3 million people and injure 20–50 million more every year. Nearly half (46 %) of those dying on the roads are VRUs, including pedestrians, cyclists, and motorcyclists. In the US, 28 % of the traffic crash fatalities in 2009 were VRUs, up from 20 % before 2000. This suggests that a key mission for DSRC evolution is indeed to extend vehicle safety applications from vehicle occupants to VRUs. Thus, a natural evolution step is from V2V/V2I to V2P (Vehicle to Pedestrian) communications. Towards this end, making cell phones as part of the vehicular communication network becomes a key mission for the society.

Over the last 2 years, Qualcomm Research has been collaborating with Honda R&D America to develop and demonstrate the feasibility of such a system [109]. In particular, a prototype pedestrian safety system based on vehicle and smartphone DSRC communications was developed and demonstrated, which gives a 360°, extended range, NLOS view between the vehicle and pedestrian. Both the driver and the pedestrian are warned of a possible collision, if an imminent safety threat is detected. However, to take this concept to field deployment, there are many other challenges to be addressed, some of which are listed below.

15.5.1 Spectrum and Channel Congestion

Channel 172 (5.855–5.865 GHz) of the DSRC spectrum is currently assigned for V2V safety applications. With the potential of thousands of vehicles in the 300 m DSRC transmission range, channel congestion has been a key issue of investigation

15 Insights into Possible VANET 2.0 Directions 441

over the last few years in the US and Europe [62, 97, 99]. Enabling billions of potential transmitters from mobile devices in Channel 172 can certainly complicate the congestion. We discuss a few mitigation schemes to eliminate or reduce channel congestion in the critical safety channel across the different operation modes of the smartphones.

1. **Receive-only mode**: Mobiles are *only* allowed to receive in the critical safety channel thus creating no additional burden to the safety channel's bandwidth. In this case, the safety alerts are dependent on safety applications running on the mobile phones. A device receives BSMs from neighboring vehicles and calculates the imminent threat of collisions. In this case, a warning will be triggered only on the pedestrian smartphones and not on the vehicles. There are non-safety critical benefits such operation mode can bring to consumers where users can use phones to receive information from DSRC-enabled infrastructure and other devices (e.g., signal phase and timing information (SPAT) from the traffic signal [88]).
2. **Allow transmission in service channels**: Mobiles are allowed to transmit in another DSRC channel, instead of Channel 172. Similar to option 1, this allows the existing V2V system using Channel 172 to not be affected by the introduction of DSRC-equipped smartphones. Industry and regulating agencies, i.e., the Federal Communications Commission would have to agree on assigning one service channel to be available for pedestrian use of DSRC. Vehicles may want to listen in the V2P channel to take advantage of the pedestrian transmissions and improve pedestrian detection functionality, which may add cost on the vehicle side, as a secondary DSRC radio would be required. Furthermore, it is likely that a congestion control protocol for smartphones would be required within the V2P channel. Clearly, this option does not exclude option 1 above, i.e. a mobile device can send its beacon in one of the service channels and yet listen to surrounding vehicles in Channel 172.
3. **Allow smartphone transmission in Channel 172**: In this scenario, smartphones are allowed to transmit in the critical safety channel, in addition to the receiving capability. Certain restrictions have to be applied on the mobile side to reduce channel congestion. To start with, mobile devices may be limited to transmit at lower power and lower duty cycle compared to the OBE transmission. This may be natural since pedestrians move at slower rates than vehicles. For safety applications to be effective, lower power and lower duty cycles may be sufficient. For example, mobile devices might only transmit at 10 dBm maximum power (instead of 20 dBm) with a periodicity of 1 Hz (instead of 10 Hz). This combination of reduced power and periodicity can reduce the air interface congestion from a mobile device by a factor of 40, as compared to an OBE, potentially mitigating congestion issues. In addition, situational awareness would have to be enforced in smartphones so DSRC signals are only transmitted when needed, e.g., when the devices detect the owner is walking near or along a street.

In summary, there are different ways to introduce mobile devices to DSRC with minimal impact to channel congestion in the critical safety channel. However, industry and standards groups must work closely together to enable this. New standards, performance requirements, and subsequent certification may be required to specify the role of mobile devices in the DSRC eco-system.

15.5.2 Mobile Positioning Accuracy

Effective DSRC operation between vehicles requires good relative positioning accuracy. Inaccurate GNSS (GPS, Galileo, etc.) positions can cause false positives and missed detections for V2V safety applications: however, GNSS and positioning technology is expected to improve, thus enhancing the reliability of DSRC systems. Compared to an in-vehicle implementation, a smartphone has a limited form factor and power budget for GPS: hence, smartphones may exhibit worse positioning accuracy under comparable conditions. To further improve the positioning accuracy on phones is certainly a key step towards making V2P safety systems a reality. One key enhancement might come from ranging capability embedded in Wi-Fi. With 160 MHz channelization, Wi-Fi based ranging, as defined in IEEE802.11mc [12], is expected to reach sub-meter accuracy. With the large-scale penetration of Wi-Fi technologies into both phones and cars, this capability will aid the GPS-based positioning approach to obtain much better relative positioning.

15.5.3 Security Design

Vehicle security is envisioned to be based on the public-key infrastructure and applied at four stages: bootstrapping, certificate provisioning, misbehavior reporting, and revocation [106]. A vehicle may not necessarily have a permanent connection to the infrastructure. However, it is assumed, once powered, that smartphones are always connected to the infrastructure, allowing for an easier security management. Nevertheless, smartphone security must be compatible with the vehicle security design.

Security will also depend on the smartphones' operation mode. For example, there will be no need for certificate provisioning to the phone if the phone is in receive-only mode. On the other hand, if a smartphone transmits awareness messages, then certificates can be provisioned through the infrastructure.

15.5.4 Certification Process

Certification of a DSRC-enabled smartphone is contingent on development of performance requirements and objective tests. If the V2P safety application is an important outgrowth of the safety proposition offered by V2V, that certification of

15 Insights into Possible VANET 2.0 Directions

communications and application performance should be an outgrowth as well. Is the smartphone's V2P function a supplemental alert or is it a safety-critical warning? Answers to such questions will dictate the certification process.

We realize that, while certifications are important, over certifying may prevent the introduction of a smartphone into what might be the largest portion of the DSRC ecosystem. Therefore, to what extent V2P applications are certified is a question which will certainly feed an important future discussion.

15.6 Complementary Solutions to Unload the Channel

The two main debates on the possible future evolution of VANETs concern novel solutions to improve them and, more in general, its perspectives (market and research). The two topics have been discussed in a workshop and in a paper of the same name [30], in which the participants concluded that Field Operational Tests (here discussed in Chap. 14), new Inter-Vehicle Applications and Heterogeneous Vehicular Networks (HVN) constitute the most concrete research perspectives. HVNs are motivated by multiple reasons, including:

- the widespread availability of multiple wireless technologies;
- each wireless solution has its own points of strength and weaknesses and, as such, they could provide complementary solutions; for example:
  - LTE is better suited to messages which have to reach the Internet (or arbitrary distances) while WiFi for local broadcasting;
  on the other hand, the two can work together for the global sustainability: in [30] the authors proposed two models: (i) class A in which HVN are agnostic on the applications and just pursue network off-load; (ii) class B, in which lower layers and applications get in direct connection, so that the most suitable medium can be selected for each task;
  - in either case, VANET would get unloaded and this would be beneficial to avoid (or upper-bound) the well-known WiFi issues presented in Sect. 15.2.2;
- the integration of wireless solutions would be beneficial to the overall goal of vehicular safety:
  - during the initial deployment of VANETs, other technologies—in particular LTE—might help them to cope with the low penetration into the market of the DSRC technology...
  - ...but LTE alone could not manage an ever growing nodes' density, due to the existing base load of the human-to-human communications: LTE could not work without DSRC, while VANETs need a solution for channel unloading;
- last but not least, the integration among multiple wireless media is something which is being retrieved but that was foreseen from the very beginning, as already discussed in the International Standard Organization (ISO) Communications access for land mobiles (CALM) architecture (Chap. 2).

Altogether, HVNs are a relevant topic and is something which should not be thought as competing against VANETs. There are several possible candidates potentially complementing VANETs: the main one is, for sure, LTE, which Chap. 16 will deal with; but many others might play a role: Bluetooth for personal devices, Zigbee for sensors, ordinary WiFi and WiMax, etc.

In the following subsections we will focus on some emerging ones which are characterized by some promising and novel features which deserve some additional insights, that is: Light-Fi, White-Space communications, and TPEG. But, before moving to these solutions it is worth mentioning some words on the possible role of some consumer technologies.

- Some papers [95, 101] propose to use the *ordinary* **WiFi** (IEEE 802.11a/b/g/n/ac) also for vehicular communications. For instance, in [101] the authors propose it as a solution to test VANETs even before a sufficient penetration of IEEE 802.11 is reached and, after some promising wardriving campaigns (to assess the background interference), they investigate the feasibility on smartphones. In [95] the authors even propose to use smartphones to have also a vehicular channel over IEEE 802.11a, so that cars equipped also with an IEEE 802.11p radio may forward it to/from the VANET.
- The previous item somehow suggests the idea of using a **smartphone as a companion device** and this is confirmed by a rich literature. For instance, in [59] they propose that the smartphones might be the ideal travel planners, since they can be brought in a whole multi-modal itinerary. In case the companion device were tightly coupled with the on-board environment, this would result in a portable environment, would enable the forwarding from the VANET to the legacy WiFi (and vice-versa), would permit a capillary and continuous driver's control [23].
- Talking about consumer wireless technology and smartphone coupling, it gets almost natural to mention also Bluetooth. In particular **Bluetooth Low Energy** (BLE) is a relatively new technology.[8] Given its characteristics, BLE can be used for *Personal Area Networks* (and on-board device-to-device), but also for vehicular communications, as suggested in [35]. However, considering the profiles defined in BLE specification, this would not be sensible for vehicle-to-vehicle safety communications—since the roles of *Peripheral* and *Central* cannot be played at the same time by the same device—but rather for sharing some files among drivers or to communicate emergency information from a Road-Side Unit (RSU) to an On-Board Unit (OBU).

---

[8]BLE was established in 2010 as Bluetooth Core Specification ver. 4.0. It is not compatible with the previous versions, often referred to as *Bluetooth Classic* (BC). BLE, as opposed to BC, was designed to reduce power consumption and to simplify the pairing. It still works in the 2.4 GHz band but with a doubled channel-width (2 MHz) compared to BC: this way, in spite of the reduced transmitted power (10 mW instead of 100 mW) it covers a distance of up to 100 m.

15.6.1 Light-Fi

Li-Fi, stands for Light-Fidelity and refers to a wireless communication technology that uses either visible or non-visible (that is, *infrared*) light for bi-directional high-speed communications: for this reason you talk about VLC (Visible-Light Communications [46]) or non-VLC (non-Visible-Light Communications). Even if non-VLC could partially solve some of the problems of outdoor VLC (such as the blinding by ambient light), the literature almost exclusively refers to VLC, perhaps in the positive perspective of mounting an only device both for lighting and communicating [85]. As a result, LiFi and VLC are used in an almost interchangeable way.

VLC systems typically involve cheap transceivers: a Light Emitting Diode (LED) as transmitter, a photo-diode as receiver and on-off keying for coding. VLC are specified (PHY and MAC) by IEEE 802.15.7 which, actually, is out-of-date: for instance, the standard defines transfer rate up to 96 Mb/s, while, currently, 1.5 Gb/s has been exceeded by LiFi using optical OFDM (O-OFDM).

Fortunately, LEDs have turned very common as automotive lighting components, thus becoming appealing for V2V communications, as suggested in [18, 69], and [70]. Intuitively, VLC transceivers should benefit from the cabling of the usual lighting systems of the cars (or scooter): Figure 15.15 depicts a possible

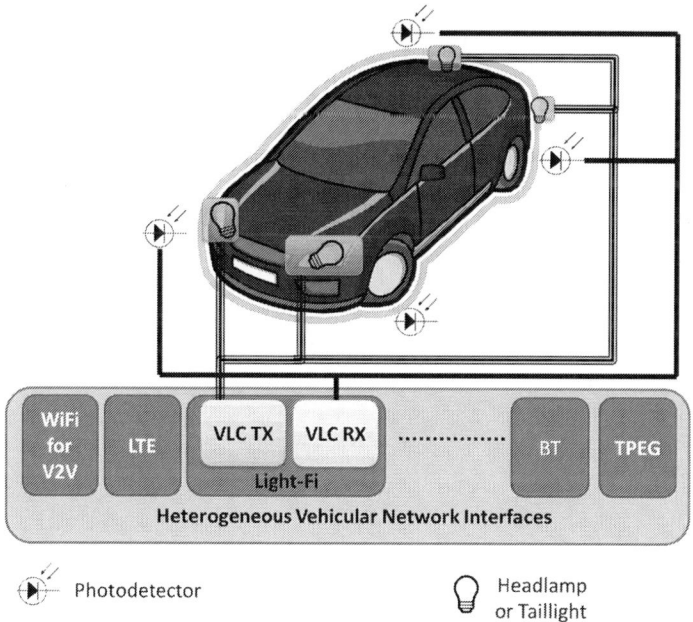

Fig. 15.15 One possible configuration of LiFi transmitters (LEDs) and receivers (photo-diodes) in an overall OBU involving multiple wireless interfaces

embodiment. According to [110], there are indeed several reasons to bet on a future on-board integration of VLC system. The authors mention:

- Low complexity and low cost. Compared to typical RF transceivers, VLC ones are much simpler, due to limited multipath phenomena and the simple modulation schemes adopted. Furthermore, in deploying on-board VLC, one can optimize the cabling ad packaging and leverage those already used for the lighting.
- No scalability issues. In RF (here including VANETs), given the characteristics of propagation (reflections, diffractions, diffusion) channel contention typically involves several nodes: this leads to well-known scalability issues. Instead, with VLC you receive only in LOS and, given the position of car lamps (headlamps or taillights), there will not be many nodes in direct LOS (even more, distinguishable by optical properties such as the incidence angle) and this will guard against scalability issues.
- Enforced vehicular safety. In [110] the authors demonstrate, by tests on prototyped VLC systems mounted on scooters, that VLC works up to 15 m apart and, at typical scooter speeds, this will allow the drivers to have as much as 0.9 s to react to an event.
- Positioning capabilities. As discussed in [10], VLC permits an accurate relative positioning among cars (with a magnitude as low as 1 cm). Notably, an accurate relative positioning may be also the key for an improved absolute positioning, exploiting tools such as particle filters [82].
- Improved security. This also is a consequence of the LOS propagation typical of LiFi: jamming[9] gets very difficult, while spoofing[10] might be prevented leveraging also positioning information.

VLCs have their weaknesses, which are basically environmental ones: rain or snow will impair reception—in the same way as they hamper the human visual perception—while direct sunlight will tend to blind photo-diodes—especially at sunrise and sunset, when the light is low in the horizon. Maybe these issues will be largely overcome in the future but, in any case, VLC will not be able to replace DSRC: in fact, VANETs are still needed to propagate information farther and to widen the so-called *electronic horizon* (the area from which the vehicle can receive information). The other way around, VLC is attractive to enforce VANETs, so that one may count on one more network interface, especially for the reception of information by the closest nodes, independently of the current congestion of VANETs.

---

[9]Jamming refers to the intentional transmission of interfering signals that disrupt a communication.

[10]A spoofing attack is a situation in which one person or program masquerades as another transmitter, either to get an advantage or to cause some damage.

15.6.2 White-Space Communications

In the plethora of wireless media potentially available for the on-board integration, the white spaces[11] are interesting to discuss because: (i) they work in a significantly different frequency range and (ii) have characteristics complementary to WiFi, but also because (iii) the topic is somehow equivocal and needs some clarification (due to the multiple standards competing in the same range, such as the TV a.k.a. Digital Video Broadcasting - Terrestrial (DVB-T), the IEEE 802.11af[12] [47] and IEEE 802.22[13] [45]); last but not least, because (iv) it extends to the challenging and high-speed vehicular environment new issues, such as the cognitive management of communications [86].

Proceeding orderly, as foreseeable, practical measurements [34] confirm that at 700 MHz the attenuation is much lower than at 5.9 GHz: the authors show that, independently of the environment (urban, suburban, rural, highway), the interpolation of the results for the two cases can happen over a same (modified) attenuation logmodel with the initial attenuation much lower at 700 MHz. Depending on the points of view, this may appear encouraging or discouraging: on the one hand, it is possible to reach longer distances; on the other hand, the broadcast and collision domains of the VANETs would widen, exacerbating the problem of scalability—and perhaps fostering a slotted solution, as from Sect. 15.4.2. For instance, in [68] it is suggested to use IEEE 802.11af just to propagate emergency messages (which constitute a limited percentage of VANETs traffic) to unload the VANETs: the results of their simulations show that, avoiding the multi-hop forwarding over the VANET of the emergency messages and using 5.9 GHz only for the periodic one-hop messages, network congestion can be prevented. The same approach is followed by other theoretical works [20] and in practical tests and demos [6].

So, between IEEE 802.11af and IEEE 802.22, the former seems to be more promising while the latter has not been, so far, met in literature—basically because it implies excessively wide areas. However things are not very simple for whitespace communications: due to the multiple network and technologies competing for the same spectrum, a cognitive approach is required to prevent the interference of communications onto TV signal and between different networks. For this purpose also some IEEE standards have been defined [45, 48]. Unfortunately, spectrum

---

[11] White spaces refer to frequencies allocated to broadcasting but not used: they may be either free for technical reasons (guard-bands) or freed by the switchover to digital TV, which compresses TV transmissions in fewer channels.

[12] IEEE 802.11af is a standard approved in February 2014, also known as White-Fi: it fosters WLAN operation in TV white space spectrum between 54 and 790 MHz. It is deigned for range of some kilometers and transfer speeds of up to 35 Mb/s, with a CSMA/CA MAC.

[13] IEEE 802.22 is a standard approved in November 2013, to build Wireless Regional Area Networks (WRAN) in TV the white space spectrum between 54 and 698 MHz. It is designed for range of up to 100 km and transfer speed of up to 19 Mb/s, with a point-to-multipoint logical architecture and a slotted access; IEEE 802.22 also embeds cognitive capabilities to dynamically adapt to the available white-space channels.

optimization at vehicular speed will be challenging, due to extreme mobile nature of vehicles and unpredictable RF channel: this is a problem which has been only approached [86] and will require an additional insight.

15.6.3 TPEG

With TPEG, the perspective on the complementary technologies varies a little. TPEG is the acronym of Transport Protocol Experts Group and was founded almost 20 years ago by the European Broadcasting Union to specify how to transmit language-independent multi-modal traffic and travel information.

Today TPEG standards are available by ISO with a set of recommendations (18,234 series) introducing the main concepts and architectures [53–55] and defining binary data formats for transmission over digital audio broadcasting (DAB) and digital multimedia broadcasting (DMB). TPEG will support several applications and, at the time of this writing, has already defined *Road Traffic Messages* (providing road traffic information, such as: accidents, obstructions, congestion [49]), *Public Transport Information* (about public transport: rail, bus, air traffic, and ferry services [50]), and Location Referencing [52]. Other applications (for parking, dynamic navigation, weather, etc.) are being developed.

In addition, there is also the so-called TpegML [51], an XML implementation designed for the Internet and DVB: it maps the TPEG binary of 18,234 series and foresees also additional tags. So TPEG might be delivered over additional channels.

In 2013 the first TPEG service over DAB was delivered; on the other hand, VANETs will soon be ready to provide drivers with road traffic information. So how may TPEG fit a VANET world? Several scenarios are envisioned: TPEG might simply unload the VANET (so to prevent congestion) or, conversely, be a redundant channel for critical information (somehow to react to congestion); otherwise, TPEG might represent a solution to have persistent information on traffic, leaving instead real-time information to VANETs; additionally, TPEG could be delivered over LTE (TpegML over Internet), over DAB or, why not, as a Wireless Access in Vehicular Environment (WAVE) service (even further loading VANETs).

The market will decide and new synergies might emerge.

15.7 Conclusions

VANETs are mature and more than ready for the practical deployment, as demonstrated by the results achieved in the Field Operational Tests; importantly, the industries and the scientific community have achieved in the meantime a deep awareness on the aspects which might require future improvements.

Fortunately, most of known issues will not become critical until a high penetration of VANET solutions will have happened, that is, when VANETs will start to get congested. As a result, we will have time to make decisions on the best measure(s) to counteract and cope with such problems. Some solutions (as in the case of DCC) have already reached a good maturity and international consensus. Additional ones, as discussed in this chapter, are being studied at physical layer, at MAC layer and by leveraging complementary technologies.

It is hard to envision today which will be more likely to meet market and/or standardization success: certainly backward compatibility will be a rewarding factor for the candidate solutions to VANET 2.0.

Acknowledgements The FP7 project GLOVE (joint GaliLeo Optimization and VANET Enhancement Grant Agreement 287175) has supported the here presented analyses on MACs. GLOVE aims at identifying VANETs weaknesses and at mitigating them by leveraging the time-space information provided by Galileo (WP1, WP2).

References

1. Abramson N (1970) The aloha system: another alternative for computer communications. In: Proceedings of the 17–19 November 1970, Fall joint computer conference, AFIPS '70 (Fall). ACM, New York, NY, pp 281–285. doi:10.1145/1478462.1478502. http://doi.acm.org/10.1145/1478462.1478502
2. Alasmary W, Zhuang W (2010) The mobility impact in IEEE 802.11p infrastructureless vehicular networks. In: 2010 IEEE 72nd vehicular technology conference fall (VTC 2010-Fall), pp 1–5. doi:10.1109/VETECF.2010.5594542
3. Alexander P, Haley D, Grant A (2011) Cooperative intelligent transport systems: 5.9-GHz field trials. Proc IEEE 99(7):1213–1235. doi:10.1109/JPROC.2011.2105230
4. Almalag M, Olariu S, Weigle M (2012) Tdma cluster-based mac for vanets (tc-mac). In: 2012 IEEE international symposium on a world of wireless, mobile and multimedia networks (WoWMoM), pp 1–6. doi:10.1109/WoWMoM.2012.6263796
5. Alsbou N, Henry D, Refai H (2010) R-aloha with priority (pr-aloha) in non ideal channel with capture effects. In: 2010 IEEE 17th international conference on telecommunications (ICT), pp 566–570. doi:10.1109/ICTEL.2010.5478849
6. Altintas O, Nishibori M, Oshida T, Yoshimura C, Fujii Y, Nishida K, Ihara Y, Saito M, Tsukamoto K, Tsuru M, Oie Y, Vuyyuru R, Al Abbasi A, Ohtake M, Ohta M, Fujii T, Chen S, Pagadarai S, Wyglinski AM (2011) Demonstration of vehicle to vehicle communications over tv white space. In: Vehicular technology conference (VTC Fall), 2011. IEEE, New York, pp 1–3. doi:10.1109/VETECF.2011.6093306
7. Amadeo M, Campolo C, Molinaro A (2012) Enhancing IEEE 802.11 p/wave to provide infotainment applications in vanets. Ad Hoc Netw 10(2):253–269
8. Autolitano A, Campolo C, Molinaro A, Scopigno RM, Vesco A (2013) An insight into decentralized congestion control techniques for vanets from etsi ts 102 687 v1.1.1. In: Wireless Days (WD), 2013 IFIP. IEEE, New York, pp 1–6
9. Autolitano A, Reineri M, Scopigno RM, Campolo C, Molinaro A (2014) Understanding the channel busy ratio metrics for decentralized congestion control in vanets. In: ICCVE 2014. IEEE, New York
10. Bai B, Chen G, Xu Z, Fan Y (2011) Visible light positioning based on led traffic light and photodiode. In: Vehicular technology conference (VTC Fall), 2011. IEEE, New York, pp 1–5. doi:10.1109/VETECF.2011.6092849

11. Bana S, Varaiya P (2001) Space division multiple access (sdma) for robust ad hoc vehicle communication networks. In: Intelligent transportation systems, 2001. Proceedings 2001. IEEE, New York, pp 962–967. doi:10.1109/ITSC.2001.948791
12. Banerji S (2013) Upcoming standards in wireless local area networks. arXiv preprint. arXiv:1307.7633
13. Batsuuri T, Bril R, Lukkien J (2010) Application level phase adjustment for maximizing the fairness in vanet. In: 2010 IEEE 7th international conference on mobile ad hoc and sensor systems (MASS), pp 697–702. doi:10.1109/MASS.2010.5663797
14. Batsuuri T, Bril R, Lukkien J (2011) Performance and fairness in vanets. In: 2011 IEEE international conference on consumer electronics (ICCE), pp 637–638. doi: 10.1109/ICCE.2011.5722782
15. Blum J, Eskandarian A (2007) A reliable link-layer protocol for robust and scalable intervehicle communications. IEEE Trans Intell Transp Syst 8(1):4–13. doi:10.1109/TITS.2006.889441
16. Borgonovo F, Capone A, Cesana M, Fratta L (2002) Rr-aloha, a reliable r-aloha broadcast channel for ad-hoc inter-vehicle communication networks. In: 1st Mediterranean ad hoc networking conference, 2002. Med-Hoc-Net
17. Borgonovo F, Capone A, Cesana M, Fratta L (2003) Adhoc: a new, flexible and reliable mac architecture for ad-hoc networks. In: 2003 IEEE wireless communications and networking, 2003. WCNC 2003, vol 2, pp 965–970. doi:10.1109/WCNC.2003.1200502
18. Cailean A, Cagneau B, Chassagne L, Topsu S, Alayli Y, Blosseville JM (2012) Visible light communications: application to cooperation between vehicles and road infrastructures. In: Intelligent vehicles symposium (IV), 2012. IEEE, New York, pp 1055–1059. doi:10.1109/IVS.2012.6232225
19. Chen Q, Jiang D, Delgrossi L (2009) Ieee 1609.4 dsrc multi-channel operations and its implications on vehicle safety communications. In: 2009 IEEE vehicular networking conference (VNC). IEEE, New York, pp 1–8
20. Chen J, Liu B, Zhou H, Wu Y, Gui L (2014) When vehicles meet tv white space: a qos guaranteed dynamic spectrum access approach for vanet. In: 2014 IEEE international symposium on broadband multimedia systems and broadcasting (BMSB), pp 1–6. doi:10.1109/BMSB.2014.6873511
21. Cheng L, Henty B, Stancil D, Bai F, Mudalige P (2007) Mobile vehicle-to-vehicle narrow-band channel measurement and characterization of the 5.9 GHz dedicated short range communication (dsrc) frequency band. IEEE J Sel Areas Commun 25(8):1501–1516. doi:10.1109/JSAC.2007.071002
22. Choi JI, Jain M, Srinivasan K, Levis P, Katti S (2010) Achieving single channel, full duplex wireless communication. In: Proceedings of the sixteenth annual international conference on Mobile computing and networking. ACM, New York, pp 1–12
23. Choudhary A, Ingole P (2014) Smart phone based approach to monitor driving behavior and sharing of statistic. In: 2014 Fourth international conference on communication systems and network technologies (CSNT), pp 279–282. doi:10.1109/CSNT.2014.61
24. CoCar feasibility study, technology, business and dissemination. Public report, CoCar Consortium - Research Project funded by the German Federal Ministry of Education and Research (BMBF) (2009). http://www.its.dot.gov/index.htm
25. Consortium G (2014) Galileo optimization with joint vehicular enhancement@ONLINE. http://www.glove-fp7.eu
26. Cozzetti H, Scopigno R (2011) Scalability and qos in ms-aloha vanets: forced slot re-use versus pre-emption. In: 2011 14th International IEEE conference on intelligent transportation systems (ITSC), pp 1759–1766. doi:10.1109/ITSC.2011.6082985
27. Cozzetti H, Campolo C, Scopigno R, Molinaro A (2012) Urban vanets and hidden terminals: evaluation through a realistic urban grid propagation model. In: 2012 IEEE international conference on vehicular electronics and safety (ICVES), pp 93–98. doi:10.1109/ICVES.2012.6294332

28. Crowther W, Rettberg R, Walden D, Ornstein S, Heart F (1973) A system for broadcast communication: reservation-aloha. In: Hawaii international conference on system science, pp 596–603
29. Davis J, Furlong J (1997) Report on the study to determine the suitability of gps disciplined oscillators as time and frequency standards traceable to the uk national time scale utc. NPL Report CTM 1 PDB: 1054
30. Dressler F, Hartenstein H, Altintas O, Tonguz O (2014) Inter-vehicle communication: Quo vadis. IEEE Commun Mag 52(6):170–177. doi:10.1109/MCOM.2014.6829960
31. ETSI (2011) Intelligent transport systems (its); on the recommended parameter settings for using stdma for cooperative its; access layer part. ETSI TR 102 861, European telecommunication standards institute, Sophia Antipolis, France
32. ETSI (2011) Intelligent transport systems (its); performance evaluation of self-organizing tdma as medium access control method applied to its; access layer part. ETSI TR 102 862, European telecommunication standards institute, Sophia Antipolis, France
33. ETSI (2011) Intelligent transport systems (its); decentralized congestion control mechanisms for intelligent transport systems operating in the 5 GHz range; access layer part. ETSI TS 102 687 v.1.1.1, European telecommunication standards institute, Sophia Antipolis, France
34. Fernandez H, Rubio L, Rodrigo-Penarrocha V, Reig J (2014) Path loss characterization for vehicular communications at 700 MHz and 5.9 GHz under los and nlos conditions. IEEE Antennas Wirel Propag Lett 13:931–934. doi:10.1109/LAWP.2014.2322261
35. Frank R, Bronzi W, Castignani G, Engel T (2014) Bluetooth low energy: an alternative technology for vanet applications. In: 2014 11th annual conference on wireless on-demand network systems and services (WONS), pp 104–107. doi:10.1109/WONS.2014.6814729
36. Gaugel T, Mittag J, Hartenstein H, Papanastasiou S (2013) Strom E In-depth analysis and evaluation of self-organizing tdma. In: 2013 IEEE vehicular networking conference (VNC), pp 79–86. doi:10.1109/VNC.2013.6737593
37. Gunter Y, Wiegel B, Grossmann H (2007) Medium access concept for vanets based on clustering. In: 2007 IEEE 66th vehicular technology conference, 2007. VTC-2007 Fall, pp 2189–2193. doi:10.1109/VETECF.2007.459
38. Hafeez K, Zhao L, Ma B, Mark J (2013) Performance analysis and enhancement of the dsrc for vanet's safety applications. IEEE Trans Veh Technol 62(7):3069–3083. doi:10.1109/TVT.2013.2251374
39. Hagenauer J (1997) The turbo principle: tutorial introduction and state of the art. In: Proceedings international symposium on turbo codes and related topics, pp 1–11
40. Helsby N (2003) Gps disciplined offset-frequency quartz oscillator. In: Frequency control symposium and pda exhibition jointly with the 17th European frequency and time forum, 2003. Proceedings of the 2003 IEEE international, pp 435–439. doi:10.1109/FREQ.2003.1275131
41. Hightower P (2008) Motion effects on gps receiver time accuracy. In: Instrumentation Technology Systems http://www.itsamerica.com/Technical_Descriptions/Dynamics%20Modes%20and%20Motion%20Effects%20on%20the%20GPS%20recei.pdf
42. Hong K, Kenney JB, Rai V, Laberteaux KP (2010) Evaluation of multi-channel schemes for vehicular safety communications. In: 2010 IEEE 71st vehicular technology conference (VTC 2010-Spring). IEEE, New York, pp 1–5
43. Hoogvelt B, Asseldonk N, Henny R (2004) Measurement technology for a calibrating vehicle for multiple-sensor weigh-in-motion system. In: 8th Heavy vehicle weights and dimensions symposium proceedings, 2004
44. Huang CL, Fallah Y, Sengupta R, Krishnan H (2010) Adaptive intervehicle communication control for cooperative safety systems. IEEE Netw 24(1):6–13. doi:10.1109/MNET.2010.5395777
45. IEEE (2011) Standard for information technology—local and metropolitan area networks–specific requirements—Part 22: Cognitive wireless ran medium access control (mac) and physical layer (phy) specifications: policies and procedures for operation in the tv bands. IEEE 802.22-2011, Institute of Electrical and Electronics Engineers-Standard Association

46. IEEE (2011) Standard for local and metropolitan area networks—Part 15.7: Short-range wireless optical communication using visible light. IEEE 802.15.7-2011, Institute of Electrical and Electronics Engineers-Standard Association
47. IEEE (2013) Standard for information technology-telecommunications and information exchange between systems - local and metropolitan area networks - specific requirements - Part 11: Wireless lan medium access control (mac) and physical layer (phy) specifications amendment 5: Television white spaces (tvws) operation. Tech. Rep. IEEE 802.11af-2013, Institute of Electrical and Electronics Engineers-Standard Association
48. IEEE (2014) Standard for information technology—telecommunications and information exchange between systems—local and metropolitan area networks—specific requirements—Part 19: Tv white space coexistence methods. IEEE 802.19.1-2014, Institute of Electrical and Electronics Engineers-Standard Association
49. ISO (2009) Traffic and travel information (tti)—tti via transport protocol expert group (tpeg) data-streams—Part 4: Road traffic message (rtm) application. ISO ISO/TS 18234-4:2006, International Standard Organization
50. ISO (2009) Traffic and travel information (tti)—tti via transport protocol expert group (tpeg) data-streams—Part 5: Public transport information (pti) application. ISO ISO/TS 18234-5:2006, International Standard Organization
51. ISO (2009) Traffic and travel information (tti)—tti via transport protocol experts group (tpeg) extensible markup language (xml)—Part 1: Introduction, common data types and tpegml. ISO ISO/TS 24530-1:2006, International Standard Organization
52. ISO (2009) Traffic and travel information (tti)-tti via transport protocol expert group (tpeg) data-streams—Part 6: Location referencing applications. ISO ISO/TS 18234-5:2006, International Standard Organization
53. ISO (2013) Intelligent transport systems—traffic and travel information via transport protocol experts group, generation 1 (tpeg1) binary data format—Part 1: Introduction, numbering and versions (tpeg1-inv). ISO ISO/TS 18234-1:2013, International Standard Organization
54. ISO (2013) Intelligent transport systems—traffic and travel information via transport protocol experts group, generation 1 (tpeg1) binary data format—Part 2: Syntax, semantics and framing structure (tpeg1-ssf). ISO ISO/TS 18234-2:2013, International Standard Organization
55. ISO (2013) Intelligent transport systems—traffic and travel information via transport protocol experts group, generation 1 (tpeg1) binary data format—Part 3: Service and network information (tpeg1-sni). ISO ISO/TS 18234-3:2013, International Standard Organization
56. ITU-R (2010) Technical characteristics for an automatic identification system using time-division multiple access in the vhf maritime mobile band. ITU ITU M.1371-4, International Telecommunication Union Radiocommunication Sector, Geneva, Switzerland
57. Jain M, Choi JI, Kim T, Bharadia D, Seth S, Srinivasan K, Levis P, Katti S, Sinha P (2011) Practical, real-time, full duplex wireless. In: Proceedings of the 17th annual international conference on mobile computing and networking. ACM, New York, pp 301–312
58. Jang HC, Feng WC (2010) Network status detection-based dynamic adaptation of contention window in IEEE 802.11p. In: 2010 IEEE 71st vehicular technology conference (VTC 2010-Spring), pp 1–5. doi:10.1109/VETECS.2010.5494183
59. Jin WL, Kwan C, Sun Z, Yang H, Qijian G (2012) Spivc: smart-phone-based intervehicle communication system. In: Proceedings of transportation research board annual meeting
60. Karamad E, Ashtiani F (2008) A modified 802.11-based {MAC} scheme to assure fair access for vehicle-to-roadside communications. Comput Commun 31(12):2898–2906. doi:http://dx.doi.org/10.1016/j.comcom.2008.01.030. http://www.sciencedirect.com/science/article/pii/S0140366408000285. Mobility Protocols for ITS/VANET
61. Katragadda S, Ganesh Murthy C, Ranga Rao M, Mohan Kumar S, Sachin R (2003) A decentralized location-based channel access protocol for inter-vehicle communication. In: The 57th IEEE semiannual vehicular technology conference, 2003. VTC 2003-Spring, vol 3, pp 1831–1835. doi:10.1109/VETECS.2003.1207140

62. Kenney JB, Bansal G, Rohrs CE (2011) Limeric: a linear message rate control algorithm for vehicular dsrc systems. In: Proceedings of the Eighth ACM international workshop on vehicular inter-networking. ACM, New York, pp 21–30
63. Kloiber B, Harri J, Strang T (2012) Dice the tx power; improving awareness quality in vanets by random transmit power selection. In: 2012 IEEE vehicular networking conference (VNC), pp 56–63. doi:10.1109/VNC.2012.6407445
64. Kloiber B, Harri J, Strang T, Sand S (2014) Bigger is better - combining contention window adaptation with geo-based backoff generation in dsrc networks. In: 2013 international conference on connected vehicles and expo (ICCVE). doi:10.1109/ICCE.2011.5722782
65. Lam RK, Kumar P (2010) Dynamic channel reservation to enhance channel access by exploiting structure of vehicular networks. In: 2010 IEEE 71st vehicular technology conference (VTC 2010-Spring), pp 1–5. doi:10.1109/VETECS.2010.5494202
66. Leng S, Fu H, Wang Q, Zhang Y (2011) Medium access control in vehicular ad hoc networks. Wiley Wireless Commun Mobile Comput 11(7):796–812. doi:10.1002/wcm.869 [Special issue on Emerging Techniques for Wireless Vehicular Communication]
67. Lenoble M, Ito K, Tadokoro Y, Takanashi M, Sanda K (2009) Header reduction to increase the throughput in decentralized tdma-based vehicular networks. In: Vehicular networking conference (VNC), 2009. IEEE, New York, pp 1–4. doi:10.1109/VNC.2009.5416383
68. Lim JH, Kim W, Naito K, Gerla M (2013) Interplay between tvws and dsrc: optimal strategy for qos of safety message dissemination in vanet. In: 2013 international conference on computing, networking and communications (ICNC), pp 1156–1161. doi:10.1109/ICCNC.2013.6504256
69. Little T, Agarwal A, Chau J, Figueroa M, Ganick A, Lobo J, Rich, T, Schimitsch, P (2010) Directional communication system for short-range vehicular communications. In: 2010 IEEE vehicular networking conference (VNC), pp 231–238. doi:10.1109/VNC.2010.5698230
70. Liu CB, Sadeghi B Knightly EW (2011) Enabling vehicular visible light communication (v2lc) networks. In: Proceedings of the eighth ACM international workshop on vehicular inter-networking, VANET '11. ACM, New York, pp 41–50. doi:10.1145/2030698.2030705. http://doi.acm.org/10.1145/2030698.2030705
71. Lombardi M (2006) Comparing loran timing capability to industrial requirements. In: Proceeding of the 2006 international loran association meeting 2006, p 14
72. Lombardi M (2008) The use of gps disciplined oscillators as primary frequency standards for calibration and metrology laboratories. Measure NCSL Int 3(3):56–65
73. Mann A, Ruckert J (1988) A new concurrent slot assignment protocol for traffic information exchange. In: IEEE 38th vehicular technology conference, 1988, pp 503–508. doi:10.1109/VETEC.1988.195408
74. Martin L (2011) Core system concept of operations (conops). Tech. rep., US Department of Transportation, Research and Innovative Technology Administration - USDOT RITA, Washington, DC. http://www.its.dot.gov/index.htm
75. Menouar H, Filali F, Lenardi M (2006) A survey and qualitative analysis of mac protocols for vehicular ad hoc networks. IEEE Wireless Commun 13(5):30–35. doi:10.1109/WC-M.2006.250355
76. Michal S (2014) Channel estimation with two-dimensional interpolation for the 802.11p communication. In: ICCVE 2014. IEEE, New York
77. MS-Aloha website. http://www.ms-aloha.eu. Accessed 30 Nov 2014
78. Mu'azu A, Jung LT, Lawal I, Shah P (2013) A qos approach for cluster-based routing in vanets using tdma scheme. In: 2013 international conference on ICT convergence (ICTC), pp 212–217. doi:10.1109/ICTC.2013.6675342
79. Nawrocki J, Dunst P (2013) Deliverable d4.1: scenario and technical feasibility of vanet synchronization by gnss. Tech. rep., Consortium of FP7 project GLOVE-Galileo integrated optimization with VANET enhancements . http://www.glove-fp7.eu
80. Ni J, Srikant R, Wu X (2011) Coloring spatial point processes with applications to peer discovery in large wireless networks. IEEE/ACM Trans Netw 19(2):575–588. doi:10.1109/TNET.2010.2090172

81. Omar H, Zhuang W, Li L (2013) Vemac: a tdma-based mac protocol for reliable broadcast in vanets. IEEE Trans Mob Comput 12(9):1724–1736. doi:10.1109/TMC.2012.142
82. Peker A, Acarman T, Yaman C, Yuksel E (2014) Vehicle localization enhancement with vanets. In: 2014 IEEE intelligent vehicles symposium proceedings, pp 661–666. doi:10.1109/IVS.2014.6856576
83. Penrod B (1996) Adaptive temperature compensation of gps disciplined quartz and rubidium oscillators. In: Proceedings of the 1996 50th IEEE international frequency control symposium, 1996, pp 980–987 doi:10.1109/FREQ.1996.560284
84. Phan MA, Rembarz R, Sories S (1996) A capacity analysis for the transmission of event and cooperative awareness messages in lte networks. In: 18th ITS world congress (2011)
85. Pujapanda K (2013) Lifi integrated to power-lines for smart illumination cum communication. In: 2013 International conference on communication systems and network technologies (CSNT), pp 875–878. doi:10.1109/CSNT.2013.189
86. Riaz F, Jalil Z, Bashir S, Imran M, Ratyal N, Sajid M (2013) Efficient spectrum optimization and mobility in cognitive radio based inter-vehicle communication system. In: 2013 Second international conference on future generation communication technology (FGCT), pp 190–195. doi:10.1109/FGCT.2013.6767195
87. Roberts LG (1975) Aloha packet system with and without slots and capture. SIGCOMM Comput Commun Rev 5(2):28–42. doi:10.1145/1024916.1024920. http://doi.acm.org/10.1145/4916.1024920
88. Robinson R, Dion F (2013) Multipath signal phase and timing broadcast project http://deepblue.lib.umich.edu/bitstream/handle/2027.42/97024/102940.pdf?sequence=1&isAllowed=y
89. Sahai A, Patel G, Sabharwal A (2011) Pushing the limits of full-duplex: Design and real-time implementation. arXiv preprint. arXiv:1107.0607
90. Sai S, Niwa E, Mase K, Nishibori M, Inoue J, Obuchi M, Harada T, Ito H, Mizutani K, Kizu M (2009) Field evaluation of uhf radio propagation for an its safety system in an urban environment. IEEE Commun Mag 47(11):120–127. doi:10.1109/MCOM.2009.5307475
91. Scopigno R, Cozzetti H, Brofferio A, Casetti C, Chiasserini C (2012) Mq-mapcast: a map-aware protocol for optimized forwarding in urban vanets. In: 2012 International conference on connected vehicles and expo (ICCVE), pp 91–98. doi:10.1109/ICCVE.2012.24
92. Scopigno R, Cozzetti H, Pilosu L, Fileppo F (2012) Advances in the analysis of urban vanets: scalable integration of radii in a network simulator. In: 2012 IEEE 8th International conference on wireless and mobile computing, networking and communications (WiMob), pp 563–570. doi:10.1109/WiMOB.2012.6379132
93. Scopigno R, et al (2014) Deliverable d2.3: time in vanets: requirements, feasibility and role of galileo/egnos. Tech. rep., Consortium of FP7 project GLOVE - Galileo integrated optimization with VANET enhancements. http://www.glove-fp7.eu, http://www.ms-aloha.eu
94. Sjoberg K, Uhlemann E, Strom E (2011) How severe is the hidden terminal problem in vanets when using csma and stdma? In: 2011 IEEE vehicular technology conference (VTC Fall), pp 1–5. doi:10.1109/VETECF.2011.6093256
95. Su KC, Wu HM, Chang WL, Chou, YH (2012) Vehicle-to-vehicle communication system through wi-fi network using android smartphone. In: 2012 International conference on connected vehicles and expo (ICCVE), pp 191–196. doi:10.1109/ICCVE.2012.42
96. Subramanian S, Werner M, Liu S, Jose J, Lupoaie R, Wu X (2012) Congestion control for vehicular safety: synchronous and asynchronous mac algorithms. In: Proceedings of the ninth ACM international workshop on vehicular inter-networking, systems, and applications, VANET '12, pp 63–72. ACM, New York. doi:10.1145/2307888.2307900. http://doi.acm.org/10.1145/2307888.2307900

97. Subramanian S, Werner M, Liu S, Jose J, Lupoaie R, Wu X (2012) Congestion control for vehicular safety: synchronous and asynchronous mac algorithms. In: Proceedings of the ninth ACM international workshop on vehicular inter-networking, systems, and applications. ACM, New York, pp 63–72
98. Tan I, Tang W, Laberteaux K, Bahai A (2008) Measurement and analysis of wireless channel impairments in dsrc vehicular communications. In: IEEE international conference on communications, 2008. ICC 08. IEEE, New York, pp 4882–4888
99. Tielert T, Jiang D, Chen Q, Delgrossi L, Hartenstein H (2011) Design methodology and evaluation of rate adaptation based congestion control for vehicle safety communications. In: 2011 IEEE vehicular networking conference (VNC). IEEE, New York, pp 116–123
100. Torrent-Moreno M, Mittag J, Santi P, Hartenstein H (2009) Vehicle-to-vehicle communication: fair transmit power control for safety-critical information. IEEE Trans Veh Technol 58(7):3684–3703. doi:10.1109/TVT.2009.2017545
101. Vandenberghe W, Moerman I, Demeester P (2011) On the feasibility of utilizing smartphones for vehicular ad hoc networking. In: 2011 11th International conference on ITS telecommunications (ITST), pp 246–251. doi:10.1109/ITST.2011.6060061
102. Vesco A, Masala E, Scopigno R (2013) Efficient support for video communications in wireless home networks. In: 2013 International conference on computing, networking and communications (ICNC), pp 599–604. doi:10.1109/ICCNC.2013.6504154
103. Wang Z, Hassan M (2008) How much of dsrc is available for non-safety use? In: VANET '08 Proceedings of the fifth ACM international workshop on vehicular inter-networking, pp 23–29. doi:10.1145/1410043.1410049. http://doi.acm.org/10.1145/1410043.1410049
104. Wang Y, Ahmed A, Krishnamachari B, Psounis K (2008) Ieee 802.11p performance evaluation and protocol enhancement. In: IEEE international conference on vehicular electronics and safety, 2008. ICVES 2008, pp 317–322. doi:10.1109/ICVES.2008.4640898
105. Wang S, Chou C, Liu K, Ho T, Hung W, Huang C, Hsu M, Chen H Lin C (2009) Improving the channel utilization of IEEE 802.11p/1609 networks. In: Wireless communications and networking conference, 2009. WCNC 2009. IEEE, New York, pp 1–6. doi:10.1109/WCNC.2009.4917753
106. Whyte W, Weimerskirch A, Kumar V, Hehn T (2013) A security credential management system for v2v communications. In: IEEE vehicular networking conference (VNC), pp 1–8
107. woo Chang S, Cha J, sun Lee S (2012) Adaptive edca mechanism for vehicular ad-hoc network. In: 2012 International conference on information networking (ICOIN), pp 379–383. doi: 10.1109/ICOIN.2012.6164404
108. Wu X, Subramanian S, Guha R, White RG, Li J, Lu KW, Bucceri A, Zhang T (2013). Vehicular communications using DSRC: challenges, enhancements, and evolution. IEEE J Sel Areas Commun 31(9-Suppl):399–408. doi:10.1109/JSAC.2013.SUP.0513036. http://dx.doi.org/10.1109/JSAC.2013.SUP.0513036
109. Wu X, Miucic R, Yang S, Al-Stouhi S, Misener J, Bai S, Chan Wh (2014) Cars talk to phones: a dsrc based vehicle-pedestrian safety system. In: 2014 IEEE vehicular technology conference (VTC Fall)
110. Yu SH, Shih O, Tsai HM, Wisitpongphan N, Roberts R (2013) Smart automotive lighting for vehicle safety. IEEE Commun Mag 51(12):50–59. doi:10.1109/MCOM.2013.6685757
111. Zhu W, Hellmich T, Walke B (1991) Dcap, a decentral channel access protocol: performance analysis. In: 41st IEEE vehicular technology conference, 1991. Gateway to the future technology in motion, pp 463–468. doi:10.1109/VETEC.1991.140532

Chapter 16
LTE for Vehicular Communications

Christian Lottermann, Mladen Botsov, Peter Fertl, Robert Müllner,
Giuseppe Araniti, Claudia Campolo, Massimo Condoluci, Antonio Iera,
and Antonella Molinaro

Abstract The cellular communication networks standard 3rd Generation Partnership Project (3GPP) Long Term Evolution (LTE) offers low latencies and high throughputs simultaneously, thus enabling more bandwidth-demanding and real-time critical services for end-users. This is of particular interest for vehicle manufacturers who in the future intend to offer a huge variety of cooperative driver assistance services with manifold quality of service requirements. This chapter analyzes the suitability of LTE as a wireless transmission technology for future vehicular services of the categories *Infotainment, Comfort, Traffic Efficiency*, and *Safety*. The investigations are based on extensive LTE system-level simulations under different load conditions and network deployments as well as on a theoretical delay analysis. Focus is set on transmission delays and reliability aspects under various quality of service settings. The results show that an accurate selection of the LTE quality of service parameters is crucial in order to meet the delay and reliability requirements of future automotive applications, especially in high-load network conditions.

Keywords VANET • LTE • LTE A • Safety applications • Comfort applications • Traffic-efficiency applications • Infotainment applications • Cooperative Awareness Messages (CAMs) • Decentralized Environmental Notification Messages (DENMs) • Floating Car Data (FCD) • Periodic Driver Assistance Service (PDAS) • Voice over LTE (VoLTE) • Voice Recognition (VR) • Resource Block (RB) • Quality of Service Class Identifier (QCI) • Evolved Multimedia

---

C. Lottermann (✉) • M. Botsov • P. Fertl
BMW Group Research and Technology, Munich, Germany
e-mail: christian.lottermann@bmw.de; mladen.botsov@bmw.de; peter.fertl@bmw.de

R. Müllner
Telefónica Germany, Munich, Germany
e-mail: robert.muellner@telefonica.com

G. Araniti • C. Campolo • M. Condoluci • A. Iera • A. Molinaro
University Mediterranea of Reggio Calabria, Reggio Calabria, Italy
e-mail: araniti@unirc.it; claudia.campolo@unirc.it; massimo.condoluci@unirc.it; antonio.iera@unirc.it; antonella.molinaro@unirc.it

Broadcast Multicast Service (eMBMS) • Vehicle-to-vehicle (V2V) • Vehicle-to-infrastructure (V2I) • Vehicle-to-X (V2X) • Vehicle-to-device (V2D) • Quality of Service (QoS)

16.1 Introduction

Improving traffic safety, efficiency, and driver's comfort becomes more and more important for modern vehicles. At the same time, the demands for high data rate information and entertainment services grow.

Modern vehicles are increasingly equipped with on-board advanced driver assistance services (ADAS) which process data from numerous on-board vehicle sensors. However, to further improve ADAS systems, it is required to enlarge the range of the sensors mounted at the vehicle by incorporating also information from the outside world. This can be obtained from cooperation with other vehicles or road infrastructure, known as V2V, vehicle-to-infrastructure (V2I), infrastructure-to-vehicle (I2V), or vehicle-to-X (V2X) communications. The IEEE 802.11p [36] standard specifies the communication technology for ITS applications in Vehicular Ad Hoc Networks (VANETs). Its advantages are easy deployment, low costs, mature technology, and the capability to natively support V2V communications in ad hoc mode. Nonetheless, this technology suffers from scalability issues and low penetration, unbounded delays, and lack of deterministic quality of service (QoS) guarantees [15]. Due to its ad hoc connectivity focus, its limited radio range and without a pervasive roadside communication infrastructure, IEEE 802.11p can only offer intermittent and short-lived V2I connectivity. These concerns motivate the investigation of wireless access technologies to support advanced V2I and V2V communications in vehicular environments. LTE [9] is the most promising wireless broadband technology that provides high throughput and low latency for mobile services. Like all cellular systems it benefits from a large coverage area, high penetration rate providing the economical basis for short development cycles, and high velocity terminal support.

LTE particularly meets the high-bandwidth demands and QoS requirements of a category of vehicular applications known as *Infotainment* (information and entertainment). This category includes traditional and emerging Internet applications. Moreover, the delivery of driving context-related information for *Traffic Efficiency* applications and applications of the *Comfort* class implies less stringent delay requirements without exhausting the LTE system resources. Nevertheless, the LTE capability to support applications specifically conceived for the vehicular environment to provide *Safety* services is still an open issue. Both *event-triggered warnings* (e.g., generated in case of accidents) and *periodic messages* (exchanged among vehicles for cooperative driving applications) belong to this category. The main concern comes from the centralized LTE architecture. Even for a localized V2V data exchange, communications always cross infrastructure nodes with negative consequences on message latency, especially for *Safety* applications. In addition, in dense traffic areas, the heavy load generated by periodic message transmissions from several vehicles strongly challenges the LTE capacity and may penalize the delivery of traditional applications.

In this chapter the suitability of LTE as a wireless transmission technology for future automotive off-board services is analyzed. In Sect. 16.2 the typical V2V and V2I service classes are introduced and their QoS requirements are defined. Section 16.3 explains the LTE architecture and shows the application in an automotive environment. Its suitability for non-safety services is evaluated in Sect. 16.4. In particular, the impact of specific LTE quality of service class identifier (QCI) settings on the transmission delays and packet discard rates are analyzed under various load conditions. Simulations to assess the performance of LTE for *Safety* services, along with an overview of literature investigations and their limitations are presented in Sect. 16.5. The results allow a consolidated view on the suitability of LTE for vehicular communications, its strengths and critical aspects. In Sect. 16.6, a comparison between IEEE 802.11p and LTE for automotive services is given. An outlook to future research topics is presented in Sect. 16.7. Finally, conclusions are drawn in Sect. 16.8.

16.2 Characterization of the Applications and Their Requirements

The considered off-board services are classified into four groups: *Infotainment*, *Comfort*, *Traffic Efficiency*, and *Safety*. However, these service categories vary considerably in their QoS requirements, such as end-to-end (E2E) transmission delay, jitter, and the overall required throughput.

16.2.1 Infotainment Applications

Infotainment applications incorporate entertainment and information related applications, such as Internet audio streaming, video streaming, and information services, which are already part of modern premium vehicle systems. Content download, media streaming, web-browsing, social networking, blog uploading, gaming, and access to different cloud services are typical *Infotainment* applications. In the future, this class will be extended to video-related services that offer advanced video-based information, such as bandwidth-consuming video surveillance (VS) [33] or video traffic information services. As of today, streaming applications of this class typically have fixed source rates and, hence, constant throughput requirements on the transmission chain that range from data rates of 92 kbit/s for low-rate audio streaming to 3 Mbit/s for high-definition video information services. In addition, the one-way E2E delay (δ_{E2E}) requirements typically vary between 100 and 600 ms.

16.2.2 Comfort Applications

Applications of the *Comfort* class aim to ease the driver's daily life. They include context-related driver information, such as voice recognition (VR), information about the current traffic situation (live traffic information), and location-based services, as well as remote software updates for the vehicle. As the majority of today's applications are run on consumer electronic (CE) devices, applications of the *Comfort* class are often based on Internet web-services which use reliable transport layer protocols for the data transmission, such as TCP/IP (transmission control protocol, Internet protocol). By definition, such web-services are less time-critical with delay constraints in the order of seconds and classified as interactive or background services based on best-effort traffic patterns with no stringent QoS requirements. Here, the application is fully responsible for the successful transmission of the data packets.

16.2.3 Traffic Efficiency Applications

Traffic Efficiency applications aim to optimize flows of vehicles by reducing travel time and traffic congestion. Similar to *Comfort* class services, applications of the *Traffic Efficiency* class are based on Internet web-services and, hence, have less stringent QoS requirements. However, their quality gradually degrades with increasing packet loss and delay. In this class, the decentralized floating car data (FCD) service and its extension, extended floating car data (XFCD) [19, 35], require periodic transmissions of information collected by vehicles from internal and external sensors (e.g., information provided at the CAN bus, in-vehicle camera, environmental monitoring sensors) to remote management servers. They process the collected data, monitor and predict traffic congestion, and send up-to-date traffic information back to the vehicle's navigation system in order to suggest alternative routes. Besides the collection of the floating car data, vehicles are able to request traffic related information with respect to their actual context from a service provider. However, two different kinds of services for such information are defined: *event-driven* services and *periodic* services, referred to as periodic driver assistance services (PDAS) in the following. Event-driven services encompass traffic and navigation related information, such as alternative route suggestions, construction site information, or traffic light cycle information. For PDAS, aggregated and relevant off-board information about the surroundings of the vehicle is sent to the vehicle's on-board component periodically, such as periodic information about the average velocity in a certain route section.

16.2.4 Safety Applications

Safety services aim at reducing the risk of car accidents and have timeliness and reliability as the major requirements. Two main types of safety messages have been standardized. Their transmission can be periodic or event-triggered. In the European Telecommunications Standards Institute (ETSI) documents [25] they are referred to as Cooperative Awareness Messages (CAMs) [22] and Decentralized Environmental Notification Messages (DENMs) [23], respectively. Basic Safety Message (BSM) is the terminology used in [53] for both periodic and event-triggered messages.[1]

CAMs, also known as *beacons* or *heartbeat* messages, are short messages periodically broadcast by each vehicle to its neighbors to provide information on presence, position, kinematics, and basic status. DENMs are event-triggered short messages broadcast to alert road users of a hazardous event.

Both CAM and DENM messages are delivered to vehicles in a particular geographic region: the immediate neighborhood (*awareness range*) for CAMs, and the area potentially affected by the notified event (*relevance area*) for DENMs, such as congestion or hazard warning. The relevance area might span over several hundred meters for DENM message distribution. The capability of transmitting a message to nodes satisfying a set of geographical criteria is called *geocast* and represents together with reliability and low-latency delivery a crucial requirement of the typical temporal- and spatial-relevant vehicular applications.

16.2.5 Summary of QoS Requirements

Table 16.1 summarizes the QoS requirements of the different application classes in terms of E2E delay, throughput, reliability, and required connectivity type. The E2E delay δ_{E2E} is the overall transmission delay which is composed of the delay contributions of the on- and off-board application units, i.e., in the vehicle and the backend server, and the overall communication path between the transmitter (e.g., vehicle) and the receiver (e.g., roadside unit). Reliability represents the requirement for a high successful packet delivery ratio (successful and in-time message transmission) which is crucial for *Safety* applications and for some *Traffic Efficiency* applications. Applications of the *Comfort* domain and *Traffic Efficiency* applications typically rely on best effort traffic patterns, since the information contained in the messages is rather informal compared to services of the *Safety* domain and, hence, has no strict time constraint.

---

[1] See Chap. 5 for a more detailed overview of those messages.

Table 16.1 Overview of the QoS requirements of the different application classes

Application class	Main requirements	Connectivity type	Examples
Infotainment	High throughput, medium-to-low latency	V2V, V2I, I2V	Web-browsing, VS, file sharing, gaming, e-mail
Comfort	Medium-to-low reliability	V2I, I2V	VR, live traffic information, remote software updates
Traffic efficiency	Medium-to-high reliability	V2I, I2V	XFCD, PDAS
Safety	High reliability, low latency ($10\,\text{ms} < \delta_{E2E} < 1\,\text{s}$)	V2V, V2I, I2V	*CAM-based:* Emergency vehicle warning, intersection collision warning, slow vehicle indication *DENM-based:* Emergency electronic brake light, collision risk warning, visibility

16.3 LTE Technology Overview

In the following section, first a comparison of LTE to other radio access technologies for vehicular connectivity is given, followed by an overview of the core features of LTE, the LTE architecture, and the QoS system concept. Table 16.2 compares the traditional vehicular technologies Wi-Fi and 802.11p with the 3GPP technologies Universal Mobile Telecommunications System (UMTS), LTE, and LTE Advanced (LTE-A). Although initially not designed for automotive applications, the enhancements in LTE may support a huge variety of novel vehicle services at acceptable QoS. LTE and LTE-A represent the most promising cellular systems for vehicular connectivity. Commercial LTE deployments have started in Europe in 2010 and LTE networks are successfully operated in many countries all over the world.

16.3.1 Overview

The overall LTE system is characterized by a flat all-IP architecture with a reduced number of network entities and a separation of the control plane and user plane traffic. IP-based data, voice, and signaling transmissions simplify extendibility with respect to previous cellular networks (UMTS and Global System for Mobile Communications (GSM)). Thanks to its flat architecture, LTE can provide round trip times (RTTs) theoretically lower than 10 ms, and transfer latency in the radio access of up to 100 ms. Measurements in current live LTE networks typically show RTTs between 15 and 60 ms. Providing low latencies also in high traffic load situations is essential for delay-sensitive vehicular applications.

Table 16.2 Main candidate wireless technologies for the vehicular connectivity

Feature	Wi-Fi	802.11p	UMTS	LTE	LTE-A
Channel bandwidth (MHz)	20	10	5	1.4, 3, 5, 10, 20	up to 100
Frequency bands (GHz)	2.4, 5.2	5.86, 5.92	0.7–2.6	0.7–2.69	0.45–4.99
Bit rate (Mbit/s)	6–54	3–27	2[a]	Up to 300	Up to 1,000
Coverage	Intermittent	Intermittent	Ubiquitous	Ubiquitous	Ubiquitous
Capacity	Medium	Medium	Low	High	Very high
Mobility support	Low	Medium	High	Very high (up to 350 km/h)	Very high (up to 350 km/h)
QoS support	Enhanced Distributed Channel Access (EDCA)	EDCA	QoS classes and bearer selection	QCI classes and bearer selection	QCI classes and bearer selection
Broadcast support/multicast support	Native broadcast	Native broadcast	Through Multimedia Broadcast Multicast Service (MBMS)	Through evolved MBMS (eMBMS)	Through evolved MBMS (eMBMS)
V2I support	Yes	Yes	Yes	Yes	Yes
V2V support	Native (ad hoc)	Native (ad hoc)	No	No	Yes (through device-to-device (D2D))
Market penetration	High	Low	High	High	Potentially high
Transmission costs	Low	Low	High	High	High

[a] Higher data rates can be achieved with the UMTS enhancements HSPA and HSPA+

The overall LTE architecture is composed of the evolved UMTS terrestrial radio access network (E-UTRAN), which is the radio access part of LTE and the Evolved Packet Core (EPC) which encompasses all core network entities (cf. Fig. 16.1).

16.3.2 LTE Air Interface

The radio access network is composed of evolved NodeBs (eNodeBs), which are responsible for radio resource and handover management. The LTE air interface has the flexibility to support frequency division duplex (FDD), time division duplex (TDD), and half-duplex FDD schemes. It provides scalable channel width ranging from 1.4 to 20 MHz.

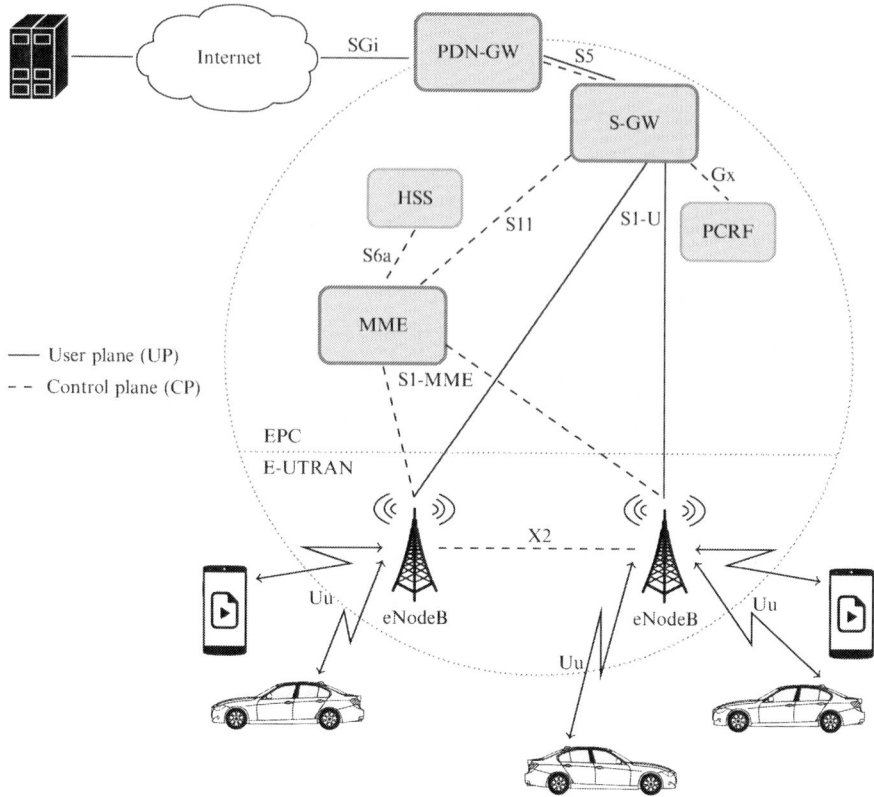

Fig. 16.1 LTE architecture: access network (E-UTRAN) and core network (EPC) entities

Orthogonal frequency division multiple access (OFDMA) is used in the downlink (DL) to fulfill the E-UTRAN performance requirements. With the use of orthogonal frequency division multiplexing (OFDM), frequency-selective fading of the multi-path channel can be exploited and low-complexity receivers can be used. Furthermore, due to a sub-carrier spacing of 15 kHz, degradations from phase noise and Doppler (for 250 km/h at 2.6 GHz) can successfully be avoided even with 64 quadrature amplitude modulation (QAM). LTE uplink (UL) uses single carrier frequency division multiple access (SC-FDMA) due to its peak-to-average power ratio (PAPR) allowing low power consumption and higher efficiency at the User Equipment (UE). Apart from a lower PAPR, the orthogonality inherited from OFDM reduces intra-cell interference. As a result, the role of power control becomes crucial to provide the required signal-to-interference-plus-noise ratio (SINR) according to QoS requirements while controlling the interference caused to neighboring cells at the same time [20, 47].

Multiple-input and multiple-output (MIMO) techniques are used to improve the spectral efficiency by a factor of 3-4 compared to 3.5 generation (3.5G) systems even at high terminal speeds, making LTE particularly efficient in challenging and dynamic propagation environments like the vehicular one.

Radio resources are centrally managed by an eNodeB at every transmission time interval of 1 ms duration, with the aim of satisfying QoS requirements while increasing channel utilization. The packet scheduler at the eNodeB plays a key role, since it selects the traffic flow based on the related QoS requirements as specified by the QCI. LTE supports quadrature phase-shift keying (QPSK), 16 QAM, and 64 QAM. For the uplink, the last one is optional. Adaptive modulation and coding (AMC) decides the appropriate modulation and coding scheme based on feedback from the mobile terminals on the channel quality [47].

16.3.3 Evolved Packet Core

The LTE core network, EPC, is responsible for the authentication, mobility management, bearer control, charging, and QoS control. It is composed of three main entities: the Mobility Management Entity (MME), the Serving Gateway (S-GW), and the Packet Data Network Gateway (PDN-GW), cf. Fig. 16.1.

The MME is the key control entity for the LTE network. It is mainly responsible for tracking, paging and storing the position information of the users, as well as authentication of the users in collaboration with the Home Subscriber Server (HSS). Furthermore, it is involved in the bearer activation/deactivation procedure and responsible for selecting the S-GW.

The task of the S-GW is routing, data forwarding, and charging by coupling with the policy and charging rules function (PCRF). It also acts as an anchor for mobility during inter-eNodeB handover and mobility between the other 3GPP technologies.

The PDN-GW is the outgoing entity that allows communication with IP and circuit-switched networks. It is responsible for packet filtering of each user, policy enforcement and charging support. The mobile terminal can have connections to multiple PDN-GWs for accessing multiple packet data networks (PDNs).

LTE also supports high-quality multicast and broadcast transmissions through the eMBMS [11], in the core and in the radio access network. It offers the possibility of sending the data only once to a set of users registered for the offered service, instead of sending it to every node separately.

16.3.4 LTE QoS Classes and Mapping for Vehicular Services

In the LTE standard the different QoS requirements of multiple applications are supported by establishing different bearers within the evolved packet system (EPS) [34]. Here, an EPS bearer is defined as the virtual connection between the terminal and the PDN-GW. Each EPS bearer is associated with a certain QoS setting defined by the QCI as well as an allocation and retention priority (ARP). QCI refers to a set of packet forwarding treatments, for example resource type, priority, acceptable packet loss rate (PLR), and delay budget. The priority, packet delay budget, and packet loss rate define how the bearer shall be handled in terms of scheduling policy, queue management, and rate shaping policy. All packet flows that are based on one bearer are treated in the same manner.

The resource type is categorized into two groups: guaranteed bit rate (GBR) bearers and non-GBR bearers. A GBR bearer is used for services that require a certain minimum bit rate which is achieved by permanently allocating dedicated bandwidth resources. Higher bit rates may only be allowed if resources are available. Conversational services like voice over LTE (VoLTE) or video-telephony using semi-persistent scheduling are examples for this group. For non-GBR bearers no specific bit rate is guaranteed, i.e., no bandwidth resources are permanently allocated. This is the case for applications such as web-browsing or file transfer protocol (FTP) services. In total, nine different QCIs with specific QoS requirements have been specified in 3GPP LTE Release 8 (cf. Table 16.3). In addition operator specific QCIs can be defined. An ARP is used by admission control to decide whether a bearer creation or modification request can be accepted or needs to be rejected due to resource limitations.

Although the QCI classes have initially not been designed for vehicular applications, an assignment of the vehicular application classes introduced in Sect. 16.2 to the QCI classes defined by 3GPP based on their QoS requirements can be performed. Applications of the *Comfort* and *Traffic Efficiency* domain are usually based on web services using a TCP-based transmission of the information. Depending on the priority, a non-GBR bearer with a QCI class 6, 8, and 9 can be selected for these applications. For *Infotainment* services, the QCI class selection depends on the traffic patterns of the applications. Streaming services with fixed bit rate requirements (such as video streaming) demand a GBR bearer with a QCI class 2 or 4. Dynamic streaming applications, such as adaptive Hypertext Transfer Protocol (HTTP) streaming, are more relaxed in their QoS requirements and can be allocated to a non-GBR bearer with a QCI class 6. In contrast to the previously mentioned

Table 16.3 Standardized QCI characteristics, [6]

QCI	Bearer type	Priority	Packet delay (ms)	PLR	3GPP sample application
1	GBR	2	100	10^{-2}	VoLTE call
2		4	150	10^{-3}	Video call
3		3	50		Online gaming (real-time)
4		5	300	10^{-6}	Video streaming
5	Non-GBR	1	100		IP Multimedia Subsystem (IMS) signaling
6		6	300		Video, TCP-based services
7		7	100	10^{-3}	Voice, video, interactive gaming
8		8	300	10^{-6}	Video, TCP-based services
9		9			

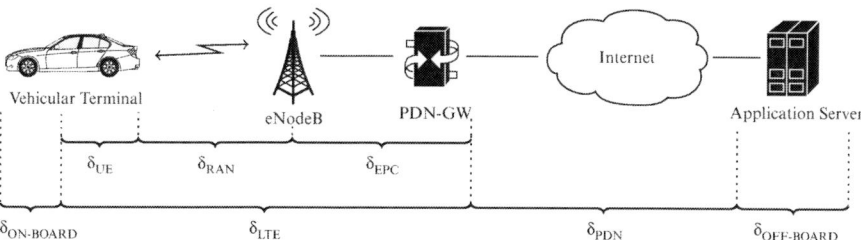

Fig. 16.2 Overall transmission chain

classes, most applications of the *Safety* class rely on the exchange of CAM and DENM messages and thus have strict end-to-end delay requirements of less than 100 ms. However, CAM and DENM messages need to pass the radio access link in the uplink and downlink direction, as well as the core network components, which may result in a total guaranteed delay of more than 100 ms. Hence, none of the defined QCI classes fulfills the requirements of the *Safety* services.

16.3.5 Overall Transmission Chain

A general overview of the overall E2E user plane transmission chain of automotive off-board services is given in Fig. 16.2. It consists of an off-board component located at an external backend server in the Internet and an on-board component in the vehicle. Both service components exchange data via the LTE system and an external PDN. The latter is usually the general Internet transmission path. Therefore, the one-way E2E transmission delay δ_{E2E} is composed of

$$\delta_{E2E} = \delta_{ON\text{-}BOARD} + \delta_{LTE} + \delta_{PDN} + \delta_{OFF\text{-}BOARD}.$$

where $\delta_{ON\text{-}BOARD}$ and $\delta_{OFF\text{-}BOARD}$ are the delays contributed by the on- and off-board application and δ_{LTE} and δ_{PDN} are the transmission delays induced by the LTE system and due to Internet routing, respectively. The latter has been thoroughly investigated in [62]. The delay of the LTE system can be further split into the terminal (UE) component δ_{UE}, the Radio Access Network (RAN) component δ_{RAN}, and the EPC component δ_{EPC} as

$$\delta_{LTE} = \delta_{UE} + \delta_{RAN} + \delta_{EPC}. \tag{16.1}$$

The delay of the LTE system δ_{LTE} also depends on the traffic load and increases if packets cannot be scheduled immediately and have to be buffered. This aspect relates to the whole chain of network nodes and transport lines. Note that in the following the QoS performance analysis will focus only on the LTE system component.

16.4 LTE for Non-safety Applications

In this section, the suitability and the performance of LTE as a wireless transmission technology for non-safety applications of the *Infotainment, Comfort,* and *Traffic Efficiency* class are investigated. Safety applications entailing different traffic patterns and requirements are discussed in Sect. 16.5. In particular, the impact of specific LTE QCI settings on the transmission delays and packet discard rates for various automotive services is analyzed. Furthermore, the overall load and performance of the LTE system in typical deployments is investigated. This performance analysis is based on extensive LTE system-level simulations under various load conditions and network deployments as well as on thorough theoretical investigations.

16.4.1 Simulation Assumptions

The simulations have been performed by using a real-time network simulator proposed in [60]. The simulator models a cellular LTE system with the number of cells depending on the scenario and concurrent users with realistic traffic patterns based on a system-level simulation paradigm. In order to guarantee meaningful statistical data, ten simulation runs per scenario with a simulated time of 1 h have been performed.

16.4.1.1 Terminal Classes

Different terminal classes have been defined to simulate realistic load scenarios. The class of *mobile terminals* represents all typical CE devices equipped with an LTE broadband modem, such as smartphones or portable computers. Additionally, *vehicular* terminals cover a class of terminals supporting the applications of automotive service classes. In particular, two types of vehicular terminals have been introduced: *fully equipped* vehicles that are able to request all services of the *Infotainment, Comfort,* and *Traffic Efficiency* classes, and *basic-equipped* vehicles that are able to request all services from the *Comfort* class.

The basic assumption is that the public LTE network serves vehicular as well as non-vehicular (referred to as mobile) terminals. A base traffic load has been assumed that represents the traffic originated from standard subscribers. On top of that base load, additional traffic from vehicular users has been simulated.

16.4.1.2 Deployment Scenarios and Traffic Load

Three characteristic scenarios in a typical LTE network deployment with specific road types have been considered: an *urban*, *rural*, and *highway* scenario. Figure 16.3 shows the simulation cell layout restricted to the coverage area of one eNodeB in each scenario. The total number of terminals, i.e., the sum of *mobile* and *vehicular terminals*, in each scenario has been derived from measurements of GSM, UMTS, and LTE live networks, cf. [43]. The simulated vehicle density has been calculated from the average vehicle frequency per day on German roads (cf. d_{20} in [18]). The velocity of each terminal is individual with a standard deviation of 10 %. The mean velocity values, introduced in the following, depend on the local scenario as well as on the road types and have been derived from [45].

The *urban* scenario (cf. Fig. 16.3a) represents a typical German inner city scenario, which is constructed by seven eNodeBs deployed in a hexagonal layout with an inter-site distance of 150 m. In accordance with current LTE network deployments in Germany, the short-range 2.6 GHz LTE spectrum has been applied. Due to the small cell-size, the overall number of simulated terminals is rather low and consists of 14 *fully-* and 14 *basic-equipped* vehicles, and 56 *mobile terminals* generating the base load. Vehicular terminals move in a bounce-back Manhattan grid model at a mean velocity of 30 km/h, whereas the *mobile terminals* are following a random walk movement model at a mean velocity of 5 km/h.

For the *rural* scenario, three eNodeBs operating in the long-range 800 MHz LTE frequency band with an inter-site distance of 3.3 km have been considered. The simulated area contains two types of roads (cf. Fig. 16.3b): urban roads and rural roads. Urban roads are deployed in a Manhattan grid model similar to the *urban* scenario. Vehicles on rural roads drive with a mean velocity of 80 km/h, while vehicles on urban roads travel with a mean velocity of 30 km/h. *Mobile terminals* move with a pedestrian velocity of 5 km/h. In total, 123 *fully-* and 102 *basic-equipped* vehicles as well as 60 *mobile terminals* for base load generation have been simulated in the rural environment.

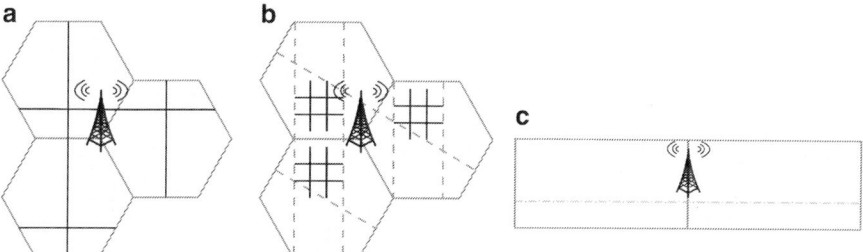

Fig. 16.3 Simulated scenario setups: Urban roads (———), rural roads (— — —), highway roads (— · — · —).

In the *highway* scenario (cf. Fig. 16.3c), the simulation area is constructed by two eNodeBs operating on the long-range 800 MHz LTE frequency band with an inter-site-distance of 25 km. The long inter-site distance in this scenario results from live network planning tools with LTE propagation models for flat environments and represents, with such a large cell size, a worst case scenario from a capacity perspective. In hilly terrains, the resulting inter-site distance is smaller, typically around 10 km. For capacity demands or the mitigation of higher in-car penetration loss due to shielded windows, further reduction of the inter-site distance is necessary. The eNodeBs are deployed in a straight line with their main antenna beams along the highway. In this scenario all terminals, i.e., mobile and vehicular, travel at a mean velocity of 130 km/h, which is a typical assumption for German highways. Here, the *mobile terminals* are assumed to be located inside the vehicle and coupled to the roof-top antenna of the vehicle. This is an approach that is considered in modern vehicles, in order to avoid mobile signal degradation of CE devices inside vehicles due to the penetration loss of the vehicle body. The simulated terminals are partitioned in 112 *fully-* and 148 *basic-equipped* vehicle terminals as well as 52 *mobile terminals*.

16.4.1.3 System Model

The simulations have been performed in FDD mode. FDD is the mode that is typically used in commercial European LTE networks. The physical layer model is simplified but still exact; it provides frequency-selective and time varying SINR values for every resource block (RB) consisting of a transmission time interval (TTI) of 1 ms duration and a bandwidth of 180 kHz. A total bandwidth of 10 MHz for each direction, uplink and downlink, has been used with frequency reuse factor of 1. In order to simulate realistic channel characteristics, interference from the surrounding cells has been taken into account via the mirror-and-shift technique. In this model the signals within the cells of interest are copied, shifted in time and frequency and introduced as interference signals from outside the simulated area. In all simulation scenarios a proportional fair scheduler has been applied [34]. The essential network parameter settings are summarized in Table 16.4.

16.4.2 Simulated Services and Their Requirements

In order to evaluate the QoS performance of the LTE network under realistic load conditions, a total number of 14 DL and 11 UL automotive services of the *Infotainment*, *Comfort*, and *Traffic Efficiency* domain have been simulated [43]. These services include a variety of current and future automotive off-board services as well as standard applications, such as web-browsing or e-mail. Since most of the

Table 16.4 Network parameters for the different simulation scenarios

Parameter	Urban	Rural	Highway
Duplexing mode	FDD		
System bandwidth	10 MHz UL/10 MHz DL		
Frequency band	2.6 GHz	800 MHz	800 MHz
Inter-site distance	150 m	3.3 km	25 km
eNodeB antenna gain	14 dBi		
eNodeB antenna beamwidth H/V	70°/10°		
eNodeB antenna height	10 m	35 m	50 m
eNodeB antenna downtilt	15°	6°	3°
UE antenna height	1.5 m		
UE antenna gain	0 dBi		
UE antenna pattern	Omni-directional		
Antenna configuration	2 × 2 (receive (Rx)/transmit (Tx))		
DL scheduling	Proportional fair with QoS support		
UL scheduling	Proportional fair with QoS support		
DL MIMO mode	Rank–adaptive closed–loop spatial multiplexing		
DL channel quality indicator reporting	Sub–bandwidth 3 RB, period 5 TTIs		
UL user bandwidth	Adaptive		
DL Tx power	46 dBm		
Max. UL Tx power	24 dBm		
UL power control	Open loop fractional power control [8, 47]		
Target uplink Rx power level P_0 [8]	−66 dBm	−85 dBm	
Uplink Pathloss Compensation factor α [8]	0.8		
Noise figure eNodeB/UE	3 dB/7 dB		

services follow similar characteristics and show similar behavior, focus has been set on a representative composition of selected applications. These services and their corresponding traffic patterns are summarized in Table 16.5.

The traffic properties of the different services are characterized by the packet size p_{size} as well as their transmission pattern. Note that most of the considered non-streaming applications trigger data transmissions in random time intervals t with the exponential probability density function $f(t) = \lambda e^{-\lambda t}$, where $t \geq 0$ and $\lambda = t_{\text{request}}^{-1}$. Here, t_{request} denotes the mean inter-request time of the respective service. Moreover, streaming services are characterized by a specific source rate R_{data}. Additionally, each service is mapped to a specific QoS setting given by a QCI and a corresponding packet delay budget. The latter specifies how long a packet remains in the transmission queue before timers expire and it is discarded. The QCIs are chosen in such a way that the corresponding packet delay budget matches the

Table 16.5 Analyzed services with QCI mappings (cf. [6]) and traffic properties

Application	Service class	Link	QCI	Packet Delay Budget (ms)	p_{size} (kbit)	$t_{request}$ (s)	R_{data} (kbit/s)
VS [43]	Infotainment	UL, GBR	2	150	200	–	1,600
		UL, non-GBR	10	60,000	200	–	1,400
VoLTE [48]	Infotainment	UL, DL	1	100	4	–	12.2
FTP [48]	Infotainment	DL	10	60,000	≤42,000	30	–
VR [30]	Comfort	UL	8	300	65	600	–
		DL	10	60,000	3.8	600	–
XFCD [43]	Traffic efficiency	UL	8	300	3.8	300	–
PDAS [43]	Traffic efficiency	DL	8	300	107	60	–

delay requirements of the corresponding service. For the simulations one additional QCI (i.e., QCI 10) has been introduced that can be one of the operator specific QCIs representing a best-effort traffic pattern with no strict delay requirements.

16.4.2.1 Infotainment Services

For the *Infotainment* domain the investigations have been focused on a quite resource-demanding application, thus indicating a worst-case scenario. VS represents a service that monitors the environment of the vehicle by means of camera systems. Off-board services like off-board traffic sign recognition use this video data for processing. Furthermore, the transmission of video streams between vehicles has been investigated to enhance safety while driving [33]. Note that VS services are only implemented in *fully-equipped* vehicles. The overall traffic flow has been separated into two bearers with different QoS characteristics. The first bearer represents a GBR bearer of QCI 2, which is the QCI class for live streaming services according to [6], with a constant source data rate of $R_{data} = 1.6$ Mbit/s. This represents 8 (fps) of an I-frame only video stream encoded with H.264, which is the minimum required number of frames in the presumed off-board VS service. The second bearer is a non-GBR bearer serving additional traffic of 7 fps with a data rate of $R_{data} = 1.4$ Mbit/s and the same encoding. This bearer is based on QCI 10 with relaxed delay requirements, since packets transmitted by this bearer contain additional frames that are used to improve the quality of the off-board services, but not necessarily required for a basic functionality of the off-board services.

Additionally, two services with different QoS demands have been analyzed: VoLTE and FTP. VoLTE represents a two-way service with stringent delay requirements. According to [6] VoLTE services are based on QCI 1, featuring the most stringent delay requirements in these investigations. FTP presents a typical background service that is not delay-sensitive and thus mapped on QCI 10. In order to accurately model this application, FTP packets of variable size p_{size} have been simulated according to the truncated log-normal distribution [48] with a maximum size of 42 Mbit.

16.4.2.2 Comfort Services

In the *Comfort* domain the results for one typical service are presented: VR, a hybrid, two-way communication service representing a speech-to-text application, which is implemented in most modern premium-vehicles [49]. For this service an on-board component records the driver's voice command and transmits the voice pattern in small data chunks of $p_{size} = 65$ kbit to the off-board component (cf., [30]). The off-board component performs the voice recognition and sends the identified text message back to the on-board component of the application. Since the VR service is based on a TCP/IP web-service, the uplink transmission path is based on QCI class 8 with a packet delay budget of 300 ms. The downlink information is not time-critical, thus relaxed delay boundaries of QCI 10 can be assumed for the identified message sent in downlink.

16.4.2.3 Traffic Efficiency Services

For the *Traffic Efficiency* class, two typical applications have been considered: XFCD and PDAS, which are applications that periodically request context-dependent information from a service provider. Both applications represent simple *one-way* uplink and downlink TCP/IP web-services, respectively [50], and are triggered in fixed periods of $t_{request}^{XFCD} = 300$ s and $t_{request}^{PDAS} = 60$ s, respectively.

16.4.3 Radio Access Network Performance Analysis

Focus has been set on three different key performance indicators (KPIs): application-specific one-way transmission delay (δ_{RAN}) and application-specific packet discard rate. The interpretation of the results of these two KPIs is supported by a third KPI, the uplink and downlink cell load.

In the first step the results of the system-level simulations are presented that have been conducted to evaluate the performance of the E-UTRAN [the RAN part in (16.1)].

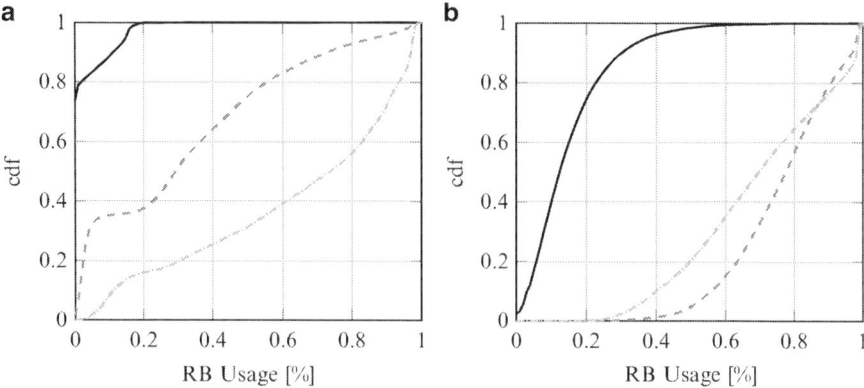

Fig. 16.4 RB usage in time for uplink and downlink data transmissions; urban (——), rural (- - -), highway (· · · ·). (**a**) Uplink. (**b**) Downlink

16.4.3.1 Network Load

The number of allocated RBs is an indicator for the load situation within the E-UTRAN. The term RB is used for the smallest scheduling unit of 1 ms interval and 180 kHz bandwidth in the time/frequency domain. Figure 16.4 shows the cumulative distribution functions (cdfs) of the percentage of allocated RBs in uplink and downlink for all TTIs in the different scenarios, respectively. Please note that the statistics involve all deployed services and terminal classes for all scenarios introduced in Sects. 16.4.1.2 and 16.4.2, respectively. They also include the traffic that is generated from the base load. In the *urban* scenario the 95th percentile shows an allocation of only 15 % of the available uplink and 37 % of the downlink resources. This means that in 95 % of the time not more than 15 % and 37 %, respectively, of the available RBs have been occupied. The network load is thus very low in this scenario. However, for the *rural* and *highway* scenario the cells are heavily loaded, which leads to occasional network congestion situations. The 95th percentile RB allocations show a downlink occupation ratio of 99 % in the *highway* and *rural* scenario. In uplink, the 95th percentile RB allocations are 87 % in the *rural* and 98 % in the *highway* scenario. Note that the average number of allocated uplink RBs in the *highway* scenario equals 65 % compared to the *rural* scenario with 33 %. However, in downlink on average more RBs have been assigned in the *rural* scenario (77 %) compared to the *highway* scenario (70 %). It can be concluded that the *urban* scenario represents a low-load scenario, whereas both, the *rural* and *highway* scenario, represent high-load scenarios. The reason for the different load situations is due to the diverse density of terminals within the cells. Because of the larger cell sizes in the *rural* and *highway* scenario, the total number of vehicles and

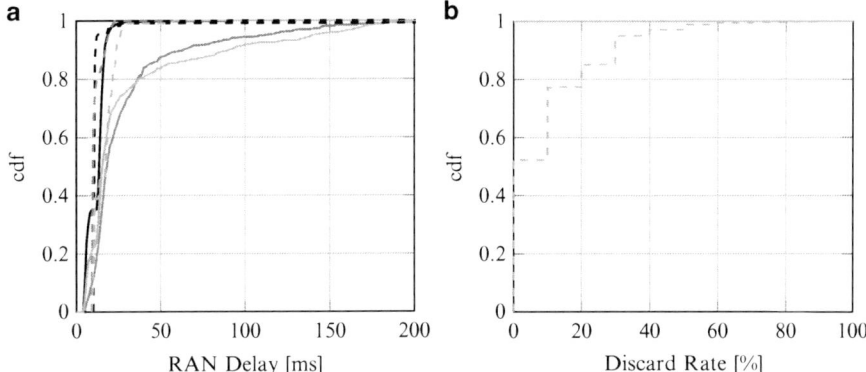

Fig. 16.5 RAN results for XFCD and PDAS; urban PDAS (——), urban XFCD (- - -), rural PDAS (——), rural XFCD (- - -), highway PDAS (——), highway XFCD (- - -). (**a**) Transmission delay. (**b**) Discard rate

mobile terminals is significantly higher than in the urban scenario. In consequence, network planning has to be adjusted to handle the high traffic demand. All in all, the selected scenarios represent examples for low and high load situations.

16.4.3.2 Delay and Packet Discard Rate Statistics

In the following, the cdfs of the delay and discard rate of packets in the radio access network (E-UTRAN) for services of the classes *Traffic Efficiency*, *Comfort*, and *Infotainment* are evaluated.

Traffic Efficiency Services The E-UTRAN delay (cf. δ_{RAN} in (16.1)) and discard rate results for the two automotive off-board services XFCD and PDAS are shown in Fig. 16.5, respectively. In the *urban* scenario the 95th percentile RAN delay is equally small for both services, i.e., 15 ms for XFCD and 19 ms for PDAS, and no packets have been discarded. For the *rural* and *highway* scenarios, the delays for XFCD and PDAS increase significantly. The 95th percentile RAN delay for PDAS in downlink is 146 ms in the *highway* scenario and 100 ms in the *rural* scenario. For the uplink service XFCD, the 95th percentile transmission delay is below 30 ms for all three scenarios. However, as a result of poor radio channel conditions and occasional network congestions in the *highway* scenario, significant amounts of XFCD packets have been discarded, with a 95th percentile discard rate of 40%, cf. Fig. 16.5b. This indicates that the scheduler has not been able to allocate enough resources within the corresponding packet delay budget of 300 ms under the prevailing conditions of this scenario.

Comfort Services The QoS performance of the voice recognition service for the E-UTRAN part is shown in Fig. 16.6. For downlink transmissions the 95th

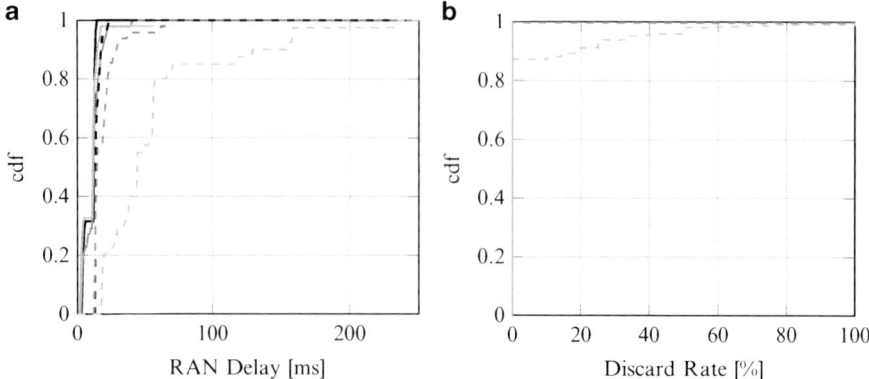

Fig. 16.6 RAN results for voice recognition; urban downlink (——), urban uplink (- - -), rural downlink (——), rural uplink (- - -), highway downlink (——), highway uplink (- - -). (**a**) Transmission delay. (**b**) Discard rate

percentile delay is around 25 ms in all scenarios, which can be explained by the small packet size. In only 5 % of the measurements, a delay higher than 25 ms has been observed. The 95th percentile uplink transmission delays are much higher in the *highway* scenario (i.e., 184 ms) compared to a delay of 52 ms in the *rural* and 20 ms in the *urban* scenario. Again, this can be attributed to the high network load and the poor radio channel conditions, which leads to packet loss, retransmissions, and discards. In the *highway* scenario the 95th percentile discard rate of uplink VR packets is 33 %, i.e., packets have not been scheduled in time and have been discarded (cf. Fig. 16.6b).

Infotainment Services The results of the video surveillance application are shown in Fig. 16.7. The cdfs show the statistics of both bearers jointly, i.e. the GBR and non-GBR bearer. In the *urban* scenario, the 95th percentile E-UTRAN delay is well below 80 ms. Nevertheless, the 95th percentile VS packet discard rate is 6 % (i.e., packets from the non-GBR bearer). In the *rural* and *highway* scenario a serious performance degradation has been observed with a 95th percentile transmission delay of 422 ms (cf. the *highway* scenario in Fig. 16.7a) as well as significant packet discards for both bearers. The 95th percentile discard rate is 50 % in the *highway* scenario and 94 % in the *rural* scenario. Yet, the 90th percentile for the *rural* scenario is below 20 %. Therefore, the average QoS performance for the VS application is much more degraded in the *highway* scenario with a mean discard rate of 42 % as opposed to the *rural* environment with 13 %. Since even packets from the GBR bearer get discarded occasionally, the VS application is rendered unreliable in these high-load scenarios.

Note that these results are related to the assumed high inter-site distance of 25 km in the *highway* scenario. The requirement from this analysis is a denser network deployment to provide more resources for the given number of users. It is more an issue of network optimization rather than a technology driven weakness.

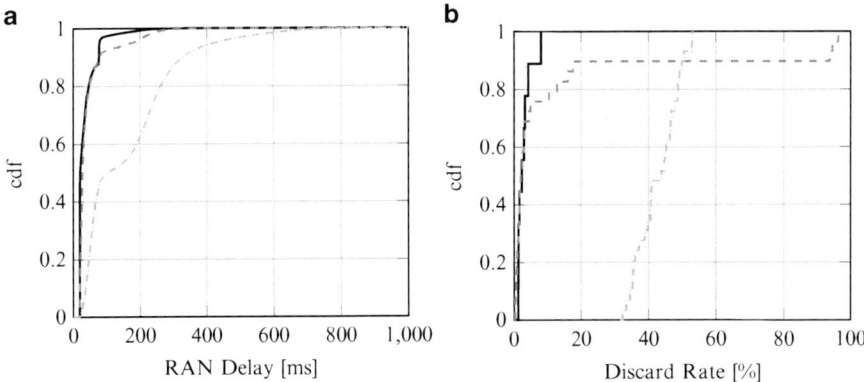

Fig. 16.7 RAN results for video surveillance; urban (———), rural (- - -), highway (-·-·-).
(**a**) Transmission delay. (**b**) Discard rate

The cdf of transmission times for the FTP service (Fig. 16.8a) shows 95th percentile transfer times of more than 25 s for the *highway* and *rural* scenario. For the interpretation of this figure, the various simulated data volumes of up to 42 Mbit have to be considered. Under good radio channel conditions in a low load situation this data volume can be transmitted within a short time. An important aspect is how radio resource management (RRM) handles services with different priorities in a high load situation. The high transmission times for FTP can be attributed to the low QoS settings for this service and, hence, to its low scheduling priority.

Figure 16.8b shows the E-UTRAN transmission delays for the VoLTE service. In all three scenarios for uplink and downlink the 95th percentile delay remains below 20 ms. The step-wise nature of the cdf with a step-width of about 8 ms is due to occasional hybrid automatic repeat request (HARQ) retransmissions (cf. [34]). No packets have been discarded for both services. For VoLTE this is due to the strict QoS requirements and high scheduling priority, while in the case of FTP this is related to a large packet delay budget of 60 s (see Table 16.5).

For validation of the simulations, the results have been compared with real LTE network measurements presented in [41]. The one-way radio access network delay measurements presented in that work have been performed with packet sizes between 10 and 5,000 bytes in a live network under low traffic load conditions. When comparing the real network measurements with the simulation results of the low-load *urban* scenario, the 95th percentile delays are almost identical. In uplink the delay measured in [41] is 14 ms and almost equals the calculated delay in the simulations of 13 ms. In downlink a delay of 35 ms has been measured in [41], whereas a delay of 31 ms has been calculated from simulations.

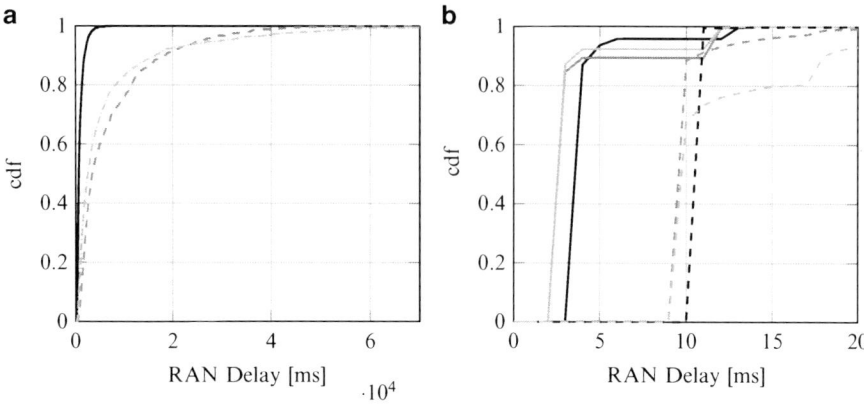

Fig. 16.8 RAN transmission delays for mobile terminal applications: FTP [urban (——), rural (- - -), highway (······)], VoLTE [urban downlink (——), rural downlink (——), highway downlink (——), urban uplink (- - -), rural uplink (- - -), highway uplink (- - -)]. (**a**) FTP transmission delay. (**b**) VoLTE transmission delay

16.4.4 Delay Analysis

Core Network Delay Analysis As defined in (16.1), the overall LTE transmission chain consists of the E-UTRAN delay as well as the core network (or EPC) and terminal (UE) delays. The EPC one-way delay (denoted as δ_{EPC} in (16.1)) in the user-plane is defined as the propagation time of a packet between the eNodeB and the PDN-GW. As estimated by 3GPP (cf. [6]), the overall core network delay, which depends on the virtual distance between the E-UTRAN and PDN-GW, can be assumed to take values between 10 ms and 50 ms. In the typical case that the radio E-UTRAN and the PDN-GW belong to the same public land mobile network (PLMN), the connection between them is fast and, thus, the average delay is around 20 ms. However, in a roaming scenario, the data packets will be routed from the foreign RAN to the home PDN-GW. In the case of large-distance communications (e.g., between the USA and Europe), the EPC delay might be 50 ms. Taking into account that the roaming case is less likely for the simulated scenarios, an EPC delay of $\delta_{EPC} = 20$ ms has been assumed.

Terminal Processing Delay Analysis Eight categories for LTE terminals are defined in [10], where each category represents different capabilities. Among others, the number of spatial multiplexing layers, modulation sizes, buffer sizes, and supported peak data rates drive the complexity and costs of the required signal processing algorithms. The results of an extensive measurement campaign in [55] show that processing time at the terminal contributes to the E2E delay with 1.5 to 5 ms [cf. δ_{UE} in (16.1)].

Further Delay Aspects Besides the E-UTRAN and EPC transmission delays, further delays corresponding to mobility management processes have to be taken into account. In [7], the handover (HO) interruption time has been estimated by 3GPP as 15–25 ms for the uplink and 13–23 ms for the downlink. During this period, no data can be sent on the respective links. However, given such short interruption times, the scheduler tries to compensate for the lost time by favoring the affected packets. In this way, the QoS requirements will be fulfilled in most cases. Furthermore, only a few packets will suffer from the HO process. In the case of network deployments with small cell sizes (i.e., the *urban* scenario), it can be calculated that a vehicle terminal potentially has to perform one HO every 10 s. Based on this assumption, 1.2 packets/h in the case of PDAS are estimated to be affected, which is almost negligible. It should also be noted that in the unlikely event of a HO failure, an additional delay of up to 130 ms has to be taken into account. Note that in this work, no additional delay due to HOs or HO failures has been considered.

Further delays can arise at the initiation phase of a bearer, if the terminal has to perform idle-to-active state transition. According to [7], this process requires 61–115.5 ms. However, in this investigation, it is assumed that the vehicle terminals are always residing in connected mode in order to support premium connectivity for the automotive services.

Depending on the network operator's policies as well as on the terminal capabilities, an additional delay caused by discontinuous reception (DRX) also needs to be considered [34]. Different DRX cycles have been specified by 3GPP ranging from 0 to 2.56 s. The main goal of DRX is to increase the limited battery life time of mobile devices. Since battery life time does not pose a problem for vehicular terminals, the DRX mode has been disabled and, thus, does not need to be considered in the E2E delays for automotive services. From a network point of view, discontinuous transmission (DTX) provides the benefit of reducing interference in the network, which is of particular importance considering the fact that typical LTE deployments have a frequency reuse of 1.

16.4.5 Performance Discussion

In the following an overview of the overall perceived QoS performance is provided based on the one-way-delay of the total LTE (EPS) transmission chain, as defined in (16.1). The following assumptions have been made: $\delta_{UE} = 5$ ms and $\delta_{EPC} = 20$ ms.

16.4.5.1 Infotainment Services

Requirements for the VS service have only been met in the *urban* scenario with a 95th percentile $\delta_{LTE}^{VS} = 105$ ms, which leaves room for additional delays such as for handovers. However, in the *rural* and *highway* scenarios the observed

E-UTRAN 95th percentile delay of 422 ms already exceeds the packet delay budget significantly. Furthermore, the 95th percentile discard rate of all VS packets is 94 % rendering this application highly unreliable in high-load environments. Note that this applies for both bearers, i.e., the GBR and non-GBR bearer. As a result, it is not feasible to deploy high-demand video streaming services in high-load scenarios with insufficient network capacity. However, a bearer based on a buffered-streaming service with QCI 6, as stated by 3GPP in [6], might be an appropriate choice for off-board video surveillance based services. Note that in high-load conditions QoS strategies can only be efficiently operated if a mix of services of different priorities is simultaneously served, i.e., also services of lower QoS priority have to be available that can be temporarily downgraded.

16.4.5.2 Traffic Efficiency Services

In the *rural* scenario, PDAS shows a 95th percentile delay of $\delta_{\text{LTE}}^{\text{PDAS}} = 205$ ms, where $\delta_{\text{RAN}}^{\text{PDAS}} = 180$ ms, $\delta_{\text{UE}} = 5$ ms, and $\delta_{\text{EPC}} = 20$ ms. Since this value is well below the delay budget (300 ms), even packets that suffer from additional handover or idle-to-active transition delays (which have not been considered in this work) would be delivered in time. In the uplink, XFCD shows even better delay behavior in the *urban* and *rural* scenarios with total LTE transmission delays below 55 ms. However, in the *highway* scenario, the bad channel conditions and corresponding high network load show 95th percentile packet discard rates of 40 %, which indicates that the quality expectations in such an environment are not fulfilled for the assumed network layout. The requirement is a denser network deployment to increase capacity.

16.4.5.3 Comfort Services

The performance of the VR service is similar to that of *Traffic Efficiency* services. In downlink, a 95th percentile delay of $\delta_{\text{LTE}}^{\text{VR,DL}} = 50$ ms has been calculated and almost no packets have been discarded. Uplink transmissions show a 95th percentile delay of $\delta_{\text{LTE}}^{\text{VR,UL}} = 209$ ms. However, in the *highway* scenario again a non-negligible amount of packets has been discarded (i.e., a 95th percentile discard rate of 33 %) and, thus, the quality expectations for this lower priority service have not been met. Overall, in order to improve the perceived QoS for the TCP/IP-based web-services, especially for the uplink based VR transmission, the bearers should be deployed under QCI 6, rather than QCI 8 or 10. According to 3GPP this class is also defined for TCP/IP-based web-services with a packet delay budget of 300 ms but with a higher priority for scheduling, which in turn leads to a prioritized handling of the corresponding bearers compared to QCI 8 or 10.

The QoS requirements for the mobile terminal services (FTP and VoLTE) can be met. The VoLTE service is based on QCI 1 with a delay budget of 100 ms. In the uplink and downlink, the 95th percentile delay is $\delta_{\text{LTE}}^{\text{VoLTE}} \leq 45$ ms. This means that even additional delays due to handovers would not exceed the QoS delay budget.

16.5 LTE for Safety Services

In this section, the suitability of LTE to deliver messages of *Safety* applications is investigated. Features and requirements of *Safety* applications are presented and an overview of related work from standardization bodies and academia is given. Finally, a simulative analysis for CAM message distribution in a typical LTE deployment is presented.

16.5.1 *Safety Applications: Features and Requirements*

Safety applications require periodic V2V data exchanges in the neighborhood of a vehicle in the case of CAMs or event-triggered V2V and V2I communications in the case of DENMs. ETSI and International Organization for Standardization (ISO) are currently investigating the ability of LTE to support data exchanges for these cooperative applications. Preliminary results are reported in [24].

CAM and DENM exchanges in LTE involve transmissions from vehicles to infrastructure nodes and successive traffic distribution to the concerned vehicles. Unicast is always used for uplink transmission. In the uplink case, the problem is to select the most appropriate channel type without congestion risks. The random access channel (RACH) is a common uplink transport channel usually selected for signaling and to transmit small data amounts, such as CAM and DENM.

Unicast, multicast, and broadcast modes can be used on the downlink by leveraging eMBMS capabilities. Although eMBMS and its predecessor MBMS developed for UMTS deployments are not implemented in most current cellular networks, broadcast mode is more resource-efficient than unicast mode. Hence, it is strongly encouraged to use eMBMS for vehicular safety applications, although it could imply longer delays due to the eMBMS session setup.

ETSI specifications foresee the presence of a special-purpose backend server that supports geocasting, by intercepting CAM and DENM traffic from vehicles and processing it before redistributing it only to the concerned vehicles in a given geographical area [25]. In order to identify the concerned vehicles in a given area [42], the backend server has to know the list of geographic areas, their coordinates, and the position and IP addresses of all vehicles in any area at all times. According to the ETSI specifications [25], each time vehicles cross over to a new area, the server informs them about the coordinates of their current geographical area. The area size is application dependent, thus affecting the signaling load. Then data is distributed to the concerned vehicles through eMBMS or via multiple unicast connections.

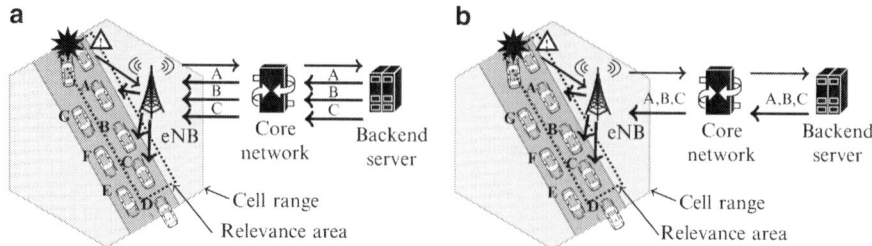

Fig. 16.9 DENM delivery in LTE. Only vehicles within the relevance area (*rectangular dotted area*) are addressed. (**a**) Unicast. (**b**) Multicast

The structure of the network deployment and location of the backend servers has an impact on the signaling procedures and latencies, as discussed in [42]. To reduce latencies in the centralized LTE architecture, the utilization of a combination of cloud-based servers and the LTE network has been proposed in [37]. In this study servers are located at different network locations between the edge of the network, i.e., network locations close to the eNodeB, and the Internet. The results show that the delay of the messages can be reduced, as the backend server moves closer to the eNodeB. If the server is installed in the mobile operator's core network, it may directly exchange location information with the MME module in the LTE architecture. This has the major advantage that the location information of the vehicles stored in the MME can be used for location-aware applications. Hence, no determination of the location through a separate process is required. In contrast, if the server is located in the Internet and thus decoupled from the mobile operator's network, each vehicle needs to maintain a connection to the server and send regularly position updates to it.

Figure 16.9 displays an example for DENM distribution procedures augmented with a backend server. In the case of unicast distribution, vehicles are addressed individually, so that the same message is separately transmitted to all concerned vehicles. In the multicast case, all vehicles in the relevance area are collectively addressed through geo-addressing capabilities leveraging the geographical position of terminals and a message transmission is performed using eMBMS (cf. Fig. 16.9b). In both cases, the latency of the message transmission might become critical, especially for localized safety-critical V2V communications.

Even for the distribution of CAM, messages have to cross the infrastructure for multicast distribution. In Fig. 16.10 the backend server collectively addresses all vehicles in the awareness range of the sending vehicles (A and B). In contrast Fig. 16.11 shows the situation when an IEEE 802.11p network is available. A single broadcast transmission can be used to distribute the message from a vehicle in its awareness range (in the case of CAMs) or within the relevance area in the case of DENMs.

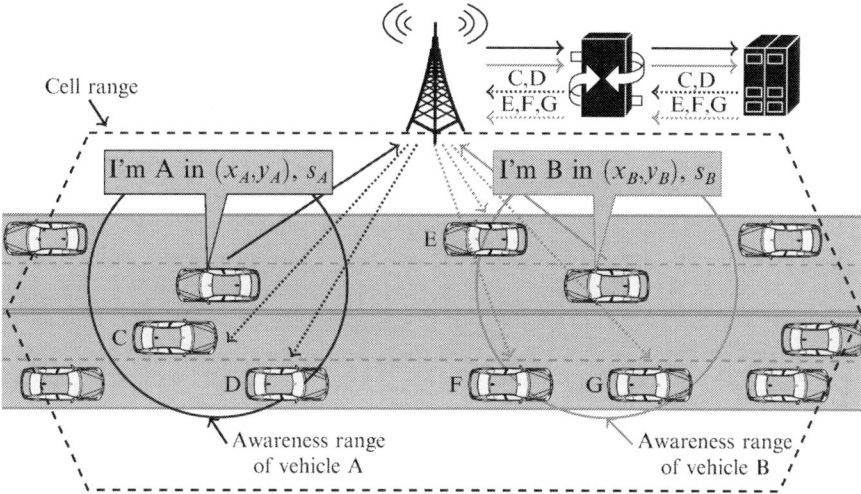

Fig. 16.10 Multicast CAM delivery. The awareness range of the vehicles does not coincide with the cell range (adapted from [16])

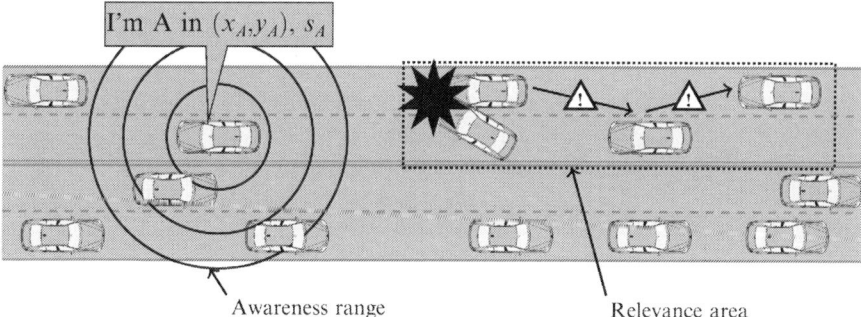

Fig. 16.11 CAM and DENM delivery in IEEE 802.11p. Messages are locally broadcasted through V2V communications (adapted from [16])

16.5.1.1 CAM Support over LTE

In addition to the outcomes of standardization bodies, the support of vehicular safety applications through LTE has been investigated in recent literature. Analytical results in [61] show that LTE is unable to satisfy the CAM delivery requirements if the eNodeB retransmits all received CAMs to every vehicle in the cell using unicast transmissions. Similar results are achieved if the eNodeB unicasts aggregated CAMs to every vehicle in the one-hop neighborhood. Improvements can be obtained through CAM broadcasting in the cell.

In [51] the authors enhance the unicast CAM downlink transport with a filtering scheme in order to reduce the load in the cells and to meet the CAM delay requirements. Filtering relies on the idea that not all vehicles in a cell need to receive all CAMs. Accordingly, based on the received vehicle's location information, the backend server selects a subset of vehicles that receive CAMs on unicast links. Results attained in urban and rural scenarios show that the selective unicast of aggregated CAMs might also overload the LTE network. A higher number of vehicles per cell can be served when decreasing the CAM rate down to 2 packets/s, instead of 10 packets/s. The conducted study has contributed to the architecture and the results reported in [25]. The authors in [25] suggest to use the eMBMS to increase the downlink capacity. Additionally, the authors of [46] advocate the complementary use of cellular systems and IEEE 802.11p to successively broadcast the received CAMs on the downlink at road intersections, where IEEE 802.11p may suffer from non-line-of-sight conditions due to the shadowing effects caused by buildings.

The impact of the remote server position on the overall performance is investigated in [38] for both eMBMS and non-eMBMS architectures. The results of the investigation using eMBMS show that up to hundrets of vehicles within each cell can be supported. Furthermore, simulation results indicate that locating the server at the edge of the network close to the vehicles reduces both the end-to-end latency and the RAN network traffic by requiring less frequent updates to achieve the intended freshness of information. The remote server plays an active role in reducing the network load, while keeping the information as fresh as possible, in the solution proposed in [63]. There, the authors suggest the server to coordinate transmissions from and to vehicles to determine an optimum transmission rate based on network observations. Such observations can be acquired at the application layer. The delay and transmission rate of the latest packet, for instance, are among the considered performance indicators. The suggested rate is transmitted in already exchanged CAM and DENM packets, without incurring additional signaling overhead.

Besides the influence of the architecture, the applied scheduling techniques in the eNodeB have a major impact on the transmission of CAM messages. In [57] it is assumed that CAM messages offer similar traffic patterns as VoLTE traffic. This is because they both foresee the frequent transmission of small-sized packets with a short information relevance. In [57] three different scheduling techniques proposed for VoLTE are discussed. First, dynamic scheduling is presented, where the UE sends a resource request message to the eNodeB for every data packet. Second, using persistent scheduling a dedicated amount of resources is statically reserved for the time of the data transmission. Third, semi-persistent scheduling applies persistent scheduling for initial transmission and dynamic scheduling for retransmissions. All schemes have their advantages and disadvantages for the CAM message distribution. Persistent and semi-persistent scheduling well match the requirements of a fixed beaconing rate, e.g., a CAM per vehicle at every 100 ms. Dynamic scheduling, on the other hand, is capable of handling adaptive beaconing schemes, where the transmission of CAMs may vary, e.g., according to the vehicle speed. The proposed solution foresees that the UE sends a scheduling request via random access for initial connection setup, then the eNodeB schedules resource blocks for subsequent CAMs. When a vehicle stops because of a red light, it stops sending CAMs and frees resources that the eNodeB can reuse.

The study in [58] has been conducted in the same scenario with the focus on a priority-based congestion control algorithm for improving CAM delivery at intersections. The eNodeB receiving CAMs is able to identify potential collision patterns and is able to send back warning messages to vehicles. No interference from other traffic is considered, since a dedicated 700 MHz Public Safety Band allocated in the USA by the Federal Communications Commission (FCC) for broadband communications is assumed to be used. However, so far it has not yet been decided how much bandwidth should be assigned for intersection assistance applications. Each user is given a priority when the bandwidth threshold is reached and low priority users (i.e., the ones further apart from the intersection) are removed to reduce the cell load.

To reduce load on the cellular network, when targeting CAM delivery at intersections, a hybrid approach is proposed in [59] that leverages both Wi-Fi and LTE technologies. According to the described solution, when a vehicle approaches an intersection, it first broadcasts beacons through its Wi-Fi interface to form a cluster. The cluster head is then responsible for sending CAMs to the base station, that, in turn, forwards CAMs to cluster heads in other road segments to keep them informed. The cluster heads forward the message sent by the base station via Wi-Fi to the cluster members. Results show that the clustering scheme significantly improves the CAMs delivery performance with respect to schemes relying on LTE and Wi-Fi only, while reducing the network load. On the other hand, the delay increases compared to a Wi-Fi only scheme where direct communications can be enforced between vehicles. In the hybrid approaches, different delay contributions should be considered, ranging from the transmission of CAMs over LTE, the processing at the eNodeB and at upper LTE layers, the identification of the set of receivers, their downlink scheduling and their forwarding from the eNodeB to other cluster heads. Overall, the delay values are below 100 ms, hence matching the application demands.

The main assumption in the above-mentioned studies is that the LTE capacity is exclusively used for CAMs, without accounting for other vehicular and non-vehicular traffic with different QoS requirements, which significantly affects performance.

16.5.1.2 DENM Support over LTE

DENMs generated as a reaction to a hazard have a limited lifetime and the number of senders is typically significantly lower compared to CAMs. Hence, DENMs generate a lower traffic load compared to CAMs.

The main challenge is related to simultaneous warning transmission attempts by all vehicles detecting a specific hazard. For example, in case of slippery roads, vehicle collision events may be detected and notified by every vehicle passing the area. In this case, the backend server plays the crucial role of a reflector and aggregator. It can filter the multiple uplink notifications of the event according to the location, time stamp, and heading field of the received messages to send only one consolidated message [24]. The latter feature allows the server to infer a better general view of the road conditions [28]. In addition, the detecting vehicle receives an implicit acknowledged notification of the same event on downlink. It has no need to repeat the same DENM transmission several times. System scalability is thus improved, channel resources are saved, and uplink congestion is avoided. As an additional benefit, the wide cellular coverage guarantees the event dissemination also when there is no nearby vehicle to relay the message, which would hinder propagation of messages when 802.11p is used instead. Therefore, DENM over LTE results in a much more reliable solution as demonstrated in [24, 51] for a system where no background load was assumed. In [40] the traffic is generated from a single vehicle transmitting a DENM to the base station, which repeatedly rebroadcasts it to all vehicles in the cell through MBMS. Different downlink scheduling schemes are compared, showing that QoS-aware schemes meet the DENM delay constraints.

16.5.2 Performance Evaluation

In this section, the delivery performance of CAMs generated by vehicles is evaluated through simulations, under different network and load conditions.

16.5.2.1 Simulation Settings and Assumptions

Simulations have been conducted using ns-3 [56] with the LENA (LTE-EPC Network Simulator) extension [44] to model the E-UTRAN and the core network modules (S-GW, PDN-GW, MME).

SUMO (Simulation of Urban Mobility) [54] is used to generate the road topology and the mobility patterns of vehicles that move over a regular 4×4 grid road topology where each road segment is 250 m long. The eNodeB is installed in the center of the road topology. Focus has been set on the scenario with one eNodeB for simulation efficiency considerations. Vehicles send CAMs to a remote server, which forwards the CAMs towards vehicles in the awareness range in downlink direction. A dedicated link is assumed for the connection between the PDN-GW and the server. To this end, the delay between the PDN-GW and the remote server (δ_{PDN}) is set to 0 ms. For the simulations, $\delta_{OFF\text{-}BOARD}$ and $\delta_{ON\text{-}BOARD}$ are also set to 0 ms.

To investigate the effect on the performance of safety messages, three traffic load cases have been analyzed. In *Case A*, only vehicles transmitting CAMs are considered in the cell. In *Case B*, fifty pedestrian users are added that generate interfering voice calls, which are mapped as VoLTE traffic and modeled with an ON/OFF Markov chain. During the ON period, the source sends 20 byte data packets every 20 ms (i.e., a source data rate of roughly 8 kbit/s). In *Case C*, the simulated interfering traffic is heavier; 40 of the 50 pedestrian users continue to generate voice calls while the remaining ten users generate video traffic with a source rate of roughly 128 kbit/s. In all three simulated cases, the number n of vehicles generating CAMs varies where $n \in \{50, 100, 150, 200\}$. Both distributed and simultaneous vehicle arrivals are considered in the three cases and separately analyzed in the following subsections.

The focus of the conducted simulations is set to the uplink direction (both data and control channels) to get quantitative insights into the LTE channel access procedure and the related congestion problem. The uplink is the more critical link of the two channel directions due to the congestion risks from massive data access. Moreover, data aggregation and geocasting (implemented through enhanced multicasting or broadcasting) can be used in downlink direction from the remote server to the intended set of receivers [51].

Similarly to [51], a round-robin MAC scheduler is used. Further LTE simulation settings are similar to the ones in [58] and are listed in Table 16.6. Focus is set on two KPIs for CAM transmissions: one-way transmission delay (δ_{RAN}), from the vehicle to the eNodeB, and packet delivery ratio. In addition, the throughput performance is computed for the interfering VoLTE and video traffic.

16.5.2.2 Scenarios with Distributed Vehicle Arrivals

The aim of the analysis reported in this subsection is to assess the suitability of LTE in supporting CAM traffic in typical scenarios (with and without interfering traffic) when several vehicles attempt to access the network at different time instants based on their mobility patterns. Specifically, a uniform vehicle arrival rate in the cell within a 2 s time interval is considered. After the arrival in the cell, each vehicle performs the random access procedure. When this procedure is successfully completed, it starts to send data. In *Cases B* and *C*, the interfering users are assumed to be already active in the cell at the start of the simulations. In all simulations, an LTE channel bandwidth of 5 MHz is considered.

The results displayed in Fig. 16.12 show the packet delivery ratio and one-way transmission delay for the three *Cases A, B*, and *C*. The amount of resources in the LTE system can ensure full reliability, i.e., CAM packet delivery ratio is 100% in all the considered cases. In other words, the results highlight that the LTE random access procedure is able to manage the arrival of hundreds of vehicles if they have different arrival time instants. As shown in Fig. 16.12, the CAM transmission delay is 12 ms independent of the number of vehicles in the cell and in all of the simulated Cases. This demonstrates that LTE is able to meet the time-constraint of CAM messages.[2]

---

[2]The delay in downlink direction is expected to be comparable with or even shorter than 12 ms, hence leading to an overall delay below 100 ms.

Table 16.6 Main simulation settings

Parameter	Value
Simulated Area	1 km × 1 km
Layout	Grid topology
Road segment length	250 m
Speed limit	50 km/h
Frequency band	2 GHz
TTI	1 ms
DL Tx Power	40 dBm, antenna gain 14 dBi, Noise figure 5 dB
UL Tx Power	20 dBm, antenna gain 0 dBi, Noise figure 9 dB
System bandwidth	5 MHz, 10 MHz (25 RB, 50 RB)
RB size	12 sub-carriers, 0.5 ms
Sub-carrier spacing	15 kHz
Data/control OFDM symbols	11/3
Scheduling algorithm	Round Robin
CAM packet size	100 Byte
CAM frequency	10 Hz
CAM QCI	8
Interfering VoLTE QCI	1
Interfering video QCI	7
Propagation model	Friis
Thermal noise	−174 dBm/Hz
Simulated time	100 s

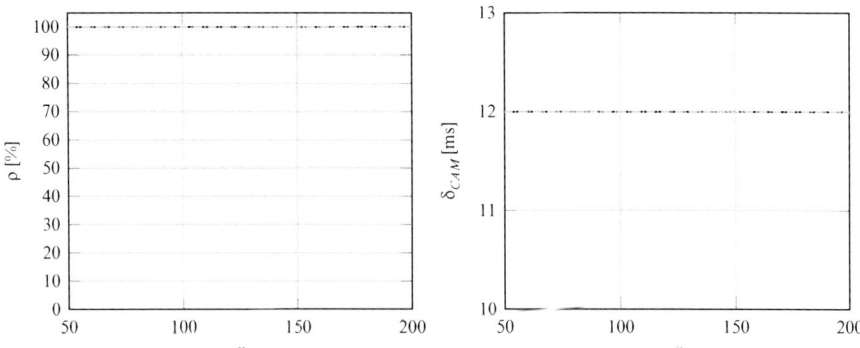

Fig. 16.12 Distributed vehicle arrivals: CAM packet delivery ratio (ρ_{CAM}) and delay (δ_{CAM}) vs. the number of vehicles transmitting CAMs (n) in *Case A* (——), *B* (– – –), and *C* (– · – · –)

The throughput of VoLTE and video interfering traffic in *Cases B* and *C* is plotted in Fig. 16.13. In *Case B*, it can be observed that VoLTE users are not

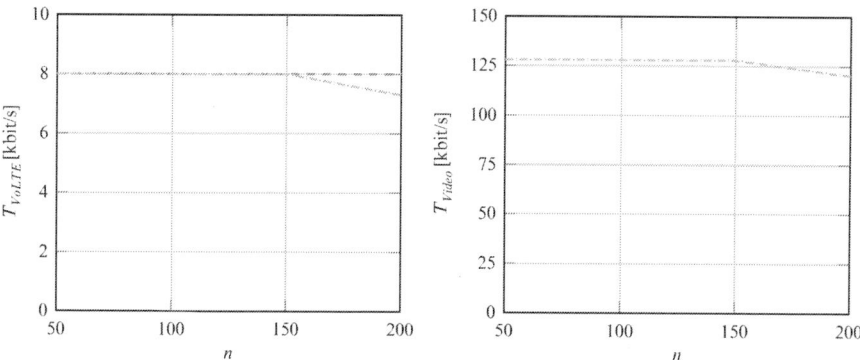

Fig. 16.13 Distributed vehicle arrivals: throughput of VoLTE (T_{VoLTE}) and video (T_{Video}) vs. number of vehicles transmitting CAMs (n) in Case B (— —) and C (— · —)

influenced by the CAM transmission, i.e., the throughput is 8 kbit/s. Hence, LTE can simultaneously handle both CAM and VoLTE traffic without any performance degradation.

In *Case C* the mean throughput of VoLTE gradually decreases when more than 150 vehicles are in the cell and it goes down to a value of 7.6 kbit/s with 200 vehicles per cell. A similar trend can be noticed for video flows, which keep a mean throughput of 128 kbit/s when up to 150 vehicles are considered. Under the heavy load of 200 active vehicles in a cell, however, their mean throughput decreases to 120 kbit/s. Under these assumptions, LTE is able to support the arrival of up to 150 vehicles in few seconds per cell without significantly affecting the performance of other traffic (i.e., VoLTE and video). Indeed, the throughput of VoLTE and video traffic is reduced by 6%, respectively, 5% when a large number of vehicles (i.e., 200) is active in the cell.

16.5.2.3 Scenarios with Simultaneous Vehicle Arrivals

In this subsection, a worst-case scenario is considered when *all* vehicles in the cell perform the random access procedure *simultaneously*. After the random access accomplishment, CAM transmission attempts by vehicles are distributed within a 2 s time interval. The performance is evaluated under different network load and deployment settings in *Cases A, B*, and *C* for two bandwidths, i.e., 5 and 10 MHz. With 10 MHz bandwidth, a higher number of resources is available with respect to the 5 MHz case, with an expected positive impact on the performance of the random access procedure.

In Fig. 16.14 the CAM delivery ratio ρ_{CAM} is displayed for all three cases and both 5 and 10 MHz bandwidths. It can be observed that full reliability (i.e., $\rho_{CAM} = 100\%$) can be achieved *(i)* for 5 and 10 MHz channels in *Case A*, and *(ii)* for 10 MHz bandwidth also in *Case C*. In the other cases, full CAM reliability

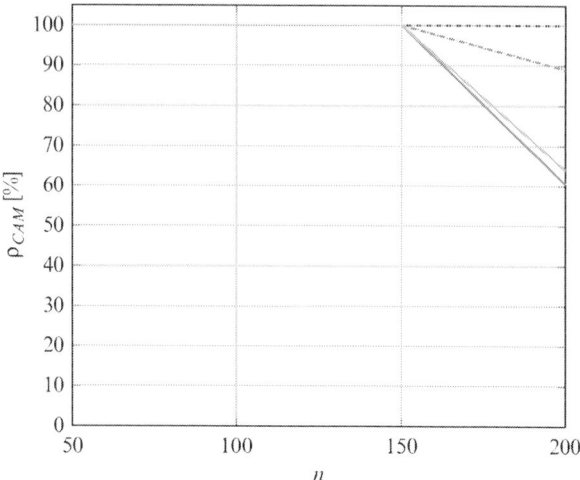

Fig. 16.14 Simultaneous vehicle arrivals: CAM packet delivery ratio (ρ_{CAM}) vs. number of vehicles transmitting CAMs in *Cases A, B* and *C* when varying the channel bandwidth. *Case A*: 5 MHz, 10 MHz (———); *Case B*: 5 MHz (———), 10 MHz (— · — · —); *Case C*: 5 MHz (———), 10 MHz (— — — —)

is achieved for a number of vehicles up to $n = 150$. For a higher number of vehicles, packet delivery ratio decreases. This results from the limitations of the random access mechanism. As a very large number of vehicles (i.e., $n \geq 150$) tries to access the network simultaneously, only a portion of them is successful. Vehicles having access to the network experience a packet delivery ratio equal to 100 % with a delay almost equal to 12 ms, while the other vehicles are blocked at the access and consequently cannot transmit CAMs.

With 10 MHz bandwidth, the number of resources available in the uplink/downlink physical channels increases and this allows a larger number of vehicles to successfully access the network. As a result, the packet delivery ratio improves compared to the cases of 5 MHz bandwidth.

It is worth noticing that in *Case C* vehicles experience better performance than in *Case B*. This is due to the fact that in *Case B* there is a higher number of VoLTE connections that frequently transmit (i.e., every 20 ms) and compete with vehicles for resource assignment. In *Case C*, VoLTE flows are fewer and the additional video flows have higher periodicity (equal to 40 ms according to the simulation settings) compared to VoLTE. This improves the scheduling efficiency in eNodeB and thus the CAM delivery (Fig. 16.15).

Similarly to CAMs, the performance of VoLTE and video traffic also deteriorates as the number of vehicles transmitting CAMs exceeds 150. This effect is more evident in scenarios with 5 MHz bandwidth.

All in all, LTE is able to support CAM traffic ensuring full reliability and meeting delay constraints under low-to-medium vehicular load conditions. If the cell is

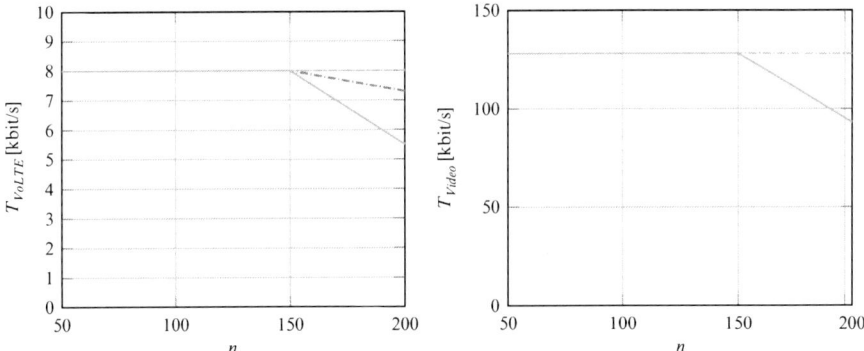

Fig. 16.15 Simultaneous vehicle arrivals: throughput of VoLTE (T_{VoLTE}) and video (T_{Video}) vs. number of vehicles transmitting CAMs (n) in *Case B* when varying the channel bandwidth. *Case B*: 5 MHz (———), 10 MHz (–·–·–·); *Case C*: 5 MHz (———), 10 MHz (–·–·–·)

highly congested, new workarounds both in the random access and scheduling techniques should be specifically conceived to meet the QoS requirements of vehicular safety applications and, at the same time, not penalize base load traffic.

16.5.3 Concluding Remarks

In this section, an overview about CAM and DENM message transmissions in LTE network deployments has been given. The following concluding remarks can be summarized:

- For DENMs, LTE can provide the ability (i) to consolidate the numerous event notifications originated from all the vehicles in a given area, and (ii) to disseminate only useful information in a specific area, with positive effects on system scalability, congestion avoidance, and delivery reliability.
- CAM delivery through LTE may suffer from poor uplink performance due to congestion under heavy load conditions. The simulated scenarios show performance degradations for a load of more than 150 vehicles per cell. However, LTE provides advantages in terms of coverage in specific hazardous areas such as road intersections, where obstacles like buildings can obstruct direct V2V communication.
- Unicast CAM delivery is less resource efficient than eMBMS delivery but it may show advantages in terms of delays, since multicast setup procedures can be avoided.
- The backend server plays a key role in V2V communications. The vehicle-to-server and in-network signaling load, which is also server-location-dependent, and the required intelligence at the server vary with the vehicular application. Besides reflecting or aggregating messages, the server may also take care of

repeating a message as long as the notified event persists so that information can be refreshed for newly arriving vehicles [28] in the relevant area. It can also regulate the transmission rate of CAMs and DENMs in response to network observations.

16.6 Comparison of 802.11p and LTE for Automotive Services

In the following the core features of LTE and IEEE 802.11p,[3] specifically conceived for vehicular environment, as potential wireless connectivity technologies for vehicular applications are compared and contrasted.

Coverage and Mobility LTE relies on a cellular deployment of eNodeBs offering a wide area coverage. This would solve the IEEE 802.11p issue of poor, intermittent, and short-range connectivity of approximately 300 m, and would particularly favor LTE for V2I communications even at high terminal speeds. The use of LTE for V2X communication also represents a viable solution to bridge the network fragmentation and extend the connectivity in those scenarios where direct V2V communications cannot be supported due to low car density (e.g., off-peak hours, rural scenarios, etc.) or due to challenging propagation conditions (e.g., corner effect due to buildings or obstructions at road intersections). Furthermore, the coverage of LTE can even be increased by incorporating other co-deployed cellular wireless technologies, such as UMTS. An inter-radio access technology handover is performed automatically by the core-network of the mobile network operator. The *centralized* nature of the cellular network has the following drawback compared to the 802.11p technology: it does not natively support direct V2V connectivity. Instead, messages require to be passed through infrastructure nodes in the core network.

Market Penetration and Transmission Costs A higher market penetration rate is expected to be achieved by LTE compared to IEEE 802.11p. This is an important aspect as ad hoc networks suffer from the typical chicken-and-egg deployment problem, since a certain penetration rate of IEEE 802.11p equipped vehicles is required before this can be considered an effective approach. An important aspect is that the LTE network interface is integrated in common user devices like smart phones and passengers are accustomed to being connected to the Internet through these devices while being on the road. A further meaningful difference between the two technologies is the cost and provider aspect. While IEEE 802.11p uses a dedicated transmission band for vehicular communication, which is free of charge and requires no further operator, the use of cellular communication systems always relies on mobile network operators, which charge for the use of the transmission system.

---

[3]See Chaps. 3 and 4 for a more detailed description of the IEEE 802.11p standard.

Capacity LTE offers high downlink and uplink capacity (up to 300 and 85 Mbit/s cell throughput, respectively, in 3GPP Rel. 8, and up to 1 Gbit/s for LTE-A in Rel. 11). Such values are much higher than the 27 Mbit/s offered by IEEE 802.11p. In case of further increasing throughput demands, additional capacity per area can be achieved by a denser LTE network deployment.

Latency Depending on the automotive service, the latency performance might have a significant influence on the application performance, especially for *Safety* services. Besides the transmission delays introduced by LTE in the E-UTRAN and the EPC [cf. Eq. (16.1)], the state mode of the mobile terminal has major impact on the overall latency. In order to save resources, cellular networks are configured to keep non-active terminals in idle mode. The transition from idle to connected state takes typically more than 50 ms, whereas within the connected mode, the transition from dormant to active state takes only around 10 ms. Therefore, it is recommended to keep vehicles that send periodic CAMs always in the connected mode. However, for the transmission of an event-triggered DENM, delays caused by state transitions from idle to connected mode are less critical.

Complementary Usage of 802.11p and Cellular Communication Systems ETSI, ISO Communications access for land mobiles (ISO CALM) and the Department of Transportation (DOT) are currently investigating the complementary roles of IEEE 802.11p, LTE, and other cellular technologies in supporting cooperative V2V and V2I applications [24, 37, 53]. In agreement with the ISO CALM guidelines, the ITS station reference architecture proposed in the ETSI specifications [25] leverages the complementary strengths of distributed short-range networks (e.g., IEEE 802.11p and its European counterpart ITS-G5, Wi-Fi) and centralized cellular technologies, among which LTE and its successor LTE-A are the most promising ones. Early indications of this trend toward heterogeneous networking in the complex vehicular environment can be found in the USA as well [53].

16.7 Open Challenges and Future Research Topics

The applicability of LTE for vehicular non-safety and safety services has been investigated in the previous sections. This evaluation shows that LTE is not able to strictly fulfill the stringent requirements for vehicular safety applications with its current settings and, therefore, requires further research and standardization efforts. This section provides an outlook to future cellular technologies such as LTE-A as well as the upcoming 5th generation (5G) wireless communication systems and highlights their applicability for future vehicular services.

16.7.1 Amendments in LTE-A

16.7.1.1 Features of LTE-A

The evolution of LTE, LTE-A, is designed to support even higher data rates, i.e., downlink rates of 1 Gbit/s for low mobility (100 Mbit/s for high mobility) and uplink rates of 500 Mbit/s. This is achieved by carrier aggregation to bandwidths of up to 100 MHz. In addition, enhanced MIMO techniques, relay nodes and an acceleration of the HARQ process have been introduced in LTE-A [32], which increases the overall capacity and reduces the delay in the radio access network significantly. The increased offered capacity reduces the transmission durations especially for high data volume transfers. It also reduces blocking and, thus, the waiting time until the start of the data transmission. In LTE-A delays are further reduced by configuration changes. The Physical Random Access Channel (PRACH) allows random access with a periodicity that can vary from 1 to 10 ms. Also the access for a scheduling request is reduced to 1 ms instead of 10 ms. Moreover, 3GPP is working on evolving LTE-A to accommodate the requirements of Machine Type Communication (MTC), that potentially involve a huge number of communication devices autonomously (i.e., without human intervention) exchanging small amounts of data traffic. Several vehicular applications, like FCD, vehicle diagnosis, and fleet management, that imply data collection from in-vehicle sensors and their transmission to a remote server, are considered for MTC in [5]. Solutions under study in 3GPP for efficient transmission of small amounts of data with minimal network impact (i.e., minimal impact on signaling load, network resources, delay, energy consumption) show promising performance [3, 4].

An important aspect for automotive applications is that LTE-A will be able to serve a higher number of users simultaneously at high quality, such as video streaming for infotainment applications. Moreover, the larger bandwidth will also provide enough network capacity for a temporary high number of CAMs to be served in parallel to the base traffic load. Although LTE-A promises latencies as low as 10 ms over the air interface, end-to-end latencies including propagation through the core network and data processing at a backend server in the cloud are expected to be in the order of several 100 ms.

16.7.1.2 Direct Device-to-Device Communication over Cellular Systems

A further reduction of latencies is expected from direct D2D communication [21]. The D2D communications paradigm enables two mobile devices in the proximity of each other to establish a direct local link and bypass the cellular infrastructure in the data plane. Among other benefits (see [29]), this leads to a hop gain referring to the usage of a single transmission link rather than two transmission links when exploiting both UL and DL resources in a conventional cellular system. Hence, the E2E delay can be reduced substantially, especially for safety applications requiring local data exchange between vehicles.

As a part of the studies conducted within the Rel. 12 activities, 3GPP has defined multiple work items considering D2D communication in cellular (i.e., LTE-A) networks [1]. In the first stage of the investigations, 3GPP focuses on a feasibility study [2] for proximity services (ProSe), i.e., services that rely on D2D communication. The objectives of this study include the investigation of use cases that benefit from the D2D paradigm and their requirements. The studies are focused on mechanisms for network-assisted device and service discovery that enable services like social networking, local advertising, or public safety applications [2]. As an example for automotive applications, 3GPP defines a use case for a parking spot finding assistant at high user density. However, none of the described use cases hint at ProSe support for two of the most important aspects of vehicular communications: high mobility[4] and strict latency requirements. Multi-operator support (i.e., enabling D2D communications between subscribers of different PLMNs) is the third major hurdle towards the realization of V2X services based on the D2D communications paradigm. This challenge is being addressed by 3GPP as it is common to virtually all of the ProSe use cases considered in [2]. While further effort on normative work regarding ProSe and a solution enabling multi-operator support should be expected from 3GPP, it remains questionable whether support for high mobility and E2E delay constraints will be part of the refinements in LTE-A. Therefore, automotive applications (or at least those with strict QoS requirements) based on the D2D communications paradigm might prove to be infeasible before the introduction of 5G wireless communication networks.

An early discussion about the usage of D2D for vehicular applications can be found in [31]. A preliminary solution leveraging D2D for broadcast dissemination of CAMs is proposed in [39].

16.7.2 Future 5G Communication Systems

To meet the expectations of the 2020 wireless communications society, future 5G mobile communication systems have to be significantly more efficient and scalable in terms of energy, costs, and spectral efficiency. Currently, several projects and initiatives investigate the requirements of 5G wireless systems [26, 27]. The technical goals highlighted in the European Union 5G flagship project METIS[5] are

- 1,000 times higher mobile data volume per area
- 10–100 times higher number of connected devices
- 10–100 times higher typical user data rate

---

[4]It should be noted that mobility has been brought up for a discussion within the 3GPP as a part of the feasibility study [12, 13], but a corresponding contribution was not adopted in the final report.

[5]*M*obile *C*ommunications *E*nablers for the *T*wenty-twenty (2020) *I*nformation *S*ociety.

- 10 times longer battery life for low power massive machine communications, and
- 5 times reduced End-to-End latency,

with similar cost and energy consumption as today's networks. Moreover, it is commonly agreed that in comparison with current legacy systems future 5G systems need to be flexible enough to support a significant diversity of applications and use cases implying different service requirements in terms of reliability, availability, and latency that current wireless communication systems typically are not able to guarantee. For example, road safety systems require very low latencies in the order of ms and high reliability (i.e., high probability of error-free packet delivery within a fixed latency deadline) even under poor radio channel conditions.

5G as Enabler for Vehicle-to-device (V2D) Communications V2X communication is currently discussed as a potential service that could be enabled by future 5G networks [27]. The potential benefit of integrating V2X communications in a future 5G communication system lies in the high market penetration, which allows to overcome the chicken-and-egg deployment problem of 802.11p based systems (referred to in Sect. 16.6). By enabling V2X communication capabilities in cellular modems, not only information between vehicles as well as between vehicles and road side units could be exchanged for safety purposes, but also between vehicles and communication devices of vulnerable road users (VRU), such as pedestrians and cyclists. This so-called V2D allows for reaching a very powerful sensor that already today almost everyone carries in his pocket: a mobile device (such as a smartphone or a tablet). In such a way, safety-relevant information can be collected directly from the VRU's devices in order to actively initiate the necessary actions for avoiding accidents. In this sense, the electronic horizon of vehicles will be significantly extended to all traffic participants. Thus, D2D communication is instrumental for the implementation of V2D safety services. However, in order to cope with their requirements, smart resource allocation and interference management schemes [17] are needed. Considering the high velocities that can be expected with V2X communication, one of the biggest challenges in this regard is the collection of reliable channel state information with a minimum amount of signaling. In addition, ultra-fast device and service discovery schemes are required. Moreover, the use of both network-controlled and pure ad-hoc (i.e., without network control) D2D communication needs to be supported in a smart and complementary manner in order to enable the exchange of information between traffic participants even in locations with insufficient network coverage. In this way, the availability of V2X safety services can be increased significantly. Finally, spectrum demand and management options for V2X communications in 5G networks require further analyses and standardization in order to provide multi-operator support among D2D/V2X devices.

5G as Enabler for Real-time Off-Board Applications The evolution of remote services allows not only the storage of data on a common entity, e.g., a server in the Internet, but also the remote execution of applications, e.g., office applications. This means that a mobile terminal can shift certain complex processing tasks to a remote server, whereas the terminal itself only serves as a user interface and therefore can relieve its own local processing units. The automotive and transportation industry will rely on remote processing to ease vehicle maintenance and to offer novel services to customers with very short time-to-market cycles. Moreover, the real-time aggregation of vehicle environment data can be used to realize an extended electronic horizon for vehicles, which can serve as an enabler for next-generation highly automated driving. The challenge to realize these services, especially when considering terminals that move at high speeds, lies not only in the provision of high data rate communication links for mobile terminals, but also in the fact that these services require low latencies and reliable transmissions. The former can be achieved by using novel waveforms, advanced modulation and coding schemes, and further diversity exploitation. These techniques enable high mobility robustness and reduced coding/decoding latency while ultra-reliable communications will ensure the reliability and availability of such services. Advanced handover optimization mechanisms [52] allow for seamless connectivity and, hence, also contribute towards the fulfillment of real-time requirements. Moreover, the utilization of context information (such as trajectory prediction) can be used as a basis for seamless content delivery and QoS control [14]. Last but not least, future 5G networks have to cope with a number of devices that is 10–100 times higher compared to a basis system of today, e.g., 3GPP LTE Rel. 11. In order to achieve this, the signaling overhead needs to be minimized.

This section has given an outlook to current and future research fields. Since new technologies are not suddenly deployed, a smooth evolution from services that can be operated in today's wireless communication networks towards more sophisticated services based on future 5G technologies is expected. On the one hand, 5G communication technologies will provide the basis for developing a huge variety of new applications, e.g., delay sensitive services. On the other hand, it is the service demands that define the requirements for future 5G technologies. Alignment of both streams will offer fantastic opportunities for vehicular applications in a full broadband wireless environment.

16.8 Conclusion

In this chapter the utilization of LTE as wireless transmission technology for vehicular applications has been analyzed. There is a wide consensus on leveraging the strengths of LTE (high capacity, wide coverage, high penetration) to mitigate the well-known drawbacks of IEEE 802.11p (poor scalability, low capacity, intermittent connectivity).

Future applications have been grouped into four service categories, *Infotainment, Comfort, Traffic Efficiency,* and *Safety*. Specific QoS settings have been chosen in order to prioritize the different services. The analysis has been carried out by extensive system-level simulations for different load scenarios and network deployments as well as theoretical investigations.

The results indicate that LTE can meet the QoS requirements of *Infotainment* and *Comfort* services in low traffic-load scenarios. However, especially for high-load scenarios and for uplink transmission the QoS settings have to be carefully selected in order to stay within the required delay budgets of the corresponding services. The additionally generated traffic also has impact on network dimensioning. For *Traffic Efficiency* services, LTE can be considered as a potential wireless transmission technology. The LTE network has to be dimensioned such that sufficient capacity is provided and radio channel quality is sufficiently high. LTE is able to support the delivery requirements of *Safety* services in terms of reliability and delay, under low-to-medium traffic conditions. However, performance decreases under heavy load.

In the initial deployment phase of vehicular networks, LTE is expected to play a crucial role in overcoming situations where no IEEE 802.11p-equipped vehicle is within the transmission range. This could be the case in rural areas where the vehicle density is low. In addition, LTE can be particularly helpful at intersections by enabling the reliable exchange of cross-traffic assistance applications, when IEEE 802.11p communications are hindered by non-line-of-sight conditions due to buildings. The wide LTE coverage can be beneficially exploited for the reliable dissemination over large areas of event-triggered safety messages with advantages for system scalability and congestion control.

Nonetheless, several challenges lie ahead before LTE can be massively exploited in vehicular environments, and a broader understanding of the performance of LTE for the wide set of relevant applications is still required. Studies should not only analyze the capacity of LTE in supporting vehicular applications, as they currently do, but also their potential impact on applications mainly conceived to benefit from this promising cellular technology (e.g., VoLTE, file sharing, video streaming). Moreover, the benefits brought by the augmented capacity and device-to-device capabilities of LTE-A should be analyzed.

Additional discussion is needed for architectural design, vehicular device deployment, and resource management. Standardization requires contributions from different stakeholders toward an integrated and synergetic networking solution leveraging the strengths of LTE, IEEE 802.11p, and emerging communication paradigms like machine-to-machine communications to match the peculiar requirements of vehicular use cases.

Furthermore, future 5G networks are expected to introduce novel technical key components that enable delay-critical services which require high reliability and ultra-low latency. Therefore, 5G will unfold a new level of vehicular connectivity and will be an enabler for even more advanced driver assistance services, such as highly automated driving. Meanwhile, effective business models should be specified to support the wide-spread use of LTE for cooperative intelligent transportation system applications. No one would agree to pay unless highly reliable safety services and attractive traffic related convenience applications can be provided.

References

1. 3GPP (2014) 3GPP active work programme. http://goo.gl/yZTp91
2. 3GPP, TR 22 803 (2013) Feasibility study for proximity services (ProSe), Rel. 12
3. 3GPP, TR 37 868 (2014) RAN improvements for machine-type communications, Rel. 11
4. 3GPP, TR 37 869 (2013) Study on enhancements to machine-type communications (MTC) and other mobile data applications; radio access network (RAN) aspects, Rel. 12
5. 3GPP, TS 22 368 (2014) Service requirements for machine-type communications (MTC), Rel. 13
6. 3GPP, TS 23 203 (2014) Policy and charging control architecture, Rel. 13
7. 3GPP, TS 25 912 (2012) Feasibility study for evolved universal terrestrial radio access (UTRA) and universal terrestrial radio access network (UTRAN), Rel. 11
8. 3GPP, TS 36 213 (2014) Evolved universal terrestrial radio access (E-UTRA); physical layer procedures, Rel. 12
9. 3GPP, TS 36 300 (2014) Evolved universal terrestrial radio access (E-UTRA) and evolved universal terrestrial radio access network (E-UTRAN), Rel. 12
10. 3GPP, TS 36 306 (2014) Evolved universal terrestrial radio access (E-UTRA); user equipment (UE) radio access capabilities, Rel. 12
11. 3GPP, TS 36 440 (2013) General aspects and principles for interfaces supporting MBMS within E-UTRAN, Rel. 11
12. 3GPP TSG-SA WG1 (2012a) S1-120067 Additional requirements for mobility of ProSe users
13. 3GPP TSG-SA WG1 (2012b) S1-120071 High speed ProSe use cases
14. Abou-zeid H, Hassanein HS, Valentin S (2013) Optimal predictive resource allocation: exploiting mobility patterns and radio maps. In: IEEE global communications conference (GLOBECOM), Atlanta
15. Amadeo M, Campolo C, Molinaro A (2012) Enhancing IEEE 802.11p/WAVE to provide infotainment applications in VANETs. Ad Hoc Netw 10(2):253–269
16. Araniti G, Campolo C, Condoluci M, Iera A, Molinaro A (2013) LTE for vehicular networking: A survey. IEEE Commun Mag 51(5):148–157
17. Botsov M, Klügel M, Kellerer W, Fertl P (2014) Location dependent resource allocation for mobile device-to-device communications. In: IEEE wireless communications and networking conference (WCNC), Istanbul, pp 1702–1707
18. Breitenberger S, Grübner B, Neuherz M (2004) Extended Floating car data: Potenziale für die Verkehrsinformation und notwendige Durchdringungsraten. In: Straßenverkehrstechnik, pp 522–531
19. Campolo C, Iera A, Molinaro A, Paratore S, Ruggeri G (2012) SMaRTCaR: an integrated smartphone-based platform to support traffic management applications. In: First international workshop on vehicular traffic management for smart cities (VTM), pp 1–6
20. Castellanos C, Villa D, Rosa C, Pedersen K, Calabrese F, Michaelsen PH, Michel J (2008) Performance of uplink fractional power control in UTRAN LTE. In: IEEE vehicular technology conference (VTC spring), pp 2517–2521
21. Doppler K, Rinne M, Wijting C, Ribeiro C, Hugl K (2009) Device-to-device communication as an underlay to LTE-Advanced networks. IEEE Commun Mag 47(12):42–49
22. ETSI EN 102 637-2 (2012) ITS; vehicular communications; basic set of applications; Part 2: Specification of cooperative awareness basic service
23. ETSI EN 102 869-X (2012) ITS; decentralized environmental notification messages (DENM)
24. ETSI EN 302 665 (2012) Intelligent transport systems (ITS); framework for public mobile network
25. ETSI TR 102 962 (2010) Intelligent transportation system (ITS); communications architecture
26. EU FP7-ICT Project 5GNOW (2014) 5th Generation non-orthogonal waveforms for asynchronous signaling. http://www.5gnow.com/
27. EU FP7 Project METIS (2014) Mobile and wireless communications enablers for the twenty-twenty information society (METIS). https://www.metis2020.com/

28. Festag A, Wiecker M, Zahariev N (2012) Safety and traffic efficiency applications for geomessaging over cellular mobile networks. In: Proceedings of the ITS world congress, pp 1–8
29. Fodor G, Dahlman E, Mildh G, Parkvall S, Reider N, Miklós G, Turányi Z (2012) Design aspects of network assisted device-to-device communications. IEEE Commun Mag 50(3):170–177
30. Franz A, Milch B (2002) Searching the web by voice. In: International conference on computational linguistics (COLING), vol 2, pp 1–5
31. Gallo L, Härri J (2013) A LTE-direct broadcast mechanism for periodic vehicular safety communications. In: IEEE vehicular networking conference (VNC), pp 166–169
32. Ghosh A, Ratasuk R, Mondal B, Mangalvedhe N, Thomas T (2010) LTE-advanced: next-generation wireless broadband technology. IEEE Wireless Commun 17(3):10–22
33. Gomes P, Vieira F, Ferreira M (2012) The see-through system: from implementation to test-drive. In: IEEE vehicular networking conference (VNC), pp 40–47
34. Holma H, Toskala A (2009) LTE for UMTS - OFDMA and SC-FDMA based radio access. Wiley, New York
35. Huber W, Lädke M, Ogger R (1999) Extended floating-car data for the acquisition of traffic information. In: Proceedings of the ITS world congress, pp 1–9
36. IEEE (2010) IEEE Standard for Information technology– local and metropolitan area networks – specific requirements – Part 11: Wireless LAN medium access control (MAC) and physical layer (PHY) specifications amendment 6: wireless access in vehicular environments
37. ISO 17515 (under development (as of August 2014)) Intelligent transport systems (ITS) – communications access for land mobiles (CALM) – LTE cellular systems
38. Kato S, Hiltunen M, Joshi K, Schlichting R (2013) Enabling vehicular safety applications over LTE networks. In: International conference on connected vehicles (ICCVE)
39. Khelil A, Soldani D (2014) On the suitability of device-to-device communications for road traffic safety. In: IEEE world forum on internet of things (WF-IoT)
40. Kihl M, Bur K, Mahanta P, Coelingh E (2012) 3GPP LTE downlink scheduling strategies in vehicle-to-infrastructure communications for traffic safety applications. In: IEEE symposium on computers and communications (ISCC), pp 448–453
41. Laner M, Svoboda P, Romirer-Maierhofer P, Nikaein N, Ricciato F, Rupp M (2012) A comparison between one-way delays in operating HSPA and LTE networks. In: International symposium on modeling and optimization in mobile, ad hoc and wireless networks (WiOpt), pp 286–292
42. Le L, Festag A, Mäder A, Baldessari R, Sakata M, Tsukahara T, Kato M (2011) Infrastructure-assisted communication for car-to-x communication. In: Proceedings of the ITS world congress
43. Lottermann C, Botsov M, Fertl P, Müllner R (2012) Performance evaluation of automotive off-board applications in LTE deployments. In: IEEE vehicular networking conference (VNC), pp 211–218
44. LTE-EPC Network Simulator (LENA) (2012). http://networks.cttc.es/mobile-networks/software-tools/lena/
45. Mangel T, Hartenstein H (2011) An analysis of data traffic in cellular networks caused by inter-vehicle communication at intersections. In: Proceedings of the IEEE intelligent vehicles symposium (IV), pp 473–478
46. Mangel T, Kosch T, Hartenstein H (2010) A comparison of UMTS and LTE for vehicular safety communication at intersections. In: IEEE vehicular networking conference (VNC), pp 293–300
47. Müllner R, Ball C, Ivanov K, Lienhart J, Hric P (2009) Contrasting open-loop and closed-loop power control performance in UTRAN LTE uplink by UE trace analysis. In: IEEE international conference on communications (ICC), pp 1–6
48. NGMN-Alliance (2008) NGMN radio access performance evaluation methodology
49. Nuance Dragon Voice in BMW Cars (2012). http://goo.gl/3Yyx42
50. Paßmann C, Schaaf G, Naab K (2009) SimTD Deliverable D11.2 Ausgewählte Funktionen. http://www.simtd.de

51. Phan MA, Rembarz R, Sories S (2011) A capacity analysis for the transmission of event and cooperative awareness messages in LTE networks. In: Proceedings of the ITS world congress
52. Ren Z, Fertl P, Liao Q, Penna F, Stanczak S (2013) Street-specific handover optimization for vehicular terminals in future cellular networks. In: IEEE vehicular technology conference (VTC Spring), pp 1–5
53. RITA, Intelligent Transportation Systems Joint Program Office (2011) Core system concept of operations (ConOps)
54. Simulation of Urban MObility (SUMO) (2012). http://sumo-sim.org/
55. Szczesny D, Showk A, Hessel S, Bilgic A, Hildebrand U, Frascolla V (2009) Performance analysis of LTE protocol processing on an ARM based mobile platform. In: International symposium on system-on-chip (SoC), pp 56–63
56. The Network Simulator-3 (ns-3) (2010). http://www.nsnam.org/
57. Tung LC, Gerla M (2013) LTE resource scheduling for vehicular safety applications. In: Annual conference on wireless on-demand network systems and services (WONS), pp 116–118
58. Tung LC, Lu Y, Gerla M (2013) Priority-based congestion control algorithm for cross-traffic assistance on LTE networks. In: IEEE vehicular technology conference (VTC Fall), pp 1–5
59. Tung LC, Mena J, Gerla M, Sommer C (2013) A cluster based architecture for intersection collision avoidance using heterogeneous networks. In: IFIP/IEEE annual Mediterranean ad hoc networking workshop (med-hoc-net), Ajaccio, Corsica
60. Viering I, Buchner C, Seidel E, Klein A (2007) Real-time network simulation of 3GPP long term evolution. In: IEEE international symposium on a world of wireless, mobile and multimedia networks (WoWMoM), pp 1–3
61. Vinel A (2012) 3GPP LTE Versus IEEE 802.11p/WAVE: which technology is able to support cooperative vehicular safety applications? IEEE Wireless Commun Lett 1(2):125–128
62. Wang Z (2001) Internet QoS: architectures and mechanisms for quality of service. Morgan Kaufmann, San Francisco
63. Wang S, Le L, Zahariev N, Leung K (2013) Centralized rate control mechanism for cellular-based vehicular networks. In: IEEE global communications conference (GLOBECOM)

Chapter 17
Information-Centric Networking for VANETs

Peyman TalebiFard, Victor C.M. Leung, Marica Amadeo, Claudia Campolo, and Antonella Molinaro

Abstract The peculiarities of the vehicular environment, characterized by dynamic topologies, unreliable broadcast channels, short-lived and intermittent connectivity, call into the question the capabilities of existing IP-based networking solutions to support the wide set of initially conceived and emerging vehicular applications. The research community is currently exploring groundbreaking approaches to transform the Internet. Among them, the *Information-Centric Networking* (ICN) paradigm appears as a promising solution to tackle the aforementioned challenges. By leveraging innovative concepts, such as *named content, name-based routing*, and *in-network content caching*, ICN well suits scenarios in which applications specify *what* they search for and not *where* they expect it to be provided and all that is required is a *localized* communication exchange. In this chapter, solutions are presented that rely on *Content-Centric Networking* (CCN), the most studied ICN approach for vehicular networks. The potential of ICN as the key enabler of the emerging *vehicular cloud computing* paradigm is also discussed.

Keywords VANET • Information-Centric Networking (ICN) • Named Data Networking (NDN) • Content-Centric Networking (CCN) • Vehicular cloud computing • In-network caching • Named data • Name-based routing • Broadcasting

17.1 Introduction

Vehicular ad-hoc networks (VANETs) are getting closer and closer to reality by providing vehicles and roadside nodes with communication capabilities. VANETs are expected to provide a wide range of crucial services, aimed at improving

---

P. TalebiFard • V.C.M. Leung
The University of British Columbia, Vancouver, BC, Canada
e-mail: peymant@ece.ubc.ca; vleung@ece.ubc.ca

M. Amadeo • C. Campolo (✉) • A. Molinaro
University Mediterranea of Reggio Calabria, Reggio Calabria, Italy
e-mail: marica.amadeo@unirc.it; claudia.campolo@unirc.it; antonella.molinaro@unirc.it

road safety and traffic efficiency, and additional commercial, informative, and entertainment services to drivers and passengers, providing revenues to the car manufacturers and service providers.

The unique features of VANETs like the fast changing topology, the short-lived intermittent connectivity, the wide set of conceived applications with heterogeneous requirements, and the harsh propagation conditions heavily challenge the traditional *end-to-end host-centric* Internet paradigm. The TCP/IP protocol stack is notoriously ineffective in mobile wireless environments and problems are further exacerbated in VANETs.

The Wireless Access in Vehicular Environments (WAVE) protocol stack (see Chap. 2) has been specifically conceived to overstep mentioned issues and support the exchange of time-sensitive safety-critical short messages and traffic management messages *without* the IP overhead through the new *lightweight* WAVE Short Message Protocol (WSMP) [15]. At the Medium Access Control (MAC) layer, IEEE 802.11p [12] also adds a new operational mode that facilitates communications in the hostile vehicular environment by allowing nodes that are not member of a Basic Service Set (BSS) to transmit data without preliminary authentication and association. This mode, referred to as *outside the context of a BSS* (OCB) [15], is worthy in a VANET as it significantly reduces access delay and signaling overhead.

Granted these preliminary attempts from the standard to cope with the high dynamicity of VANETs, the WAVE stack nevertheless leverages the traditional TCP/UDP/IPv6 protocols for the exchange of non-safety data.

Rather, many vehicular services, due to time and space relevance (e.g., location-based services, road congestion information), would benefit from *consumer-driven* protocols for *information dissemination* that exploit in-network data *caching* and *replication* [6].

The latter concepts are among the main pillars of *Information-Centric Networking* (ICN), a new paradigm conceived for future Internet architectures that allows applications, services, and networks to interact using information as the main primitive [1]. ICN focuses on finding and delivering named contents, instead of maintaining end-to-end communications between hosts identified by IP addresses.

ICN is envisioned as a basis for a future networking solution to work in both infrastructure and ad hoc situations to adapt to extreme levels of dynamism, and to serve applications ranging from content distribution to Machine-to-Machine (M2M) and Internet of Things (IoT) applications such as vehicular ones. ICN enables these through features like topology independent name-based routing, multicast/anycast, in-network caching, native mobility support, and content-level security.

Specifically, ICN can enhance the scalability and effectiveness of data dissemination in vehicular networks for the following reasons. First, ICN can cope with the dynamics of changes in time and space independently of location identities such as IP addresses. Second, the caching capability can enhance data retrieval and replication performance, which can lead to addressing challenges associated with the intermittent connectivity. Furthermore, the broadcast nature of the wireless medium can be advantageous to support multi-path capabilities that ICN can enable for transport of information in VANETs.

Several works have been published targeting ICN as a networking paradigm to support typical applications in VANETs, ranging from data dissemination and collection, e.g., [4, 11, 25, 26, 31], to safety messages delivery [5]. They address forwarding, naming and transport mechanisms, by proposing enhancements and improvements to the Content-Centric Networking (CCN) paradigm [14], the ICN solution mainly considered for VANETs in the literature.

In addition, ICN has been recognized as an enabling networking technology for an emerging paradigm in VANETs: vehicular cloud computing (VCC) [10, 19]. VCC can be considered as an instance of Mobile Cloud Computing (MCC), that will move cloud services away from individual devices and remote servers and instantly use resources, such as computing, storage and connectivity, hosted by the distributed and dynamic community of vehicles. VCC will play a significant role in shaping the future of VANETs, by providing a distributed, more scalable and reliable management and offering of useful and attractive services. Designing vehicular clouds based on a *clean slate* ICN paradigm, virtualized over the main elements of elastic and scalable computing, storage and networking, can enhance the design of a system at large by realizing the commonality among the design principles that will govern content-oriented services and applications in future vehicular clouds.

The remainder of the chapter is organized as follows. Section 17.2 introduces the main ICN concepts and architectures. Focus is on the CCN paradigm [14], mainly considered for VANETs in the literature. Section 17.3 debates the main advantages, potential weaknesses and open issues of CCN for VANETs. A comprehensive overview of CCN literature solutions for VANETs is provided in Sect. 17.4. Section 17.5 introduces the VCC paradigm and the potentialities of ICN to support it, while scanning relevant literature. Section 17.6 concludes the chapter.

17.2 Information-Centric Networking: An Overview

The new forms of web interaction offered by social networks, today engaging the interest of millions of active users worldwide, the tremendous popularity of video sharing websites like YouTube, and of peer-to-peer (P2P) file sharing protocols like BitTorrent are pushing towards a paradigm shift in the way users communicate and surf the Internet. People have become active information producers and consumers, and the current usage of the Internet is mainly for search and distribution of information to static and mobile users.

This paradigm shift from the *host-centric* conversational model of the traditional Internet to the *information-centric* model pushed by new applications, and the increasing demand for highly scalable, efficient and effective distribution of contents over wired and wireless connections have triggered an intense research activity for the design of new networking architectures.

ICN is a new paradigm for the future Internet where information is the first class network element and communication is based on *unique, persistent, and location independent content names* which are directly used by applications for search and retrieval [1]. Every piece of information that can be stored and accessed through an ICN network is therefore a *Named Data Unit* (NDU). Security is embedded with the NDU: *authenticity* and *integrity* are ensured by a verifiable binding between the data and its name. This makes every NDU a self-consistent object and facilitates caching and replication operations.

Some of the advantages of ICN are:

- Improving efficiency in data distribution and energy efficiency.
- Reducing congestion and latency.
- Better reliability since information can be delivered using any available network.
- Reduction in set-up time, manual configuration, and operating costs.

Deploying ICN at a large scale is a promising solution not only to improve performance in the global Internet, but also to enable simpler and more efficient mobile wireless networking. In fact, by relying on named data, ICN ignores location-based host-centric communications and distributes data in a receiver-driven connectionless mode, without the need of creating and maintaining stable end-to-end sessions [24].

There are many active research initiatives around the world focusing on the ICN paradigm. Among them, we can mention the Named Data Networking (NDN)[1] [33] and CCNx[2] projects in the USA, the Publish Subscribe Internet Routing Protocol (PSIRP) [18] (now PURSUIT[3]) and SAIL[4]/NetInf [7] projects in Europe.

NDN and CCNx initiatives are based on the Content Centric Networking (CCN) architecture proposed by Van Jacobson et al. in [14]. Conversely, PSIRP/PURSUIT and SAIL/NetInf borrow some of their main features from the Data Oriented Network Architecture (DONA) [16]. All these architectures share the same ICN principles (i.e., the use of self-authenticating and self-identifying data units), but they implement different naming and security schemes and different data dissemination methods. Specifically, we can identify four common building blocks of an ICN network, which are summarized in the following.

Naming and Security The ICN naming system can be hierarchical or flat. The *hierarchical* namespace, as specified by the CCN architecture [14], has a structure similar to current Uniform Resource Identifiers (URIs) and may originate user-friendly names. The hierarchy enables aggregation of routing information, thus improving scalability of the routing system. Content authentication is obtained by leveraging a public key infrastructure (PKI) and NDUs must be transferred with a digital signature. Conversely, the *flat* namespace, as in DONA, PSIRP and NetInf, is associated with self-certifying (non human-readable) names. Name-data integrity can be verified without needing a PKI to first establish trust in the key.

---

[1] http://named-data.net/.

[2] http://www.ccnx.org/.

[3] http://www.fp7-pursuit.eu.

[4] http://www.sail-project.eu/.

Content Discovery The discovery operation in ICN consists of forwarding the content request to one (or more) node(s) that generated or cached the NDU. This can be performed by leveraging: (i) routers with a Forwarding Information Base (FIB) that plays the same role of a routing table in the traditional Internet, but it is populated with content names instead of IP addresses, as in CCN; or (ii) an ad hoc defined rendezvous or resolution system, as in DONA, PSIRP and NetInf. Specifically, in DONA, entities called *Resolution Handlers* process the requests received by local subscribers and identify the content storage location. Similarly, in PSIRP, a complex *rendezvous* network is defined: any time a subscription is processed by a rendezvous point, a logical forwarding path is created to allow subsequent simple and fast forwarding back to the subscribers. NetInf, instead, employs a hierarchical multilevel DHT-based resolution system.

Content Delivery The delivery operation consists in forwarding the requested NDU from the storage node back to the consumer. This can be performed in different ways: by (i) traditional IP-based routing protocols, like in DONA and NetInf; (ii) *source-routing*, i.e., the route information is encoded in the packet's header, like in PSIRP; (iii) soft-state delivery, i.e., nodes keep track of each forwarded request and temporarily store information about the link-layer interface where the request has been received from, like in CCN.

Source routing schemes generate an extra per-packet overhead, but little information status must be kept in the network nodes. To limit such overhead, the forwarding mechanism in PSIRP uses a space-efficient probabilistic data structure, the Bloom Filter, to encode the entire route (expressed as a sequence of unique link identifiers instead of node identifiers) into the packet's header. Vice versa, the soft-state delivery in CCN does not require the computation of any overall route and does not generate any additional packet overhead. The price to pay is the maintenance of a state in every node that has forwarded a content request, but proper requests aggregation mechanisms are foreseen to limit per-node load.

Caching The use of self-consistent NDUs allows easy in-network caching operations that can speed up information retrieval and reduce traffic congestion. CCN nodes, ranging from mobile terminals to routers, potentially have a content store that caches incoming NDUs and is managed depending on local policy and constraints. Therefore, CCN combines caching at the network edge, as in P2P networks, with in-network caching. Conversely, in DONA, PSIRP and NetInf, a specific subset of nodes is required to cache NDUs. Information about the storage locations is managed by the Resolution Handlers in DONA, the Rendezvous system in PSIRP and the DHT-based resolution system in NetInf.

17.2.1 What ICN Model for VANETs?

All ICN approaches provide the basic routines to perform data retrieval under mobility conditions.

DONA [16] handles consumer mobility by changing the Resolution Handler associated with the node when this latter changes the network attachment point. Mobility of producers is managed by re-registering the content with a new Handler. Similarly, in NetInf, consumer mobility is easily supported by the DHT-based resolution system, which identifies the best content source(s) on the basis of the new location of the requester. Of course, when a producer moves, the resolution system needs to be updated.

Consumer mobility in PSIRP [18] is managed by issuing a new subscription to the content. This results in the computation of a new route towards the new location of the consumer. Producer mobility, instead, requires the update of the routing information in the Rendezvous system.

Consumer mobility in CCN [14] is intrinsically supported: when a consumer moves, it can simply re-issue any unsatisfied request without the need to perform any other operation. Producer mobility, instead, may require routing updates in content routers. However, CCN inherently supports multi-sourcing and in-network caching, thus alleviating the problems induced by producer mobility.

So far, CCN has been the most studied ICN approach for dynamic wireless ad hoc environments like VANETs. Compared to PSIRP and DONA, the CCN architecture does not require any additional rendezvous network for data discovery and delivery. Every CCN node is basically a content router and also a possible storage location of contents. This also implies that data retrieval and dissemination can be easily performed in stand-alone environments, where nodes communicate in a multi-hop peer-to-peer fashion by simply broadcasting requests and content packets and following ad hoc defined forwarding strategies. In such a context, delay tolerant and opportunistic networking paradigms can be natively supported with proper caching policies and routing protocols.

But these are only a few of the benefits that CCN can offer in a VANET. In the following, we discuss more in detail the advantages and the main open issues related to the deployment of CCN in vehicular networks, by also scanning related works.

17.2.2 CCN in a Nutshell

CCN presents an hourglass model in which the narrow waist leverages data names instead of IP addresses for data delivery, Fig. 17.1. Specifically, the original content is typically segmented in chunks, which are individually named and secured.

Communication is based on two packet types: the *Interest*, used to request an NDU by name, and the *Data*, which represents the NDU and carries the data payload together with the content name and additional information for security services. CCN refers to as *face* any medium for transmitting and receiving packets: both upper layers (application processes) and lower layers (hardware network interfaces) interact with the core of the CCN system using the face abstraction.

17 Information-Centric Networking for VANETs

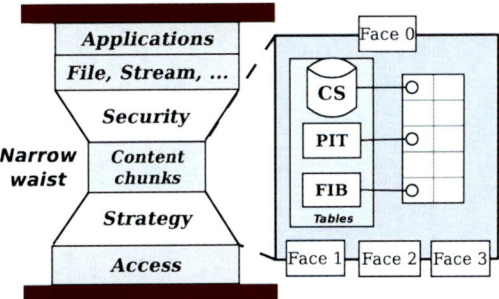

Fig. 17.1 CCN node model [14]

The so-called Strategy Layer in the CCN hourglass model manages the forwarding and transport schemes that vary with respect to the face(s) packets are transmitted and received.

A consumer device requests a named content by broadcasting an Interest packet over its available network faces. The Interest is forwarded hop-by-hop in the network until a provider replies with Data.

As shown in Fig. 17.1, each CCN node maintains three data structures: (i) a Content Store (CS) to cache incoming Data packets; (ii) a routing table named Forwarding Information Base (FIB), which stores the outgoing interface(s) to forward the Interests; (iii) a Pending Interest Table (PIT), which keeps track of the forwarded Interest(s), so that received Data can be sent back to the requester(s).

Each node N receiving an Interest runs the following algorithm, as sketched in Fig. 17.2:

- First, looks at its CS for a prefix-based longest-match lookup on the content name. If N finds a matching, it sends the Data back to the same interface the Interest arrived from.
- Otherwise, if there is a matching PIT entry, the Interest's arrival interface is added to the PIT and the Interest is discarded (since an equal request has been already forwarded).
- Otherwise, if there is a matching FIB entry, the Interest is sent towards the data source and a new PIT entry is created.
- If there is no match for the Interest, it is discarded.

The retrieved Data packets follow the chain of PIT entries back to the requester(s): CCN assumes that an Interest can consume a single NDU to have flow control in the network.

CCN is mainly considered as a clean-slate approach to be developed on top of layer two technologies and replacing IP, but it can be also incrementally deployed as an overlay, i.e., on top of an IP network, in order to be available to applications without requiring universal adoption.

Fig. 17.2 CCN Interest processing

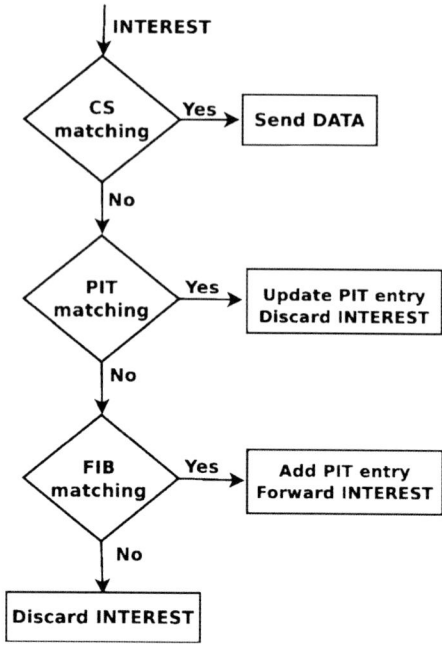

17.3 CCN Benefits and Open Issues for VANETs

Differently from the current *host-centric* Internet model, which poorly fits the dynamicity of the vehicular environment, the described CCN paradigm naturally matches various VANETs requirements and features.

Matching the Nature of VANET Applications. CCN is especially appropriate for the set of location-relevant and time-relevant applications, which are unique to the vehicular environment and usually target vehicles in a given area regardless of their identity or IP address (e.g., road information, advertisements about nearby points-of-interests).

Lightweight Configuration Networking. By leveraging location-independent content names, CCN would allow a network node to communicate without any a priori need for network parameters configuration (i.e., IP address, network mask, default router, name server). This is especially useful in dynamic environments where static configuration is not possible due to node mobility and where solutions like Mobile IP would be unsatisfactory [25].

Easy and Straightforward In-Network Caching. Since each Data packet is a self-consistent unit not bound to a specific location, packets can be cached in any network node; this also enables data storing closer to the consumer, thus reducing the content access delay. In-network caching techniques can be applied at low cost in VANETs, thanks to the capabilities of vehicular nodes not limited by battery, memory, or computation constraints. Although the benefits of caching in mobile and vehicular networks have been largely investigated, e.g., in [9] and the references therein, the novelty of CCN is the coupling of caching and named-data. In fact, the use of named-data makes the content accessible in an application-independent manner, so a content request can be satisfied by any matching data regardless of its location.

Service over Sporadically Connected Links. CCN natively supports *asynchronous* data exchange between end-nodes. By using its cached data, a mobile node can serve as a link between disconnected areas and enable communications even under intermittent connectivity, typical of vehicular environments with sparse roadside infrastructure and short-lived connectivity among nodes.

Broadcast Communications. Another positive aspect of deploying CCN in vehicular networks is related to the possibility of exploiting the broadcast nature of the radio channel. A node can indeed take advantage from overhearing the contents requested by other nodes; in case the content is also of its own interest, the node does not need to explicitly require the overheard content.

Multipath Forwarding. By allowing multipath forwarding of Interests, the consumer can discover more than one content source. Several sources indeed may exist that can satisfy the same request, either because the data has been cached due a previous request, or because several nodes may provide the requested data, e.g., by generating on-the-fly information about the traffic conditions in a given road segment. Depending on application and user constraints, the consumer can select the provider that guarantees the best performance or fetch the content from different directions at same time. Multipath retrieval is particularly beneficial in highly dynamic vehicular networks because it can mitigate the service disruption periods and limit the overhead caused by continuous routing updates.

Native Multicast Support. The majority of VANETs applications consists of information addressed to more than one recipient. Such information can be created explicitly for public dissemination (e.g., road, traffic, weather information), or it can involve restricted groups of recipients (e.g., a video streaming). CCN natively supports multicast data delivery, thanks to the possibility of Interests aggregation in the PIT, according to which intermediate nodes avoid forwarding multiple requests for the same Data packet while the first one is pending.

Despite the potential benefits of CCN in vehicular environments, its adoption is not straightforward, and a CCN solution should address the following main challenging issues:

Coping with Channel Unreliability. The wireless medium is characterized by frequent packet losses induced by adverse propagation conditions and interference, e.g., path loss, fading, capture effect, hidden and exposed terminals. Interests and/or Data packets may be lost or corrupted in transit, or Data may be temporarily unavailable.

Coping with Dynamic Network Topologies. Vehicular network topologies are characterized by frequent and often unpredictable changes, both consumers and providers may move. Although the CCN communication model is connectionless, lightweight, and based on content names, the delivery service can suffer from potential delays and losses due to time-and space-varying connectivity conditions.

Broadcast Storm Mitigation. Broadcast transmissions, which CCN relies on for Interest delivery, are unreliable in a VANET that uses carrier sense multiple access with collision avoidance (CSMA-CA). No explicit acknowledgement can be supported, so the transmitter is not aware of the success or failure of the packet transmission. Moreover, each time a vehicle broadcasts an Interest, potentially all nearby vehicles may rebroadcast the same Interest, inducing unnecessary redundant packet transmissions (i.e., the broadcast storm phenomenon [29]), with consequent scalability issues. The CCN model assumes that nodes implement some packet suppression mechanisms based on randomizing the response times and aborting transmission when detecting that another node has already transmitted a response. However, CCN packet suppression rules are not yet standardized.

All these issues motivate research aiming at design efficient and robust CCN solutions for VANETs, as discussed in the following section.

17.4 CCN for VANETs: A Literature Overview

The intrinsic features of CCN not only can improve the basic mechanisms of content discovery and delivery in VANETs, but they also open new perspectives in the design and development of novel applications for different stakeholders: road authorities, local agencies, car manufacturers and users *on wheels*.

For instance, in [25] a data collection system from mobiles via named data is designed for manufacturers to collect information from vehicles for monitoring and alert purposes.

In [17] a solution is proposed, *CarSpeak*, that enables cars to request and access sensory information from other cars, as well as static infrastructure sensors, in a manner similar to how they access their own sensory information, by following content-centric principles.

In this section, we scan the literature and focus on practical implementations of CCN VANETs. Due to the costs of deploying large-scale test-beds with on-board units empowered with a CCN stack, most of the related literature is based on simulations to evaluate the benefits of CCN in vehicular environments, with the exception of a few works (e.g., [11]) that instead address real deployment in cars.

So far, the majority of works applying CCN principles for data dissemination in VANETs has focused on improving forwarding routines, while only a few of

17 Information-Centric Networking for VANETs 513

Table 17.1 ICN approaches for vehicular ad hoc networks

Solution	Main contribution	Access Technology	Evaluation platform
DMND [25]	Single hop Data collection from vehicles	802.16	Qualnet
NDN naming [26]	Data names for traffic applications	Not specified	ndnSIM
V-NDN [11]	Real implementation of NDN VANET	802.11p, WiMax	UCLA testbed
CCVN [4]	V2V and V2I forwarding	802.11p	NS-2
HBFR [31]	Hierarchical Bloom-Filter routing	Not specified	Qualnet
Hybrid VANETs [5]	Safety messaging with multi-face selection	Multiple 802.11b chan.	NS-2
V2V NDN [27]	Traffic info dissemination with collision avoidance and data pushing	802.11a	ndnSIM
CRoWN [2]	V2V and V2I forwarding	802.11p	NS-2
NDN Transport [3]	Interest Retransmission and content segmentation	802.11p	NS-2
LER [32]	Geographical opportunistic routing with encounter information	Not specified	ndnSIM

them have addressed naming and transport design. Table 17.1 summarizes the main features of CCN[5] solutions proposed for VANETs in the literature, which are discussed in detail in the following.

17.4.1 Naming and Security

There is a wide consensus on claiming that there is a tight relationship between content-centric namespaces and applications. CCN defines some basic conventions for the hierarchical name structure, while the name semantics and the number of substrings can be customized on the basis of applications and conventions. The naming system is still under active research, and some naming proposals in the context of VANETs have recently started to appear.

In [26] the potentialities of data naming in single-hop V2V and V2I are discussed. The following name structure is proposed for a traffic information

---

[5] Since the NDN framework relies on the CCN architecture [14], in the following we use the names CCN and NDN interchangeably.

dissemination application: */traffic/geolocation/timestamp/datatype*, in which the name components identify the temporal and geographical scopes of traffic information, and the application data type.

In [31] the following naming convention for each data chunk is assumed */Category/ServiceName/AdditionalInfo/*, where *Category* identifies the type of data according to the popularity, the size, etc. that drives routing decisions, the *ServiceName* identifies a service, that could be provided by multiple nodes, and *AdditionalInfo* includes the content identifier. An example name for a map of Westwood requested to a local shareable map service can be */typeA/map/losAngeles/westwood*.

The idea of content-based security, where protection and trust are embedded in the content instead of securing the communication channel, is particularly beneficial in VANETs.

In [25], a secure application is presented for data collection from vehicles that allows manufacturers to verify integrity and authenticity of incoming data and to protect privacy of mobile users. Data packets from vehicles are tagged with their signature and encrypted using the public key of a reference database server. The authors assume that the data collector has access to each mobile's public key. This is reasonable since manufacturers could record public keys of vehicles and also store the public key of the database server inside vehicles before release.

In [21] a Name-Based Cryptography approach is proposed that is a variation of the Identity-Based Cryptography (IBC) approach to secure data communication in content-centric networks. In this work, a dynamic private key is used and each entity refreshes its own private key periodically by itself, without any need to the Private Key Generator (PKG) to reproduce the private key.

Using IBC to maintain the trust and to secure the CCN-based system is also proposed in [34] where preserving trust is achieved only by source authentication, which limits the trust concept in an Internet based communication. This approach uses content and publisher names as sources for the IBC encryption system.

17.4.2 Content Discovery and Delivery

Content delivery in vehicular networks has been deeply studied with the concept of IP networking in mind, and several routing protocols have been proposed and analysed [20].

In IP networks, the responsibility of data discovery and delivery is only on the routing scheme, while the forwarding operation is stateless. Conversely, in CCN, both routing and forwarding are responsible for an effective content discovery procedure. Although a routing protocol can be used to populate FIB(s), the forwarding operation is stateful and adaptive to the network conditions. CCN forwarding implies the Interest and Data packets processing, which is done hop-by-hop according to the decisions of the Strategy Layer in each node. The latter one leverages information stored in PIT and FIB but also accounts for requirements of applications, devices and users, network features and constraints.

17.4.2.1 Routing Schemes

According to the literature on VANETs, routing schemes are broadly classified in proactive, reactive and hybrid approaches. This subdivision can be easily extended in CCN environments.

A proactive approach requires the periodic exchange of advertisements so that all nodes may maintain fresh routing information, which is then used at the reception of a content request. Conversely, in a reactive approach, advertisements are not considered. Content discovery starts with the Interest transmission.

Typically, routing solutions conceived for vehicular environments do not address the proactive content advertisement and rely on a reactive content discovery via Interest flooding. Indeed, the proactive transmission of content announcements to populate FIBs is not suitable for VANETs due to the large control overhead, which is even exacerbated by the highly dynamic topology. In addition, in such distributed environments contents are commonly time- and location-relevant and can be generated on the fly, making the FIB population a heavy task. Flooding-based content discovery has the advantage of quickly finding the nearest cached data objects. However, frequent large-scale flooding may cause network congestion and broadcast storm issues. This is why solutions in literature try to improve the reactive routing scheme with additional features, which sometimes directly involve the forwarding plane, or with hybrid strategies.

In Content-Centric Vehicular Networking (CCVN) [4], only the first Interest is flooded to discover the reachable content producer(s) in the VANET. Then, a smart forwarding based on a distance metric is enabled.

In [31], a hybrid routing framework is presented that adaptively performs proactive or reactive content discovery based on content characteristics. The authors identify three major content classes: (i) popular sharable data, such as emergent announcements, map services; (ii) popular non-sharable/non-cacheable data, such as large-size, popular recorded video files; and (iii) unpopular data services, such as private messaging. The class of the content is included in the hierarchy of its name, thus routing can perform differently on the basis of it. Specifically, popular non-sharable/non-cacheable data require proactive routing, while the other two content classes are managed with a reactive scheme. The proposed Hierarchical Bloom Filter Routing (HBFR) uses Bloom Filters (BF) to announce only the popular prefixes and defines hierarchically organized geographical partitions to limit the distribution of advertisements, and therefore the overhead. Vehicles are partitioned in corresponding geographical clusters and they may advertise their contents by including the prefixes in the BF. Each node should have the full view of all the BFs in all the partitions to which it belongs and it is responsible for the distributed aggregation of BF.

A different approach for CCN routing in VANET is LER (Last Encounter Content Routing) [32], which integrates last encounter content discovery and geographical opportunistic forwarding to achieve low overhead and congestion level. In LER, every time two vehicles encounter each other, they exchange their content lists and the content locations known to them. The Interest is only flooded when the provider location is unknown, and only until a relay finds a matching location information. When this happens, the Interest is forwarded by geo-routing instead of flooding, i.e., a node is a candidate forwarder only if it is closer to the destination than the last hop.

17.4.2.2 Forwarding Strategies

The main target of a forwarding strategy for CCN VANETs is to counteract broadcast storm effects and limit useless transmissions while speeding up data delivery. Therefore, the Strategy Layer is usually provided with (i) packet collision avoidance schemes, (ii) *face(s)* selection algorithms, when different access technologies are available, and (iii) provider selection and switching routines, when more than one content source is available in the VANET.

Collision Avoidance To overstep broadcast issues, Zhang et al. in [27] propose a set of timers to coordinate Data transmissions and reduce the chances for packet collisions in a multihop highway scenario. In particular, a *collision-avoidance* timer is randomly selected by neighbouring cars that receive an Interest at the same instant to schedule Data broadcasts at different times.

A similar approach coupled with a counter-based suppression technique is followed in CCVN [4], where two different defer timers for Interest and Data transmissions are used in order to minimize the collision probability and prioritize Data over Interest forwarding. During the defer time, a node listens to ongoing transmissions: if it overhears the same packet transmitted by other nodes for a number of times C_t, then it aborts its broadcast forwarding.

In LER [32], the transmission of Interests is performed with a collision avoidance timer, calculated with a specific target in mind: to expand the content search range. Therefore, the timer computation takes into account the distance to the previous hop, so that farther neighbours have higher priority in transmission.

Multi-Face Scenarios When multiple faces are available, CCN forwarding may leverage the information stored in PITs and FIBs to select the best outgoing interface(s) at each node. For example, the FIB may keep track of the delivery performance (e.g., in terms of latency, throughput, round-trip times) of each outgoing interface, so that packets are transmitted via the best performing interface.

In [5] vehicles may have access to several network interfaces at the same time, such as IEEE 802.11p, WiMAX and UMTS, to disseminate safety messages. Compared to [33], a few modifications are introduced accounting for (i) event-triggered safety packets dealt through a newly introduced *unsolicited* content packet, called *Event Packet*, and (ii) a modified FIB which stores information about the latency of each radio interface, so that *Event Packets* get priority over other packets, if transmitted over the low latency node interface.

Packets could be also simultaneously transmitted over all available interfaces to cope with disruption in connectivity, as investigated in a prototype for a vehicular NDN network (V-NDN) [11], where nodes are equipped with multiple network

17 Information-Centric Networking for VANETs

interfaces, such as 3G/ LTE, WiMAX, Wi-Fi, IEEE 802.11p, and IEEE 1901 power line communication. Implemented at the University of California, Los Angeles, and tested using the UCLA Vehicular Testbed, the V-NDN prototype consists of a group of cars, which can utilize any of these interfaces to communicate with each other and with infrastructure servers, as needed by applications.

Provider Selection and Switching In case the consumer discovers more than one content source in the VANET, it is possible selecting the best performing provider based on some criteria, as suggested in CCVN [4].

The selected provider P can be advertised in subsequent Interests so that intermediate nodes can decide if broadcasting the packet or not, by considering their proximity to P, i.e., intermediates nodes forward a request only if they are closer to the provider than the previous sender. Light path-state information (the identifier of the selected provider and its hop distance to the consumer) are included in Interest and Data packets and left as "bread crumbs" in an additional data structure kept by CCN nodes, the so-called Provider Table. The objective is to forward Interests towards the selected provider and to route Data over the shortest identified path. In doing so, the twofold benefit is achieved of keeping the channel load under control, and reducing download time and energy consumption.

It is worth noticing that the selection of a provider does not mean neither single-path nor unicast delivery. In fact, packet transmission is always broadcast and can be overheard by any neighbour node. This also implies that if there is a cache hit in an intermediate node (different from P), it can immediately answer the Interest without forwarding the Interest farther. This sort of seamless provider handover, which is also referred to as provider switching, can be intrinsically supported by the CCN packet processing fabric.

17.4.2.3 Transport

A transport strategy in a content-centric network has to control Interests transmissions to effect Data delivery, according to the available resources and network dynamics.

Depending on the application, transport may offer reliability and flow control. To provide a reliable service in CCN, the Interest must be retransmitted if it is not satisfied in a given period of time. Retransmission is usually in charge of the original consumer, however, in some advanced forwarding schemes [30], also intermediate nodes can perform this operation. This is especially true when a node has different available outgoing interfaces and explores alternative paths after network problems are detected.

The Interest retransmission timeout setting is critical to quickly recover packet losses while limiting useless retransmissions. CCN foresees that a consumer maintains a timer on each unsatisfied Interest, and retransmits it when the timer expires. Although a standard algorithm for the computation of the retransmission timeout (RTO) is not detailed, it is common use to adopt estimations based on the round-trip-time (RTT), which is the time elapsed from the Interest transmission and the reception of the Data.

In [3] an Interest retransmission policy is proposed, according to which each CCN node tracks the time when an Interest has been forwarded and records an RTT sample when the requested Data is received. RTT estimation is done according to Exponential Weighted Moving Average (EWMA), similarly to TCP RTT estimation [13], to filter out transient effects. The work in [3], however, is only a preliminary study, which requires fundamental improvements. It is worth noticing, in fact, that the presence of in-network caching can produce high RTT fluctuations, which could lead to a wrong RTO estimation. In addition, in (multi-hop) wireless environments, the RTT measurement becomes much more unreliable due to errors, channel rate variability and vehicle mobility.

Due to the complex vehicular scenario, advanced transport layer solutions for content-centric VANETs are still not investigated and the topic of flow control has never been considered. Currently, solutions in literature rely on very simple strategies to guarantee reliability. In [27], a node broadcasts a packet several times with a preconfigured retransmission timeout, $T_{rtx} = 50$ ms. When it hears that the packet has been successfully re-broadcast farther, it will cancel subsequent retransmission. The same strategy is applied in [31], but the value of T_{rtx} is set to 100 ms.

17.5 ICN for Vehicular Cloud Computing

17.5.1 From MCC to VCC

VCC [10] is emerging as an instance of Mobile Cloud Computing with several new facets and will play a significant role in shaping the future of VANETs.

MCC is typically defined as the infrastructure where both the data storage and the data processing happen outside of a mobile device with limited resources [28].

Relying on the Internet cloud would not be convenient for several vehicular application where all that is required is the exchange of local and time-relevant contents and services. Moreover, vehicles themselves can play the role of a distributed cloud within which services are produced, maintained and consumed. Indeed, vehicles are becoming powerful multi-faceted sensors. They host processing and storage resources, are instrumented with cameras and sensors, and enabled with embedded or plug-in interaction and communication capabilities with in-vehicle telematics, other cars, the road-side infrastructure, and the surrounding environment. Most of such resources in a vehicle are typically underutilized and could be pooled with those of other vehicles to produce advanced services an individual cannot make alone. By moving services and facilities away from individual remote devices and servers a more scalable and reliable service provisioning can be assured.

In [8] it is argued that vehicular cloud services can be provided by parked cars organized into independent networks or clusters. Parked cars have the advantage of being almost ubiquitously available in city and urban environments, i.e., at exactly the locations where RSUs would have to be deployed, so they can provide an efficient and effective spatio-temporal storage infrastructure. The formed network will be able to store data and to provide connectivity between the clusters. Creation and maintenance of such clusters of parked vehicles, to which moving cars passing by can connect to upload and download information, is the objective of the study in [8].

Prospective applications that would benefit from VCC range from the ones improving road safety and reducing traffic congestion, to new services exploiting the underutilized resources of vehicles when parked either in an airport or in the city and offering unprecedented business opportunities [28].

One of the major beneficiaries of the vehicular cloud architecture will be Cooperative Automated-driving vehicles. Autonomous driving would extensively rely on vehicle's local sensors, such as Radar, global positioning system (GPS), light detection and ranging (LiDAR), cameras, and would benefit from information from neighboring vehicles to build maps of the surrounding environment (cars, trucks, pedestrians, motorcycles, and other nearby objects) aimed at improving the accuracy and safety of the driving. The vehicular cloud would serve the demands of future autonomous vehicles, to run large complex problems, such as sensor data collection, aggregation, fusion and processing.

17.5.2 How ICN Can Support VCC?

The increasing demand to use networks to provide services, applications and information in vehicular clouds makes the impact of networking significant. However, traditional networking solutions hardly match the unique VCC requirements.

There are several motivations behind the shift in the networking paradigm towards ICN to enable VCC. Since VCC goes beyond the today's recognized scope of VANETs applications (i.e., data dissemination), a separate discussion (in addition to Sect. 17.3) is provided in the following to introduce the potential of ICN.

First, the end-to-end networking solution, unaware of the type of content being transported, could not appropriately support the inherent dynamic properties of VANETs in the massive production and consumption of typically location-relevant services and contents in vehicular clouds. IP-based solutions neither exploit the semantics of information to utilize the available contextual information, whose availability is crucial for vehicular services.

Second, in the cloud computing paradigm, typically, shared resources, software and information are provided to computers and other devices as a utility in a manner that do not require end-user knowledge of the physical location and configuration of the system that delivers the services. Moreover, in a vehicular cloud, the provider of information is not relevant since vehicles and devices can collaborate using their resources towards a participatory approach.

Finally, ICN can achieve an effective content distribution that VCC requires to support its content-oriented applications.

Overall, ICN can be the motivator that utilizes the cloud beyond data centre and enable virtualization of network functions as well as more intelligent strategies to meet the latency and reliability demands of vehicular services.

17.5.2.1 Preliminary Works

Discussion and analysis about the joint potential of ICN and VCC can be found in [19, 22].

In [19] the authors dub a new concept, namely *vehicular cloud networking* (VCN). The latter one couples two successful computing and networking models provided by VCC and ICN, respectively, to support emerging VANETs applications and services. The envisioned vehicle cloud is temporarily created by interconnecting resources available in the vehicles and road-side nodes. VCC and ICN contribute together to create the cloud and running the common virtual platform over such networked resources. A high-level description of how VCN should work is provided in [19]. Unlike in the conventional cloud, three types of resources (i.e., storage, sensing and computing) are provided by vehicles. Resources are requested through the broadcast transmission of *resource request* messages. Nodes willing to share their resources reply with a *resource reply* message. The cloud leader then creates the cloud and allocates the tasks among the cloud members, according to their available resources. Cloud maintenance and release operations are also envisioned to better manage available resources while vehicles move.

Several open issues should be tackled in VCN. Incentive mechanisms should be deviced to properly reward nodes sharing their resources. Proactive, reactive and hybrid solutions for resource discovery, as introduced for content discovery in Sect. 17.4.2, need to be defined. In addition to content, other envisioned resources should be adequately named to reflect their types and capabilities and to facilitate cooperation.

In such a scenario, based on massive amount of information and where nodes may provide *similar* resources, it is crucial to define meaningful and scalable naming schemes. The distributed nature and diversity of the nodes that participate in vehicular clouds, by producing and consuming services, make the interoperability with other entities a challenging task.

The authors in [22] argue that semantically rich applications can benefit from a higher level view of topology that can manifest the connectivity of objects such as services, content items and users to enhance the control and forwarding planes. Possible directions are efficient caching methodologies and facilitating management and configuration such as path optimization at the strategy layer (as per CCN implementation). By bringing a new level of semantic meaning towards the notion of *information topology* at the network layer, ICN represents the key networking solution to enrich VCC with semantics, as proposed in [22, 23]. Since services are not merely based on IP address of source and destination, as in traditional networking approaches, other dynamics and dimensions (like the context of consumers and providers of information, quality of information) play a crucial role.

The proposed method makes use of predicates and attributes in the CCN Interest and Data packets to deliver a content item to the users who are interested in receiving it. Such a demand can be inferred based on the context of the users, e.g., their location, preferences, and device capabilities, etc. The routine involves a decision making approach that is based on the semantics of content items along with the collected information from other sources, such as the user's devices, sensors, and network infrastructure. The collected information should be analysed to yield the best set of data that reflect the most important attributes for decision making. Learning-based methods can be used to this purpose.

Furthermore, in addition to the *logical* and *physical* connectivity of the elements, the concept of *information topology* is introduced by the authors. This higher level vision of the topology should also consider the connectivity of information objects and services that mimics the social connectivity of the objects. It is based on commonality of interest among entities, such as contextual information, i.e., how nodes are connected to the neighbouring clusters and what characteristics make them a better alternative as candidate seed nodes for a specific type of content as described by the prefixes. One of the determinant factors is the spectral characteristics of the graph representation of nodes based on the type of content. For example, centrality, betweenness and other expansion properties can be considered as the spectral characteristics. Leveraging the contextual information with modelling and inferring the information topology can assist in identifying popular services and content items that leads to reduction of redundant traffic and maximum utilization resources. Elasticity of resources in the cloud computing model demands a dynamic virtual infrastructure mapping at the cloud edge that can be benefitted from predictions based on user behaviours.

17.6 Discussion and Conclusion

The unique and challenging features of the vehicular environment and of applications tailored to it heavily challenge traditional end-to-end host-centric networking solutions.

The research community recently started to investigate innovative concepts laying their foundations into future Internet research. Among them, information-centric networking that leverages in-network caching, data replication, and an interaction models decoupling senders and receivers by focusing on the content well suit the demands of vehicular scenarios.

Recent studies have explored and proven the feasibility of applying ICN in vehicular networks. Most of them focused on the CCN paradigm, while targeting the design of forwarding, transport and naming strategies for highly dynamic vehicular networks.

Despite the achieved efforts so far, several research challenges still lie ahead to provide efficient and effective content distribution.

There is a wide consensus on leveraging some kind of awareness in the forwarding fabric. However, the trade-off between the overhead of transferring and/or keeping awareness in every node (e.g., additional information about providers and/or neighbours) and the improved delivery performance should be pursued by accounting for the requirements of the applications and the network conditions. Moreover, the routing design should be tighten to both caching and transport routines.

Transport issues pose several concerns related to the regulation of the Interest rate and the estimation of the retransmission interval, which are especially critical in presence of dynamic topologies and high node mobility.

Meaningful naming schemes, accounting for the spatial and time-relevant nature of dynamically generated contents, should be designed.

ICN as the key enabler of upcoming vehicular clouds is also debated, as recently argued in some preliminary works. Vehicular clouds will go beyond traditional applications for VANETs and offer new attractive services benefiting from a participatory distributed sharing of in-vehicle storage, sensing, communication and processing capabilities.

The peculiarities of ICN, enriching the network layer with a new semantic meaning, allow for the definition of such a virtualized platform, that could additionally take advantage of semantic relationships and inference of context information to facilitate interoperability among different entities.

References

1. Ahlgren B, Dannewitz C, Imbrenda C, Kutscher D, Ohlman B (2012) A survey of information-centric networking. IEEE Commun Mag 50(7):26–36
2. Amadeo M, Campolo C, Molinaro A (2012) CRoWN: content-centric networking in vehicular ad hoc networks. IEEE Commun Lett 16(9):1380–1383
3. Amadeo M, Campolo C, Molinaro A (2013) Design and analysis of a transport-level solution for content-centric VANETs. In: IEEE ICC workshops, Budapest
4. Amadeo M, Campolo C, Molinaro A (2013) Enhancing content-centric networking for vehicular environments. Elsevier Comput Netw 57(16):3222–3234
5. Arnould G, Khadraoui D, Habbas Z (2011) A self-organizing content centric network model for hybrid vehicular ad hoc networks. In: First ACM international symposium on design and analysis of intelligent vehicular networks and applications (DIVANet'11), Miami, FL
6. Bai F, Krishnamachari B (2010) Exploiting the wisdom of the crowd: localized, distributed information-centric VANETs [topics in automotive networking]. IEEE Commun Mag 48(5):138–146

7. Dannewitz C, Kutscher D, Ohlman B, Farrell S, Ahlgren B, Karl H (2013) Network of information (NetInf)—an information-centric networking architecture. Elsevier Comput Commun 36(7):721–735
8. Dressler F, Handle P, Sommer C (2014) Towards a vehicular cloud—using parked vehicles as a temporary network and storage infrastructure. In: ACM international workshop on wireless and mobile technologies for smart cities (WiMobCity), Philadelphia, PA, August 2014, pp 11–18
9. Fiore M, Casetti C, Chiasserini C (2011) Caching strategies based on information density estimation in wireless ad hoc networks. IEEE Trans Veh Technol 60(5):2194–2208
10. Gerla M (2012) Vehicular cloud computing. In: The 11th annual Mediterranean ad hoc networking workshop (Med-Hoc-Net). IEEE, New York, pp 152–155
11. Grassi G, Pesavento D, Pau G, Vuyyuru R, Wakikawa R, Zhang L (2014) VANET via named data networking. In: IEEE INFOCOM workshop on name oriented mobility (NOM), Toronto
12. IEEE 802.11p (2010) Amendment 6: wireless access in vehicular environments. IEEE, Washington, DC
13. Jacobson V (1988) Congestion avoidance and control. ACM SIGCOMM Comput Commun Rev 18:314–329
14. Jacobson V, Smetters DK, Thornton JD, Plass M, Briggs N, Braynard RL (2009) Networking named content. In: Proceedings of 5th ACM international conference on emerging networking experiments and technologies (ACM CoNEXT), Rome
15. Kenney JB (2011) Dedicated short-range communications (DSRC) standards in the United States. Proc IEEE 99(7):1162–1182
16. Koponen T, Chawla M, Chun BG, Ermolinskiy A, Kim KH, Shenker S, Stoica I (2007) A data-oriented (and beyond) network architecture. In: Proceedings of the 2007 conference on applications, technologies, architectures, and protocols for computer communications (SIGCOMM'07), Kyoto, pp 181–192
17. Kumar S, Shi L, Ahmed N, Gil S, Katabi D, Rus D (2012) CarSpeak: a content-centric network for autonomous driving. ACM SIGCOMM Comput Commun Rev 42(4):259–270
18. Lagutin D, Visala K, Tarkoma S (2010) Publish/Subscribe for internet: PSIRP perspective. In: Tselentis G, Galis A, Gavras A, Krco S, Lotz V, Simperl E, Stiller B, Zahariadis T (eds) Towards the future internet—emerging trends from European research. IoS Press, Amsterdam, pp 75–84
19. Lee E, Lee EK, Gerla M, Oh SY (2014) Vehicular cloud networking: architecture and design principles. IEEE Commun Mag 52(2)148–155
20. Li F, Wang Y (2007) Routing in vehicular ad hoc networks: a survey. IEEE Veh Technol Mag 2(2):12–22
21. Nicanfar H, TalebiFard P, Zhu C, Leung V (2013) Efficient security solution for information-centric networking. In: Green computing and communications (GreenCom), 2013 IEEE Internet of things (iThings/CPSCom), IEEE international conference on IEEE cyber, physical and social computing, pp 1290–1295
22. TalebiFard P, Leung V (2012) A content centric approach to dissemination of information in vehicular networks. In: Proceedings of the second ACM international symposium on design and analysis of intelligent vehicular networks and applications, pp 17–24
23. TalebiFard P, Nicanfar H, Hu X, Leung V (2013) Semantic based networking of information in vehicular clouds based on dimensionality reduction. In: Proceedings of the third ACM international symposium on design and analysis of intelligent vehicular networks and applications, pp 69–76
24. Tyson G, Sastry N, Rimac I, Cuevas R, Mauthe A (2012) A survey of mobility in information-centric networks: challenges and research directions. In: ACM NoM'12, Hilton Head, SC, pp 1–6
25. Wang J, Wakikawa R, Zhang L (2010) DMND: collecting data from mobiles using named data. In: IEEE vehicular networking conference (VNC 2010)
26. Wang J, Wakikawa R, Kuntz R, Vuyyuru R, Zhang L (2012) Data naming in vehicle-to-vehicle communications. In: IEEE INFOCOM'12 workshop on emerging design choices in name-oriented networking, March 2012

27. Wang L, Afanasyev A, Kunts R, Vuyyuru R, Wakikawa R, Zhang L (2012) Rapid traffic information dissemination using named data. In: First ACM mobihoc workshop on emerging name-oriented mobile networking design (NoM'12), Hilton Head Island, SC
28. Whaiduzzaman M, Sookhak M, Gani A, Buyya R (2013) A survey on vehicular cloud computing. J Netw Comput Appl 40:325–344
29. Wisitpongphan N, Tonguz OK, Parikh J, Mudalige P, Bai F, Sadekar V (2007) Broadcast storm mitigation techniques in vehicular ad hoc networks. IEEE Wirel Commun 14(6):84–94
30. Yi C, Afanasyev A, Moiseenko I, Wang L, Zhang B, Zhang L (2013) A case for stateful forwarding plane. Comput Commun Inf-Centric Netw 36:779–791
31. Yu YT, Li X, Gerla M, Sanadidi M (2013) Scalable VANET content routing using hierarchical bloom filters. In: Wireless communications and mobile computing conference (IWCMC). IEEE, New York, pp 1629–1634
32. Yu YT, Li Y, Ma X, Shang W, Sanadidi M, Gerla M (2013) Scalable opportunistic VANET content routing with encounter information. In: International conference on network protocols (ICNP), pp 1–6
33. Zhang L, et al (2010) Named data networking (NDN) project. Technical Report NDN-0001, PARC
34. Zhang X, Chang K, Xiong H, Wen Y, Shi G, Wang G (2011) Towards name-based trust and security for content-centric network. In: 19th IEEE international conference on network protocols (ICNP), pp 1–6

Chapter 18
Future Applications of VANETs

Cristofer Englund, Lei Chen, Alexey Vinel, and Shih Yang Lin

Abstract Current transportation systems face great challenges due to the increasing mobility. Traffic accidents, congestion, air pollution, etc., are all calling for new methods to improve the transportation system. With the US legislation in progress over vehicle communications and EU's finalization of the basic set of standards over cooperative intelligent transportation systems (C-ITS), vehicular ad hoc network (VANET) based applications are expected to address those challenges and provide solutions for a safer, more efficient and sustainable future intelligent transportation systems (ITS). In this chapter, transportation challenges are firstly summarized in respect of safety, efficiency, environmental threat, etc. A brief introduction of the VANET is discussed along with state of the art of VANET-based applications. Based on the current progress and the development trend of VANET, a number of new features of future VANET are identified, together with a set of potential future ITS applications. The on-going research and field operational test projects, which are the major enabling efforts for the future VANET-based C-ITS, are presented. The chapter is of great interest to readers working within ITS for current development status and future trend within the C-ITS area. It is also of interest to general public for an overview of the VANET enabled future transportation system.

Keywords VANET • ITS • Cooperative systems • Traffic management • Traffic coordination • Video transmission • Basic set of applications • C-ITS • Next generation C-ITS • Cloud services • Vehicle cloud • Platooning • Cooperative intersection • Negotiation • Cooperative vehicle behaviour • Automated vehicle behaviour • Behaviour standardization

---

C. Englund (✉) • L. Chen
Viktoria Swedish ICT, Lindholmspiren 3A, SE-417 56 Gothenburg, Sweden
e-mail: cristofer.englund@viktoria.se; lei.chen@viktoria.se

A. Vinel • S.Y. Lin
Halmstad University, Box 301, SE-301 18 Halmstad, Sweden
e-mail: alexey.vinel@hh.se; shih.yang_lin@hh.se

© Springer International Publishing Switzerland 2015
C. Campolo et al. (eds.), *Vehicular ad hoc Networks*,
DOI 10.1007/978-3-319-15497-8_18

18.1 Introduction

The transportation sector is facing great challenges, e.g. ever increasing passenger mobility, strong demand on freight transportation, environmental issues. The urbanization process is creating large cities with very high population density. Sixty-four percent of the developing world and 86 % of the developed world are predicted to be urbanized by 2050. This comes with significant challenges. Because of the high exposure of vulnerable road users (VRUs), such as pedestrians and cyclists, in urban areas, traffic safety is of the utmost importance. In 2012, EU road traffic accidents claim 28,000 fatalities, where urban fatalities account for 40 %, and half of them are pedestrians and cyclists.

The past two decades have witnessed advancements of the vehicle safety systems from the early types, such as seat belts, airbags, to the more recent advanced sensors and driver assistant systems. However, the benefits of those technologies gradually reach the limits. The vision of vehicle sensors is limited to their surrounding area and the sensing is limited to specific information. Thus sensors can only have a certain information for the immediate neighbourhood. Even with powerful fusion components within the vehicle, vehicles can still only adapt to the local environment.

Together with the urbanization, congestion has become a daily routine, which causes an annual cost of EURO 80 billion and brings big social and economical challenges. The space for new roads and infrastructure is very limited for the EU urban cities, indicating that improving the efficiency is key. Furthermore, the current inefficient urban transportation system results in high CO_2 emission, accounting for 23 % of the total CO_2 emission from the transportation sector.

Similar challenges are present in the highway sector. Generally speaking, the typical maximum capacity of a highway today is about 2,200 vehicles/hour/lane [1]. When the traffic on highways becomes dense, so-called shockwaves may occur due to vehicle speed variations, which will result in reduced throughput. Meanwhile, road transportation takes the largest part for the increasing goods transportation. In 2010, 45.8 % of the intra-EU goods transportation and 72.7 % of the inland goods transportation in the EU countries were carried out by road transportation. Road transportation keeps growing and it is estimated that the freight road traffic will increase by 75 % by the year 2030 [2]. About 25 % of the total EU CO_2 emissions come from road transportation and the percentage has increased by 23 % from 1990 to 2010. Currently, transportation is the only major sector within the EU where the greenhouse gas emission keeps rising. To deal with this, the EU has set a goal for the year 2030 to reduce the EU road transportation greenhouse gas emissions to approximately 80 % of the 2008 level [3].

To deal with the above challenges new methods are required. The advancements in both vehicle automation and communications enable vehicular ad hoc networks (VANETs), which will potentially play a key role for addressing the future transportation challenges. Based on both short range and long range communications, VANETs have the capability to resolve the limitations of sensors, which help vehicles to extend the vision unlimitedly. Besides, efficient and real-time communication enables a cooperative transportation system, where all components within the system are able to communicate and cooperate with each other. Therefore, in addition to efficient information collection and exchange, vehicles and infrastructure are now able to negotiate and cooperate with the surrounding vehicles, which brings the safety to a new level.

Transportation efficiency can potentially be improved significantly by VANETs. VANETs allow vehicles to communicate with each other, and share information such as vehicle dynamics, driving intentions, thus enabling traffic management on an individual level. VANETs also connect vehicles to the infrastructure network and even to the internet, allowing real-time information distribution and traffic coordination. Together with the advancement of data analysis and control algorithms, vehicles as part of the VANET are able to adjust and coordinate their driving behaviours, such as driving routes, speed, leading to an efficient transportation network scaling to large areas.

Platooning, referred to as road trains, is a concept that a number of vehicles run in a group with synchronized driving dynamics through a VANET. All vehicles driving in a platoon constantly communicate and synchronize with each other. Vehicles are able to take actions simultaneously such as braking, accelerating, and decelerating, to avoid dangerous situations. The distance between vehicles can be shortened and the highway capacity can be improved without further investment in the infrastructure. Meanwhile, vehicle communications enable that vehicles both in and outside the platoons have the updated status of each other. This will make the highway vehicle merging and lane changing much safer and more efficient.

VANET-based cooperative transportation system has strong potential for reducing the energy consumption and lowering the CO_2 emission within the transportation sector. Through the real-time data sharing and exchange among vehicles and infrastructure, it is possible to monitor and coordinate transportation both locally and globally. Traffic lights will be much smarter or even completely removed from the system. Vehicles are able to adapt to the traffic system through cooperation with other vehicles, with infrastructure, and even with the whole traffic network. Combining with advanced reasoning algorithms, VANET-enabled transportation is able to minimize inefficient driving such as frequent braking and stops, leading to a greener urban mobility. For platooning, air drag can be reduced for vehicles, which leads to less fuel consumption and reduced CO_2 emission.

VANET-based applications attract increasing research and development. Standards over vehicular communications have been continuously published worldwide. In Japan, Dedicated Short Range Communication (DSRC) based intelligent transportation system (ITS) has been standardized in STD-T75 [4] for radio interface between a land mobile and a base station, in STD-T88 [5] for the DSRC application layer, in STD-T110 [6] for the non-IP type applications. In the USA, IEEE publishes standards over DSRC, i.e. IEEE WAVE (Wireless Access in Vehicular Environments), through the IEEE 1609 series. In the EU, standardization has focused on the cooperative intelligent transportation systems (C-ITS). A basic standardization package for C-ITS, the so-called Release 1 [7], based on ITS-G5, was finalized in 2013 and formally confirmed in 2014.[1] A thorough description of the published standards can be found in [8].

This chapter first summaries the major efforts that have been working toward the cooperative systems in Sect. 18.2. This includes worldwide projects that apply to different areas including traffic management and control, platooning, traffic efficiency, intersection management, and VRUs. Followed by the previous and on-going efforts, in view of the fast development of vehicular communication technologies, future applications based on VANETs are identified and discussed in Sect. 18.3. Those applications are based on a context of fully connected traffic system enabled by VANETs and are expected to emerge in further future after implementations of the basic set of applications (BSA) specified in C-ITS Release 1. Toward a fully connected transportation system, a number of new projects focusing on different aspects of VANETs-based ITS applications have started and are presented in Sect. 18.4. And the chapter is concluded in Sect. 18.5.

18.2 Cooperative Systems for Road Transportation: A State of the Art

The benefits of the VANETs have been proved in projects focusing on major areas of road transportation in both the context of highways and urban environments. Typical working areas include improving road safety and traffic efficiency, traffic management and control, video delivery, intersection management, and so on.

It is noticed that among the following discussed state-of-the-art research works few of the applications are commercially available. There are a few ITS application pilots that are based on VANETs in, e.g., Japan, see Sect. 18.2.1. However, the applications are mostly to assist drivers through a better information awareness, while limited VANETs capabilities have been utilized. Other applications such as platooning for high-way automation, see Sect. 18.2.2, cooperative intersection for urban traffic, see Sect. 18.2.3, as well as applications for VRU protection, see Sect. 18.2.4, are mostly in the phase of research and proof of concept. The state of the art is discussed in this section for a glimpse of the current research effort that may lead to future applications that fully explore the capabilities of VANETs, and potential future applications are presented and discussed in the next section.

18.2.1 Traffic Management and Control

Through efficient traffic management and control, ITS helps to eliminate traffic congestion, improve traffic safety and comfort, and reduce the environmental

---

[1]http://www.goo.gl/v7Mmbr.

impact. In Japan, *Vehicle Information and Communication System (VICS)* is an information and communication system that processes vehicle road traffic data at the VICS center. It enables users to receive real-time road traffic information such as congestion or regulation that are shown on the navigation screen [9]. Another Japanese traffic management system is the *Universal Traffic Management System (UTMS)* that includes: (1) Integrated Traffic Control Systems (ITCS), which provide advanced traffic management; (2) Advanced Mobile Information Systems (AMIS), which provide traffic information to on-board devices via infrared beacons; (3) Public Transportation Priority Systems (PTPS), which control traffic signals to give buses and other public transportation priorities; (4) Mobile Operation Control Systems (MOCS), which provide administrators with accurate time and locations of their vehicles for fleet management; (5) Environment Protection Management Systems (EPMS), which provide route guidance for reducing traffic pollution; (6) Driving Safety Support Systems (DSSS), which assist drivers for safe driving; (7) Help system for Emergency Life saving and Public safety (HELP), which immediately report location information to rescue organizations when an emergency occurs such as traffic accidents; (8) Pedestrian Information and Communication Systems (PICS), which provide accurate voice notification about the intersection safety; (9) Fast Emergency Vehicle Preemption Systems (FAST), which detect emergency vehicles and then control traffic signals to give priorities [9]. Another traffic system is **Smartway**, which consists of roads, vehicles, communication and processing systems. In Smartway, various services are provided using a platform that integrates elements such as DSRC, ITS on-board units, digital maps, and road-side sensors, with vehicles, drivers, road managers, service providers, and so on [10]. In the EU, **sim**^{TD} is one of the earliest and largest German field operational tests (FOT) projects on V2X communication. The project focused on realistic deployment of V2X communication with large scenarios covering safety, efficiency, and other commercial services. sim^{TD} follows closely the Car to Car Communication Consortium (C2C-CC) and ETSI draft standards for a better acceptance of the results. The system includes mainly three categories of components: the ITS vehicle station (IVS), the ITS roadside station (IRS), and the ITS control station (ICS).

18.2.2 Platooning

The concept of platooning dates back to the World's Fairs in New York 1939 when General Motors presented a vision of driverless vehicles keeping safe distance to each other maintained by automatic radio control. Since then technology has evolved and after the introduction of computers in the 1960s numerous lateral and longitudinal control systems for vehicles have been presented.

PROMETHEUS (1988–1995). Program for European Traffic with Highest Efficiency and Unprecedented Safety is the largest project so far concerning automated driving. The vision of the PROMETHEUS project was to create intelligent vehicles as a part of an overall intelligent road traffic system. The PROMETHEUS project was structured into seven sub-programs concerning driving assistance, communications, vehicle control, artificial intelligence, standards, scenario and use cases, etc. Three of the sub-programs were carried out by the motor industry and the other four represented basic research areas [11].

PATH (1986). The Californian Partners for Advanced Transportation TecHnology project stretches back until 1986. Currently it is divided into three research program areas: Transportation safety research, Traffic operations research, and Modal applications research. Historically PATH was one of the pioneers in platooning and demonstrated an Automated Highway System (AHS) with automated longitudinal control of a four-car platoon in 1994. At the National Automated Highway System Consortium Demo '97 an eight-car platoon was demonstrated in 1997. At this demo the inter-vehicle position error was maintained with a 20 cm root mean square (RMS) error creating a feeling as if the vehicles were mechanically connected, however at the same time maintaining the smooth ride quality for comfort [12].

COVER² (2006–2009) is a 3-year European Commission co-funded FP6 project. The main focus is on cooperation between vehicles and infrastructure aiming to enable intelligent control of vehicles in order to increase infrastructure efficiency. In particular applications for advanced cruise-assist on highway systems and truck platooning in order to safely and efficiently handle queues and congestions were proposed.

SARTRE (2010–2012) is a European Commission co-funded FP7 project that aims to develop strategies and technologies to allow vehicle platoons to operate on normal public highways with significant environmental, safety, and comfort benefits. Within the SARTRE project platooning with automated control in both lateral and longitudinal directions was demonstrated in 2012. With the use of vehicle-to-vehicle communication (V2V) local vehicle signals such as speed and sensor data are shared among the vehicles in the platoon. The SARTRE system was highly integrated with the vehicles' own system and can be thought of as a distributed control system. Each vehicle tries to maintain the distance to the preceding vehicle and at the same time follows the trajectory of the lead vehicle. The results from SARTRE clearly showed the benefit of platooning with measured fuel savings of up to 20 % for the platoon members, nevertheless, fuel savings for the first vehicle were also observed.

GCDC (2011). The Grand Cooperative Driving Challenge 2011 [13] was a competition where the main goal was to accelerate the development, integration, demonstration, and deployment of cooperative mobility. In GCDC both a highway and an urban scenario were demonstrated. The urban scenario was about traffic coordination in a traffic light controlled intersection where two platoons in the same lane should join. The highway part was about demonstrating how traffic shockwaves, that are common on highways, can be attenuated. To be able to

---

[2]http://www.cvisproject.org/en/links/cover.htm.

participate in the challenge, the teams had to comply to a number of rules and to implement a common communication and interaction protocol. GCDC was a competition with a multi-vendor approach. The participants developed their vehicular control systems based on different vehicle manufactures and types, and both heavy-duty and passenger vehicles took part in the challenge. In GCDC, vehicles incorporated automatic longitudinal control and the driver was responsible for the lateral control.

Energy ITS (2008) is a Japanese platooning project financed by the Ministry of Economy, Trade and Industry. The goal is to develop energy saving technology to reduce CO_2 emissions and other greenhouse gases by means of ITS technology. The project is divided into two tracks (1) automated truck platooning and (2) evaluation methods to measure the effectiveness of ITS in terms of energy savings. Demonstrations, in which three automated heavy-duty trucks driving in a platoon at 80 km/h with a gap of 4 m were made. The experimental result indicates that it is possible to reduce energy consumption by 15 % due to aerodynamic drag reductions [9, 14].

KONVOI (2005–2009). The aim of the KONVOI project is to realize and analyse the use of electronically regulated truck convoys. One of the goals is to examine, both through the use of a driving simulator and with real experimental vehicles, what effect does convoys have on traffic. The project will develop five experimental vehicles with automated longitudinal and lateral control, V2V communication, and a human machine interface (HMI) for interacting with the system. The vehicles will be tested on test-tracks and in real traffic surrounded by normal traffic. They will also be used by road carrier companies to test the functions under realistic conditions.

Platooning is the major application for the highway road automation to improve the capacity and traffic efficiency. Despite the field operational test, it has not yet reached the point for realistic deployment. The research and test of the concept of platooning is a very active area. Besides the vehicle controlling, some of the tough requirements that platooning needs include reliable and robust vehicle communications, positioning with high accuracy, as well the platooning operational logic. Two major research and testing areas involving platooning are with and without platoon leaders.

The challenges with platooning is the coordination of platoon members. This is usually done with a platoon leader as the controlling vehicle. Bergenhem et al. [15] present a set of manoeuvres to create, join, maintain, leave, and dissolve platoons and demonstrate the feasibility and duration of these manoeuvres given a specific constellation of vehicles. All manoeuvres are controlled by a manually driven lead vehicle, either a bus or a truck with a certified driver.

In the work presented in [16] a platoon coordination procedure for a vehicle to join from the side of the platoon was developed. Coordination of the manoeuvre is managed by the platoon leader. When the joining vehicle reports to the leader that it is in position to join, the leader commands the vehicles already in the platoon to open up a gap large enough for the new vehicle to enter. The joining vehicle waits for confirmation from the platoon leader and steers in to the platoon when instructed to.

Within the PATH project a five-layered control architecture is proposed where the two highest layers, situated in the infrastructure, monitor and give directives to the different platoons. The manoeuvres of the vehicles in the platoon are controlled by the currently active control law. The control layers may be changed depending on the vehicles' position, its current state in the platoon and active mode of operation. The proposed system requires the exchange of information between vehicles and infrastructure and relies on well-defined vehicle behaviour.

The other method of platooning was demonstrated in the GCDC 2011 where vehicles formed platoons with no centralized controlling vehicle. In this context, all vehicles were able to take any role in the platoon, thus there were no dedicated vehicles or driver who was leading the platoon.

Distributed platooning is promising when the autonomous vehicles become reality. The platooning operation will purely be done in a distributed way through the ego vehicle and its neighbouring vehicles. Meanwhile, all vehicles within the platoon are aware and synchronized of the platoon status through vehicle communications.

18.2.3 Cooperative Intersection

A cooperative intersection is the concept of integrating multiple elements operating at the intersections through communication and coordination for enabling a safe and smooth traffic flow at the intersection. Vehicle collision warning/avoidance systems are introduced to improve the intersection safety, while traffic management methods, such as Cooperative Intersection Management (CIM) [17–19], are introduced for a smooth and comfort intersection passing.

It is generally believed that with the introduction of VANETs, future intersections will be traffic light free. Vehicles can be controlled through sensors and automated algorithms, or driven by humans with assistance from on-board systems. Vehicles communicate with infrastructure and/or vehicles through vehicle-to-infrastructure communication (V2I)/V2V techniques for negotiating the optimal passing sequence of the intersection. Each vehicle will individually receive the personalized right of way to cross the intersection after negotiation.

For improving the intersection safety, a *Smart Intersection*[3] project was launched in 2006, where Ford, General Motors Corp., Honda Motor Co. Ltd., Daimler AG, Toyota Motor Corp., Virginia Tech Polytechnic Institute and State University and the federal government collaborated. Smart Intersection is an active safety technology that warns drivers about possible hazards at an intersection. It also transmits a detailed digital map of the signalized intersection, six additional maps of the surrounding stop sign intersections or crosswalks, lane-specific global positioning system (GPS) information (e.g. lane changing, next exit, speed limit)

---

[3] http://www.goo.gl/DNKUFP.

and signal status information to vehicles. The collision avoidance system within the vehicle is able to determine whether it will safely cross the intersection or if it needs to stop before reaching the intersection. A visual warning on the windshield and an audio alarm to the driver will be issued in case of a detected potential collision.

Cooperative intersection collision avoidance systems (CICAS)[4] is one of the US Department of Transportation (DOT) programs that mainly addresses intersection crash problems related to stop sign violations, traffic signal violations, stop sign movements, and unprotected signalized left turn movements. Three major technologies of CICAS are (1) Vehicle-based technologies and systems-sensors, processors, and driver interfaces that equipped within each vehicle, (2) Infrastructure-based technologies and systems-roadside sensors and processors to detect vehicles and identify hazards and signal systems, messaging signs, and/or other interfaces to communicate various warnings to drivers, and (3) Communications systems-DSRC to communicate warnings and data between the infrastructure and equipped vehicles.

Another project focusing on intersection safety is *INTERSAFE-2*,[5] which is a European Commission co-funded FP7 project. The project aims to develop and demonstrate a Cooperative Intersection Safety System (CISS) that can reduce accidents at intersections. CISS introduces cooperative sensor data fusion techniques based on data from on-board sensors and V2X communications for a better situational awareness [20].

BMW and Volkswagen have developed prototypes of smart intersection assistant systems[6] for improving intersection safety through V2X communication. The BMW's intersection assistant system is activated by an array of sensors and fuses highly precise location data that come from the navigation system for determining potentially dangerous situations. The intersection assistant system[7] developed by Volkswagen can recognize critical situations at intersection and warn the driver. Direct communication between the vehicles and the infrastructures allows vehicles to take actions. On-board devices fuse data received from V2X communication with the vehicles sensor data, and provide driving recommendations to the driver.

The energy consumption aspects have also been addressed for the intersection passing. Concept of Green Cooperative Intersection (GCI) has been presented by the EU project **eCoMove**, which is a 3-year integrated project focusing on the introduction of cooperative ITS for improving the energy consumption and reducing the CO_2 emission. The project conducts field operational test in mainly three areas, e.g., decrease inefficient driving, reducing fuel consumption, and improving traffic efficiency. In addition to the message sets that have been included in the C-ITS standards (see Chap. 5), eCoMove proposed message sets such as eCoMessage

---

[4] http://www.its.dot.gov/cicas/.

[5] http://www.cvisproject.org/en/links/intersafe2.htm.

[6] https://www.observatorio.iti.upv.es/resources/new/12912.

[7] http://www.goo.gl/VGdO4N.

specifically for addressing the energy aspects. Vehicles can receive speed advice through those message sets while approaching the intersection, thus coordinating and adjusting the passing for minimizing the CO_2 emission [21].

18.2.4 Vulnerable Road Users

VANET applications also include the communication with other road users e.g., VRUs. In [22] researchers from Panasonic in Japan presented a device for pedestrian-to-vehicle (P2V) communication. The system was tested by pedestrians crossing a road intersection with oncoming traffic from all directions. It was demonstrated both in simulation and real-life experiments that the P2V is capable of communicating with other road users in order to disseminate information about its presence. Another example of VANETs within road safety for VRUs is the collaboration between two Swedish companies, i.e. Volvo Cars and POC, for the communication between vehicles and bicyclists. Other projects include the EU funded WATCH OVER, SAFESPOT, etc., and German project Aktiv-AS. Honda R&D recently demonstrated how DSRC could be used to detect pedestrians with a DSRC enabled smartphone [23]. They also extended this to include Vehicle-to-Motorcycle technology along with the University of Michigan Transportation Research Institute [24].

18.3 VANET Applications for Future Cooperative Mobility

ETSI has standardized a BSA within ITS and published it in [25] (see Chap. 5). BSA are considered as Day 1 applications and are selected with considerations of, e.g., the technology status, the stakeholders' requirements, the societal and economical factors. The applications are expected to be deployed within a 3-year time frame after the release of standards. The applications are categorized into (a) active road safety, (b) cooperative traffic efficiency (c) cooperative local services (d) global internet services.

Day 1 applications focus on the driver awareness, i.e. to provide necessary information or certain warnings to the driver, so that drivers will take actions accordingly. The currently available message sets are also for this purpose. The information is simple and can be easily interpreted by the applications and thus it is straightforward to create an interface to deliver the message. Among the Day 1 use cases, no manipulation of the vehicle control is allowed, and it is the drivers' responsibility to act appropriately, therefore the applications require a HMI to fulfil its purpose. In [26] it was found that currently, there is no harmonization within the BSA on how to express desired behaviour of the road users. Different approaches are used within different applications, thus it is proposed that custom logic has to be created to handle the behavioural aspects associated with each individual message. In general, communications at this stage are still considered as additional sensors, and VANETs-based benefits are far from being fully explored. VANETs will form a scalable vehicular network that is able to collect and process large amounts of vehicle and traffic data in real-time. Future VANETs-based applications will go beyond the BSA and introduce new functions such as situation awareness, negotiation, coordination. A number of features for the VANETs are summarized in the following section and potential future applications enabled by VANETs are identified and discussed.

18.3.1 VANETs for Enabling a Safer, More Efficient, and Sustainable Road Transportation

Automakers have already started to equip their vehicles with communication capabilities, and the number of communicating vehicles grow fast. When more and more vehicles with communication capabilities populate the traffic system, VANETs become reality and advanced cooperative applications can be implemented and benefits of vehicle coordination will appear. VANET-enabled traffic systems have new features that can be utilized for enabling advanced cooperative applications. Some of the key enabling concepts are identified as follows:

- Cooperative context awareness [27–30]—Vehicles will have full awareness of local and global environment, such as the traffic information, congestion level, tunnel and bridge ahead, road states. This goes beyond the current deployment and extends vehicles' perception from local to global level. With the combination of local and global updated information, traffic management can be implemented from macro level, e.g. traffic flow management, to micro level, e.g. traffic manoeuvres control, simultaneously depending on the involving scenarios. Vehicles are able to take both pro-active and passive actions well ahead of time, contributing to a globally effective transportation system.
- Virtual Infrastructure [31–33]—Vehicles negotiate with each other and with infrastructure to improve safety, efficiency and contribute to reduced environmental impact. Road infrastructure will gradually become virtualized, contributing to reducing the cost of installation and maintenance. Potentially, there will be no road signs, marks, traffic lights, etc., in the future transportation system. All information will be provided virtually and vehicles have full awareness through the VANET. Beyond the fixed physical infrastructure, future virtual infrastructure will be dynamic and adjusted constantly according to the traffic situation, e.g. lanes can be changed based on the congestion level or when accidents happen.
- Internet of Vehicles [34–36]—VANETs are part of the evolution of smart cities. A connected transportation system, enabled by VANETs, is in parallel with the Internet of Things (IoT) and consists of an indispensable part of the future smart cities. VANETs take care of the vehicle networks and connect to the generic Internet, forming a part of the networked and connected society. This indicates

that the transportation system will contribute to other city functions automatically with minimum requirements on cooperation between different city authorities. For example, a transportation system automatically optimizes itself, e.g. driving lanes, traffic lights, to accommodate a big event in the city.

- VANET big data [37, 38]—VANETs will generate large amounts of data in real-time. The utilization of VANET data will enable disruptive applications for improving the traffic on all aspects from safety, efficiency, CO_2 emission to comfort, for both road users and administrators. VANET data go well beyond the traditional road data collected from, e.g., road cameras, loop detectors, and provide much more detailed information. Vehicle dynamics including, e.g., real-time acceleration/deceleration status, lighting, braking become available for data analysis and real-time control. The rich data, with many new features to be explored, together with the emerging big data analysis, potentially will leap the traffic management to a whole new level. A typical example is congestion prediction and dynamic traffic redirection through real-time analysis of the traffic situation data and online optimization. Travel directions can be issued directly through virtual infrastructure, leading to a much more efficient and potentially congestion free transportation system.
- Energy aspects [39–42]—VANETs will enable a much more sustainable transportation system. With real-time information from VANETs, together with the fusion methods and intelligent decision modules in vehicles, inefficient driving behaviour can be avoided. Eco-driving, where drivers are advised with the most eco-friendly information, e.g. route suggestion, driving schedule, even accelerating and braking operations, will become common practices. The prevalence of electric vehicles contributes to reducing the greenhouse gas emission. VANETs will bridge the gap between the transportation network of electric vehicles and the grid—two of the biggest parts of the future smart cities and consequently open doors to innovative applications in both of the areas.

18.3.2 Application Scenarios and Use Cases

VANETs are still emerging and the Day 1 applications to be implemented in the near future mainly focus on providing information for drivers, i.e. driver assistance. Day 2 applications are expected to further improve the road safety, transportation efficiency and potentially provide methods that have not been possible without VANETs. Meanwhile, they will also address the global challenges on the environment and help to reduce the emission from the transportation sector. The highway capacity will be further explored by advanced platooning concepts and transportation management and control methods enabled by VANETs. The Zero vision, e.g., no fatal traffic accidents, are to be realized in the road transportation sector. Also, being connected anytime and anywhere, road mobility will be much more comfortable for travellers. Day 2 applications will mostly be based on the initial deployment of Day 1 applications with further enhancement. Meanwhile, the

18 Future Applications of VANETs   537

fast development of mobile broadband, in parallel with VANETs, is expected to introduce emerging services for future mobility. Based on the current status and the connectivity introduced by VANETs, as well as the fast development of mobile Internet industry, some of the key applications are identified that target the future challenges of transportation.

18.3.2.1 Enhancements via Live Video Delivery

In spite of the above projects successes, there are still many open issues (like policy, liability and acceptance) to be solved before platooning can become a reality on our public roads. An acceptance of joining a platoon by drivers where vehicles will have a time gap of 0.5 s may become an issue. For the driver and passengers of an ordinary car which is following a truck in only 10 m distance, the platooning application may be perceived as uncomfortable. Live video streaming between the vehicles could be seen as one way of increasing the acceptance. For instance, the platoon leader with a camera installed in its windshield can distribute its current view of a road ahead to all other vehicles in the platoon. Live video transmission between vehicles in the platoon for entertainment purposes is another possibility. Potential applications within the area include [43, 44]:

- Pedestrian crossing assistance: Live video is exchanged between the vehicles approaching a pedestrian crossing for protecting VRUs.
- Public transportation assistance: Live video is delivered from the bus at the stop to the vehicles passing by for protecting disembarking passengers.
- In-vehicle video surveillance: Public transportation is monitored in real-time by the control center to help counteract vandalism and other crimes.
- Traffic condition video surveillance: The current situation at a given road section, an intersection or even a lane is transmitted from the nearest vehicle to the management center.
- Overtaking assistance: Live video information is delivered from the truck vehicle to the vehicles behind on rural roads.
- Video conferencing in platoons: Group video conferencing is organized between vehicles in a platoon for a pleasant journey.
- Police assistance at a crime scene: Live video is exchanged between the emergency vehicles at the crime scene.

18.3.2.2 Enhanced Road Safety

VANETs extend the vision of the vehicle beyond the sensors, and enable fully context awareness for the drivers. The future driving assistance system with VANETs will be enhanced significantly. Information of both the surrounding environments and that further away will be utilized for providing safety related information and warnings for drivers to avoid potential dangerous situations. Platooning will be further enhanced, enabling further cooperative applications based on automation and VANET technology.

Potential applications within the area include:

- Advanced danger warning: Road dangers can be sent far away before the vehicles arrive on sites, allowing vehicles to take actions ahead of time. Warning messages based on situations detected by passing vehicles, such as slippery road, uneven road, bridges, tunnels, can be created and cancelled with exact location information and sent out to warn other vehicles.
- Cooperative overtaking and merging. VANETs will allow safe and efficient road overtaking and merging. Future road manoeuvring such as overtaking and merging will be done cooperatively. When vehicles are planning any manoeuvres that may introduce safety issues, they will communicate and coordinate with the relevant vehicles. In such cases, vehicles trying to overtake or merge into the traffic will negotiate with vehicles that will be part of the process, and the operation will be done only when they agree that it is safe.
- Dangerous goods warning. Dangerous goods transportations may cause hazards for other road users. With context awareness from VANETs, information of vehicles with dangerous good will be known ahead of time for all involved vehicles. Manoeuvres involving vehicles with dangerous goods will be prioritized and will be subject to extra safety measures.
- Enhanced VRU protection: Future road safety will cover various VRUs, e.g. pedestrians, two-wheelers, bikers, mopeds, motorcycles or electrical light weight vehicles. Wearable devices, wireless tags, communications such as DSRC, WiFi-Direct, will integrate VRUs into VANET. Together with information fusion from infrastructure data and accurate positioning technologies, VRUs can be identified ahead of time, thus preventing potential dangerous situations from occurring.

18.3.2.3 Enhanced Traffic Control, Planning and Guidance

Various sensors including positioning, road monitoring, environmental, vehicle interior and exterior provide large amount of real-time data. The integration of VANETs provides efficient ways to transportation and utilization of data. Vehicle big data analysis can be used to assist the traffic control centres for more efficient traffic planning and real-time optimization. Platooning methods will be further enhanced with intensive communications enabled by VANETs.

- Intelligent traffic management: Traffic control centres of the future will be data driven and automated. They will directly access all infrastructures, both physical and virtual, and adjust traffic policies for certain purposes in real time. Generally, all management on infrastructure, such as dynamic traffic light control, dynamic lanes, dynamic speed limitation, can be realized without too

much human intervention. Both long-term and short-term traffic prediction are possible. Potential traffic problems can be identified through mining and learning algorithms and precautions can be initialized ahead of time.
- Cooperative manoeuvring: Enhanced from the current lane changing warning application, future traffic lane changing/merging will be made in a cooperative way. Besides the guarantee of safety, vehicles intending to change lanes or merge with other traffic will communicate and negotiate the most efficient way. The advantage is that both the involved vehicles and vehicles nearby have the information of the activities and are able to adjust their behaviour with an agreed efficiency goal, such as to minimize the speed reduction, or to maximize the comfort. Since the operation is based on cooperation, the procedure is monitored automatically and uncertainties are potentially to be reduced.
- Advanced platoon operations: Platoon applications will be further enhanced, where platoon operations, such as platoon merge, platoon split, join a platoon, leave a platoon, will be available with VANETs. VANETs will enable fully distributed platoons, where platoon operations will happen only with the involving vehicles, while the rest of the platoon vehicles will know the information through context awareness.
- Cooperative navigation and travel guidance: Traffic guidance applications will be enhanced by cooperative positioning with implementation of local dynamic maps. Sensor fusion combining with detailed digital maps will improve the positioning accuracy and enable high precision manoeuvres. Cooperative navigation with hybrid navigation systems combining different data sources and the real-time computation at backbone networks, will be able to provide the most efficient travel guidance from the network level, e.g. route guidance with purposes on saving time and reducing emission, to vehicle level, e.g. vehicle manoeuvres to maximize comfort and minimize emission.
- Cooperative intersections: VANETs enable fully cooperative intersections. Future intersection may have no traffic lights, and passing intersections will be based on the coordination between involved vehicles. Vehicles start coordinating and planning in good time before reaching the intersections and pass through smoothly according to their negotiated passing plan. This cooperative intersection passing will also be extended to future platoons where a group of vehicles running in a platoon incorporate the intersection passing application. In this case, vehicles from other directions might adjust the speed further so that the platoon is able to pass the intersection without splitting.

18.3.2.4 Vehicle Cloud

The integration of VANETs with the transportation network, together with the fast development of cloud computing, make vehicle cloud a promising method to provide services for both traditional and emerging purposes. Vehicle cloud provides a platform for enhancing e.g. the conventional traffic management, control, real-time information distribution. Meanwhile, since all vehicles are considered as computing units in the cloud system, all resources within the vehicle can be utilized, if available, by the cloud, e.g., vehicles as a service providers. This potentially will join the effort with the evolution of Internet cloud and create emerging services. Potential applications includes:

- Vehicle temporary data center: Future vehicles have powerful computing power, when they are parked, they may provide data services through the VANET. For example, parked vehicles may provide services such as map updating, music and film streaming, to neighbouring devices.
- Vehicle torrent network: Peer-to-Peer based information distribution, e.g. vehicle torrent networks, will be available with VANETs. Infotainment content, especially data of large volume, can be distributed among vehicles through vehicle torrent networks. Even environmental perception data, such as live video data, can be distributed effectively with vehicle torrent networks.
- Grid balancing: Electric vehicles will dominate the future transportation, and will be key components of future grids. When they are parked and plug in, it is possible they can contribute to balance the grid with their battery power. In this case, VANETs will be closely integrated with the smart grid.
- Personal travel management: Electric vehicle (EV) charging will be managed automatically according to the owners personal schedules. VANETs, together with advanced control algorithms, will make sure that travel will never be disturbed by charging the EV with enough amount of power ahead of time or providing routes with charging stations.

18.3.2.5 Green Transportation

Reducing greenhouse emission is one of the key missions for future transportation planning, and VANETs will play a key role in this process through data provision and advanced traffic management and guidance. Example applications include:

- Eco-driving: Environmental friendly driving guidance can be either done at the traffic management center or vehicles themselves. Besides the above-mentioned cooperative navigation and travel guidance, vehicles can optimize themselves through data fetched from VANETs. An individual vehicle is able to compare itself with other vehicles, of the same kind, driving the same path, both currently and in the past, for achieving the most environmentally efficient driving. This may be combined with travel guidance from control centers for even better performance.
- Green goods transportation: VANETs will enable advanced fleet management systems, thus green goods transportation. With real-time information about the goods delivery status, as well as global traffic information such as congestion level, fleet management is able to perform on-line route planning and optimization to adjust the goods distribution strategy with a goal of minimizing the energy consumption or greenhouse gas emission.

18.3.2.6 Emergency Scenarios

Emergency situations may happen any time in the cities or regions. VANETs under this context may provide efficient information distribution, advanced traffic guidance in case of emergency situations. Example scenarios include:

- Emergency vehicle: With VANETs, emergency vehicles are able to warn vehicles and notify the preferred driving route so that other vehicles are able to prepare the way ahead of time [26].
- Immediate emergency assistance: When accidents occur, VANETs may access public networks for requesting immediate help from its proximity before emergency vehicles arrive. People nearby may receive asking-for-help information with certain basic information. Those who have experiences may perform first aid or provide professional help, thus maximizing the chances of survival of the wounded.
- Emergency evacuation: In case of predicted or unpredicted natural or other disasters, especially when infrastructure-based communications, e.g. fixed line, cellular, etc., are lost, VANETs will play a critical role. It will help drivers to be fully aware of the situation by real-time information exchange. It will also help the evacuation with emergent travel guidance, either from authorities or from individuals.

18.4 On the Way Towards the Cooperative Automated Driving (Plans, Strategies)

Following the release of the BSA, i.e. Day 1 applications, future applications of VANETs are also under investigation. A number of projects have been started towards this direction.

i-GAME (2013–2016). The Interoperable GCDC AutoMation Experience, i-GAME, is a European Commission co-funded FP7 project and a successor project of GCDC 2011. The objective is to develop technologies that speed-up the real-life implementation of automated driving, supported by V2X communication. Automated systems must be safe and to some extent be able to cope with unanticipated events. Different scenarios, both on highway and in urban environment, will be demonstrated in i-GAME. Since GCDC is a competition the organizers must define requirement for, e.g., technology and communication for the teams to conform to. The requirements will keep from being too specific to allow the creativities of the teams. Also, the requirements will be kept at a minimum level, which allow vehicles from different teams to communicate and negotiate in order to resolve the complicated challenge scenarios. Whereas GCDC 2011 was about platooning supported by V2V communication, i-GAME is focusing on interoperability and on demonstrating the benefit of coordination supported by V2V communication.

ACDC (2014–2016) is a Swedish national project (Scania, Volvo GTT, Volvo Cars, Kapsch TrafficComAB, Halmstad University, Qamcom) funded by the Knowledge Foundation, which aims at supporting autonomous cooperative driving with dependable wireless real-time communications. There are two application scenarios considered: (a) Platooning (road trains), including special cases like joining in the middle of the platoon to arrange for best fuel saving; (b) Fully autonomous driving in a confined area like a construction site, a harbour, or a mine. The main research questions are: How can wireless communication enable/enhance autonomous cooperative driving? What requirements on the communication for applications? How to design and configure communication protocols and methods to fulfil the requirements on wireless real-time communications?

COMPANION[8] (2013–2016) is an FP7 project and is motivated by the environmental challenge. The project is building knowledge about truck platooning in order to increase common interest and awareness about platooning in Europe. The project investigates how information should be presented to the drivers on where they can join and leave platoons in a safe manner. The project also investigates the possibility to allow shorter distances between trucks in a platoon. VANETs play an important role when the inter-vehicular distance is decreased. However, the closer the distance the greater the fuel savings hence the need of a reliable wireless connection.

18.5 Conclusion and Outlook

VANETs, enabled by vehicle automation and the introduction of V2V and V2I communications, will enable the development of an efficient and safe future transportation system. The EU publishes a BSA targeting applications for the near future and focusing on driving assistance. In this chapter, future transportation challenges are discussed and motivate that VANETs may play a key role for addressing future transportation issues. A state-of-the-art development of VANETs-based applications involving traffic management and control, platooning, intersections, VRUs, etc., is then presented. Based on the current status, a number of potential future VANETs-based applications are discussed. Those applications involve major areas of transportation, e.g. safety, efficiency, sustainability, as well as comfort. Some of the applications are based on the context of other emerging technologies, such as IoT, smart cities, smart grids. The chapter serves as both an overview of the current transportation system and an outlook for the future VANETs-based applications. The presented future VANETs applications may provide guidance towards a fully automated and cooperative transportation system.

---

[8]http://www.companion-project.eu.

References

1. iMobility Forum (2013) Automation in road transport, May 2013
2. Björn Stigson (2004) Mobility 2030: meeting the challenges to sustainability. World business council for sustainable development, 2004
3. EU Commission (2011) com(2011) 144 Roadmap to a single European transport area: towards a competitive and resource efficient transport system
4. Association of Radio Industries and Businesses (ARIB) (2001) Dedicated short-range communication system. ARIB STD-T75, 2001
5. Association of Radio Industries and Businesses (ARIB) (2004) DSRC application sub-layer. ARIB STD-T88, 2004
6. Association of Radio Industries and Businesses (ARIB) (2012) Dedicated short-range communication (DSRC) basic application interface. ARIB STD-T110, 2012
7. ETSI. TR 101 607 intelligent transport systems (ITS); cooperative ITS (C-ITS); release 1
8. Chen L, Englund C (2014) Cooperative ITS - EU standards to accelerate cooperative mobility. In: The 3rd international conference on connected vehicles & expo (ICCVE 2014), 2014
9. Hanai T (2013) Intelligent transport systems. Society of Automotive Engineers of Japan, 2013
10. Infrastructure Ministry of Land and Transport (2007) ITS introduction guide-shift from legacy systems to smartway. Chapter 2, smartway. pp 25–28, 2007
11. Gillan WJ (1989) PROMETHEUS and DRIVE: their implications for traffic managers, 1989
12. Rajesh R, Han-Shue T, Kait LB, Zhang W-B (2000) Demonstration of integrated longitudinal and lateral control for the operation of automated vehicles in platoons. IEEE Trans Control Syst Technol 8(4):695–708
13. van Nunen E, Kwakkernaat M, Ploeg J, Netten BD (2012) Cooperative competition for future mobility. IEEE Trans Intell Transp Syst 13:1018–1025
14. Tsugawa S, Kato S (2010) Energy ITS: another application of vehicular communications. IEEE Commun Mag 48(11):120–126
15. Bergenhem C, Huang Q, Benmimoun A (2010) Challenges of platooning on public motorways. In: Proceedings of the 17th ITS world congress, pp 1–12, 2010
16. McGurrin M (2012) Vehicle information exchange needs for mobility applications: version 2.0, FHWA-JPO-12-021. Technical report
17. Wu J, Abbas-Turki A, El Moudni A (2012) Cooperative driving: an ant colony system for autonomous intersection management. Appl Intell 37(2):207–222
18. Perronnet F, Abbas-Turki A, Buisson J, El Moudni A, Zéo R, Ahmane M (2012) Cooperative intersection management: real implementation and feasibility study of a sequence based protocol for urban applications. In: 2012 15th international IEEE conference on intelligent transportation systems, Sept 2012. IEEE, New York, pp 42–47
19. Abbas-Turki A, Perronnet F, Buisson J, El-Moudnia A, Ahmane M, Zéo R (2012) Cooperative intersections for emerging mobility systems. In: 15th edition of European working group on transportation, volume: compendium of papers, pp 1–11, 2012
20. Shooter C, Reeve J (2009) INTERSAFE-2 architecture and specification. In: Proceedings of IEEE 5th international conference on intelligent computer communication and processing, pp 379–386, 2009
21. Alesiani F, Lykkja OM, Festag A, Baldessari R (2012) Cooperative ITS messages for green mobility: an overview from the eCoMove project. In: Proceedings of 19th ITS world congress, pp 1–8, 2012
22. Nagai M, Nakaoka K, Doi Y (2012) Pedestrian-to-vehicle communication access method and field test results. In: 2012 international symposium on antennas and propagation (ISAP). IEEE, New York, pp 712–715
23. Fujita M, Watanabe T, Kaneko Y, Hamaguchi M, Satomura M, Yamamura M (2010) Development of 'DSRC safety mobile phone' for pedestrian using 5.8GHz DSRC inter-vehicle communication technology. In: 17th ITS world congress, 2010

24. Bai S (2015) Improving motorcycle safety through DSRC motorcycle-to-vehicle communication. SAE Technical Paper, No. 2015-01-0291
25. ETSI (2010) Intelligent transport systems (ITS); vehicular communication, basic set of applications, Part 1: functional requirements. ETSI TS 102 637-1 V1.1.1, 2010
26. Englund C, Lidström K, Nilsson J (2013) On the need for standardized representations of cooperative vehicle behavior. In: Second international symposium on future active safety technology toward zero-traffic-accident, Nagoya, Sept 2013. Society of Automotive Engineers of Japan, Nagoya, pp 1–6
27. Lytrivis P, Thomaidis G, Tsogas M, Amditis A (2011) An advanced cooperative path prediction algorithm for safety applications in vehicular networks. IEEE Trans Intell Transp Syst 12(3):669–679
28. Wan J, Zhang D, Zhao S, Yang L, Lloret J (2014) Context-aware vehicular cyber-physical systems with cloud support: architecture, challenges, and solutions. IEEE Commun Mag 52(8):106–113
29. Rico J, Sancho J, Cendon B, Camus M (2013) Parking easier by using context information of a smart city: enabling fast search and management of parking resources. In: 2013 27th international conference on advanced information networking and applications workshops, Mar 2013, pp 1380–1385
30. Wang Y, Jiang J, Mu T (2013) Context-aware and energy-driven route optimization for fully electric vehicles via crowdsourcing. IEEE Trans Intell Transp Syst 14(3):1331–1345
31. Ferreira M, D'Orey PM (2012) On the impact of virtual traffic lights on carbon emissions mitigation. IEEE Trans Intell Transp Syst 13(1):284–295
32. Neudecker T, An N, Hartenstein H (2013) Verification and evaluation of fail-safe virtual traffic light applications. In: 2013 IEEE vehicular networking conference, Dec 2013, pp 158–165
33. Sinha R, Roop PS, Ranjitkar P (2013) Virtual traffic lights+. Transp Res Rec J Transp Res Board 2381(1):73–80
34. Whaiduzzaman M, Sookhak M, Gani A, Buyya R (2013) A survey on vehicular cloud computing. J Netw Comput Appl:1–20
35. Gerla M, Lee E-K, Pau G, Lee U (2014) Internet of vehicles: from intelligent grid to autonomous cars and vehicular clouds. In: 2014 IEEE world forum on internet of things (WF-IoT), Mar 2014. IEEE, New York, pp 241–246
36. Lu N, Cheng N, Zhang N, Shen X, Mark JW (2014) Connected vehicles: solutions and challenges. IEEE Internet Things J 1(4):289–299
37. Zhang J, Wang F-Y, Wang K, Lin W-H, Xu X, Chen C (2011) Data-driven intelligent transportation systems: a survey. IEEE Trans Intell Transp Syst 12(4):1624–1639
38. Lee E, Lee E-K, Gerla M, Oh S (2014) Vehicular cloud networking: architecture and design principles. IEEE Commun Mag 52(2):148–155
39. Alsabaan M, Alasmary W, Albasir A, Naik K (2013) Vehicular networks for a greener environment: a survey. IEEE Commun Surv Tutorials 15(3):1372–1388
40. Barth M, Boriboonsomsin K (2012) ECO-ITS: intelligent transportation system applications to improve environmental performance. Technical report
41. Cheng X, Hu X, Yang L, Husain I, Inoue K, Krein P, Lefevre R, Li Y, Nishi H, Taiber JG, Wang F-Y, Zha Y, Gao W, Li Z (2014) Electrified vehicles and the smart grid: the ITS perspective. IEEE Trans Intell Transp Syst 15(4):1388–1404
42. Wang M, Liang H, Zhang R, Deng R, Shen X (2014) Mobility-aware coordinated charging for electric vehicles in VANET-enhanced smart grid. IEEE J Sel Areas Commun 32(7):1344–1360
43. Belyaev E, Molchanov P, Vinel A, Koucheryavy Y (2013) The use of automotive radars in video-based overtaking assistance applications. IEEE Trans Intell Transp Syst 14(3):1035–1042
44. Belyaev E, Vinel A, Jonsson M, Sjöberg K (2014) Live video streaming in ieee 802.11p vehicular networks: demonstration of an automotive surveillance application. In: 2014 IEEE conference on computer communications workshops (INFOCOM WKSHPS), Apr 2014, pp 131–132